Operator Theory
Advances and Applications
Vol. 115

Editor:
I. Gohberg

Operator Theory and Interpolation

International Workshop on Operator Theory and Applications, IWOTA 96

Hari Bercovici
Ciprian Foias
Editors

Springer Basel AG

Editors:

Hari Bercovici and Ciprian I. Foias
Department of Mathematics
Indiana University
Bloomington, IN 47405-5701
USA

1991 Mathematics Subject Classification 47-06; 47A57

A CIP catalogue record for this book is available from the
Library of Congress, Washington D.C., USA

Deutsche Bibliothek Cataloging-in-Publication Data

Operator theory and interpolation / International Workshop on
Operator Theory and Applications, IWOTA 96. Hari Bercovici ;
Ciprian Foias eds. - Basel ; Boston : Berlin : Birkhäuser, 2000
(Operator theory ; Vol. 115)
ISBN 978-3-7643-6229-4 ISBN 978-3-0348-8422-8 (eBook)
DOI 10.1007/978-3-0348-8422-8

ISBN 978-3-7643-6229-4

Originally published by Birkhäuser Verlag, Basel, Switzerland in 2000

Printed on acid-free paper produced from chlorine-free pulp. TCF ∞
Cover design: Heinz Hiltbrunner, Basel
ISBN 978-3-7643-6229-4

9 8 7 6 5 4 3 2 1

Contents

Editorial Preface

Systems and Control theories use sophisticated operator theoretical methods. They also provide new ideas and problems in Operator Theory. As a consequence, the biannual MTNS (Mathematical Theory of Networks and Systems) conference is attended by many operator theorists. At the initiative of J.W. Helton and I. Gohberg an *International Workshop in Operator Theory and Applications* (IWOTA) was organized since the early 80's, as a satellite of MTNS. We are now presenting the proceedings of the 1996 edition of IWOTA which was held at Indiana University, Bloomington, in June as a prelude to the MTNS which was held in St. Louis.

The contributions in this volume represent most of the areas that were discussed at the workshop. There is some emphasis on modern interpolation theory, a subject which has seen much progress in recent years. As usual, all the papers in this volume were subject to the refereeing process.

The workshop would not have been possible without the logistic and financial help of the Mathematics Department, the College of Arts and Sciences, and the office of Research and Graduate Development at Indiana University. For this support we thank Chairman Robert Glassey, and Deans Morton Lowengrub and George Walker. Additional funding was provided by the National Science Foundation, and by the U.S. Office of Naval Research. We also thank the staff and students of the Mathematics Department, especially Kathy Reuter (administrative secretary of the workshop) and Donna Fink.

We learned recently that Béla Sz.-Nagy, one of the founders of modern operator theory, and one of its major contributors, passed away. Much of the work in this volume is related with Sz.-Nagy's results in interpolation and dilation theory. Therefore we feel that it is particularly appropriate to dedicate this volume to his memory.

Hari Bercovici and Ciprian Foias

Operator Theory:
Advances and Applications, Vol. 115

Classical Nevanlinna-Pick Interpolation with Real Interpolation Points

Daniel Alpay, Aad Dijksma and Heinz Langer

We consider a scalar Nevanlinna–Pick interpolation problem with at most countably many interpolation points which lie in $\mathbb{C}^+ \cup \mathbb{R}$. Questions about the existence and uniqueness of the solutions are considered. In the case of nonuniqueness a description of all solutions is given which generalizes Potapov's formula. Our results are obtained in two ways: in Sections 1–5 through the theory of selfadjoint extensions of a symmetric relation in a Hilbert space, including Kreĭn's formula and the socalled u-resolvent matrix, and in Sections 6 and 7 via the theory of reproducing kernel Hilbert spaces.

Introduction

Recall that a Nevanlinna function (or function of the class **N**) is a function N which is holomorphic in the upper half plane $\mathbb{C}^+$ and such that

$$\operatorname{Im} N(\zeta) \geq 0 \text{ if } \operatorname{Im} \zeta > 0.$$

Often Nevanlinna functions are extended by symmetry to the lower half plane $\mathbb{C}^-$: $N(\zeta^*) = N(\zeta)^*$ if $\zeta \in \mathbb{C}^+$ and also by holomorphy into those points of the real axis $\mathbb{R}$ where this is possible. A function $N \in \mathbf{N}$ admits a Herglotz integral representation

$$N(\zeta) = \beta\zeta + \alpha + \int_{-\infty}^{+\infty} \left(\frac{1}{t-\zeta} - \frac{t}{1+t^2} \right) d\sigma(t), \tag{0.1}$$

where $\beta \geq 0$, $\alpha \in \mathbb{R}$, and σ is a nondecreasing function on $\mathbb{R}$ such that $\int_{-\infty}^{+\infty} \frac{d\sigma(t)}{1+t^2} < \infty$. It follows that in each real point x the possibly improper limit

$$d_x^{(N)} := \lim_{\zeta \hat{\to} x} \frac{\operatorname{Im} N(\zeta)}{\operatorname{Im} \zeta} = \beta + \int_{-\infty}^{+\infty} \frac{d\sigma(t)}{(t-x)^2}$$

exists: $0 \leq d_x^{(N)} \leq \infty$, where $\lim_{\zeta \hat{\to} x}$ stands for the nontangential limit (it is the limit $\zeta \to x$ where ζ is restricted to a sector of the form $\frac{\pi}{2} - \gamma \leq |\arg(\zeta - x)| \leq \frac{\pi}{2} + \gamma$ for some $\gamma \in (0, \frac{\pi}{2})$). It follows that $d_x^{(N)} = 0$ for some $x \in \mathbb{R}$ implies that $N(\zeta) \equiv \alpha$, and if $d_x^{(N)} < \infty$ then the limit $\lim_{\zeta \hat{\to} x} N(\zeta)$ exists and is real; it will be denoted by $N(x)$. For further properties of Nevanlinna functions, see the Appendix.

Now the Nevanlinna-Pick interpolation problem considered in this note can be formulated as follows:

Given is an at most countable nonempty set $\mathcal{Z} \subset \mathbb{C}^+ \cup \mathbb{R}$ without accumulation point in $\mathbb{C}^+$; $\mathcal{Z}$ is called the set of interpolation points. For each $z \in \mathcal{Z} \cap \mathbb{C}^+$ there is given a complex number v_z, and for each $x \in \mathcal{Z} \cap \mathbb{R}$ there are given real numbers v_x and d_x, $0 < d_x < \infty$. Find conditions such that there exists at least one or exactly one function $N \in \mathbf{N}$ with the properties

$$N(z) = v_z \text{ if } z \in \mathcal{Z} \cap \mathbb{C}^+, \tag{0.2}$$

$$\lim_{\zeta \hat{\to} x} \frac{\operatorname{Im} N(\zeta)}{\operatorname{Im} \zeta} \leq d_x, \quad N(x) = v_x \text{ if } x \in \mathcal{Z} \cap \mathbb{R}, \tag{0.3}$$

and if there is more than one function N with these properties give a description of all of them. The Nevanlinna-Pick interpolation problem is said to be determined, if there exists exactly one solution N, and indetermined if there is more than one solution. In the case considered here, that also real interpolation points are given, this problem is sometimes called the boundary Nevanlinna-Pick interpolation problem.

We suppose that $\mathcal{Z}$ does not have an accumulation point in $\mathbb{C}^+$, as otherwise the solution of the problem, if it exists, is uniquely determined. If $x \in \mathcal{Z} \cap \mathbb{R}$ is such that there exists a sequence $(z_n) \subset \mathcal{Z} \cap \mathbb{C}^+$ with $z_n \hat{\to} x$, then, if the problem has a solution, $\lim_{n\to\infty} \frac{\operatorname{Im} v_{z_n}}{\operatorname{Im} z_n} =: d$ is finite and $v_{z_n} \to v_x$ if $n \to \infty$. Hence a necessary condition for the existence of a solution is $v_{z_n} \to v_x$ and $d \leq d_x$, and if the point x is excluded from $\mathcal{Z}$, we get the same set of solutions.

Fractional linear transformations of the dependent and independent variables turn the above Nevanlinna-Pick problem into an interpolation problem for Schur functions. These are holomorphic functions on the open unit disk $\mathbb{D}$ with values in the closed unit disk. Pioneers such as Carathéodory, Pick, Schur, Nevanlinna, and Denjoy, and many others later, also considered Schur interpolation problems. For example, Denjoy and Nevanlinna gave necessary and sufficient conditions for the existence of one or infinitely many solutions in terms of the divergence or convergence of a series. The summands in these series, like Schur parameters, are defined by solving the interpolation problem step by step. In case of infinitely many solutions, these solutions can be described by a fractional linear transformation

$$s(\zeta) = \frac{a(\zeta)t(\zeta) + b(\zeta)}{c(\zeta)t(\zeta) + d(\zeta)}, \quad \zeta \in \mathbb{D},$$

where $a(\zeta), b(\zeta), c(\zeta)$, and $d(\zeta)$ are functions on $\mathbb{D}$ determined by the data. For each Schur function $t(\zeta)$, $s(\zeta)$ is a solution and conversely each solution is of this form. For details, see, for example, Nevanlinna [N]. For finitely many interpolation points, Potapov [KP] gave a particularly nice formula for the 2×2 matrix

$$\begin{pmatrix} a(\zeta) & b(\zeta) \\ c(\zeta) & d(\zeta) \end{pmatrix}.$$

If the points in $\mathcal{Z}$ are written as a (finite or infinite) sequence: $\mathcal{Z} = (z_k)$, $k = 0, 1, \ldots$, the conditions for the existence and the uniqueness of the solutions of the interpolation problem are formulated in terms of the sequence of Pick matrices $\mathbb{P}_n = \left(p_{z_j z_k}\right)_{j,k=0}^{n-1}$, where for $z, w \in \mathcal{Z}$,

$$p_{zw} = \begin{cases} \dfrac{v_z - v_w^*}{z - w^*} & \text{if } z \neq w^*, \quad z, w \in \mathcal{Z}, \\ d_z & \text{if } z = w^* \in \mathcal{Z} \cap \mathbb{R}. \end{cases} \tag{0.4}$$

In fact, a necessary and sufficient condition that the interpolation problem has at least one solution is the nonnegativity of all the Pick matrices $\mathbb{P}_n$. The proof of the necessity is easy if the integral representation (0.1) is used, hence we can assume that $\mathbb{P} \geq 0$, $n = 1, 2, \ldots$ For the proof of the sufficiency, which also yields a description in the indetermined case, we use two approaches.

The **extension method** in Sections 1–5 is based on the fact that a Nevanlinna function N admits a representation in terms of a selfadjoint relation A: $N = N_A$ (see the Appendix). We associate with the data a Hilbert space $\mathcal{H}$, with inner product defined by means of the Pick matrices $\mathbb{P}_n$, and a symmetric relation S in $\mathcal{H}$ with defect numbers less than or equal to 1. The pair $\mathcal{H}$, S is called the model associated with the interpolation problem. We prove in Section 2 that the solutions N of the interpolation problem are precisely those of the form $N = N_A$ where A is a minimal selfadjoint extension of S. In Section 3 we show that S can have unequal defect numbers and that the defect index (0, 1) cannot occur. Further we characterize the socalled canonical solutions $N = N_A$ of the interpolation problem, for which the selfadjoint extension A acts in the Hilbert space $\mathcal{H}$ of the model. If one of the Pick matrices is degenerate, S is selfadjoint and hence the problem has a unique solution (Theorem 4.1). Therefore, in order to formulate a criterion for the existence of infinitely many solutions in Section 4 we can suppose that all the Pick matrices $\mathbb{P}_n$ are invertible. The criterion is given in Theorem 4.3 in terms of the convergence of sequences $(\alpha_n(\zeta))$ for some $\zeta \in \mathbb{C}^+$ and some $\zeta \in \mathbb{C}^-$, ensuring that S has defect index (1, 1). Finally, in Section 5 a parametrization of the set of all solutions N in the indetermined case is derived from M.G. Kreĭn's formula describing the set of all u-resolvents of S by means of a fractional linear transformation. In the case of a finite set $\mathcal{Z}$ and $\mathcal{Z} \subset \mathbb{C}^+$ this description coincides with the formula given by V.P. Potapov [KP].

We mention that the extension theory of linear operators was first applied to the Nevanlinna-Pick problem by B. Sz.-Nagy and A. Koranyi [SzK1], [SzK2]. They proved necessary and sufficient conditions for the existence of solutions, also in the case of operator valued Nevanlinna functions. In [KL2] this extension method was used in the study of the Q–function of a π–symmetric operator, in [ABDS2], [Br1] and [Br2] it was applied to interpolation problems for matrix valued Nevanlinna pairs with finitely many nonreal interpolation points. Finally, this method is also used in the general scheme for interpolation problems based on the socalled Fundamental Matrix Inequality; see [KKY], where further references can be found.

In the **reproducing kernel approach** in Sections 6 and 7 various reproducing kernel Hilbert spaces of holomorphic functions play a role. One is the space $\mathcal{L}(N)$ which can be associated with each Nevanlinna function N through its integral representation (see the Appendix). From the interpolation data a linear space $\mathcal{M}$ of 2-vector valued functions is constructed together with an inner product determined by the Pick matrices. This space is isomorphic to the model space $\mathcal{H}$ in the extension method, but different representations are being used. In Section 6 we show that the Nevanlinna function N is a solution of the interpolation problem if and only if a certain linear map τ from $\mathcal{M}$ to $\mathcal{L}(N)$ is a contraction. To explain the method further we consider the interpolation problem associated with the first n points and denote the corresponding space $\mathcal{M}$ of 2-vector functions by $\mathcal{M}_n$. It is finite dimensional and its inner product is defined by $\mathbb{P}_n$. Denote by J the 2×2 matrix

$$J = \begin{pmatrix} 0 & -1 \\ 1 & 0 \end{pmatrix}.$$

If the inner product on $\mathcal{M}_n$ is nondegenerate then by a structure theorem of de Branges (applied to a finite dimensional situation; see [B1], [AD1]), there is a rational iJ-inner 2×2 matrix function θ_n such that $\mathcal{M}_n = \mathcal{H}(\theta_n)$, the reproducing kernel Hilbert space with reproducing kernel

$$\frac{\theta_n(\zeta)J\theta_n(\eta)^* - J}{\zeta - \eta^*}.$$

Let N be a solution of the interpolation problem. Applying a theorem of de Branges and Rovnyak [BR] which characterizes the contractivity of $\tau : \mathcal{M}_n = \mathcal{H}(\theta_n) \to \mathcal{L}(N)$ in terms of θ_n and N, we find that N is a linear fractional transformation of a Nevanlinna function with coefficient matrix θ_n. It now follows that θ_n is a parametrization matrix for the interpolation problem with interpolation points $z_0, z_1, \ldots, z_{n-1}$. For more details and further references, see [ABDS3], [Br1] and [Br2].

If for some n the Pick matrix $\mathbb{P}_n$ and hence the inner product on the space $\mathcal{M}_n$ are degenerate, the unique solution of the problem is determined by the nonzero neutral elements in $\mathcal{M}_n$; see formula (6.4). If all $\mathbb{P}_n$'s are invertible, a solution can be obtained as a limit of a subsequence of the sequence (N_n) of Nevanlinna functions where N_n is a solution of the interpolation problem with interpolation points $z_0, z_1, \ldots, z_{n-1}$. In Section 7 we show that the problem has infinitely many solutions if the inner product space $\mathcal{M}$ can also be written as an $\mathcal{H}(\theta)$ space, and that this is the case precisely if the limits of the sequences $(\alpha_n(\zeta))$ mentioned above are finite; recall that in the extension method this condition means that S has defect index $(1, 1)$. It turns out that the θ here is a parametrization matrix for the interpolation problem.

The reproducing kernel method in interpolation was originally developed and promoted by H. Dym; see [Dy], where other applications can be found. The relation between the u–resolvent matrix and the de Branges spaces is explained in [AD2].

Other approaches to interpolation can be found, for example, in [BH] and in the monograph [BGR].

1 Existence of Solutions and the Model

We recall the definition (0.4) of the numbers p_{zw}, $z, w \in \mathcal{Z}$:

$$p_{zw} = \begin{cases} \dfrac{v_z - v_w^*}{z - w^*} & \text{if } z \neq w^*,\ z, w \in \mathcal{Z}, \\ d_z & \text{if } z = w^* \in \mathcal{Z} \cap \mathbb{R}. \end{cases}$$

Theorem 1.1 *The Nevanlinna Pick problem has a solution if and only if for any n and interpolation points $z_0, \dots, z_{n-1} \in \mathcal{Z}$ the $n \times n$ matrix $\mathbb{P}_n = (p_{z_j z_k})_{j,k=0}^{n-1}$ is nonnegative.*

Proof of the only if part of Theorem 1.1: The only if part follows from the integral representation (0.1) of a function $N \in \mathbf{N}$:

$$N(\zeta) = \beta\zeta + \alpha + \int_{-\infty}^{+\infty} \left(\frac{1}{t - \zeta} - \frac{t}{1 + t^2} \right) d\sigma(t).$$

Then

$$K_N(\zeta, \eta) := \frac{N(\zeta) - N(\eta)^*}{\zeta - \eta^*} = \beta + \int_{-\infty}^{+\infty} \frac{1}{(t - \zeta)(t - \eta^*)} d\sigma(t).$$

It follows that for $y > 0$ the matrix $(K_N(z_j + iy, z_k + iy))_{j,k=0}^{n-1}$ is nonnegative. If N is a solution of the interpolation problem, the limit matrix

$$\mathbb{K}_n = \left(\lim_{y \downarrow 0} (K_N(z_j + iy, z_k + iy)) \right)_{j,k=0}^{n-1}$$

exists and is nonnegative also. Finally, the difference $\mathbb{P}_n - \mathbb{K}_n$ is a diagonal matrix with nonnegative entries. Hence the matrix $\mathbb{P}_n$ is nonnegative. □

The if part of the theorem will be proved by constructing a symmetric relation S in a Hilbert space $\mathcal{H}$ from the interpolation data. The pair $S, \mathcal{H}$ is called the model associated with the given data. To begin, we denote by $\mathcal{L}$ the linear space of all finite formal sums $\sum_z f_z \varepsilon_z$ where $z \in \mathcal{Z}$, $f_z \in \mathbb{C}$ and ε_z is a symbol associated with each $z \in \mathcal{Z}$, and we equip $\mathcal{L}$ with the inner product

$$\left(\sum_z f_z \varepsilon_z, \sum_w g_w \varepsilon_w \right)_{\mathcal{L}} = \sum_{z,w} g_w^* p_{wz} f_z.$$

In $\mathcal{L}$ we define the operator S by:

$$\operatorname{dom} S = \left\{ \sum_z f_z \varepsilon_z : \sum_z f_z = 0 \right\}, \quad S \sum_z \varepsilon_z f_z = \sum_z z^* f_z \varepsilon_z.$$

Straightforward calculation shows that S is Hermitian, that is, if $\sum_z f_z\varepsilon_z$ and $\sum_w g_w\varepsilon_w$ belong to dom S, then

$$\left(S\sum_z f_z\varepsilon_z, \sum_w g_w\varepsilon_w\right)_{\mathcal{L}} = \left(\sum_z f_z\varepsilon_z, S\sum_w g_w\varepsilon_w\right)_{\mathcal{L}}.$$

If $z, w \in \mathcal{Z}$ then, evidently,

$$(S - z^*)(\varepsilon_z - \varepsilon_w) = (z^* - w^*)\varepsilon_w. \tag{1.1}$$

It follows that ran $(S - z^*)$ contains all elements ε_w for all $w \neq z$, and therefore

$$\dim \mathcal{L}/\text{ran}\,(S - z^*) \leq 1, \quad z \in \mathcal{Z}. \tag{1.2}$$

We make two more observations:

(i) Let $x_0 \in \mathcal{Z} \cap \mathbb{R}$, and assume there exists a sequence of interpolation points (z_n) in $\mathcal{Z} \cap \mathbb{C}^+$ such that as $n \to \infty$,

$$z_n \longrightarrow x_0, \quad v_{z_n} \longrightarrow v_{x_0}, \quad \frac{v_{z_n} - v_{x_0}}{z_n - x_0} \longrightarrow d_{x_0}, \quad \frac{v_{z_n} - v_{z_n}^*}{z_n - z_n^*} \longrightarrow d_{x_0}. \tag{1.3}$$

Then ran $(S - x_0)$ is dense in $\mathcal{L}$. Indeed,

$$\begin{aligned} \|\varepsilon_{z_n} - \varepsilon_{x_0}\|_{\mathcal{L}}^2 &= (\varepsilon_{z_n} - \varepsilon_{x_0}, \varepsilon_{z_n} - \varepsilon_{x_0})_{\mathcal{L}} \\ &= (\varepsilon_{z_n}, \varepsilon_{z_n})_{\mathcal{L}} - (\varepsilon_{x_0}, \varepsilon_{z_n})_{\mathcal{L}} - (\varepsilon_{z_n}, \varepsilon_{x_0})_{\mathcal{L}} + (\varepsilon_{x_0}, \varepsilon_{x_0})_{\mathcal{L}} \\ &= \frac{v_{z_n} - v_{z_n}^*}{z_n - z_n^*} - \frac{v_{x_0} - v_{z_n}^*}{x_0 - z_n^*} - \frac{v_{z_n} - v_{x_0}}{z_n - x_0} + d_{x_0} \longrightarrow 0, \end{aligned}$$

and, according to (1.1), $\varepsilon_{z_n} \in$ ran $(S - x_0)$.

(ii) Assume there exists a sequence $(z_n) \subset \mathcal{Z}$ such that as $n \to \infty$,

$$\frac{\text{Im}\, v_{z_n}}{\text{Im}\, z_n} \longrightarrow 0.$$

Then dom S is dense in $\mathcal{L}$. Indeed, the assumption is equivalent to

$$\|\varepsilon_{z_n}\|_{\mathcal{L}}^2 = \frac{\text{Im}\, v_{z_n}}{\text{Im}\, z_n} \longrightarrow 0$$

if $n \to \infty$. Now, for arbitrary $\sum_z f_z\varepsilon_z$ in $\mathcal{L}$, the element $\sum_z f_z\varepsilon_z - (\sum_z f_z)\varepsilon_{z_n}$ belongs to dom S and

$$\sum_z f_z\varepsilon_z - \left(\sum_z f_z\right)\varepsilon_{z_n} \longrightarrow \sum_z f_z\varepsilon_z.$$

These results are taken from [KL1]-[KL3]; in [SzK1] and [SzK2] a similar model is constructed but for S^{-1} instead of S.

If $\mathcal{L}_0 = \{\sum_z f_z \varepsilon_z : (\sum_z f_z \varepsilon_z, \sum_z f_z \varepsilon_z)_{\mathcal{L}} = 0\}$ is the isotropic part of $\mathcal{L}$ then the factor space $\mathcal{L}/\mathcal{L}_0$ is a pre-Hilbert space, whose completion is a Hilbert space which we denote by $\mathcal{H}$. Denote by τ the canonical embedding from $\mathcal{L}$ to $\mathcal{H}$. Then the Hermitian operator S in $\mathcal{L}$ induces a relation in $\mathcal{H}$ given by

$$\left\{ \left\{ \tau \sum_z f_z \varepsilon_z, \tau \sum_z z^* f_z \varepsilon_z \right\} : \sum_z f_z = 0 \right\}.$$

Its closure is a symmetric relation which will also be denoted by S. If, for example, the assumption in **(ii)** holds, dom S is dense in $\mathcal{H}$ and S is an operator. If $\mathcal{Z}$ is finite and the Pick matrix is nondegenerated, then S is a nondensely defined operator. In the following the elements of $\mathcal{L}$ and their images under τ are denoted by the same symbols.

Remark 1.2 There are various ways to make the space $\mathcal{L}$ and the operator S more explicit. For example, one can view the symbol ε_z, $z \in \mathcal{Z}$, as the function $\varepsilon_z(\zeta) = \frac{1}{z^* - \zeta}$. Then $\mathcal{L}$ is the linear space of functions on $\mathbb{C}\backslash\mathcal{Z}$ of the form $\zeta \mapsto \sum_z \frac{f_z}{z^* - \zeta}$ ($z \in \mathcal{Z}$, finite sum) with inner product

$$\left(\sum_z \frac{f_z}{z^* - \zeta}, \sum_w \frac{g_w}{w^* - \zeta} \right)_{\mathcal{L}} = \sum_{z,w} g_w^* p_{wz} f_z.$$

The operator S in $\mathcal{L}$ now becomes multiplication by the independent variable ζ. Indeed, from $\sum_z f_z = 0$, we get

$$S \sum_z \frac{f_z}{z^* - \zeta} = \sum_z \frac{z^* f_z}{z^* - \zeta} = \sum_z \frac{(z^* - \zeta + \zeta) f_z}{z^* - \zeta} = \zeta \sum_z \frac{f_z}{z^* - \zeta}.$$

In sequel we shall encounter other realizations of $\mathcal{L}$ and S. See, for example, in the proofs of Theorem 2.1 and Theorem 4.1.

We mention here already, that the solution of the interpolation problem is uniquely determined if the isotropic part $\mathcal{L}_0$ of $\mathcal{L}$ is not trivial, that is, if the inner product degenerates on $\mathcal{L}$. Indeed, if $\mathcal{L}_0$ degenerates, there exists a finite number of interpolation points $z_0, z_1, \dots, z_{n-1}$ such that the corresponding Pick matrix $\mathbb{P}_n$ is singular. Then, according to Theorem 4.1, the restricted problem with the interpolation points $z_0, z_1, \dots, z_{n-1}$ has a unique solution and the same is trivially true for the original problem. Therefore, if the interpolation problem has infinitely many solutions the isotropic part $\mathcal{L}_0$ is trivial and there is no need to form the factor space.

Consider the upper and lower defect numbers $n_\pm(S) := \dim(\ker(S^* - \zeta))$, $\zeta \in \mathbb{C}^\pm$, and the defect index $(n_-(S), n_+(S))$ of the symmetric relation S. From (1.2) it follows that S has defect numbers ≤ 1. Hence there are four possibilities for the defect index of S: $(0,0)$, $(1,0)$, $(0,1)$ or $(1,1)$. In the first case S itself is selfadjoint. In the last case S has infinitely many minimal nonisomorphic selfadjoint extensions, including the canonical extensions. To explain the last notions, let $\widetilde{A}$ be a selfadjoint extension of S in the Hilbert space $\widetilde{\mathcal{H}}$. Then the orthogonal complement $\widetilde{\mathcal{H}} \ominus \mathcal{H}$ is called the exit space. The extension $\widetilde{A}$ is called canonical if $\widetilde{\mathcal{H}} = \mathcal{H}$, that is, the exit space consists of the zero element only; it is called minimal if

$$\overline{\operatorname{span}}\,\{\varepsilon_z, (\widetilde{A}-\zeta)^{-1}\varepsilon_z : \zeta \in \rho(\widetilde{A}), z \in \mathcal{Z}\} = \widetilde{\mathcal{H}}.$$

Note that some of the selfadjoint extensions may be linear relations; see the Appendix. If the defect numbers of S are $(1,0)$ or $(0,1)$ then S is maximal symmetric and has a minimal selfadjoint extension; it is unique up to isomorphisms and has an infinite dimensional exit space. In Section 3 we shall see that for a symmetric relation S associated with a Nevanlinna-Pick problem as above the case of a defect index $(0,1)$ does not occur.

Proof of the if part of Theorem 1.1: Let $\widetilde{A}$ be an arbitrary selfadjoint extension of S in a Hilbert space $\widetilde{\mathcal{H}}$. We show that there is a solution of the interpolation problem of the form

$$N(\zeta) = N(\mu)^* + (\zeta - \mu^*)((I + (\zeta - \mu)(\widetilde{A} - \zeta)^{-1})u, u)_{\widetilde{\mathcal{H}}}, \tag{1.4}$$

where μ is any fixed point in $\mathbb{C}^+$ (see (9.5) in the Appendix):

If $z_0 \in \mathcal{Z} \cap \mathbb{C}^+$, we consider the function

$$N(\zeta) = v_{z_0}^* + (\zeta - z_0^*)((I + (\zeta - z_0)(\widetilde{A} - \zeta)^{-1})\varepsilon_{z_0}, \varepsilon_{z_0})_{\widetilde{\mathcal{H}}}. \tag{1.5}$$

It is of the form (1.4) with $u = \varepsilon_{z_0}$, $\mu = z_0$ as $N(z_0) = v_{z_0}$; the last equality implies that N satisfies (0.2) for $z = z_0$.

If $z_0 \in \mathcal{Z} \cap \mathbb{R}$, the function N in (1.5) takes the form (1.4) if we set, for example,

$$u = (I + (\mu - z_0)(\widetilde{A} - \mu)^{-1})\varepsilon_{z_0}.$$

Moreover, it satisfies (0.3) for $x = z_0$. Indeed, from $\|(\widetilde{A} - \zeta)^{-1}\| \leq 1/|\mathrm{Im}\,\zeta|$ we get

$$\lim_{\zeta \hat{\to} z_0} N(\zeta) = v_{z_0}.$$

If $\widetilde{E}(t)$ is the spectral function of $\widetilde{A}$ (see the Appendix) then

$$\frac{\mathrm{Im}\,N(\zeta)}{\mathrm{Im}\,\zeta} = d_{z_0} + \int_{-\infty}^{+\infty} \left(\frac{(t - z_0)^2}{|t - \zeta|^2} - 1\right) d(\widetilde{E}(t)\varepsilon_{z_0}, \varepsilon_{z_0}),$$

and hence

$$\begin{aligned}\lim_{\zeta \hat{\to} z_0} \frac{\mathrm{Im}\,N(\zeta)}{\mathrm{Im}\,\zeta} &= d_{z_0} - ((\widetilde{E}(z_0 + 0) - \widetilde{E}(z_0 - 0))\varepsilon_{z_0}, \varepsilon_{z_0}) \\ &= \|\widetilde{P}_0 \varepsilon_{z_0}\|^2 \leq \|\varepsilon_{z_0}\|^2 = d_{z_0},\end{aligned}$$

where $\widetilde{P}_0$ is the projection in $\widetilde{\mathcal{H}}$ onto $\overline{\operatorname{ran}}\,(\widetilde{A} - z_0)$. Hence $N(\zeta)$ satisfies the interpolation condition(s) for $\zeta = z_0 \in \mathcal{Z}$.

Assume that $\mathcal{Z}$ contains more than one point, and consider $z \in \mathcal{Z}$, $z \neq z_0$. By (1.1) with $w = z_0$,

$$\{\varepsilon_z - \varepsilon_{z_0}, (z^* - z_0{}^*)\varepsilon_{z_0}\} \in \widetilde{A} - z^*. \tag{1.6}$$

If $z \in \mathcal{Z} \cap \mathbb{C}^+$, it follows that

$$\varepsilon_z = \varepsilon_{z_0} + (z^* - z_0^*)(\widetilde{A} - z^*)^{-1}\varepsilon_{z_0}.$$

Taking the inner product with ε_{z_0}, we find

$$v_z = v_{z_0}^* + (z - z_0^*)((I + (z - z_0)(\widetilde{A} - z)^{-1})\varepsilon_{z_0}, \varepsilon_{z_0})_{\widetilde{\mathcal{H}}}.$$

Hence $N(\zeta)$ in (1.5) also solves the Nevanlinna–Pick problem for all $z \in \mathcal{Z} \cap \mathbb{C}^+$.

Now assume $z \in \mathcal{Z} \cap \mathbb{R}$. First we consider the case that $z \in \sigma_p(\widetilde{A})$. Denote by $\widetilde{P}_1$ the orthogonal projection in $\widetilde{\mathcal{H}}$ onto $\overline{\operatorname{ran}}\,(\widetilde{A} - z)$ and by $\widetilde{A}_1$ the restriction of $\widetilde{A}$ to $\overline{\operatorname{ran}}\,(\widetilde{A} - z)$. From (1.6) it follows that ε_{z_0} belongs to ran $(\widetilde{A} - z)$. Then by Lemma 9.1 in the Appendix applied to N given by (1.5),

$$\widetilde{P}_1\varepsilon_z = \varepsilon_{z_0} + (z - z_0^*)(\widetilde{A_1} - z)^{-1}\varepsilon_{z_0}, \tag{1.7}$$

and

$$\lim_{\zeta \hat{\to} z} N(\zeta) = v_{z_0}^* + (z - z_0^*)((I + (z - z_0)(\widetilde{A_1} - z)^{-1})\varepsilon_{z_0}, \varepsilon_{z_0})_{\widetilde{\mathcal{H}}}.$$

Writing out the equality $(\varepsilon_{z_0}, \widetilde{P}_1\varepsilon_z)_{\widetilde{\mathcal{H}}} = (\varepsilon_{z_0}, \varepsilon_z)_{\widetilde{\mathcal{H}}}$ with the help of (1.7) we obtain

$$v_{z_0}^* + (z - z_0^*)((I + (z - z_0)(\widetilde{A_1} - z)^{-1})\varepsilon_{z_0}, \varepsilon_{z_0})_{\widetilde{\mathcal{H}}} = v_z,$$

and hence

$$\lim_{\zeta \hat{\to} z} N(\zeta) = v_z.$$

Finally, again by (1.7),

$$\begin{aligned}\lim_{\zeta \hat{\to} z} \frac{\operatorname{Im} N(\zeta)}{\operatorname{Im} \zeta} &= \lim_{\zeta \hat{\to} z} \|(I + (\zeta - z_0)(\widetilde{A_1} - \zeta)^{-1})\varepsilon_{z_0}\|_{\widetilde{\mathcal{H}}}^2 \\ &= \|(I + (z - z_0)(\widetilde{A_1} - z)^{-1})\varepsilon_{z_0}\|_{\widetilde{\mathcal{H}}}^2 = \|\widetilde{P}_1\varepsilon_z\|_{\widetilde{\mathcal{H}}}^2 \le \|\varepsilon_z\|_{\widetilde{\mathcal{H}}}^2 = d_z.\end{aligned}$$

This implies that N given by (1.5) is a solution of the interpolation problem in the point z. If $z \notin \sigma_p(\widetilde{A})$, then $\widetilde{\mathcal{H}} = \overline{\operatorname{ran}}\,(\widetilde{A} - z)$ and the same arguments with $\widetilde{P}_1$ replaced by the identity operator yield that also in this case N in (1.5) interpolates in z. □

2 Characterization of the Solutions by Selfadjoint Extensions

In the previous section we have shown that every selfadjoint extension of the symmetric relation S in the model gives rise to a solution of the Nevanlinna–Pick interpolation problem, see (1.5). That is, we have proved part of the following theorem.

Theorem 2.1 *Let $z_0 \in \mathcal{Z}$. The formula*

$$N(\zeta) = v_{z_0}^* + (\zeta - z_0^*)((I + (\zeta - z_0)(\widetilde{A} - \zeta)^{-1})\varepsilon_{z_0}, \varepsilon_{z_0})_{\widetilde{\mathcal{H}}} \tag{2.1}$$

establishes a bijective correspondence between all solutions N of the interpolation problem and all minimal selfadjoint extensions $\widetilde{A}$ of (a unitary copy of) the symmetric relation S in the model.

Proof: It suffices to show that if $N \in \mathbf{N}$ is a solution of the interpolation problem, then N admits the representation (2.1), where $\widetilde{A}$ is a minimal selfadjoint extension of the relation S of the model $S, \mathcal{H}$, which is uniquely determined up to unitary equivalence. Let

$$N(\zeta) = N(\mu)^* + (\zeta - \mu^*)((I + (\zeta - \mu)(A_1 - \zeta)^{-1})u, u)_{\mathcal{H}_1} \tag{2.2}$$

be the u–minimal representation of N, where A_1 is a selfadjoint relation in the Hilbert space $\mathcal{H}_1$ (see the Appendix). We define for $\zeta \in \rho(A_1)$,

$$e(\zeta) = (I + (\zeta - \mu)^*(A_1 - \zeta^*)^{-1})u.$$

These elements are dense in $\mathcal{H}_1$, and by the resolvent identity,

$$(e(\zeta), e(\eta))_{\mathcal{H}_1} = \frac{N(\zeta)^* - N(\eta)}{\zeta^* - \eta}, \quad \zeta, \eta \in \rho(A_1). \tag{2.3}$$

Since N is a solution, for each $x \in \mathcal{Z} \cap \mathbb{R}$ the weak limit

$$e(x) := \text{w}-\lim_{\zeta \hat{\to} x} e(\zeta)$$

exists, and $(e(x), e(x))_{\mathcal{H}_1} \le d_x$. We denote by $\mathcal{Z}_s$ those $x \in \mathcal{Z} \cap \mathbb{R}$ for which this inequality is strict. With each $x \in \mathcal{Z}_s$ we associate a symbol e_x and the closed linear span $\mathcal{E}$ of these symbols becomes a Hilbert space under the inner product determined by

$$(e_x, e_y) = 0, \ x \ne y, \quad (e_x, e_x) = d_x - ((e(x), e(x))_{\mathcal{H}_1}, \quad x, y \in \mathcal{Z}_s.$$

In the Hilbert space $\widetilde{\mathcal{H}}$ which is the orthogonal sum $\widetilde{\mathcal{H}} = \mathcal{H}_1 \oplus \mathcal{E}$ the relation

$$\widetilde{A} = A_1 \oplus \overline{\text{span}}\,\{\{e_x, xe_x\} : x \in \mathcal{Z}_s\}$$

is selfadjoint. We set

$$\tilde{e}(z) = \begin{cases} e(z), & z \in \mathcal{Z}\backslash\mathcal{Z}_s, \\ e(z) + e_z, & z \in \mathcal{Z}_s, \end{cases}$$

and we consider in the subspace

$$\mathcal{P} = \overline{\text{span}}\,\{\tilde{e}(z) : z \in \mathcal{Z}\}$$

of the space $\widetilde{\mathcal{H}}$ the linear relation (the sums below are sums over finite subsets of $\mathcal{Z}$)

$$S = \text{closure}\left\{\left\{\sum_z f_z\tilde{e}(z),\ \sum_z z^* f_z\tilde{e}(z)\right\} : \sum_z f_z = 0\right\},$$

where the closure is taken in the orthogonal direct sum $\mathcal{P} \oplus \mathcal{P}$. Then S is a symmetric relation in $\mathcal{P}$ and a unitary copy of the relation in the model. We claim that $\widetilde{A}$ is a minimal extension of S.

(a) First we show it is an extension: For all $\zeta \in \rho(A_1)$,

$$\{(A_1 - \zeta^*)^{-1}u,\ \zeta^*(A_1 - \zeta^*)^{-1}u + u\} \in A_1,$$

hence $\{e(\zeta) - u, \zeta^* e(\zeta) - \mu u\} \in A_1 \subset \widetilde{A}$. It follows that

$$\{\tilde{e}(z) - u,\ z^*\tilde{e}(z) - \mu u\} \in \widetilde{A}, \quad z \in \mathcal{Z}, \tag{2.4}$$

and hence every finite linear combination of such elements belongs to $\widetilde{A}$:

$$\left\{\sum_z f_z\tilde{e}(z) - u\sum_z f_z,\ \sum_z z^* f_z\tilde{e}(z) - \left(\mu\sum_z f_z\right)u\right\} \in \widetilde{A}.$$

If $\sum_z f_z = 0$, this linear combination belongs to S, which implies that $S \subset \widetilde{A}$.

(b) Minimality means that

$$\widetilde{\mathcal{P}} := \overline{\text{span}}\,\{\tilde{e}(z),\ (\widetilde{A} - \zeta)^{-1}\tilde{e}(z) : \zeta \in \rho(\widetilde{A}), z \in \mathcal{Z}\} = \widetilde{\mathcal{H}}. \tag{2.5}$$

From (2.4) it follows that for $\zeta \in \rho(\widetilde{A})$,

$$(\widetilde{A} - \zeta)^{-1}\tilde{e}(z) = \frac{1}{z^* - \zeta}(\tilde{e}(z) - u) + \frac{\mu - \zeta}{z^* - \zeta}(\widetilde{A} - \zeta)^{-1}u.$$

Letting $\zeta \to \mu$, we find that $\tilde{e}(z) - u$ belongs to $\widetilde{\mathcal{P}}$; since also $\tilde{e}(z)$ belongs to this set, $u \in \widetilde{\mathcal{P}}$. The u-minimality of A implies $\mathcal{H}_1 \subset \widetilde{\mathcal{P}}$. Now since $\tilde{e}(z)$, $z \in \mathcal{Z}_s$, are elements of $\widetilde{\mathcal{P}}$, so are their projections onto the orthogonal complement of $\mathcal{H}_1$, hence $\mathcal{E} \subset \widetilde{\mathcal{P}}$. This proves (2.5).

We have yet to show that $N(\zeta)$ given by (2.2) is of the form (2.1) with $\varepsilon_{z_0} = e(z_0)$. For that we note that by (2.3)

$$N(\zeta) = N(\eta)^* + (\zeta - \eta^*)(e(\eta), e(\zeta))_{\mathcal{H}_1}.$$

Set $\eta = z_0$ if $z_0 \in \mathcal{Z} \cap \mathbb{C}^+$ or let $\eta \xrightarrow{\wedge} z_0$ if $z_0 \in \mathcal{Z} \cap \mathbb{R}$, to obtain

$$N(\zeta) = v_{z_0}^* + (\zeta - z_0^*)(\tilde{e}(z_0), e(\zeta))_{\widetilde{\mathcal{H}}}.$$

Here we use that $e(\zeta)$ belongs to $\mathcal{H}_1 \subset \widetilde{\mathcal{H}}$. In this equality we have that, by (2.4) with $z = z_0$,

$$e(\zeta) = (I + (\zeta^* - \mu)(\widetilde{A} - \zeta^*)^{-1})u = (I + (\zeta^* - z_0^*)(\widetilde{A} - \zeta^*)^{-1})\tilde{e}(z_0),$$

which readily yields (2.1).

Finally we must show that $\widetilde{A}$, because of its minimality, is, up to unitary equivalence, uniquely determined by the solution N and the data. In order to see this, assume that we have two relations $\widetilde{A}$ and $\widetilde{A}_0$ such that the representation (2.1) of N holds also with $\widetilde{A}$ replaced by $\widetilde{A}_0$ and $\widetilde{\mathcal{H}}$ replaced by $\widetilde{\mathcal{H}}_0$. Then $\widetilde{\mathcal{H}}_0$ is the closed linear span of the elements $\tilde{e}_0(z)$, $(\widetilde{A}_0 - \zeta)^{-1}\tilde{e}_0(z)$, $z \in \mathcal{Z}$, $\zeta \in \rho(\widetilde{A}_0)$. The elements ε_z, $z \in \mathcal{Z}$, of the space $\mathcal{H}$ of the model correspond to the elements $\tilde{e}(z)$ of $\widetilde{\mathcal{H}}$ and $\tilde{e}_0(z)$ of $\widetilde{\mathcal{H}}_0$. Further, from (2.3) and the relations which follow it we get for $z, \eta \in \mathcal{Z}$ and also for all nonreal z, η

$$(e(z), e(\eta))_{\widetilde{\mathcal{H}}} = (e_0(z), e_0(\eta))_{\widetilde{\mathcal{H}}_0}, \quad (\tilde{e}(z), \tilde{e}(\eta))_{\widetilde{\mathcal{H}}} = (\tilde{e}_0(z), \tilde{e}_0(\eta))_{\widetilde{\mathcal{H}}_0}.$$

Since we have that

$$(\widetilde{A} - \zeta^*)^{-1}\tilde{e}(z) = \frac{\tilde{e}(\zeta) - \tilde{e}(z)}{\zeta^* - z^*}$$

and that a corresponding relation with $\widetilde{A}_0$ and $\tilde{e}_0(z)$ holds, the inner products between two generating elements of $\widetilde{\mathcal{H}}$ in (2.5) and the corresponding generating elements of $\widetilde{\mathcal{H}}_0$ coincide. Thus, $\widetilde{A}$ and $\widetilde{A}_0$ are unitarily equivalent. □

Corollary 2.2 *The interpolation problem has either a unique or infinitely many solutions.*

Indeed, the first case prevails if and only if S is selfadjoint or maximal symmetric, that is, at least one of its defect numbers is zero. Otherwise S has infinitely many selfadjoint extensions, which give infinitely many solutions of the interpolation problem according to Theorem 2.1.

The solution N of the interpolation problem is called canonical if the corresponding selfadjoint relation $\widetilde{A}$ from Theorem 2.1 is a canonical extension of S. A more explicit characterization of the canonical solutions of the interpolation problem will be given in Theorem 3.1.

Theorem 2.3 *Let $x_0 \in \mathcal{Z} \cap \mathbb{R}$ and suppose that the interpolation problem has infinitely many solutions. Then either there exists a number d'_{x_0}, $d'_{x_0} \leq d_{x_0}$, such that for all solutions N of the interpolation problem,*

(2.6) $$\lim_{\zeta \xrightarrow{\wedge} x_0} \frac{\operatorname{Im} N(\zeta)}{\operatorname{Im} \zeta} = d'_{x_0},$$

or there exists a unique canonical solution N_0 such that

$$\min\left(\lim_{\zeta \hat{\to} x_0} \frac{\operatorname{Im} N(\zeta)}{\operatorname{Im}\zeta}\right) = \lim_{\zeta \hat{\to} x_0} \frac{\operatorname{Im} N_0(\zeta)}{\operatorname{Im}\zeta} < d_{x_0},$$

where the minimum is taken with respect to all solutions N of the interpolation problem, and for the canonical solutions $N \neq N_0$ we have

$$\lim_{\zeta \hat{\to} x_0} \frac{\operatorname{Im} N(\zeta)}{\operatorname{Im}\zeta} = d_{x_0}.$$

Proof: Additionally to the notations at the end of Section 1 denote by P_1 the orthogonal projection onto $\overline{\operatorname{ran}}\,(S - x_0)$ in $\mathcal{H}$. Then if the solution N corresponds to the selfadjoint extension $\widetilde{A}$ of S we have

$$(2.7) \qquad \|\varepsilon_{x_0}\|^2 = d_{x_0} \geq \lim_{\zeta \hat{\to} x_0} \frac{\operatorname{Im} N(\zeta)}{\operatorname{Im}\zeta} = \|\widetilde{P}_1\varepsilon_{x_0}\|^2_{\widetilde{\mathcal{H}}} \geq \|P_1\varepsilon_{x_0}\|^2_{\mathcal{H}}$$

(observe that ran $(\widetilde{A} - x_0) \supset$ ran $(S - x_0)$, hence $\|\widetilde{P}_1 v\|^2 \geq \|P_1 v\|^2_{\mathcal{H}}$ for $v \in \mathcal{H}$). If $\overline{\operatorname{ran}}\,(S - x_0) = \mathcal{H}$, then $P_1 = I$ and the righthand side of (2.7) equals $\|\varepsilon_{x_0}\|^2_{\mathcal{H}} = d_{x_0}$, which implies that the case (2.6) holds with $d'_{x_0} = d_{x_0}$. If $x_0 \in \sigma_p(S)$, we have dim (ran $(S - x_0))^{\perp} = 1$ and $\widetilde{P}_1 v = P_1 v$ for all $v \in \mathcal{H}$. Evidently, $P_1\varepsilon_{x_0}$ is independent of the extension $\widetilde{A}$, and also the case (2.6) prevails. If $\overline{\operatorname{ran}}\,(S - x_0) \neq \mathcal{H}$ and $x_0 \notin \sigma_p(S)$, then among the canonical extensions there is exactly one with $\widetilde{P}_1 = P_1$, namely the extension $\widetilde{A}_0$ with the property

$$\ker(\widetilde{A}_0 - x_0) \neq \{0\}.$$

For all other canonical extensions $\widetilde{A}$ we have $\overline{\operatorname{ran}}\,(\widetilde{A} - x_0) = \mathcal{H}$, hence $\widetilde{P}_1 = I$. In this case the second case of the alternative prevails. □

3 The Defect Numbers of S, Canonical Extensions

We start by recalling some well known results. Let $N \in \mathbf{N}$ and consider its integral representation (0.1):

$$N(\zeta) = \beta\zeta + \alpha + \int_{-\infty}^{+\infty} \left(\frac{1}{t - \zeta} - \frac{t}{1 + t^2}\right) d\sigma(t).$$

It leads to the Hilbert space $\mathbb{C}_\beta \oplus L^2(d\sigma)$ of all pairs $(\beta\tau; \varphi) : \tau \in \mathbb{C}, \varphi \in L^2(\sigma)$, with inner product

$$[(\beta\tau; \varphi), (\beta\omega; \psi)] := \beta\tau\omega^* + \int_{-\infty}^{+\infty} \phi\psi^*\, d\sigma,$$

$$(\beta\tau; \phi), (\beta\omega; \psi) \in \mathbb{C}_\beta \oplus L^2(\sigma).$$

This space is isomorphic with the space $\mathcal{L}(N)$ which we consider in the reproducing kernel approach; see, for example, [ABDS1]. Here, of course, if $\beta = 0$ the first component $\mathbb{C}_\beta$ does not appear. We also introduce a Hilbert space $\mathcal{H}_N$ as follows: Let $\mathcal{L}_N$ be the linear space of all formal finite sums $\sum_z f_z \varepsilon_z$ where $z \in \mathbb{C}^+$ (as in the definition of $\mathcal{H}$ in Section 2 but with $\mathcal{Z}$ replaced by $\mathbb{C}^+$) equipped with the inner product

$$\left(\sum_z f_z \varepsilon_z, \sum_w g_w \varepsilon_w\right)_{\mathcal{L}} = \sum_{z,w} g_w^* \frac{N(z)^* - N(w)}{z^* - w} f_z$$

$$= \beta \sum_w g_w^* \sum_z f_z + \int_{-\infty}^{+\infty} \sum_{z,w} \frac{g_w^* f_z}{(t - z^*)(t - w)} d\sigma(t)$$

and let $\mathcal{H}_N$ be the Hilbert space obtained by factoring and completion. With the identification

$$\varepsilon_z \longmapsto \left(\beta; \frac{1}{t - z^*}\right),$$

where the expression $\frac{1}{t-z^*}$ denotes the corresponding function of t for $t \in \mathbb{R}$, $\mathcal{H}_N$ can be considered as the subspace $\mathbb{C}_\beta \oplus L^2_+(d\sigma)$ of $\mathbb{C}_\beta \oplus L^2(d\sigma)$ where $L^2_+(d\sigma)$ is the subspace of $L^2(d\sigma)$ generated by the functions $\frac{1}{t-z^*}$, $z \in \mathbb{C}^+$. Here $L^2(d\sigma) \neq L^2_+(d\sigma)$ if and only if the nonnegative measure σ has a nontrivial absolutely continuous component σ_{ac} such that

$$\int_{-\infty}^{+\infty} \frac{\ln \sigma'_{ac}(t)}{1 + t^2} dt > -\infty, \tag{3.1}$$

or equivalently,

$$\int_{-\infty}^{+\infty} \frac{|\ln \sigma'_{ac}(t)|}{1 + t^2} dt < \infty.$$

For a proof see [A, pp. 261–266] and [DyM, Chapter 4]. Analogously, the subspace $L^2_-(d\sigma)$ of $L^2(d\sigma)$ is the closed linear span of the functions $\frac{1}{t-z^*}$, $z \in \mathbb{C}^-$, in $L^2(d\sigma)$, and we have $L^2(d\sigma) \neq L^2_-(d\sigma)$ if and only if (3.1) holds. In $\mathcal{H}_N$ the symmetric relation S_N is defined analogously to S in $\mathcal{H}$, and given by:

$$S_N = \{\{(0; f(t)), (\beta\tau; tf(t))\} : \tau \in \mathbb{C},\ f(t), tf(t) \in L^2_+(d\sigma)\},$$

hence S_N is the direct sum of a selfadjoint linear relation in $\mathbb{C}_\beta$ and the operator of multiplication by the independent variable in $L^2_+(d\sigma)$, or is just multiplication by the independent variable when $\beta = 0$. The defect numbers of S_N are ≤ 1. We claim that S_N either is selfadjoint, or is maximal symmetric and has defect index $(1, 0)$, and the latter occurs if and only if (3.1) holds. To prove the claim we suppose without loss of generality that $\beta = 0$. S_N has upper defect number

$n_+(S_N) = 0$, as ran $(S - z^*)$ for $z \in \mathbb{C}^+$ contains all the generating elements ε_w, $w \in \mathbb{C}^+$. If S_N is selfadjoint, or equivalenly, $n_-(S_N) = 0$, then

$$\frac{1}{t-z} = (I + (z - z^*)(S_N - z)^{-1})\frac{1}{t - z^*} \in L^2_+(d\sigma), \quad z \in \mathbb{C}^+,$$

hence $L^2_-(d\sigma) \subset L^2_+(d\sigma)$, which implies that $L^2_+(d\sigma) = L^2(d\sigma)$, that is, (3.1) does not hold. If (3.1) does not hold, then $L^2_+(d\sigma) = L^2(d\sigma)$ and since multiplication by the independent variable in $L^2(d\sigma)$ is a selfadjoint operator, S_N is selfadjoint in $L^2_+(d\sigma)$. This completes the proof of the claim.

Now let N be a solution of the interpolation problem having the integral representation (0.1). The selfadjoint relation $\widetilde{A}$ corresponding to this solution N can be considered in the Hilbert space $\mathbb{C}_\beta \oplus L^2(d\sigma)$ and is given by

$$\widetilde{A} = \{\{(0; f(t)), (\beta\tau; tf(t))\} : \tau \in \mathbb{C},\ f(t), tf(t) \in L^2(d\sigma)\},$$

see, for example, [ABDS1, just after Corollary 4 and Theorem 5]. Thus we get immediately the following characterization of canonical solutions.

Theorem 3.1 *The solution N of the Nevanlinna–Pick interpolation problem is canonical if and only if the functions $\frac{1}{t-z^*}$, $z \in \mathcal{Z}$, form a total subset of the Hilbert space $L^2(d\sigma)$.*

To only thing to check is that $\frac{1}{t-x_0}$ belongs to $L^2(d\sigma)$ for $x_0 \in \mathcal{Z} \cap \mathbb{R}$. But this follows from

$$\lim_{\zeta \hat{\to} x_0} \frac{\operatorname{Im} N(\zeta)}{\operatorname{Im} \zeta} = \lim_{\zeta \hat{\to} x_0} \beta + \int_{-\infty}^{+\infty} \frac{d\sigma(t)}{|t-\zeta|^2} = \beta + \int_{-\infty}^{+\infty} \frac{d\sigma(t)}{(t-x_0)^2} \le d_{x_0}.$$

Next we give an example of an interpolation problem where the symmetric relation S has unequal defect numbers. We start with the following simple lemma, the proof of which is left to the reader. Let $W := \{z = i + z_1 : \beta \le \arg z_1 \le \pi - \beta\}$ for some β with $0 < \beta < \frac{\pi}{2}$.

Lemma 3.2 *If $\widehat{F} \in \mathbf{N}$, then for all points $z \in W$ it holds*

$$|\widehat{F}(z)| \le C_1 + C_2|z|$$

with some constants C_1, C_2.

Now let $N \in \mathbf{N}$. We choose a sequence of points $z_n = i\eta_n$, $n = 0, 1, \dots$, such that $2 \le \eta_0 < \eta_1 < \cdots$, $\lim_{n\to\infty} \eta_n = \infty$, and

$$\lim_{R\to\infty} \frac{1}{\ln R} \int_0^R \frac{\nu(t)}{t^2}\, dt = \infty, \tag{3.2}$$

where $\nu(t)$ for $t > 0$ denotes the number of points z_n with $|z_n| \leq t$. Then the Nevanlinna-Pick interpolation problem with data set $\mathcal{Z}_0 = \{z_n : n = 0, 1, \ldots\}$ and $v_{z_n} = N(z_n)$, $n = 0, 1, \ldots$, is determined.

Indeed, assume that N_1 is a solution of this problem which is different from N. Then $N(i\eta_0) \neq N_1(i\eta_0)$ for some η_0 with $0 < \eta_0 < 2$ and without loss of generality we can suppose that $\eta_0 = 1$. The function $F := N - N_1$ has the property $|F(z)| \leq C_1 + C_2|z|$ for all $z \in W$. Consider the function $G(z) = F(z+i)$ in $W' := W - i$. We introduce the new variable $\zeta = z^{\gamma} e^{-i\frac{\pi}{2}}$, where $\gamma > 1$ is such that W' is mapped onto the closed right half plane. Then the function $\widehat{G}(\zeta) = G(z)$ is defined and analytic on the closed right half plane, and it holds

$$|\widehat{G}(\zeta)| \leq C_1 + C_2 |\zeta|^{\frac{1}{\gamma}} \tag{3.3}$$

with some constants C_1, C_2. With $\alpha(R) = \arcsin \frac{d}{R}$ for some $d > 0$ the formula (4.20) from [L] implies

$$\begin{aligned} \int_0^R \frac{\nu(t)}{t^2} dt &= \frac{1}{2\pi} \int_{-\frac{\pi}{2}+\alpha(R)}^{\frac{\pi}{2}-\alpha(R)} \frac{\ln |\widehat{G}(R\cos\theta e^{i\theta})|}{R} \frac{d\theta}{\cos^2\theta} + O(1) \\ &\leq \frac{1}{2\pi} \int_{-\frac{\pi}{2}+\alpha(R)}^{\frac{\pi}{2}-\alpha(R)} \frac{\ln^+ |\widehat{G}(R\cos\theta e^{i\theta})|}{R} \frac{d\theta}{\cos^2\theta} + O(1). \end{aligned}$$

Using the relation (3.3) the integral on the righthand side can easily be estimated from above by $D_1 + D_2 \ln R$ with positive constants D_1, D_2, hence the expression on the lefthand side of (3.2) remains bounded if $R \to \infty$, a contradiction to (3.2).

Now choose a function $N \in \mathbf{N}$ such that in its representation (0.1) we have $\beta = 0$ and for the measure σ the relation (3.1) holds. We construct the model $\mathcal{H}$, S for the interpolation problem with the set $\mathcal{Z}_0$ introduced above according to Section 1, and also the space $\mathcal{H}_N$ and the relation S_N. Then with the natural identification it evidently holds $\mathcal{H} \subset \mathcal{H}_N$ and $S \subset S_N$. The assumption (3.1) implies that the symmetric operator S_N has lower defect number $n_-(S_N) = 1$. As the interpolation problem has a unique solution, S is maximal symmetric. If S would be selfadjoint, then also S_N would be selfadjoint, a contradiction.

Finally we show that the relation S in the model cannot have the defect index $(0, 1)$. In this case the interpolation problem would have a unique solution, say N, and we consider the corresponding Hilbert space $\mathbb{C}_\beta \oplus L^2(d\sigma)$. Then the space $\mathcal{H}$ of the model is the closed linear span in $\mathbb{C}_\beta \oplus L^2(d\sigma)$ of the elements $\varepsilon_z = (\beta; \frac{1}{t-z^*})$, $z \in \mathcal{Z}$. If S has defect index $(0, 1)$, then for $z \in \mathbb{C}^+$ the functions

$$\begin{aligned} (I + (z - z_0^*)(S - z)^{-1})\varepsilon_{z_0} &= \left(\beta; \frac{1}{t - z_0^*}\right) \\ + \left(0; \frac{z - z_0^*}{(t - z)(t - z_0^*)}\right) &= \left(\beta; \frac{1}{t - z}\right) \end{aligned}$$

belong to $\mathcal{H}$. Hence we have $\mathbb{C}_\beta \oplus L^2_-(d\sigma) \subset \mathcal{H}$. On the other hand $\mathcal{H} \subset \mathbb{C}_\beta \oplus L^2_+(d\sigma)$. It follows that $\int_{-\infty}^{+\infty} \frac{\ln \sigma'_{ac}(t)}{1+t^2} dt = -\infty$. But in this case also the functions $\frac{1}{t-z^*}$, $z \in \mathbb{C}^+$, generate the space $L^2(d\sigma)$, hence $\mathcal{H} = \mathbb{C}_\beta \oplus L^2(d\sigma)$ and $S \subset \widetilde{A}$, therefore $S = S^*$, a contradiction to the assumption that S has defect index $(0, 1)$.

4 Uniqueness and Nonuniqueness of Solutions (I)

The following result is well known. We repeat the proof (see [W], and also [Br2]) here for the convenience of the reader.

Theorem 4.1 *Let the set $\mathcal{Z}$ of interpolation points be finite: $\mathcal{Z} = \{z_0, z_2, \ldots, z_{n-1}\}$. Then the Nevanlinna-Pick interpolation problem* (0.2), (0.3) *is determined if and only if the corresponding Pick matrix $\mathbb{P}_n$ degenerates.*

Proof: If $\mathbb{P}_n$ degenerates there exists an element $f^0 = \sum_{j=0}^{n-1} f^0_{z_j} \varepsilon_{z_j}$ in $\mathcal{L}_0$ such that $\sum_{j=0}^{n-1} |f^0_{z_j}| \neq 0$. We denote by e^* the functional $e^*(\sum_{j=0}^{n-1} f_{z_j}\varepsilon_{z_j}) := \sum_{j=0}^{n-1} f_{z_j}$ on $\mathcal{L}$ and by Z^* the operator

$$Z^* \sum_{j=0}^{n-1} f_{z_j}\varepsilon_{z_j} = \sum_{j=0}^{n-1} z_j^* f_{z_j}\varepsilon_{z_j}$$

in $\mathcal{L}$. Then $\operatorname{ran}(S-\zeta) = \{f \in \mathcal{L} : e^*((Z^*-\zeta)^{-1} f) = 0\}$ and for arbitrary $g \in \mathcal{L}$ the element

$$g - \frac{e^*((Z^*-\zeta)^{-1} g)}{e^*((Z^*-\zeta)^{-1} f^0)} f^0$$

belongs to $\operatorname{ran}(S-\zeta)$. Therefore, if we denote the linear relation which is induced by S in $\widehat{\mathcal{L}} := \mathcal{L}/\mathcal{L}_0$ by $\widehat{S}$, it follows that with

$$\widehat{x} = (Z^*-\zeta)^{-1}\left(g - \frac{e^*((Z^*-\zeta)^{-1} g)}{e^*((Z^*-\zeta)^{-1} f^0)} f^0\right)$$

and the element $\widehat{g}$ which corresponds to g in $\widehat{\mathcal{L}}$ we have $\{\widehat{x}; \widehat{g}\} \in \widehat{S} - \zeta$, hence $\operatorname{ran}(\widehat{S} - \zeta) = \widehat{\mathcal{L}}$ and the operator $\widehat{S}$ is selfadjoint.

Conversely, if $\mathbb{P}_n$ does not degenerate, we have $\mathcal{H} = \mathcal{L}$ for the model space and the domain of S is of codimension one, hence S has equal defect numbers $n_+(S) = n_-(S) = 1$. The theorem now follows from Theorem 2.1. □

Remark 4.2 If the Pick matrix $\mathbb{P}_n$ degenerates the unique solution N of the interpolation problem is given by the formula (see [W])

$$N(\zeta) = v_{z_k} + (\zeta - z_k)\frac{\sum_{j=0}^{n-1} p_{z_k z_j} \frac{\varphi_j}{z_j^*-\zeta}}{\sum_{j=0}^{n-1} \frac{\varphi_j}{z_j^*-\zeta}},$$

where

$$\begin{pmatrix} \varphi_0 \\ \varphi_1 \\ \vdots \\ \varphi_{n-1} \end{pmatrix}$$

is any nonzero vector in $\ker \mathbb{P}_n$. This formula can also be obtained from Theorem 2.1 by calculating $((\widehat{S} - \zeta)^{-1}\widehat{e}_k, \widehat{e}_k)$, where $\widehat{S}$ is defined in the proof of Theorem 4.1 above.

We now assume that $\mathcal{Z}$ consists of infinitely many points: $\mathcal{Z} = \{z_j : j = 0, 1, \ldots\}$. Then the model concerns a symmetric relation S in an infinite dimensional Hilbert space. The defect numbers of S are at most one, and hence S has infinitely many nonisomorphic minimal selfadjoint extensions if and only if both defect numbers are one. In this case and in this case only, by Theorem 2.1, there are infinitely many solutions to the interpolation problem. That one of the defect numbers is not zero, can be characterized by the convergence of a nondecreasing sequence. We use the following notation. For $n \in \mathbb{N}$, $\mathbb{P}_n$ is the $n \times n$ Pick matrix $(p_{z_j z_k})_{j,k=0}^{n-1}$, $a_n(\zeta)$ stands for the n-vector function

$$a_n(\zeta) = \begin{pmatrix} \frac{1}{z_0 - \zeta} \\ \frac{1}{z_1 - \zeta} \\ \vdots \\ \frac{1}{z_{n-1} - \zeta} \end{pmatrix}, \quad \zeta \in \mathbb{C} \backslash \mathcal{Z},$$

and $\alpha_n(\zeta)$ is the scalar function

$$\alpha_n(\zeta) = \sup \left\{ \frac{|a_n(\zeta^*)^* x|^2}{x^* \mathbb{P}_n x} : x \in \mathbb{C}^n, x \neq 0 \right\}, \quad \zeta \in \mathbb{C} \backslash (\mathbb{R} \cup \mathcal{Z}^*),$$

where $\mathcal{Z}^* = \{z^* : z \in \mathcal{Z}\}$. For each ζ, the sequence $(\alpha_n(\zeta))$ is nondecreasing, because, if n increases, the set of all quotients

$$\frac{|a_n(\zeta^*)^* x|^2}{x^* \mathbb{P}_n x}, \quad x \in \mathbb{C}^n, x \neq 0,$$

becomes larger. Here we assume that all $\mathbb{P}_n$ are nonnegative and invertible, for otherwise, by Theorem 1.1 and Theorem 4.1, there is at most one solution. The number $\alpha_n(\zeta)$ can also be written as

$$\alpha_n(\zeta) = a_n(\zeta^*)^* \mathbb{P}_n^{-1} a_n(\zeta^*).$$

Theorem 4.3 *Assume that* $\mathbb{P}_n > 0$ *for all* $n \in \mathbb{N}$. *The interpolation problem* (0.2), (0.3) *has infinitely many solutions if and only if for one (and hence all)* $\zeta \in \mathbb{C}^+$

and for one (and hence all) $\zeta \in \mathbb{C}^-\backslash\mathcal{Z}^*$,

$$\lim_{n\to\infty} \alpha_n(\zeta) < \infty. \tag{4.1}$$

Proof: One of the defect numbers of S is not equal to zero if for one and then for all ζ in one of the open halfplanes (but not in $\mathcal{Z}^*$):

$$\operatorname{ran}(S-\zeta) = \left\{ \sum_k f_{z_k}\varepsilon_{z_k} \in \mathcal{L} : \sum_k \frac{f_{z_k}}{z_k^* - \zeta} = 0 \right\}$$

is not dense in $\mathcal{L}$. The sums here are always over finite subsets of $\mathbb{N}$. This range is not dense if and only if the functional F_ζ defined by

$$F_\zeta\left(\sum_k f_{z_k}\varepsilon_{z_k}\right) = \sum_k \frac{f_{z_k}}{z_k^* - \zeta}$$

is continuous, that is, if and only if there exists a number $\gamma > 0$ such that

$$\left|\sum_k \frac{f_{z_k}}{z_k^* - \zeta}\right|^2 \leq \gamma \left(\sum_j f_{z_j}\varepsilon_{z_j}, \sum_k f_{z_k}\varepsilon_{z_k}\right)_{\mathcal{L}} = \gamma \sum_{j,k} f_{z_k}^* p_{z_k z_j} f_{z_j}.$$

This is equivalent to

$$\gamma \mathbb{P}_n - (\,\cdot\,, a_n(\zeta^*))_{\mathbb{C}^n} a_n(\zeta^*) \geq 0, \quad n = 1, 2, \ldots,$$

and this inequality is equivalent to

$$(\,\cdot\,, \mathbb{P}_n^{-1/2} a_n(\zeta^*))_{\mathbb{C}^n} \mathbb{P}_n^{-1/2} a_n(\zeta^*) \leq \gamma I,$$

or

$$\|\mathbb{P}_n^{-1/2} a_n(\zeta^*)\|_{\mathbb{C}^n}^2 = (\mathbb{P}_n^{-1} a_n(\zeta^*), a_n(\zeta^*))_{\mathbb{C}^n} \leq \gamma.$$

The lefthand side of this inequality is equal to $\alpha_n(\zeta)$. □

5 Parametrization of Solutions

In this section we suppose that the operator S of the model has defect index (1, 1). We shall apply the description of all u-resolvents of S (see, for example, [KL4], [LT], [Br1], [ABDS3], [DLS1]) and Theorem 2.1 in order to get a description of all solutions of the interpolation problem (0.2), (0.3). Below we use the following notation: If $W = (w_{jk})_{j,k=1}^2$ is a 2×2 matrix function, then for $T \in \widetilde{\mathbf{N}} := \mathbf{N}\cup\{\infty\}$, we denote by $W(\zeta)_{T(\zeta)}$ the fractional linear transform

$$W(\zeta)_{T(\zeta)} = \frac{w_{11}(\zeta)T(\zeta) + w_{12}(\zeta)}{w_{21}(\zeta)T(\zeta) + w_{22}(\zeta)}.$$

For $T(\zeta) \equiv \infty$ this expression reduces to $w_{11}(\zeta)/w_{21}(\zeta)$.

Recall that given a closed symmetric operator S with defect index $(1, 1)$ in a Hilbert space $\mathcal{H}$, an element $u \in \mathcal{H}$ is called a module for S if the set $r_u(S)$ of all $\zeta \in \mathbb{C}$ for which

$$\mathcal{H} = \operatorname{ran}(S - \zeta) \dot{+} \operatorname{span}\{u\} = \operatorname{ran}(S - \zeta^*) \dot{+} \operatorname{span}\{u\}, \quad \text{direct sum in } \mathcal{H}, \tag{5.1}$$

is nonempty. Then each element $f \in \mathcal{H}$ can be decomposed as

$$f = f_\zeta + \lambda_\zeta u, \tag{5.2}$$

where $f_\zeta \in \operatorname{ran}(S - \zeta)$ and $\lambda_\zeta \in \mathbb{C}$. We denote by $P(\zeta)$ and $Q(\zeta)$ the linear mappings from $\mathcal{H}$ to $\mathbb{C}$ defined by

$$P(\zeta)f = \lambda_\zeta, \qquad Q(\zeta)f = ((\overset{o}{A} - \zeta)^{-1} f_\zeta, u)_{\mathcal{H}}, \quad f \in \mathcal{H}. \tag{5.3}$$

Here $\overset{o}{A}$ is an arbitrary canonical selfadjoint extension of S. Since $f_\zeta \in \operatorname{ran}(S - \zeta)$, $Q(\zeta)f$ is independent of the special choice of $\overset{o}{A}$. The functionals $P(\zeta)$ and $Q(\zeta)$ are holomorphic at all points $\zeta \in r_u(S)$ of regular type of S.

If $\tilde{A}$ is a selfadjoint extension of S in some Hilbert space $\tilde{\mathcal{H}} \supset \mathcal{H}$, then

$$((\tilde{A} - \zeta)^{-1} u, u)_{\tilde{\mathcal{H}}}$$

is called a u-resolvent of S. The set of all u–resolvents of S can be parametrized over the Nevanlinna functions with parametrization matrix $W(\zeta)$ given by, for example, [KL4, formula (3.12)]. That is, the relation

$$((\tilde{A} - \zeta)^{-1} u, u)_{\tilde{\mathcal{H}}} = W(\zeta)_{T(\zeta)} \tag{5.4}$$

establishes a bijective correspondence between all u-resolvents of S and all $T \in \tilde{\mathbf{N}}$. If $r_u(S)$ contains a real point a, then the parametrization matrix $W(\zeta)$ can be chosen as the 2×2 matrix function

$$W^a(\zeta) = I - (\zeta - a) \begin{pmatrix} Q(\zeta) \\ -P(\zeta) \end{pmatrix} \begin{pmatrix} Q(a)^* & -P(a)^* \end{pmatrix} J, \quad J = \begin{pmatrix} 0 & -1 \\ 1 & 0 \end{pmatrix}. \tag{5.5}$$

The mappings $P(a)^*$ and $Q(a)^*$ are the adjoints of $P(a)$ and $Q(a)$ and map $\mathbb{C}$ to $\mathcal{H}$, where $\mathbb{C}$ is equipped with the usual inner product; we have, for example, $(P(a)^*\xi, f)_{\mathcal{H}} = \xi(P(a)f)^*$.

We apply these formulas to the operator S of the model. Let $\mathcal{Z} = \{z_j : j = 0, 1, 2, \ldots\}$; if $\mathcal{Z}$ contains only finitely many points, the following considerations simplify considerably and the obvious changes are left to the reader. As a module we choose the element $u = \varepsilon_{z_0}$ from Theorem 2.1. To show that this element is a module for S in the model we consider the symmetric relation S in $\mathcal{H}$ of the model as an extension of S_n in $\mathcal{L}_n = \mathbb{C}^n$ provided with the inner product $(x, y) = y^*\mathbb{P}_n x$. We denote by P_n the projection in $\mathcal{H}$ onto $\mathcal{L}_n$ and by $\overset{o}{\varphi}(\zeta)$ the element in $\mathcal{H}$ which spans $\ker(S^* - \zeta)$. This element is orthogonal to $\operatorname{ran}(S - \zeta^*)$ and therefore $P_n \overset{o}{\varphi}(\zeta)$ is orthogonal to $\operatorname{ran}(S_n - \zeta^*)$ in $\mathcal{L}_n$, that is, $P_n \overset{o}{\varphi}(\zeta) \in \ker(S_n^* - \zeta)$. On the other hand it is easy to see that in $\mathcal{L}_n$ the kernel $\ker(S_n^* - \zeta)$ is spanned by the vector $\mathbb{P}_n^{-1}(Z_n - \zeta)^{-1}e$, where $Z_n = \operatorname{diag}(z_0, z_1, \ldots, z_{n-1})$ and e is the n-column vector

$$e = \begin{pmatrix} 1 \\ 1 \\ \vdots \\ 1 \end{pmatrix}.$$

It follows that there is a complex number $\gamma_n(\zeta)$ such that

$$P_n \overset{o}{\varphi}(\zeta) = \gamma_n(\zeta)\mathbb{P}_n^{-1}(Z_n - \zeta)^{-1}e, \quad \zeta \notin \mathcal{Z}.$$

As P_n tends to the identity and $\overset{o}{\varphi}(\zeta) \neq 0$, this complex number is not zero for n sufficiently large, and we find for these n

$$(\overset{o}{\varphi}(\zeta), u)_{\mathcal{H}} = (P_n \overset{o}{\varphi}(\zeta), \varepsilon_{z_0})_{\mathcal{L}_n} = \gamma_n(\zeta)\frac{1}{z_0 - \zeta} \neq 0.$$

From this it follows that $r_u(S)$ contains all $\zeta \in \mathbb{C}\backslash\mathbb{R}$ for which ζ and ζ^* do not belong to $\mathcal{Z}$.

In order to find explicit expressions for $P(\zeta)$ and $Q(\zeta)$ we consider (5.2) for the element $f \in \mathcal{L}$. Observe that the inner product is nondegenerate on $\mathcal{L}$ (see Section 1) and that $\mathcal{L}$ is a dense subset of $\mathcal{H}$. It is easy to check that with $f = \sum_{j=0}^{\infty} f_j \varepsilon_{z_j}$ (these sums are actually finite sums), the decomposition (5.2) is

$$f = \left(f - (z_0^* - \zeta)\sum_{j=0}^{\infty} \frac{f_j}{z_j^* - \zeta}\varepsilon_{z_0}\right) + (z_0^* - \zeta)\sum_{j=0}^{\infty} \frac{f_j}{z_j^* - \zeta}\varepsilon_{z_0},$$

hence $P(\zeta)f = (z_0^* - \zeta)\sum_{j=0}^{\infty} \frac{f_j}{z_j^* - \zeta}$. According to the decomposition (5.2) this expression and hence also $\sum_{j=0}^{\infty} \frac{f_j}{z_j^* - \zeta}$ defines a continuous linear functional on $\mathcal{L}$, which can be extended by continuity to all of $\mathcal{H}$, and therefore, according to a theorem of F. Riesz, the functional $\sum_{j=0}^{\infty} \frac{f_j}{z_j^* - \zeta}$ can be represented by an element $e(\zeta^*) \in \mathcal{H}$:

$$f \longmapsto \sum_{j=0}^{\infty} \frac{f_j}{z_j^* - \zeta} = (f, e(\zeta^*))_{\mathcal{H}}, \quad f \in \mathcal{L}. \tag{5.6}$$

Evidently, for $f \in \mathcal{L}$, the function $\zeta \mapsto P(\zeta)f$ is holomorphic at all points outside the closure $\overline{\mathcal{Z}^*}$ of $\mathcal{Z}^*$, and its only singularities are simple poles. It follows from a basic result of M.G. Kreĭn (see [K], [GG], compare also [KL4]) that for all $f \in \mathcal{H}$ the function $P(\zeta)f$ is holomorphic outside $\overline{\mathcal{Z}^*}$ and that its singularities which are not accumulation points of $\overline{\mathcal{Z}^*}$ are poles of first order. The real points ζ where all functions $P(\zeta)f$, $f \in \mathcal{H}$, are holomorphic are points of regular type of S and belong to $r_u(S)$; see [K, Theorem 8]. Hence if $a \in \mathbb{R}\backslash\overline{\mathcal{Z}}$, then $a \in r_u(S)$. From its definition in (5.3) it follows that $Q(\zeta)f$ is holomorphic at all points ζ of regular type of S, where $P(\zeta)f$ is holomorphic, $f \in \mathcal{H}$. Indeed, for each such point ζ there exists a selfadjoint canonical extension $\overset{o}{A}$ of S such that $\zeta \in \rho(\overset{o}{A})$: This is true if $\zeta \neq \zeta^*$, and a real point ζ belongs to the spectrum of only one canonical extension $\overset{o}{A}$. Further, for $f = \sum_{j=1}^{\infty} f_j \varepsilon_{z_j} \in \mathcal{L}$,

$$
\begin{aligned}
&Q(\zeta)f\\
&= \left((\overset{o}{A} - \zeta)^{-1} \left(f - (z_0^* - \zeta) \sum_{j=0}^{\infty} \frac{f_j}{z_j^* - \zeta} \varepsilon_{z_0} \right), \varepsilon_{z_0} \right)_{\mathcal{H}}\\
&= \left((\overset{o}{A} - \zeta)^{-1} \left(\sum_{j=1}^{\infty} f_j \varepsilon_{z_j} - (z_0^* - \zeta) \sum_{j=1}^{\infty} \frac{f_j}{z_j^* - \zeta} \varepsilon_{z_0} \right), \varepsilon_{z_0} \right)_{\mathcal{H}}\\
&= \sum_{j=1}^{\infty} p_{0j} \frac{f_j}{z_j^* - \zeta} - p_{00} \sum_{j=1}^{\infty} \frac{f_j}{z_j^* - \zeta}.
\end{aligned}
$$

We introduce the infinite matrices

$$
\mathbb{P} = (p_{z_j z_k})_{j,k=0}^{\infty}, \quad p_{z_j z_k} = (\varepsilon_{z_k}, \varepsilon_{z_j})_{\mathcal{L}}; \quad Z = \operatorname{diag}(z_0, z_1, z_2, \ldots),
$$

and the column vectors

$$
e = \begin{pmatrix} 1 \\ 1 \\ 1 \\ \vdots \end{pmatrix}, \quad e_0 = \begin{pmatrix} 1 \\ 0 \\ 0 \\ \vdots \end{pmatrix}.
$$

Then the formulas for $P(\zeta)f$ and $Q(\zeta)f$ can be written as

$$
P(\zeta)f = (z_0^* - \zeta)e^*(Z^* - \zeta)^{-1}f, \quad Q(\zeta)f = (e_0^*\mathbb{P} - p_{00}e^*)(Z^* - \zeta)^{-1}f, \tag{5.7}
$$

where now f stands for the column vector

$$\begin{pmatrix} f_0 \\ f_1 \\ f_2 \\ \vdots \end{pmatrix}$$

of the coefficients f_j of ε_{z_j} in the representation of f; we frequently identify f with its corresponding column vector. Note that since $f \in \mathcal{L}$ all the sums which appear in these relations are finite. Again, $Q(\zeta)f$ is a continuous linear functional on $\mathcal{L}$ and hence on $\mathcal{H}$, and can therefore be written as

$$Q(\zeta)f = (f, q(\zeta^*))_{\mathcal{H}}$$

for some $q(\zeta^*) \in \mathcal{H}$.

According to Theorem 2.1 the solutions of the interpolation problem (0.2), (0.3) have the form

$$N(\zeta) = v_0^* + (\zeta - z_0^*)(\varepsilon_{z_0}, \varepsilon_{z_0})_{\mathcal{L}} + (\zeta - z_0^*)(\zeta - z_0)((\tilde{A} - \zeta)^{-1}\varepsilon_{z_0}, \varepsilon_{z_0})_{\tilde{\mathcal{H}}}$$

Inserting the expression for the u-resolvents from (5.4) into the righthand side of this relation we obtain

$$N(\zeta) = (X(\zeta)W^a(\zeta))_{\hat{T}(\zeta)} \tag{5.8}$$

with $W^a(\zeta)$ from (5.5) and

$$X(\zeta) = \begin{pmatrix} (\zeta - z_0^*)(\zeta - z_0) & v_0^* + p_{00}(\zeta - z_0^*) \\ 0 & 1 \end{pmatrix}. \tag{5.9}$$

That is, the formula (5.8) establishes a bijective correspondence between all $\widehat{T} \in \widetilde{\mathbf{N}}$ and all solutions of the interpolation problem (0.2), (0.3). In order to replace the matrix XW^a on the righthand side of (5.8) by a matrix which is more closely related to the data, we replace the functional $Q(\zeta)$ defined in (5.7) by the functional

$$f \mapsto v^*(Z^* - \zeta)^{-1}f, \qquad f \in \mathcal{L}, \tag{5.10}$$

with

$$v = \begin{pmatrix} v_0 \\ v_1 \\ v_2 \\ \vdots \end{pmatrix}.$$

Lemma 5.1 *The functional defined on $\mathcal{L}$ by* (5.10) *is continuous on $\mathcal{L}$ and hence on $\mathcal{H}$.*

Proof: We claim

$$(\zeta - z_0)e_0^*\mathbb{P} + (z_0 - z_0^*)p_{00}e^* + v_0^*e^* = -e_0^*\mathbb{P}(Z^* - \zeta) + v^*.$$

Indeed, the lefthand side equals

$$\begin{aligned}
&(\zeta - z_0)(p_{00}\ p_{01}\ p_{02}\ \ldots) + (v_0 - v_0^*)e^* + v_0^*e^* \\
&= (p_{00}p_{01}p_{02}\ldots)(\zeta - Z^*) \\
&+ (p_{00}\ p_{01}\ p_{02}\ \ldots)(Z^* - z_0) + v_0e^* \\
&= -e_0^*\mathbb{P}(Z^* - \zeta) + \left(\frac{v_0 - v_0^*}{z_0 - z_0^*}\ \frac{v_0 - v_1^*}{z_0 - z_1^*}\ \frac{v_0 - v_2^*}{z_0 - z_2^*}\ \cdots\right)(Z^* - z_0) + v_0e^* \\
&= -e_0^*\mathbb{P}(Z^* - \zeta) + v^*.
\end{aligned}$$

It follows that

$$\begin{aligned}
&(\zeta - z_0)e_0^*\mathbb{P}(Z^* - \zeta)^{-1} + (z_0 - z_0^*)p_{00}e^*(Z^* - \zeta)^{-1} + v_0^*e^*(Z^* - \zeta)^{-1} \\
&= -e_0^*\mathbb{P} + v^*(Z^* - \zeta)^{-1},
\end{aligned}$$

and hence

$$\begin{aligned}
&(z_0 - \zeta)(e_0^*\mathbb{P} - p_{00}e^*)(Z^* - \zeta)^{-1}f \\
&+ (z_0^* - \zeta)p_{00}e^*(Z^* - \zeta)^{-1}f - v_0^*e^*(Z^* - \zeta)^{-1}f \\
&= e_0^*\mathbb{P}f - v^*(Z^* - \zeta)^{-1}f.
\end{aligned} \tag{5.11}$$

The three terms on the lefthand side of this equality define continuous linear functionals on $\mathcal{L}$, see (5.7). The first term on the righthand side equals $-(f, \varepsilon_{z_0})_{\mathcal{H}}$, which is obviously a continuous linear functional on $\mathcal{L}$. It follows that also (5.10) defines a continuous linear functional on $\mathcal{L}$. □

The functional in (5.10) is generated by an element of $\mathcal{H}$, say $v(\zeta^*)$:

$$v^*(Z^* - \zeta)^{-1}f = (f, v(\zeta^*))_{\mathcal{H}}, \qquad f \in \mathcal{L}.$$

Under the assumption that there exists an $a \in \mathbb{R}\backslash\overline{\mathcal{Z}}$, we introduce the 2×2 matrix function θ^a:

$$\begin{aligned}
\theta^a(\zeta) = &\begin{pmatrix} 1 & 0 \\ 0 & 1 \end{pmatrix} \\
&- (\zeta - a)\begin{pmatrix} (v(a), v(\zeta^*))_{\mathcal{H}} & (v(a), e(\zeta^*))_{\mathcal{H}} \\ (e(a), v(\zeta^*))_{\mathcal{H}} & (e(a), e(\zeta^*))_{\mathcal{H}} \end{pmatrix}\begin{pmatrix} 0 & -1 \\ 1 & 0 \end{pmatrix}.
\end{aligned} \tag{5.12}$$

Theorem 5.2 *Suppose that the interpolation problem has infinitely many solutions and that the real point a does not belong to the closure $\overline{\mathcal{Z}}$. Then the relation*

$$N(\zeta) = \theta^a(\zeta)_{T(\zeta)}$$

establishes a bijective correspondence between all $T \in \widetilde{\mathbf{N}}$ and all solutions N of the interpolation problem (0.2), (0.3). *The entries of the matrix function θ^a are holomorphic functions on $\mathbb{C}\backslash\overline{\mathcal{Z}^*}$ with simple poles at the isolated points of $\overline{\mathcal{Z}^*}$.*

In the general case where $\mathbb{R} \subset \overline{\mathcal{Z}}$ is allowed, the relation

$$N(\zeta) = \theta(\zeta)_{T(\zeta)},$$

where $\theta(\zeta) = X(\zeta)W(\zeta)$ with $X(\zeta)$ given by (5.9) and $W(\zeta)$ by [KL4, formula (3.12)] (it also appears in the formula (5.4) above), gives a bijective correspondence between all $T \in \widetilde{\mathbf{N}}$ and all solutions N of the interpolation problem.

Proof of Theorem 5.2: Without loss of generality we suppose that $a = 0$, otherwise the interpolation points have to be shifted by a real number. From the definition of $X(\zeta)$ in (5.9):

$$X(0)^{-1} = \frac{1}{|z_0|^2}\begin{pmatrix} 1 & p_{00}z_0^* - v_0^* \\ 0 & |z_0|^2 \end{pmatrix}.$$

Since $p_{00}z_0^* - v_0^* \in \mathbb{R}$, the mapping $T(\zeta) \mapsto \hat{T}(\zeta) = (X(0)^{-1})_{T(\zeta)}$ is a bijection from $\widetilde{\mathbf{N}}$ onto itself:

$$\hat{T}(\zeta) = \frac{1}{|z_0|^2}(T(\zeta) - v_0^* + p_{00}z_0^* - v_0^*).$$

Now we use the description (5.8) of the solutions of the interpolation problem:

$$N(\zeta) = (X(\zeta)W^0(\zeta))_{\hat{T}(\zeta)} = (X(\zeta)W^0(\zeta)X(0)^{-1})_{T(\zeta)}.$$

Therefore the theorem follows if we show that

$$\frac{1}{z_0^*}(z_0^* - \zeta)\theta^0(\zeta) = X(\zeta)W^0(\zeta)X(0)^{-1}$$

and observe that the matrices $\theta^0(\zeta)$ and $\frac{1}{z_0^*}(z_0^* - \zeta)\theta^0(\zeta)$ define the same fractional linear transformation. We start with the expression on the righthand side:

$$\begin{aligned} & X(\zeta)W^0(\zeta)X(0)^{-1} \\ & = X(\zeta)X(0)^{-1} - \zeta X(\zeta)\begin{pmatrix} Q(\zeta) \\ -P(\zeta) \end{pmatrix}\begin{pmatrix} -P(0)^* & -Q(0)^* \end{pmatrix} X(0)^{-1} \end{aligned} \tag{5.13}$$

and introduce the following mappings from $\mathcal{L}$ to $\mathbb{C}$:

$$\begin{aligned} E(\zeta)f &= e^*(Z^* - \zeta)^{-1}f = (f, e(\zeta^*))_{\mathcal{H}}, \\ V(\zeta)f &= v^*(Z^* - \zeta)^{-1}f = (f, v(\zeta^*))_{\mathcal{H}}. \end{aligned}$$

Then, on account of (5.13) and (5.11), $P(\zeta) = (z_0^* - \zeta)E(\zeta)$ and

$$(z_0 - \zeta)Q(\zeta) = -(z_0^* - \zeta)p_{00}E(\zeta) + v_0^*E(\zeta) + e_0^*\mathbb{P} - V(\zeta).$$

With these relations it follows easily that

$$X(\zeta)\begin{pmatrix} Q(\zeta) \\ -P(\zeta)\end{pmatrix} = (z_0^* - \zeta)\begin{pmatrix} e_0^*\mathbb{P} - V(\zeta) \\ -E(\zeta)\end{pmatrix} \tag{5.14}$$

and

$$\begin{pmatrix} P(0)^* & Q(0)^* \end{pmatrix} X(0)^{-1} = \frac{1}{z_0^*}\begin{pmatrix} E(0)^* & e_0 - V(0)^* \end{pmatrix}.$$

Now the relation (5.13) implies

$$\begin{aligned}
&X(\zeta)W^0(\zeta)X(0)^{-1}\\
&= X(\zeta)X(0)^{-1} + \frac{1}{z_0^*}(z_0^* - \zeta)\begin{pmatrix} e_0^*\mathbb{P} - V(\zeta) \\ -E(\zeta)\end{pmatrix}\begin{pmatrix} E(0)^* & e_0 - V(0)^* \end{pmatrix}\\
&= -\frac{1}{z_0^*}\zeta(z_0^* - \zeta)\begin{pmatrix} V(\zeta) \\ E(\zeta)\end{pmatrix}\begin{pmatrix} V(0)^* & E(0)^* \end{pmatrix}\begin{pmatrix} 0 & -1 \\ 1 & 0\end{pmatrix}\\
&+ X(\zeta)X(0)^{-1} + \frac{1}{z_0^*}\zeta(z_0^* - \zeta)\Bigg[\begin{pmatrix} e_0^*\mathbb{P} \\ 0\end{pmatrix}\begin{pmatrix} E(0)^* & -V(0)^* \end{pmatrix}\\
&- \begin{pmatrix} V(\zeta) \\ E(\zeta)\end{pmatrix}\begin{pmatrix} 0 & e_0 \end{pmatrix} + \begin{pmatrix} e_0^*\mathbb{P} \\ 0\end{pmatrix}\begin{pmatrix} 0 & e_0 \end{pmatrix}\Bigg].
\end{aligned} \tag{5.15}$$

If we observe the relations

$$\begin{aligned}
e_0^*\mathbb{P}V(\eta)^* &= \frac{v_0}{z_0 - \eta^*}, & V(\zeta)e_0 &= \frac{v_0^*}{z_0^* - \zeta},\\
e_0^*\mathbb{P}E(\eta)^* &= \frac{1}{z_0 - \eta^*}, & E(\zeta)e_0 &= \frac{1}{z_0^* - \zeta},
\end{aligned} \tag{5.16}$$

and $e_0^*\mathcal{P}e_0 = p_{00}$, the expression in the square brackets becomes

$$\begin{pmatrix} \frac{1}{z_0} & -\frac{v_0}{z_0} \\ 0 & 0\end{pmatrix} - \begin{pmatrix} 0 & \frac{v_0^*}{z_0^* - \zeta} \\ 0 & \frac{1}{z_0^* - \zeta}\end{pmatrix} + \begin{pmatrix} 0 & p_{00} \\ 0 & 0\end{pmatrix},$$

and it can be checked by a straightforward calculation that the expression on the last two lines of (5.15) equals $\frac{1}{z_0^*}(z_0^* - \zeta)\begin{pmatrix} 1 & 0 \\ 0 & 1\end{pmatrix}$. □

Evidently, with the mappings $E(\zeta)$, $V(\zeta)$ the parametrization matrix $\theta^a(\zeta)$ in (5.12) can be written as

$$\theta^a(\zeta) = \begin{pmatrix} 1 & 0 \\ 0 & 1\end{pmatrix} - (\zeta - a)\begin{pmatrix} V(\zeta) \\ E(\zeta)\end{pmatrix}\begin{pmatrix} V(a)^* & E(a)^* \end{pmatrix} J.$$

Theorem 5.3 *The parametrization matrix $\theta^a(\zeta)$ in (5.12) satisfies the following identity:*

$$\frac{\theta^a(\zeta)J\theta^a(\eta)^* - J}{\zeta - \eta^*} = \begin{pmatrix} V(\zeta) \\ E(\zeta) \end{pmatrix} \begin{pmatrix} V(\eta)^* & E(\eta)^* \end{pmatrix}. \tag{5.17}$$

Proof: We use the fact that in general the u-resolvent matrix in (3.16) of [KL4] satisfies the relation (3.15) in [KL4] with $\beta(z) = 1$, that is, we have

$$\frac{W(\zeta)JW(\eta)^* - J}{\zeta - \eta^*} = \begin{pmatrix} Q(\zeta) \\ -P(\zeta) \end{pmatrix} \begin{pmatrix} Q(\eta)^* & -P(\eta)^* \end{pmatrix}. \tag{5.18}$$

In order to see this we observe that according to Theorem 3.9 of [KL4] the resolvent matrix W satisfies a relation (3.15) of [KL4] with some function $\beta(z)$. If in (3.15) we choose $\zeta = a$ and compare this relation with (3.16) it follows that $\beta(z)\beta(a)^* = 1$ and hence $\beta(z)\beta(\zeta)^* = 1$.

Now we consider the kernel on the lefthand side of (5.17) (as in the proof of Theorem 5.2 we assume $a = 0$):

$$\frac{\theta^a(\zeta)J\theta^a(\eta)^* - J}{\zeta - \eta^*}$$
$$= \frac{1}{\zeta - \eta^*}\left\{\frac{|z_0|^2}{(z_0^* - \zeta)(z_0 - \eta^*)} X(\zeta)W(\zeta)X(0)^{-1}JX(0)^{-*}W(\zeta)^*X(\eta) - J\right\}.$$

It is easy to see that

$$X(0)^{-1}JX(0)^{-*} = |z_0|^{-2}J.$$

If we now observe (5.18) and (5.14), a straightforward calculation yields

$$\frac{\theta^a(\zeta)J\theta^a(\eta)^* - J}{\zeta - \eta^*}$$
$$= \begin{pmatrix} V(\zeta) \\ E(\zeta) \end{pmatrix} \begin{pmatrix} V(\eta)^* & E(\eta)^* \end{pmatrix} + \frac{X(\zeta)JX(\eta)^*}{(\zeta - \eta^*)(z_0^* - \zeta)(z_0 - \eta^*)} - \frac{J}{\zeta - \eta^*}$$
$$- \begin{pmatrix} e_0^*\mathbb{P} \\ 0 \end{pmatrix} \begin{pmatrix} V(\eta)^* & E(\eta)^* \end{pmatrix} - \begin{pmatrix} V(\zeta) \\ E(\zeta) \end{pmatrix} \begin{pmatrix} e_0 & 0 \end{pmatrix}$$
$$+ \begin{pmatrix} e_0^*\mathbb{P} \\ 0 \end{pmatrix} \begin{pmatrix} e_0 & 0 \end{pmatrix}$$

and it follows easily (use (5.16) that the last five terms on the righthand side add up to zero. □

We mention that in the case where $\mathcal{Z}$ contains only finitely many points the description of the solutions in Theorem 5.2 coincides with the description of V.P. Potapov (see [KP]). Indeed, if we use the vector representation for $\mathcal{L}$, we have for $f \in \mathcal{L}$

$$e^*(Z^* - \zeta)^{-1}f = \left(f, (Z - \zeta^*)^{-1}e\right)_{\mathbb{C}^n} = \left(f, \mathbb{P}^{-1}(Z - \zeta^*)^{-1}e\right)_{\mathcal{H}},$$

hence $e(\zeta) = \mathbb{P}^{-1}(Z-\zeta)^{-1}e$ and, in the same way, $v(\zeta) = \mathbb{P}^{-1}(Z-\zeta)^{-1}v$. Now, the formula (5.12) becomes

$$\theta^a(\zeta) = \begin{pmatrix} 1 & 0 \\ 0 & 1 \end{pmatrix} - (\zeta - a)\begin{pmatrix} v^* \\ e^* \end{pmatrix}(Z^* - \zeta)^{-1}\mathbb{P}^{-1}(Z-a)^{-1}\begin{pmatrix} e & -v \end{pmatrix}$$

which is the representation of the parametrization matrix given by Potapov. By letting $a \to \infty$ we obtain another parametrization matrix:

$$\theta^\infty(\zeta) = \begin{pmatrix} 1 & 0 \\ 0 & 1 \end{pmatrix} - \begin{pmatrix} v^* \\ e^* \end{pmatrix}(Z^* - \zeta)^{-1}\mathbb{P}^{-1}\begin{pmatrix} e & -v \end{pmatrix}.$$

That this is a parametrization matrix can be seen by showing that $\theta^a(\zeta)^{-1}\theta^\infty(\zeta)$ is an iJ–unitary matrix which is independent of ζ (see [Br1, formula (1.35)]), but also as follows: Choose $a \in \mathbb{R}\backslash\mathcal{Z}$, transform the ζ–plane into a ζ'–plane by $\zeta' = -\frac{1}{\zeta} + a$, then write the parametrization matrix for the new interpolation problem and finally transform it back to the variable ζ.

We mention that P. Bruinsma (see [Br1], [ABDS3]) has considered a more general interpolation problem but with only finitely many data. He proved corresponding formulas by essentially the same method as in this section. However, he used a more complicated model, but in this model the derivation of the formulas (in the finite dimensional case) was easier.

6 Existence of a Solution and Reproducing Kernels

Throughout this section we assume that all Pick matrices associated with the interpolation problem (0.2), (0.3) are nonnegative; some may be singular. By the only if part of Theorem 1.1 (the proof of which has nothing to do with the extension method or the reproducing kernel method), this is a necessary condition for the existence of a solution of the interpolation problem. In the reproducing kernel approach, we take a concrete choice for the ε_z, $z \in \mathcal{Z}$, namely ε_z is chosen to be the function

$$f_z(\zeta) = \frac{\begin{pmatrix} v_z{}^* \\ 1 \end{pmatrix}}{z^* - \zeta}, \quad \zeta \in \mathbb{C}\backslash\mathbb{R}.$$

We denote by $\mathcal{M}$ the linear span of these functions provided with the inner product

$$(f_{z_1}, f_{z_2})_{\mathcal{M}} = p_{z_2 z_1}, \quad z_1, z_2 \in \mathcal{Z},$$

where $p_{z_i z_j}$ is given by (0.4). This inner product is well defined on $\mathcal{M}$ since the f_z's are linearly independent functions; it is nonnegative, and possibly degenerate.

For $N \in \mathbf{N}$ with integral representation (0.1) recall that $\mathcal{L}(N)$ is the reproducing kernel Hilbert space with reproducing kernel

$$K_N(\zeta, \eta) = \frac{N(\zeta) - N(\eta)^*}{\zeta - \eta^*}.$$

We refer to the Appendix for a review on these spaces, which were introduced by de Branges; see [B2]. Spaces $\mathcal{L}(N)$ play an important role in inverse scattering and in operator models; see [BR]. Their role in the present work is given by the following theorem:

Theorem 6.1 *The Nevanlinna function N is a solution of the interpolation problem* (0.2), (0.3) *if and only if the linear map τ which to f_z associates the function*

$$\tau f_z(\zeta) = (1 \quad - N(\zeta))f_z(\zeta) = \frac{N(\zeta) - v_z{}^*}{\zeta - z^*} \tag{6.1}$$

is a contraction from $\mathcal{M}$ into $\mathcal{L}(N)$.

The following simple lemmas are the key to the preceding result.

Lemma 6.2 *Let v_0 and z_0 be real numbers such that the function $\zeta \mapsto \frac{N(\zeta)-v_0}{\zeta - z_0}$ belongs to $\mathcal{L}(N)$. Then the function $t \mapsto \frac{1}{t-z_0}$ belongs to $L^2(d\sigma)$ and*

$$\frac{N(\zeta) - v_0}{\zeta - z_0} = \beta + \int_{-\infty}^{+\infty} \frac{1}{(t - z_0)} \frac{d\sigma(t)}{(t - \zeta)}. \tag{6.2}$$

Lemma 6.3 *Let v_0 and z_0 be numbers in $\mathbb{C}^+$ such that the function $\zeta \mapsto \frac{N(\zeta)-v_0{}^*}{\zeta - z_0{}^*}$ belongs to $\mathcal{L}(N)$. Then $N(z_0) = v_0$.*

Proof of Lemma 6.2: By hypothesis there are an $x \in \mathbb{C}$ and a function $f \in L^2(d\sigma)$ such that

$$\frac{N(\zeta) - v_0}{\zeta - z_0} = \beta x + \int_{-\infty}^{+\infty} \frac{f(t)d\sigma(t)}{t - \zeta},$$

and hence

$$N(\zeta) = v_0 + (\zeta - z_0)\beta x + (\zeta - z_0) \int_{-\infty}^{+\infty} \frac{f(t)d\sigma(t)}{t - \zeta}.$$

Take $\eta \in \mathbb{C} \setminus \mathbb{R}$ and apply the difference quotient operator R_η (this operator maps a function h to the function $R_\eta h(\zeta) = \frac{h(\zeta)-h(\eta)}{\zeta - \eta}$ to the functions on both sides of this equality. We obtain:

$$\begin{aligned}
\beta + & \int_{-\infty}^{+\infty} \frac{d\sigma(t)}{(t - \zeta)(t - \eta)} \\
&= \beta x + \frac{(\zeta - z_0) \int_{-\infty}^{+\infty} \frac{f(t)d\sigma(t)}{t-\zeta} - (\eta - z_0) \int_{-\infty}^{+\infty} \frac{f(t)d\sigma(t)}{t-\eta}}{\zeta - \eta} \\
&= \beta x + \int_{-\infty}^{+\infty} \frac{f(t)d\sigma(t)}{t - \eta} + (\zeta - z_0) \int_{-\infty}^{+\infty} \frac{f(t)d\sigma(t)}{(t - \zeta)(t - \eta)}.
\end{aligned}$$

Therefore

$$\beta + \int_{-\infty}^{+\infty} \left(\frac{1}{t-\zeta} - f(t) \right) \frac{d\sigma(t)}{t-\eta} = \beta x + \int_{-\infty}^{+\infty} \left(\frac{\zeta - z_0}{t-\zeta} f(t) \right) \frac{d\sigma(t)}{t-\eta}.$$

Hence, by the uniqueness of the representation of elements in $\mathcal{L}(N)$, $\beta = \beta x$ (so $x = 1$ when $\beta \neq 0$), and

$$\frac{1}{t-\zeta} - f(t) = \frac{\zeta - z_0}{t-\zeta} f(t),$$

that is, $f(t) = \frac{1}{t-z_0} \in L^2(d\sigma)$. □

Proof of Lemma 6.3: Since the functions in $\mathcal{L}(N)$ are analytic on $\mathbb{C}\backslash\mathbb{R}$, the function $\zeta \mapsto \frac{N(\zeta)-v_0{}^*}{\zeta - z_0{}^*}$ is analytic at $z_0{}^*$. Therefore, $N(z_0{}^*) = v_0{}^*$, which is equivalent to $N(z_0) = v_0$ since $N(\zeta^*)^* = N(\zeta)$. □

Proof of Theorem 6.1: Assume first that the map τ is a contraction. By Lemma 6.3, we have $N(z_0) = v_{z_0}$ for $z_0 \in \mathcal{Z} \cap \mathbb{C}^+$, and so the interpolation condition (0.2) at z is satisfied. Now let $z_0 \in \mathcal{Z} \cap \mathbb{R}$. By Lemma 6.2, the function $\frac{N(\zeta)-v_{z_0}}{\zeta - z_0}$ belongs to $\mathcal{L}(N)$ and has norm less or equal to $\sqrt{d_{z_0}}$. Using (6.2) we find that

$$\lim_{\zeta \hat{\to} z_0} \frac{\operatorname{Im} N(\zeta)}{\operatorname{Im} \zeta} = \lim_{\zeta \hat{\to} z_0} \beta + \int_{-\infty}^{+\infty} \frac{d\sigma(t)}{|t-\zeta|^2}$$

$$= \beta + \int_{-\infty}^{+\infty} \frac{d\sigma(t)}{(t-z_0)^2} = \left\| \frac{N(\zeta) - v_{z_0}}{\zeta - z_0} \right\|^2_{\mathcal{L}(N)} \leq d_{z_0}.$$

Thus N satisfies the first condition in (0.3). We now show that it satisfies the second condition of (0.3), that is, the interpolation condition. Let (ζ_j) be a sequence of points in $\mathbb{C}^+$ converging nontangentially to z_0 as $j \to \infty$. The functions

$$F_j(\zeta) = K_N(\zeta, \zeta_j) = \frac{N(\zeta) - N(\zeta_j)^*}{\zeta - \zeta_j{}^*} \tag{6.3}$$

belong to $\mathcal{L}(N)$ and their norms are uniformally bounded by some finite number M_{z_0} since

$$\int_{-\infty}^{+\infty} \frac{d\sigma(t)}{(t-z_0)^2} < \infty.$$

It follows that

$$\left| \frac{N(\zeta_j) - v_{z_0}}{\zeta_j - z_0} \right|^2 = \left| \left(\frac{N(\zeta) - v_{z_0}}{\zeta - z_0}, F_j(\zeta) \right)_{\mathcal{L}(N)} \right|^2 \leq M_{z_0}^2 \left\| \frac{N(\zeta) - v_{z_0}}{\zeta - z_0} \right\|^2_{\mathcal{L}(N)}$$

and so $\lim_{j\to\infty} N(\zeta_j) = v_{z_0}$, that is, $\lim_{\zeta \hat{\to} z_0} N(\zeta) = v_{z_0}$. We have shown that N is a solution of the interpolation problem.

Conversely, assume now that N is a solution. For $z \in \mathcal{Z} \cap \mathbb{C}^+$,

$$\tau f_z(\zeta) = \frac{N(\zeta) - v_z{}^*}{\zeta - z^*} = \frac{N(\zeta) - N(z)^*}{\zeta - z^*} = K_N(\zeta, z)$$

and so τf_z belongs to $\mathcal{L}(N)$.

Now let $z \in \mathcal{Z} \cap \mathbb{R}$, and let (ζ_j) denote a sequence of points in $\mathbb{C}^+$ converging nontangentially to z. The first condition in (0.3) forces the norms of the functions F_j (defined in (6.3)) to be uniformly bounded. Since $\mathcal{L}(N)$ is a Hilbert space, the sequence (F_j) has a weak limit, say g. By the reproducing kernel property, $\lim F_j(w) = g(w)$ for every $w \in \mathbb{C}^+$. On account of the second condition in (0.3), $g = \tau f_z$. Furthermore, by the properties of the weak limit,

$$\|g\|^2_{\mathcal{L}(N)} \leq \lim \|F_{\eta_i}\|^2_{\mathcal{L}(N)} \leq d_z.$$

Hence the map τ sends $\mathcal{M}$ into $\mathcal{L}(N)$. It remains to show that it is a contraction. From Lemmas 6.2 and 6.3 we have that for any $z \in \mathcal{Z}$,

$$\tau f_z(\zeta) = \beta + \int_{-\infty}^{+\infty} \frac{d\sigma(t)}{(t-\zeta)(t-z^*)} = K_N(\zeta, z).$$

Thus, for $n \in \mathbb{N}$, $z_0, \ldots, z_{n-1} \in \mathcal{Z}$ and $c_0, \ldots, c_{n-1} \in \mathbb{C}$,

$$\left\| \tau\left(\sum_{j=0}^{n-1} c_j f_{z_j}\right)\right\|^2_{\mathcal{L}(N)} = \beta \left|\sum_{j=0}^{n-1} c_j\right|^2 + \int_{-\infty}^{+\infty} \left|\sum_{j=0}^{n-1} \frac{c_j}{t - z_j{}^*}\right|^2 d\sigma(t)$$

$$= (c_0{}^* \cdots c_{n-1}{}^*)\, \mathbb{K}_n \begin{pmatrix} c_0 \\ \vdots \\ c_{n-1} \end{pmatrix} \leq (c_0{}^* \cdots c_{n-1}{}^*)\, \mathbb{P}_n \begin{pmatrix} c_0 \\ \vdots \\ c_{n-1} \end{pmatrix},$$

where the $n \times n$ matrices $\mathbb{K}_n$ and $\mathbb{P}_n$ are given by

$$\mathbb{K}_n = \left(\lim_{y \downarrow 0} K_N(z_j + iy, z_k + iy)\right)_{j,k=0}^{n-1}, \qquad \mathbb{P}_n = (p_{z_j z_k})_{j,k=0}^{n-1}.$$

It follows that τ is a contraction. □

We now show that under the assumption that all Pick matrices (defined by (0.4)) are nonnegative there is at least one solution. We first consider the case where (at least) one of them is singular.

Corollary 6.4 *If one of the Pick matrices is singular, then the interpolation problem* (0.2), (0.3) *has a unique solution. This solution is rational, and the corresponding measure σ in the Herglotz integral representation is either identically equal to* 0 *or is locally constant except at finitely many points where it has a jump.*

Proof: Write the elements of $\mathcal{Z}$ as a sequence $\mathcal{Z} = (z_j)$, $j = 0, 1, \ldots$, and assume that for some n and $z_0, z_1, \ldots, z_{n-1} \in \mathcal{Z}$ the Pick matrix $\mathbb{P}_n = (p_{z_j z_k})_{j,k=0}^{n-1}$ is singular. Then there is a nonzero function h of the form

$$h(\zeta) = \sum_{j=0}^{n-1} c_j f_{z_j}(\zeta) = \begin{pmatrix} h_1(\zeta) \\ h_2(\zeta) \end{pmatrix}$$

which is neutral in $\mathcal{M}$. Here $h_2(\zeta) = \sum_{j=0}^{n-1} \frac{c_j}{\zeta - z_j^*}$ is not identically equal to zero. Assume that N is a solution of the interpolation problem. By Theorem 6.1, the corresponding operator τ is a contraction, and therefore $\|\tau h\|_{\mathcal{L}(N)} = 0$. The space $\mathcal{L}(N)$ is a Hilbert space, so $\tau h(\zeta) \equiv 0$, that is,

$$N(\zeta) = \frac{h_1(\zeta)}{h_2(\zeta)}. \tag{6.4}$$

It follows that the solution N is unique (that is, does not depend on the choice of the element h) and rational. The last claim of the corollary about σ concerns general results on rational Nevanlinna functions. □

We now consider the case where all Pick matrices are strictly positive. As before, we assume some ordering $z_0, z_1, \ldots$ of the elements in $\mathcal{Z}$ and study the sequence of interpolation problems corresponding to the points $z_0, z_1, \ldots, z_{n-1}$ for $n = 1, 2, \ldots$. We recall the definitions of the $n \times n$ matrix Z_n and the n-column vectors v and e in $\mathbb{C}^n$:

$$Z_n = \operatorname{diag}(z_0, z_1, \ldots, z_{n-1}), \quad v = \begin{pmatrix} v_{z_0} \\ v_{z_1} \\ \vdots \\ v_{z_{n-1}} \end{pmatrix}, \quad e = \begin{pmatrix} 1 \\ 1 \\ \vdots \\ 1 \end{pmatrix}.$$

Furthermore we define the 2×2 matrix function $\theta_n(\zeta) = \theta_n^a(\zeta) = (\theta_{ij}(\zeta))_{j,k=1}^2$ by

$$\theta_n(\zeta) = \begin{pmatrix} 1 & 0 \\ 0 & 1 \end{pmatrix} - (\zeta - a) \begin{pmatrix} v^* \\ e^* \end{pmatrix} (Z_n^* - \zeta)^{-1} \mathbb{P}_n^{-1} (Z_n^* - a)^{-*} (e \quad -v), \tag{6.5}$$

where $a \in \mathbb{R} \backslash \mathcal{Z}_n$, where $\mathcal{Z}_n = \{z_0, z_1, \ldots, z_{n-1}\}$.

Theorem 6.5 *The formula*

$$N(\zeta) = (\theta_n)_T(\zeta) := \frac{\theta_{11}(\zeta)T(\zeta) + \theta_{12}(\zeta)}{\theta_{21}(\zeta)T(\zeta) + \theta_{22}(\zeta)} \tag{6.6}$$

gives a one-to-one correspondence between all solutions of the interpolation problem associated with $\mathcal{Z}_n$ and the parameters $T \in \widetilde{\mathbf{N}} = \mathbf{N} \cup \{\infty\}$.

If $\mathcal{Z}_n \subset \mathbb{C}^+$, this result is a special case of [ABDS2, Theorem 0.1]; see in particular formula (0.10) there. In the proof of the above theorem we use the following theorem of de Branges and Rovnyak; see [BR, Theorem 13]. Recall that J is the 2×2 matrix

$$J = \begin{pmatrix} 0 & -1 \\ 1 & 0 \end{pmatrix}.$$

Theorem 6.6 *Let θ be an iJ–inner function and $N \in \mathbf{N}$. Denote by $\mathcal{H}(\theta)$ and $\mathcal{L}(N)$ the reproducing kernel Hilbert spaces with reproducing kernels*

$$\frac{\theta(\zeta)J\theta(\eta)^* - J}{\zeta - \eta^*} \quad \text{and} \quad \frac{N(\zeta) - N(\eta)^*}{\zeta - \eta^*}$$

respectively. The map which to f associates the function

$$\tau f(\zeta) = (1 \quad -N(\zeta))f(\zeta) \tag{6.7}$$

is a contraction from $\mathcal{H}(\theta)$ into $\mathcal{L}(N)$ if and only if the kernel

$$-\frac{(1 \quad -N(\zeta))\theta(\zeta)J\theta(\eta)^* \begin{pmatrix} 1 \\ -N(\eta)^* \end{pmatrix}}{\zeta - \eta^*}$$

is nonnegative in $\mathbb{C}\backslash\mathbb{R}$, or equivalently, if and only if for some $T \in \widetilde{\mathbf{N}}$,

$$N = \theta_T(\zeta). \tag{6.8}$$

Sketch of the proof of Theorem 6.5: First of all, the space $\mathcal{M}_n$ spanned by the functions f_{z_j}, $j = 0, 1, \ldots, n-1$, with inner product

$$(f_{z_k}, f_{z_j})_{\mathcal{M}_n} = p_{z_j z_k},$$

is a finite dimensional reproducing kernel space with reproducing kernel

$$\frac{\theta(\zeta)J\theta(\eta)^* - J}{\zeta - \eta^*},$$

with $\theta = \theta_n$ given by (6.5). This is well known and comes from the fact that the Pick matrix $\mathbb{P} = \mathbb{P}_n$ is a solution of the Lyapunov equation

$$\mathbb{P}Z_n^* - Z_n\mathbb{P} = ev^* - ve^*.$$

For references and more details, see for example [ABDS3, Theorem 2.3]. Secondly, formula (6.6) describes the set of all solutions. Indeed, by Theorem 6.1, a Nevanlinna function N is a solution if and only if the map τ is a contraction from $\mathcal{M}_n$ into $\mathcal{L}(N)$ and since $\mathcal{M}_n = \mathcal{H}(\theta)$ with $\theta = \theta_n$, we may invoke Theorem 6.6. □

The function θ_n is rational and iJ–inner. It has McMillan degree n, and is a product of n elementary factors of McMillan degree 1. Such a factor is either a Blaschke factor, that is, a 2×2 matrix of the form

$$\theta(\zeta) = \begin{pmatrix} 1 & 0 \\ 0 & 1 \end{pmatrix} - (\zeta - a) \begin{pmatrix} v^* \\ 1 \end{pmatrix} \frac{v - v^*}{(z^* - \zeta)(z - z^*)(z - a)} (1 \quad - v),$$

or a Brune factor, that is, of the form

$$\theta(\zeta) = \begin{pmatrix} 1 & 0 \\ 0 & 1 \end{pmatrix} - \begin{pmatrix} v \\ 1 \end{pmatrix} \frac{1}{(\zeta - z)d_z} (1 \quad - v).$$

Brune factors are also called Blaschke Potapov factors of the third kind. (The factors of the first kind are Blaschke factors which are analytic in the upper open half plane, those of the second kind are analytic in the lower open half plane.) For other expressions, see Dym [Dy]. For every n, $\theta_n^{-1}\theta_{n+1}$ is a Blaschke factor of degree 1 if $z_n \in \mathcal{Z} \cap \mathbb{C}^+$ or a Brune factor of degree 1 if $z_n \in \mathcal{Z} \cap \mathbb{R}$.

We use a limit process to show that the interpolation problem has at least one solution, when $\mathcal{Z}$ consists of infinitely many points and all Pick matrices are strictly positive.

Corollary 6.7 *For every n, let N_n be a Nevanlinna function which solves the problem for $z_0, \ldots, z_{n-1}$. Then there is a subsequence of the sequence (N_n) converging to a Nevanlinna function N. This N is a solution of the interpolation problem.*

Proof: For every n, we set $R_n = \frac{1+iN_n}{1-iN_n}$. Then (R_n) is a sequence of Schur functions and hence by Montel's theorem, has a subsequence which converges uniformly on compact subsets of $\mathbb{C}_+$ to a Schur function R, say. All functions R_n have at $\zeta = z_1$ the value $\frac{1+iv_1}{1-iv_1}$. This number is different from -1 and so the limit function R is not the constant function identically equal to -1. The function $N = i\frac{1-R}{1+R}$ is a well defined Nevanlinna function and is a solution of the interpolation problem. □

7 Uniqueness and Nonuniqueness of Solutions (II)

We assume that $\mathcal{Z}$ has infinitely many points and that all Pick matrices are strictly positive. The question at hand is whether the interpolation problem has an infinite number of solutions, or a unique one. As we shall see there are no other possibilities. We begin with the following lemma.

Lemma 7.1 *For every natural number n, let $\mathcal{H}_n$ be a reproducing kernel Hilbert space of functions defined on a common set Ω with reproducing kernel $K_n(\zeta, \eta)$. Assume that for every n, $\mathcal{H}_n \subset \mathcal{H}_{n+1}$ and that the inclusion map is isometric. Let $\overset{\circ}{\mathcal{H}} = \cup_n \mathcal{H}_n$ be endowed with the inner product*

$$[u, v]_{\overset{\circ}{\mathcal{H}}} = [u, v]_{\mathcal{H}_n}, \qquad u, v \in \mathcal{H}_n.$$

Then $\overset{\circ}{\mathcal{H}}$ is a pre-Hilbert space. Denote by $\mathcal{H}$ its completion. Then $\mathcal{H}$ is a reproducing kernel Hilbert space if and only if for all $\zeta \in \Omega$,

$$\lim_{n \to \infty} K_n(\zeta, \zeta) < \infty. \tag{7.1}$$

The limit in (7.1) exists or equals ∞ because $\mathcal{H}_n \subset \mathcal{H}_{n+1}$ implies that

$$K_n(\zeta, \zeta) \leq K_{n+1}(\zeta, \zeta), \qquad \zeta \in \Omega.$$

A similar result can be found in Schwartz [Sc, Proposition 18].

Proof of Lemma 7.1: Suppose first that $\mathcal{H}$ is a reproducing kernel Hilbert space. Then the point evaluation $f \to f(\zeta)$ from $\overset{\circ}{\mathcal{H}} \to \mathbb{C}$ extends to a continuous map from $\mathcal{H} \to \mathbb{C}$. Thus there is a constant $k > 0$ such that

$$|f(\zeta)| \leq k \|f\|_{\overset{\circ}{\mathcal{H}}}, \qquad f \in \overset{\circ}{\mathcal{H}}.$$

In particular,

$$\|K_n(\cdot, \zeta)\|_{\mathcal{H}_n} = \sup_{0 \neq f \in \mathcal{H}_n} \frac{(f, K_n(\cdot, \zeta))_{\mathcal{H}_n}}{\|f\|_{\mathcal{H}_n}} = \sup_{0 \neq f \in \mathcal{H}_n} \frac{|f(\zeta)|}{\|f\|_{\mathcal{H}_n}} \leq k,$$

from which it follows that for all n,

$$K_n(\zeta, \zeta) = (K_n(\cdot, \zeta), K_n(\cdot, \zeta))_{\mathcal{H}_n} = \|K_n(\cdot, \zeta)\|^2_{\mathcal{H}_n} \leq k^2.$$

Conversely, assume that (7.1) holds and that $\lim_{n \to \infty} K_n(\zeta, \zeta) = k^2$. Then since the sequence $(K_n(\zeta, \zeta))$ is nondecreasing, we have for $f \in \mathcal{H}_n$,

$$|f(\zeta)|^2 \leq K_n(\zeta, \zeta) \|f\|^2_{\mathcal{H}_n} \leq k^2 \|f\|^2_{\mathcal{H}_n}.$$

Hence the map $f \to f(\zeta)$ extends by continuity to $\mathcal{H}$ and $\mathcal{H}$ is a reproducing kernel Hilbert space. □

The proof of the lemma shows that point evaluation is bounded at all ζ for which (7.1) holds. The set of these points may be smaller than Ω. For example, if

$$K_n(\zeta, \eta) = \frac{1 - \zeta^{n+1} \overline{\eta}^{n+1}}{1 - \zeta \overline{\eta}}, \qquad \Omega = \mathbb{C} \backslash \partial \mathbb{D},$$

then (7.1) holds if and only if $\zeta \in \mathbb{D}$. On the other hand if

$$K_n(\zeta,\eta) = \begin{pmatrix} \frac{1-\zeta^{n+1}\overline{\eta}^{n+1}}{1-\zeta\overline{\eta}} & 0 \\ 0 & \frac{\zeta^{-(n+1)}\overline{\eta}^{-(n+1)}-1}{1-\zeta\overline{\eta}} \end{pmatrix}, \qquad \Omega = \mathbb{C}\backslash(\{0\}\cup\partial\mathbb{D}),$$

then (7.1) does not hold for all $\zeta \in \mathbb{C}$.

We denote by f_j the vector function

$$f_j(\zeta) = \begin{pmatrix} f_{j1} \\ f_{j2} \end{pmatrix} = f_{z_j}(\zeta) = \frac{\begin{pmatrix} v_{z_j}^* \\ 1 \end{pmatrix}}{z_j{}^* - \zeta}, \qquad \zeta \in \mathbb{C}\backslash\mathbb{R},$$

and we provide the linear spaces $\mathcal{M}_n = \operatorname{span}\{f_j : j = 0, 1, \ldots, n-1\}$ and $\mathcal{M}_{n2} = \operatorname{span}\{f_{j2} : j = 0, 1, \ldots, n-1\}$ with the same inner product:

$$(f_k, f_j)_{\mathcal{M}_n} = (f_{k2}, f_{j2})_{\mathcal{M}_{2n}} = p_{z_j z_k}, \quad j, k = 0, 1, \ldots, n-1.$$

Recall that $\mathbb{P}_n = (p_{z_j z_k})_{j,k=0}^{n-1}$ is assumed to be positive, and that $\mathcal{M}_n$ concides with the reproducing kernel Hilbert space $\mathcal{H}(\theta_n)$ with reproducing kernel

$$\frac{\theta_n(\zeta)J\theta_n(\eta)^* - J}{\zeta - \eta^*},$$

where θ_n given by (6.5); see the proof of Theorem 6.5. The space $\mathcal{M}_{n2}$ is also a reproducing kernel Hilbert space: the reproducing kernel is given by

$$M_n(\zeta,\eta) := a_n(\zeta^*)^*\mathbb{P}_n^{-1}a_n(\eta^*),$$

where $a_n(\zeta)$ is the n-vector

$$a_n(\zeta) = \begin{pmatrix} \frac{1}{z_0-\zeta} \\ \frac{1}{z_1-\zeta} \\ \vdots \\ \frac{1}{z_{n-1}-\zeta} \end{pmatrix} = (Z-\zeta)^{-1}e, \quad \zeta \in \mathbb{C}\backslash\mathcal{Z}.$$

We set

$$\alpha_n(\zeta) = M_n(\zeta,\zeta) = a_n(\zeta^*)^*\mathbb{P}_n^{-1}a_n(\zeta^*).$$

Theorem 7.2 *Let $\mathcal{M}$ be the completion of the linear span of the reproducing kernel Hilbert spaces $\mathcal{M}_n = \mathcal{H}(\theta_n)$ as in Lemma 7.1. Then $\mathcal{M}$ is a reproducing kernel Hilbert space if and only if*

(7.2) $$\lim_{n\to\infty} \alpha_n(\zeta) < \infty, \qquad \zeta \in \mathbb{C}\backslash(\mathbb{R}\cup\overline{\mathcal{Z}}).$$

Proof: Assume first that the limit in (7.2) is finite. Let N be a solution of the interpolation problem. Such a solution exists since all Pick matrices $\mathbb{P}_n$ are positive. Then the map $\tau f(\zeta) = (1 \quad -N(\zeta))f(\zeta)$ from $\mathcal{H}(\theta_n)$ to $\mathcal{L}(N)$ is a contraction. Thus for every $f \in \mathcal{H}(\theta_n)$ there exist an $x_f \in \mathbb{C}$ and a function $h_f \in L^2(d\sigma)$ such that

$$f_1(\zeta) - N(\zeta)f_2(\zeta) = \beta x_f + \int_{-\infty}^{+\infty} \frac{h_f(t)d\sigma(t)}{t-\zeta}.$$

Here we assume that N has the integral representation (0.1). It follows that there exists a constant $c_\zeta \geq 0$ such that

$$|f_1(\zeta)| \leq |N(\zeta)||f_2(\zeta)| + \beta|x_f| + c_\zeta \|h_f\|_{L^2(d\sigma)}.$$

By Lemma 7.1 applied to the family of spaces $\mathcal{M}_{n2}$, condition (7.2) implies that there is a constant $k_\zeta \geq 0$ such that

$$\begin{aligned}|f_2(\zeta)| &= |(f_2, M_n(\cdot,\zeta))_{\mathcal{M}_n}| \leq \|f_2\|_{\mathcal{M}_n} M_n(\zeta,\zeta)^{1/2} \leq k_\zeta \|f_2\|_{\mathcal{M}_n}\\ &= k_\zeta \|f\|_{\mathcal{M}},\end{aligned}$$

and so

$$|f_1(\zeta)| \leq k_\zeta |N(\zeta)| \|f\|_{\mathcal{M}} + (c_\zeta^2 + \beta)^{1/2}(\beta|x_f|^2 + \|h_f\|^2_{L^2(d\sigma)})^{1/2}.$$

Moreover,

$$\beta|x_f|^2 + \|h_f\|^2_{L^2(d\sigma)} = \|\tau f\|^2_{\mathcal{L}(N)} \leq \|f\|^2_{\mathcal{M}_n} = \|f\|^2_{\mathcal{M}}.$$

Hence with $K_\zeta = k_\zeta|N(\zeta)| + (c_\zeta^2 + \beta)^{1/2}$, which is independent of n,

$$|f_1(\zeta)| \leq K_\zeta \|f\|_{\mathcal{M}},$$

and therefore

$$\|f(\zeta)\|_{\mathbb{C}^2} \leq (K_\zeta^2 + k_\zeta^2)^{1/2} \|f\|_{\mathcal{M}}.$$

This inequality implies that the point evaluation $f \mapsto f(\zeta)$ on span $\{\mathcal{M}_n : n \in \mathbb{N}\}$ can be extended to a continuous map from $\mathcal{M}$ to $\mathbb{C}^2$.

To prove the converse, assume $\mathcal{M}$ is a reproducing kernel space. Then $f \mapsto f(\zeta)$ is bounded and so are the maps $f \mapsto f_1(\zeta)$ and $f \mapsto f_2(\zeta)$. The latter is a mapping from span $\{\mathcal{M}_n : n \in \mathbb{N}\}$ to $\mathbb{C}$. Since $M_n(\cdot,\zeta)$ belongs to this span, there is a number $C_\zeta \geq 0$, independent of n, such that

$$\begin{aligned}M_n(\zeta,\zeta) &= (M_n(\cdot,\zeta), M_n(\cdot,\zeta))_{\mathcal{M}_n}\\ &= (M_n(\cdot,\zeta), M_n(\cdot,\zeta))_{\mathcal{M}} \leq C_\zeta \|M_n(\cdot,\zeta)\|_{\mathcal{M}} = C_\zeta M_n(\zeta,\zeta)^{1/2},\end{aligned}$$

that is, $\alpha_n(\zeta) \leq C_\zeta^2$, and this implies (7.2). □

The proof of the following theorem is an adaptation of the proof of [AD1, Theorem 2.2, p. 596].

Theorem 7.3 *If*

$$\lim_{n\to\infty} \alpha_n(\zeta) < \infty,$$

holds on an open subset Ω *of* $\mathbb{C}\backslash\mathbb{R}$ *which is symmetric with respect to the real axis, then there is a* iJ*-inner function* θ *on* Ω *such that the reproducing kernel* $K(\zeta,\eta)$ *of the completion* $\mathcal{M}$ *of the linear span of the spaces* $\mathcal{M}_n = \mathcal{H}(\theta_n)$ *is given by*

$$K(\zeta,\eta) = \frac{\theta(\zeta)J\theta(\eta)^* - J}{\zeta - \eta^*}, \qquad \zeta,\eta \in \Omega. \tag{7.3}$$

Furthermore, one can find a sequence of iJ*–unitary matrices* (U_n) *such that*

$$\lim_{n\to\infty} \theta_n U_n = \theta \tag{7.4}$$

in the symmetric open set Ω.

Proof: In the following the nonreal numbers ζ, η and μ belong to the symmetric open set Ω. Let

$$\begin{aligned} K_n(\zeta,\eta) &= \frac{\theta_n(\zeta)J\theta_n(\eta)^* - J}{\zeta - \eta^*}, \\ L_n(\zeta,\eta) &= (\zeta - \eta^*)K_n(\zeta,\eta) = \theta_n(\zeta)J\theta_n(\eta)^* - J. \end{aligned}$$

Then from the iJ-unitarity of θ_n we have that

$$L_n(\zeta,\eta) = L_n(\mu,\eta) + L_n(\zeta,\mu^*) - L_n(\zeta,\mu^*)JL_n(\mu,\eta).$$

By Theorem 7.2, the limit

$$\lim_{n\to\infty} K_n(\zeta,\eta) = K(\zeta,\eta), \qquad \zeta,\eta \in \Omega,$$

exists when $\zeta = \eta$. The case $\zeta \neq \eta$ then follows from the Cauchy–Schwarz inequality.

Let

$$L(\zeta,\eta) = \lim_{n\to\infty} L_n(\zeta,\eta).$$

Then $L(\mu,\mu^*) = 0$ and

$$L(\zeta,\eta) = L(\mu,\eta) + L(\zeta,\mu^*) - L(\zeta,\mu^*)JL(\mu,\eta),$$

and setting $M(\zeta,\eta) = J + L(\zeta,\eta)$, we get $M(\mu,\mu^*) = J$ and

$$L(\zeta,\eta) = -J - M(\zeta,\mu^*)JM(\mu,\eta).$$

It follows that

$$K(\zeta,\eta) = \frac{X_1(\zeta)JX_2(\eta)^* - J}{\zeta - \eta^*}, \tag{7.5}$$

where

$$\begin{aligned} X_1(\zeta) &= iM(\zeta,\mu^*) = iJ + i(\zeta-\mu)K(\zeta,\mu^*), \\ X_2(\zeta)^* &= iM(\mu,\zeta) = iJ + i(\mu-\zeta^*)K(\mu,\zeta). \end{aligned}$$

At this stage the proof breaks into two cases.

(1) The first case is when all the θ_n are analytic in a fixed neighbourhood of a real point. Then in the preceding analysis one can take $\mu \in \mathbb{R}$ and one obtains that $X_1 = X_2 =: \theta$. The kernel (7.5) is then of the form (7.3). Moreover, from

$$\theta_n(\zeta)(iJ) = iJ + i(\zeta-\mu)K_n(\zeta,\mu) \longrightarrow iJ + i(\zeta-\mu)K(\zeta,\mu) = \theta(\zeta),$$

and so (7.4) with $U_n = iJ$ holds.

(2) If $\mu \neq \mu^*$, then in general $X_1(\zeta) \not\equiv X_2(\zeta)$ and the kernel (7.5) does not have the symmetric form (7.3). We now show that X_1 and X_2 may be chosen to be equal. Since $K(\zeta,\eta) = K(\eta,\zeta)^*$, we have that

$$X_1(\zeta)JX_2(\eta)^* = X_2(\zeta)JX_1(\eta)^*,$$

and hence

$$E = X_1(\zeta)^{-1}X_2(\zeta)$$

is a constant matrix (independent of ζ). As $X_1(\mu) = iJ = X_2(\mu^*)^*$, we have that $\det X_1 \not\equiv 0$ and $\det X_2 \not\equiv 0$, and therefore the inverse $X_1(\zeta)^{-1}$ in the definition of E exists and E is invertible. Moreover, the matrix E satisfies $EJ = JE^*$, and so $(iJ)E^*$ is selfadjoint and

$$K(\zeta,\eta) = \frac{X_1(\zeta)JE^*X_1(\eta)^* - J}{\zeta - \eta^*}.$$

From $K(\zeta,\zeta) > 0$, we obtain

$$\begin{aligned} iJ &\geq X_1(\zeta)(iJ)E^*X_1(\zeta)^* \text{ if } \operatorname{Im}\zeta > 0, \\ & X_1(\zeta)(iJ)E^*X_1(\zeta)^* \geq iJ \text{ if } \operatorname{Im}\zeta < 0. \end{aligned}$$

As $X_1(\zeta)$ is invertible these inequalities imply that iJE^* is neither positive nor negative. Hence it has a positive and a negative eigenvalue and can therefore be written in the form $(iJ)E^* = W(iJ)W^*$ for some constant and invertible matrix W. If we set $\theta(\zeta) = X_1(\zeta)W$, then the kernel $K(\zeta,\eta)$ is of the form described in the theorem.

It remains to show that there is a sequence (U_n) of iJ–unitary matrices such that (7.4) holds. The argument is quite involved. It is somewhat easier to consider first J_0–inner functions, where J_0 is the signature matrix

$$J_0 = \begin{pmatrix} 1 & 0 \\ 0 & -1 \end{pmatrix},$$

and the class $\mathcal{S}$ of Schur functions of the upper half plane. We recall that if

$$\theta(\zeta) = \begin{pmatrix} a(\zeta) & b(\zeta) \\ c(\zeta) & d(\zeta) \end{pmatrix},$$

then the symbol $\theta(\zeta)_{T(\zeta)}$ denotes the linear fractional transformation

$$\theta(\zeta)_{T(\zeta)} = \frac{a(\zeta)T(\zeta)+b(\zeta)}{c(\zeta)T(\zeta)+d(\zeta)}.$$

In the following to simplify the notation we shall suppress the depence on ζ. It is well to recall that

$$\theta_{\mathcal{S}} := \{\theta_T \,:\, T \in \mathcal{S}\} \subset \mathcal{S},$$

and that there is equality when θ is a J_0–unitary constant. For ease of exposition, we proceed in a number of steps:

Step 1: *Let (θ_n) be a sequence of J_0–inner functions. The following are equivalent:*

(1) *There is a sequence of J_0–unitary matrices (U_n) and a J_0–inner function θ such that (maybe via a subsequence)*

$$\lim_{n\to\infty} \theta_n U_n = \theta.$$

(2) *There is a subsequence such that* $\lim_{n\to\infty} d_n^{-1}$ *is not identically zero, where we have set*

$$\theta_n U_n = \begin{pmatrix} a_n & b_n \\ c_n & d_n \end{pmatrix}.$$

(3) *There is a subsequence such that the intersection $\cap_n (\theta_n)_{\mathcal{S}}$ is not a singleton.*

Proof of Step 1: Before starting the proof, we recall the following: if

$$\theta = \begin{pmatrix} \alpha & \beta \\ \gamma & \delta \end{pmatrix}$$

is J_0–inner, then $\delta^{-1} \not\equiv 0$ and the function Σ defined by

$$\Sigma = \begin{pmatrix} \alpha - \beta\delta^{-1}\gamma & \beta\delta^{-1} \\ \delta^{-1}\gamma & \delta^{-1} \end{pmatrix},$$

the Potapov–Ginzburg transform of θ, is inner. In particular, the entries of Σ are Schur functions. Finally, $|\delta^{-1}(z)\gamma(z)| < 1$ at all points of analyticity z of δ in the upper half–plane. Still before starting the proof let us make the following preliminary computations. We fix some point ω_0 in the upper–half plane at which all the functions θ_n are analytic. Since θ_n is J_0–inner, we obtain that $K_n = -\delta_n(\omega_0)^{-1}\gamma_n(\omega_0)$ is strictly less than 1 in modulus. Let

$$U_n = H(-K_n) = \begin{pmatrix} \frac{1}{\sqrt{1-|K_n|^2}} & \frac{K_n}{\sqrt{1-|K_n|^2}} \\ \frac{K_n^*}{\sqrt{1-|K_n|^2}} & \frac{1}{\sqrt{1-|K_n|^2}} \end{pmatrix}.$$

The matrix U_n is J_0–unitary and we have that $c_n(\omega_0) = 0$ and $d_n(\omega_0)^{-1} \neq 0$. In particular $d_n(\omega_0)^{-1}c_n(\omega_0) = 0$. (This is the main point in the proof, as will appear in the sequel). We assume that a subsequence of the sequence $(\theta_n U_n)$ is chosen such that the limits $\lim_{n\to\infty} a_n - b_n d_n^{-1} c_n$, $\lim_{n\to\infty} d_n^{-1} c_n$, $\lim_{n\to\infty} b_n d_n^{-1}$, and $\lim_{n\to\infty} d_n^{-1}$ exist, and that furthermore

$$\lim_{n\to\infty} d_n^{-1} \not\equiv 0.$$

Let $T \in \cap(\theta_n)_S$. Then there is a sequence of Schur functions (T_n) such that $T = (\theta_n U_n)_{T_n}$. This can be rewritten as

$$T = b_n d_n^{-1} + (a_n - b_n d_n^{-1} c_n) T_n (1 + c_n d_n^{-1} T_n)^{-1} d_n^{-1}. \tag{7.6}$$

Since $c_n d_n^{-1}$ vanishes at the point ω_0 we can write,

$$|c_n d_n^{-1}(z)| \le |b_{\omega_0}(z)| < 1$$

for z in the upper half plane, where we have set

$$b_{\omega_0}(z) = \frac{z - \omega_0}{z - \omega_0^*}.$$

By Schwarz' lemma,

$$|(1 - c_n d_n^{-1} \sigma_n)^{-1}| \le \frac{1}{1 - |b_{\omega_0}(z)|}. \tag{7.7}$$

We are now in a position to prove Step 1: We prove first that (3) implies (2). Assume that (2) does not hold, that is, $\lim d_n^{-1} \equiv 0$. From (7.6) and the estimate (7.7) we obtain that $T = \lim c_n d_n^{-1}$, that is, the intersection is reduced to a point. Thus (3) does not hold. The fact that (2) implies (1) is a direct consequence of the Potapov–Ginzburg transform. That (1) implies (3) is also easy.

Step 2: *Let (X_n) be a sequence of iJ–inner functions. The following are equivalent:*

1. *There is a sequence of iJ–unitary matrices (U_n) such that $(X_n U_n)$ converges.*
2. *The intersection $\cap_n \{(X_n)_T : T \in \tilde{\mathbf{N}}_0\}$ is not a singleton.*

Proof of step 2: One simply goes from J_0 to iJ via $iJ = M J_0 M^{-1}$ with

$$M = \begin{pmatrix} -i & 1 \\ 1 & -i \end{pmatrix}.$$

Step 3: *The function θ in the kernel* (7.3) *may be chosen such that $\lim_{n\to\infty} \theta_n U_n = \theta$.*

Proof of step 3: We apply step 2 with $X_n = \theta_n$. The intersection in (2) is not a singleton because we assume that the interpolation problem has infinitely many solutions. So by (1), there exists a sequence of iJ–unitary matrices (U_n) such that $\tilde{\theta} = \lim_{n\to\infty} \theta_n U_n$ exists. From

$$\theta_n(\zeta) J \theta_n(\eta)^* = \theta_n(\zeta) U_n J U_n^* \theta_n(\eta)^*$$

we obtain that

$$\tilde{\theta}(\zeta)J\tilde{\theta}(\eta)^* = \theta(\zeta)J\theta(\eta)^*.$$

Hence $\theta = \tilde{\theta}U$, where U is a iJ–unitary matrix and so $\theta = \lim_{n\to\infty}\theta_n U_n V$. □

Theorem 7.4 *The function* θ *in Theorem* 7.3 *is a parametrization matrix for the Nevanlinna–Pick problem* (0.2), (0.3).

Proof of Theorem 7.4: For every n, the function $\Psi_n(\zeta) = \theta_n(\zeta)^{-1}\theta(\zeta)$ is iJ–inner, and hence if T belongs to $\widetilde{\mathbf{N}}$, then so does $(\Psi_n)_T$. It follows that $N = \theta_T = (\theta_n)_{(\Psi_n)_T}$ is a solution of the Nevanlinna–Pick problem for every n for the set $\{z_0, \ldots, z_{n-1}\}$, and hence is a solution of the Nevanlinna–Pick problem. Conversely, let N be a solution of the Nevanlinna–Pick problem; it is of the form $N = (\theta_n)_{T_n}$ for some sequence of parameters (T_n). There is a subsequence (T_{n_j}) which converges, uniformly on compact subsets of $\mathbb{C}\backslash\mathbb{R}$, to some $T \in \widetilde{\mathbf{N}}$, say. The Nevanlinna–Pick problem has more than one solution since (7.2) holds, and by the preceding theorem, the sequence (θ_n) may be assumed to converge to a limit, which is θ. Hence, we see that

$$N = \lim_{j\to\infty}(\theta_{n_j})_{T_{n_j}} = \theta_T.$$

As $\det\theta \not\equiv 0$, the relation $N = \theta_T$ defines a bijection between the solutions N and the parameters T. □

8 Examples

We now illustrate condition (4.1) or (7.2) by a number of examples. First take $z_m = m$, $d_m = \frac{1}{m+1}$ and $v_m = 0$ for $m = 0, \ldots$ Then

$$a_n(\zeta)^*\mathbb{P}_n^{-1}a_n(\eta) = \sum_{m=0}^{n-1}\frac{m+1}{(m-\zeta^*)(m-\eta)}.$$

The series $a_n(\zeta^*)^*\mathbb{P}_n^{-1}a_n(\zeta^*)$ diverges for every ζ. There is only one Nevanlinna function N (namely the function identically equal to 0) such that $N(m) = 0$ and $\lim_{\zeta\hat{\to}m}\frac{\operatorname{Im}N(\zeta)}{\operatorname{Im}\zeta} \leq \frac{1}{m+1}$.

With the same interpolation points z_m the choice $d_m = 1$ leads to

$$a_n(\zeta)^*\mathbb{P}_n^{-1}a_n(\eta) = \sum_{m=0}^{n-1}\frac{1}{(m-\zeta^*)(m-\eta)}.$$

The corresponding functions θ_n and θ are given by

$$(8.1)\quad \theta_n(\zeta) = \begin{pmatrix} 1 & 0 \\ t_n(\zeta) & 1 \end{pmatrix}, \quad t_n(\zeta) = -\left(\zeta - \frac{1}{2}\right)\sum_{m=0}^{n-1} \frac{1}{(m-\zeta)(m-\frac{1}{2})},$$

$$(8.2)\quad \theta(\zeta) = \begin{pmatrix} 1 & 0 \\ t(\zeta) & 1 \end{pmatrix}, \quad t(\zeta) = -\left(\zeta - \frac{1}{2}\right)\sum_{m=0}^{\infty} \frac{1}{(m-\zeta)(m-\frac{1}{2})}.$$

The above case is easy since there is analyticity across the real line. This does not hold in the next case. We take $\mathcal{Z}$ to be the rational numbers: $\mathcal{Z} = (r_m)$, and fix a sequence (d_m) such that

$$\lim_{n\to\infty} \alpha_n(\zeta) = \sum_{m=1}^{\infty} \frac{1}{d_m} \frac{1}{|r_m - \zeta|^2} < \infty.$$

For instance, if $r_m = \frac{p}{q}$ with p and q relatively prime, one can take $d_m = (p^2+q^2)^2$. In these examples the interpolation problem has infinitely many solutions.

Another choice where there is analyticity at a point on the real line is given by $z_m = \frac{1}{m}$, $v_m = 0$ and $d_m = m^2$, for $m = 1, \ldots$. Then

$$\lim_{n\to\infty} \alpha_n(\zeta) = \sum_{m=1}^{\infty} \frac{1}{m^2|\frac{1}{m} - \zeta|^2},$$

and so there are nontrivial Nevanlinna functions N such that $N(\frac{1}{m}) = 0$ with nontangential derivative at $\frac{1}{m}$ less or equal to m^2, $m = 1, \ldots$.

Finally we mention an example involving one real point: $z_0 = x_0 \in \mathbb{R}$, $v_{x_0} = y_0$ and $d_{x_0} = d_0$. The Nevanlinna functions $N(\zeta)$ with $N(x_0) = y_0$ and nontangential derivative at x_0 less than or equal to d_0 are given by

$$N(\zeta) = \frac{d_0(x_0 - \zeta)T(\zeta) + y_0^2}{-T(\zeta) + d_0(x_0 - \zeta) + y_0}, \quad T(\zeta) \in \widetilde{\mathbf{N}}.$$

This formula goes back to G. Julia [J].

9 Appendix: Nevanlinna Functions

The Herglotz integral representation of a Nevanlinna function N is given in (0.1) in the Introduction. Recall that for all $x \in \mathbb{R}$, the (possibly improper) limit

$$\lim_{\zeta \hat{\to} x} \frac{\operatorname{Im} N(\zeta)}{\operatorname{Im} \zeta} = \beta + \int_{-\infty}^{+\infty} \frac{d\sigma(t)}{|t-x|^2} =: d_x$$

exists, that is, $0 \le d_x \le \infty$. If $d_x = 0$, then $\beta = 0$ and $\sigma(t) \equiv$ constant on $\mathbb{R}$ and hence $N(\zeta) \equiv \alpha$ on $\mathbb{C}$. If $d_x < \infty$, then from

$$N(\zeta) = \beta\zeta + \alpha + \int_{-\infty}^{+\infty} \frac{(1+t\zeta)(t-x)}{1+t^2} \frac{(t-x)}{t-\zeta} \frac{d\sigma(t)}{(t-x)^2}$$

and from the boundedness of $\frac{t-x}{t-\zeta}$ as $\zeta \stackrel{\wedge}{\to} x$:

$$\text{(9.1)} \qquad \left|\frac{1}{2}\pi - \arg(\zeta - x)\right| \le \gamma < \frac{1}{2}\pi \Rightarrow \left|\frac{t-x}{t-\zeta}\right| \le 1 + \frac{1}{\cos\gamma},$$

it follows that the limit

$$\lim_{\zeta \stackrel{\wedge}{\to} x} N(\zeta) =: v_x$$

exists and $v_x \in \mathbb{R}$. Therefore in the interpolation problem, as mentioned in the Introduction, we require from the beginning that for the interpolation points $x \in \mathcal{Z} \cap \mathbb{R}$, the bound d_x on the Carathéodory derivative in x is positive and the limit value v_x at x is real.

If N is a Nevanlinna function and has the integral representation (0.1), the kernel

$$\text{(9.2)} \qquad K_N(\zeta, \eta) = \frac{N(\zeta) - N(\eta)^*}{\zeta - \eta^*} = \beta + \int_{-\infty}^{+\infty} \frac{d\sigma(t)}{(t-\zeta)(t-\eta^*)}$$

is nonnegative in $\mathbb{C}\backslash\mathbb{R}$ in the sense of reproducing kernels. We denote by $\mathcal{L}(N)$ the associated reproducing kernel space:

$$\mathcal{L}(N) = \left\{F(\zeta) = \beta x + \int_{-\infty}^{+\infty} d\sigma(t)\frac{f(t)}{t-\zeta} \,:\, x \in \mathbb{C},\ f \in L^2(d\sigma)\right\}.$$

The elements of $\mathcal{L}(N)$ are analytic in $\mathbb{C}\backslash\mathbb{R}$, and the norm is given by

$$\|F\|_{\mathcal{L}(N)} = (\beta|x|^2 + \|f\|^2_{L^2(d\sigma)})^{1/2}, \qquad F \in \mathcal{L}(N).$$

The $\mathcal{L}(N)$ spaces were introduced by de Branges [B2]. $\mathcal{L}(N)$ can be identified with the orthogonal sum Hilbert space $\mathbb{C}_\beta \oplus L^2(d\sigma)$, where $\mathbb{C}_\beta$ is identified with the range of multiplication by β on $\mathbb{C}$: it is the zero space when $\beta = 0$, and otherwise $\mathbb{C}_\beta = \{\beta\tau : \tau \in \mathbb{C}\}$ with inner product $(\beta\tau, \beta\omega) = \beta\tau\omega^*$.

To explain the operator representation of a Nevanlinna function we recall that a (closed) linear relation A in a Hilbert space $\mathcal{H}$ is a (closed) linear subset A of the direct sum space $\mathcal{H}^2 = \mathcal{H} \oplus \mathcal{H}$. The adjoint of A is the closed linear relation

$$A^* = \{\{u, v\} \in \mathcal{H}^2 : (v, f)_\mathcal{H} - (u, g)_\mathcal{H} = 0 \text{ for all } \{f, g\} \in A\}.$$

A is called symmetric if $A \subseteq A^*$ and selfadjoint if equality prevails. The graph of an operator is a linear relation; a linear relation A is (the graph of) an operator if and only if its multivalued part $A(0) = \{g \in \mathcal{H} : \{0, g\} \in A\}$ is the trivial subspace of $\mathcal{H}$. The resolvent set $\rho(A)$, the spectrum $\sigma(A)$, and the point spectrum $\sigma_p(A)$ of a closed linear relation A are defined as if A were an operator. The point ∞ is a regular point of a closed linear relation A if A is a bounded and everywhere defined operator, and it is an eigenvalue of A if $A(0)$ is nontrivial. Any selfadjoint relation A in a Hilbert space $\mathcal{H}$ can be decomposed as

$$A = A_\infty \oplus A_o \tag{9.3}$$

where $A_\infty^{-1} = 0$ is the operator of muliplication by 0 in the space $A(0)$ and A_o is a selfadjoint (densely defined) operator in the Hilbert space $\mathcal{H}_o = \mathcal{H} \ominus A(0)$. A_0 is called the operator part of A. It follows, for example, that $\mathbb{C}\backslash\mathbb{R} \subseteq \rho(A)$; in fact, if P_o denotes the orthogonal projection in $\mathcal{H}$ onto $\mathcal{H} \ominus A(0)$, then for $z \in \mathbb{C}\backslash\mathbb{R}$, $(A-z)^{-1} = (A_o - z)^{-1} P_o$. We define the spectral function $E(t)$ of the selfadjoint relation (9.3) as the orthogonal sum $E(t) = 0 \oplus E_o(t)$, where $E_o(t)$ is the spectral function of A_o. Then

$$(A - \zeta)^{-1} = \int_{-\infty}^{+\infty} \frac{dE(t)}{t - \zeta}. \tag{9.4}$$

Of course, $\int_{-\infty}^{+\infty} dE(t) = 0 \oplus I = P_o$, where I is the identity on $\mathcal{H}_o$.

The operator representation can be stated as follows. The function N belongs to **N** if and only if it can be written as

$$N(\zeta) = N(\mu)^* + (\zeta - \mu^*)((I + (\zeta - \mu)(A - \zeta)^{-1})u, u)_{\mathcal{H}}, \tag{9.5}$$

where $A = A^*$ is a selfadjoint linear relation in a Hilbert space $\mathcal{H}$, μ is a point in $\rho(A)$, and u belongs $\mathcal{H}$. For proofs, see, for example, [KL2] and [DLS2]. The representation (9.5) can be chosen u–minimal which means that

$$\overline{\text{span}}\,\{(I + (\zeta - \mu)(A - \zeta)^{-1})u : \zeta \in \rho(A)\} = \mathcal{H}.$$

In that case the representation is unique up to an isomorphism. To see this, suppose that N has the u-minimal representation (9.5) and also the v-minimal representation

$$N(\zeta) = N(\lambda)^* + (\zeta - \lambda^*)((I + (\zeta - \lambda)(B - \zeta)^{-1})v, v)_{\mathcal{K}},$$

where $B = B^*$ in a Hilbert space $\mathcal{K}$, $\lambda \in \rho(B)$, and $v \in \mathcal{K}$. For elements of the form

$$\begin{aligned} e(\zeta) &= (I + (\zeta - \mu)(A - \zeta)^{-1})u \in \mathcal{H}, \\ d(\zeta) &= (I + (\zeta - \lambda)(B - \zeta)^{-1})v \in \mathcal{K}, \quad \zeta \in \mathbb{C}\backslash\mathbb{R}, \end{aligned}$$

we have that

$$(e(\zeta), e(\eta))_{\mathcal{H}} = \frac{N(\zeta) - N(\eta)^*}{\zeta - \eta^*} = (d(\zeta), d(\eta))_{\mathcal{K}}.$$

This implies that the mapping $Ue(\zeta) = d(\zeta)$ can be extended to a unitary mapping from $\mathcal{H}$ onto $\mathcal{K}$. From $U(A-\eta)^{-1}e(\zeta) = (B-\eta)^{-1}Ue(\zeta)$, $\zeta, \eta \in \mathbb{C}\backslash\mathbb{R}$, we deduce that the relations A and B are unitarily equivalent under this extended mapping.

The integral representation (0.1) of a Nevanlinna function can be obtained from its operator representation (9.5) and (9.4). One can show that if N has the u-minimal representation (9.5) then A is multivalued, that is, $A(0)$ is nontrivial, if and only if $\beta \neq 0$ in the integral representation (0.1) of N.

Let A be a selfadjoint relation in a Hilbert space $\mathcal{H}$ and let x be a point in $\mathbb{R}$. Then $\ker(A-x) = \ker(A_o - x)$ and $\operatorname{ran}(A-x) = A(0) \oplus \operatorname{ran}(A_o - x)$. We denote by P_1 the orthogonal projection in $\mathcal{H}$ onto $\overline{\operatorname{ran}}(A-x)$ and by A_1 the restriction of A to $\overline{\operatorname{ran}}(A-x)$ if A is densely defined, otherwise A_1 stands for the direct sum of A_∞ and the restriction of A_o to $\overline{\operatorname{ran}}(A_o - x)$. Evidently, if $x \in \sigma_c(A) \cup \rho(A)$, then $P_1 = I$ and $A_1 = A$.

Lemma 9.1 *Let N be a Nevanlinna function with u-minimal representation* (9.5) *and assume that for some $x \in \mathbb{R}$, $u \in \operatorname{ran}(A-x)$. Then for $\zeta \in \rho(A) \cup \{x\}$,*

$$P_1(A-\zeta)^{-1}u = (A_1 - \zeta)^{-1}u, \tag{9.6}$$

and

$$N(\zeta) = N(\mu)^* + (\zeta - \mu^*)((I + (\zeta - \mu)(A_1 - \zeta)^{-1})u, u)_{\mathcal{H}}. \tag{9.7}$$

Moreover,

$$\lim_{\zeta \hat{\to} x} N(\zeta) = N(\mu)^* + (x - \mu^*)((I + (x - \mu)(A_1 - x)^{-1})u, u)_{\mathcal{H}}. \tag{9.8}$$

Proof: We only consider the case that $x \in \sigma_p(A)$. Assume first that A is a densely defined selfadjoint operator. Then it can be written in the matrix form

$$A = \begin{pmatrix} A_1 & 0 \\ 0 & x \end{pmatrix} : \begin{pmatrix} \overline{\operatorname{ran}}(A-x) \\ \ker(A-x) \end{pmatrix} \longrightarrow \begin{pmatrix} \overline{\operatorname{ran}}(A-x) \\ \ker(A-x) \end{pmatrix}.$$

This representation implies that

$$(A-\zeta)^{-1}u = \begin{pmatrix} (A_1 - \zeta)^{-1}u \\ 0 \end{pmatrix},$$

whence (9.6) and (9.7). To prove the second part of the lemma for this case, we denote by $E_1(t)$ the spectral function of A_1. Since $u \in \operatorname{ran}(A-x) = \operatorname{ran}(A_1 - x)$, $u = (A_1 - x)v$ for a $v \in \operatorname{dom} A_1$, and we have for $z \in \mathbb{C}^+$,

$$\begin{aligned} ((A_1 - \zeta)^{-1}u, u)_{\mathcal{H}} &= \int_{-\infty}^{+\infty} \frac{(t-x)^2}{t-\zeta}\, d(E_1(t)v, v)_{\mathcal{H}}, \\ \frac{\operatorname{Im}((A_1 - \zeta)^{-1}u, u)_{\mathcal{H}}}{\operatorname{Im}\zeta} &= \int_{-\infty}^{+\infty} \left|\frac{t-x}{t-\zeta}\right|^2 d(E_1(t)v, v)_{\mathcal{H}}. \end{aligned} \tag{9.9}$$

By (9.1) and the dominated convergence theorem, the limits for $\zeta \stackrel{\wedge}{\to} x$ of the expresssions on the lefthand side of the equalities in (9.9) exist. Moreover, from

$$((A_1 - \zeta)^{-1}u, u)_{\mathcal{H}} - ((A_1 - x)^{-1}u, u)_{\mathcal{H}}$$
$$= (\zeta - x) \int_{-\infty}^{+\infty} \frac{t - x}{t - \zeta} d(E_1(t)v, v)_{\mathcal{H}}$$

it follows that the limit for $\zeta \stackrel{\wedge}{\to} x$ of the lefthand side equals zero. So the lemma is proved for this special case.

In general, however, A is not densely defined, but of the form (9.3). Because $\operatorname{ran}(A - x) = A(0) \oplus \operatorname{ran}(A_o - x)$, $P_o u \in \ker(A_o - x)$. Denote by P_{o1} the orthogonal projection in $\mathcal{H}_o$ onto $\overline{\operatorname{ran}}(A_o - x)$. Then, by what we have just proved for operators, for $\zeta \in \mathbb{C}^+ \cup \{x\}$,

$$P_{o1}(A_o - \zeta)^{-1} P_o u = (A_{o1} - \zeta)^{-1} P_o u.$$

From

$$(A - \zeta)^{-1} = \begin{pmatrix} 0 & 0 & 0 \\ 0 & (A_o - \zeta)^{-1} & 0 \\ 0 & 0 & (x - \zeta)^{-1} \end{pmatrix} : \begin{pmatrix} A(0) \\ \overline{\operatorname{ran}}(A_o - x) \\ \ker(A - x) \end{pmatrix} \longrightarrow \begin{pmatrix} A(0) \\ \overline{\operatorname{ran}}(A_o - x) \\ \ker(A - x) \end{pmatrix}$$

we find that for $\zeta \in \mathbb{C}^+ \cup \{x\}$,

$$P_1(A - \zeta)^{-1}u = P_{o1}(A_o - \zeta)^{-1} P_o u = (A_{o1} - \zeta)^{-1} P_o u = (A_1 - \zeta)^{-1}u.$$

This implies (9.6) and (9.7). The remaining assertions can be proved by rewriting them in terms of A_o and A_{o1}. We omit the details. □

The condition $u \in \operatorname{ran}(A - x)$ in Lemma 9.1 holds if and only if the Carathéodory derivative

$$\lim_{\zeta \stackrel{\wedge}{\to} x} \frac{\operatorname{Im} N(\zeta)}{\operatorname{Im} \zeta}$$

is finite. Indeed, by straightforward computation,

$$\frac{\operatorname{Im} N(\zeta)}{\operatorname{Im} \zeta} = \|(I - P_o)u\|_{\mathcal{H}}^2 + \int_{-\infty}^{+\infty} \frac{|t - \mu|^2}{|t - \zeta|^2} d(E_o(t) P_o u, P_o u)_{\mathcal{H}},$$

and hence the limit of the lefthand side as $\zeta \stackrel{\wedge}{\to} x$ exists if and only if

$$\int_{-\infty}^{+\infty} \frac{d(E_o(t) P_o u, P_o u)_{\mathcal{H}}}{|t - x|^2} < \infty.$$

The latter implies that

$$v := \int_{-\infty}^{+\infty} \frac{dE_o(t)P_o u}{t-x}$$

belongs to dom $(A_o - x)$ and then $(A_o - x)v = P_o u$. It follows that $u = (I - P_o)u + P_o u \in A(0) \oplus \operatorname{ran}(A_o - x) = \operatorname{ran}(A - x)$. The converse can be proved by tracing these arguments in the opposite direction.

References

[A] N.I. Akhiezer, *Vorlesungen über Approximationstheorie*, Akademie-Verlag, Berlin, 1953.

[ABDS1] D. Alpay, P. Bruinsma, A. Dijksma and H.S.V. de Snoo, *A Hilbert space associated with a Nevanlinna function*, in: M.A. Kaashoek, J.H. van Schuppen and A.C.M. Ran (eds.), *Signal Processing, Scattering and Operator Theory, and Numerical Methods*, Proceedings International Symposium MTNS-89, vol. III, Progress in Systems and Control Theory, Birkhäuser, Basel, 1990, 115–122.

[ABDS2] D. Alpay, P. Bruinsma, A. Dijksma and H.S.V. de Snoo, *Interpolation problems, extensions of symmetric operators and reproducing kernels* I, Oper. Theory: Adv. Appl., vol. 50, Birkhäuser, Basel, 1991, 35–82.

[ABDS3] D. Alpay, P. Bruinsma, A. Dijksma and H.S.V. de Snoo, *Interpolation problems, extensions of symmetric operators and reproducing kernels* II, Integral Equations and Operator Theory **14** (1991), 465–500; *Interpolation problems, extensions of symmetric operators and reproducing kernels* II (missing section 3), Integral Equations and Operator Theory **15** (1992), 378–388.

[AD1] D. Alpay and H. Dym, *Hilbert spaces of analytic functions, inverse scattering and operator models* I, Integral Equations and Operator Theory **7** (1984), 589–641.

[AD2] D. Alpay and H. Dym, *Hilbert spaces of analytic functions, inverse scattering and operator models* II, Integral Equations and Operator Theory **8** (1985), 145–180.

[B1] L. de Branges, *Some Hilbert spaces of analytic functions* I, Trans. Amer. Math. Soc. **106** (1963), 445–468.

[B2] L. de Branges, *Hilbert spaces of entire functions*, Prentice Hall, Englewoods Cliffs, N.J., 1968 (French transl.: *Espaces Hilbertiens des fonctions entières*, Masson et Cie, Paris, 1972).

[BGR] J.A. Ball, I. Gohberg and L. Rodman, *Interpolation of rational matrix functions*, Oper. Theory: Adv. Appl., vol. 45, Birkhäuser, Basel, 1990.

[BH] J.A. Ball and W.J. Helton, *Interpolation problems of Pick–Nevanlinna and Loewner types for meromorphic matrix functions: parametrization of the set of all solutions*, Integral Equations and Operator Theory **9** (1986), 155–203.

[Br1] P. Bruinsma, *Interpolation problems for Schur and Nevanlinna pairs*, Dissertation University of Groningen, 1991.

[Br2] P. Bruinsma, *Degenerate interpolation problems for Nevanlinna pairs*, Indag. Mathem., N.S. **2** (1991), 179–200.

[BR] L. de Branges and J. Rovnyak, *Canonical models in quantum scattering theory*, in: C. Wilcox (ed.), *Perturbation theory and its applications in quantum mechanics*, Wiley, New York, 1996, 295–392.

[DLS1] A. Dijksma, H. Langer and H.S.V. de Snoo, *Selfadjoint Π_κ–extensions of symmetric subspaces: an abstract approach to boundary value problems with spectral parameter in the boundary conditions*, Integral Equations and Operator Theory **7** (1984), 460–515.

[DLS2] A. Dijksma, H. Langer and H.S.V. de Snoo, *Eigenvalues and pole functions of Hamiltonian systems with eigenvalue parameter depending boundary conditions*, Math. Nachr. **161** (1993), 107–154.

[Dy] H. Dym, *J contractive matrix functions, reproducing kernel Hilbert spaces and interpolation*, Regional conference series in mathematics **71**, Amer. Math. Soc., Providence, R.I., 1989.

[DyM] H. Dym and H.P. McKean, *Gaussian processes, function theory, and the inverse spectral problem*, Academic Press, New York, 1976.

[GG] M.L. Gorbachuk and V.I. Gorbachuk, *M.G. Kreĭn's lectures on entire operators*, Birkhäuser, Basel, monograph in preparation.

[J] G. Julia, *Principes Géometrique d'Analyse*, Gauthiers-Villars, Paris, Volume I (1930), Volume II (1932).

[K] M.G. Kreĭn, *Fundamental aspects of the representation theory of Hermitian operators with deficiency index (m, m)*, Ukr. Mat. Zh. **1** no.2 (1949), 3–66; English transl.: Amer. Math. Soc. Transl. (2) **97** (1970), 75–143.

[KKY] V. Katsnelson, A. Kheifets and P. Yuditskii, *An abstract interpolation problem and extension theory of isometric operators*, translated from: V.A. Marchenko (ed.), *Operators in function spaces and problems in function theory*, Naukova Dumka, Kiev, 1987, 83–96.

[KP] I.V. Kovalishina and V.P. Potapov, *An indefinite metric in the Nevanlinna–Pick problem*, Akad. Nauk. Arm. SSR Dokl. **59** (1974), 17–22; English transl. in: *Seven papers translated from the Russian by I.V. Kovalishina and V.P. Potapov*, Amer. Math. Soc. Transl., vol. 138, Amer. Math. Soc., Providence, R.I., 1988, 15–19.

[KL1] M.G. Kreĭn and H. Langer, *Über die verallgemeinerten Resolventen und die charakteristische Funktion eines isometrischen Operators in Raume Π_κ*, Colloquia Mathematica Societatis János Bolyai **5**, Hilbert Space Operators, Tihany (Hungary) (1970), 353–399.

[KL2] M.G. Kreĭn and H. Langer, *Über die Q-Funktion eines π-hermiteschen Operators im Raume Π_κ*, Acta Sci. Math. (Szeged) **34** (1973), 191–230.

[KL3] M.G. Kreĭn and H. Langer, *Über einige Fortsetzungsprobleme, die eng mit der Theorie hermitescher Operatoren im Raume Π_κ zusammenhängen. I. Einige Funktionenklassen und ihre Darstellungen*, Math. Nachr. **77** (1977), 187–236.

[KL4] M.G. Kreĭn and H. Langer, *Über einige Fortsetzungsprobleme, die eng mit der Theorie hermitescher Operatoren im Raume Π_κ zusammenhängen. II. Verallgemeinerte Resolventen, u-Resolventen und ganze Operatoren*, J. Funct. Anal. **30** (1978), 390–447.

[L] B.Ja. Levin, *Distribution of zeros of entire functions*, Transl. of Math. Monographs, vol. 5, Amer. Math. Soc., Providence, R.I., revised edition, 1972.

[LT] H. Langer and B. Textorius, *On generalized resolvents and Q–functions of symmetric linear relations (subspaces) in Hilbert space*, Pacific J. Math. **72** (1977), 135–165.

[N] R. Nevanlinna, *Über beschränkte analytische Funktionen*, Ann. Acad. Sci. Fennicae Ser A **XXXII. 7** (1929), 1–75.

[Sc] L. Schwartz, *Sous–espaces hilbertiens d'espaces vectoriels topologiques et noyaux associés (noyaux reproduisants)*, J. Analyse Math. **13** (1964), 115–256.

[SzK1] B. Sz.-Nagy and A. Korányi, *Relations d'un problème de Nevanlinna et Pick avec la théorie des opérateurs de l'espace hilbertien*, Acta Math. Acad. Sci. Hung. **7** (1956), 295–303.

[SzK2] B. Sz.-Nagy and A. Korányi, *Operatortheoretische Behandlung und Verallgemeinerung eines Problemkreises in der komplexen Funktionentheorie*, Acta Math. **100** (1958), 171–202.

[W] H. Woracek, *Nevanlinna–Pick interpolation: the degenerate case*, Linear Algebra Appl. **252** (1997), 141–158.

Daniel Alpay
Department of Mathematics
Ben-Gurion University of the Negev
P.O. Box 653
84105 Beer–Sheva
Israel
dany@math.bgu.ac.il

Aad Dijksma
Department of Mathematics
University of Groningen
P.O. Box 800
9700 AV Groningen
The Netherlands
a.dijksma@math.rug.nl

Heinz Langer
Department of Mathematics
Technical University of Vienna
Wiedner Hauptstrasse 8–10
A-1040 Vienna
Austria
hlanger@mail.zserv.tuwien.ac.at

AMS Classification Numbers: 47A57, 46E22, 30E05, 42A82.

Operator Theory:
Advances and Applications, Vol. 115

The Abstract Interpolation Problem and Commutant Lifting: A Coordinate-free Approach

Joseph A. Ball and Tavan T. Trent

We present a coordinate-free formulation of the Abstract Interpolation Problem introduced by Katsnelson, Kheifets and Yuditskii in an abstract scattering theory framework. We also show how the commutant lifting theorem fits into this new formulation of the Abstract Interpolation Problem, giving a coordinate-free version of a result of Kupin.

Introduction

The Abstract Interpolation Problem (AIP) as introduced by Katsnelson, Kheifets and Yuditskii [KaKhY] (see also [KhY]) is as follows.

One is given as a data set for the problem a collection $(T_1, T_2, D(\cdot,\cdot), M_1, M_2, \mathcal{X}, \mathcal{E}_1, \mathcal{E}_2)$ where $\mathcal{X}$ is a linear space, $\mathcal{E}_1$ and $\mathcal{E}_2$ are two Hilbert spaces (playing the role of input and output space respectively), $D(\cdot,\cdot)$ is a positive-semidefinite Hermitian form on $\mathcal{X} \times \mathcal{X}$, T_1 and T_2 are two operators on $\mathcal{X}$, and $M_1 : \mathcal{X} \to \mathcal{E}_1$ and $M_2 : \mathcal{X} \to \mathcal{E}_2$ are two linear operators. It is assumed in addition that these objects satisfy a so-called Potapov identity:

$$D(T_1x, T_1x) + \|M_1x\|^2_{\mathcal{E}_1} = D(T_2x, T_2x) + \|M_2x\|^2_{\mathcal{E}_2}. \tag{0.1}$$

Denote by $\mathcal{X}_D$ the Hilbert space completion of $\mathcal{X}$ in the $D(\cdot,\cdot)$ inner-product (with elements with zero self-inner-product identified to zero). In terms of this data set one defines an "abstract interpolation problem"; the unknown in the abstract interpolation problem is a pair (F, ω) where ω is a Schur-class function with values equal to operators from $\mathcal{E}_1$ into $\mathcal{E}_2$, where F is a contraction operator from $\mathcal{X}_D$ into the de Branges-Rovnyak model space $\mathcal{H}^\omega$ associated with ω (precise definitions are given below in Section 1) and where F is required to satisfy a certain identity involving the interpolation data.

The significance of the identity (0.1) is that it enables one to define an isometry V with domain equal to im $\begin{bmatrix} T_1 \\ M_1 \end{bmatrix} \subset \begin{bmatrix} \mathcal{X}_D \\ \mathcal{E}_1 \end{bmatrix}$ and with image space equal to im $\begin{bmatrix} T_2 \\ M_2 \end{bmatrix} \subset \begin{bmatrix} \mathcal{X}_D \\ \mathcal{E}_2 \end{bmatrix}$ by

$$V : \begin{bmatrix} T_1 \\ M_1 \end{bmatrix} x \longrightarrow \begin{bmatrix} T_2 \\ M_2 \end{bmatrix} x \tag{0.2}$$

for all $x \in \mathcal{X}$. Here we use the canonical quotient map to identify the element $T_1 x \in \mathcal{X}$ with an element of $\mathcal{X}_D$ and similarly for $T_2 x$. The main result concerning the Abstract Interpolation Problem is the clean characterization of the set of all solutions (F, ω) in terms of unitary extensions $U : \begin{bmatrix} \mathcal{H} \\ \mathcal{E}_1 \end{bmatrix} \to \begin{bmatrix} \mathcal{H} \\ \mathcal{E}_2 \end{bmatrix}$ of V. Specifically, if U has block matrix representation

$$U = \begin{bmatrix} A & B \\ C & D \end{bmatrix} : \begin{bmatrix} \mathcal{H} \\ \mathcal{E}_1 \end{bmatrix} \longrightarrow \begin{bmatrix} \mathcal{H} \\ \mathcal{E}_2 \end{bmatrix}$$

then $\omega(z) = D + zC(I - zA)^{-1}B$ (the so-called *scattering matrix* associated with the unitary colligation U) is part of a solution (F, ω) of the AIP. The set of all ω's arising in this way, in turn, can be given a linear-fractional parametrization in terms of a Schur-function free-parameter via a technique due to Arov and Grossman [ArG]. The problem AIP itself, in turn, encodes the most general bitangential Nevanlinna-Pick interpolation problem, boundary Nevanlinna-Pick interpolation problem, and Hamburger moment problem (after a change of variables from the real line to the unit circle), and leads to a sharp analysis of the most degenerate cases of these problems which has proved elusive for other methods; these elaborations can be found in the series of papers by Kheifets [Kh1, Kh2, Kh3, Kh4]. An explanation of how the Sz.-Nagy-Foias commutant lifting theorem (see Theorem 2.3 of Chapter II of [Sz.-NF]) fits in as a special case of the AIP is given by Kupin in [Ku]. For a more recent treatment of the Commutant Lifting Theorem itself which includes applications to interpolation and engineering problems, we recommend the monograph of Foias and Frazho [FF].

The purpose of this paper is to reformulate the AIP in a more coordinate-free manner which eliminates the need for the introduction of de Branges-Rovnyak model spaces. In the terminology of Nikolskii and Vasyunin [NV], the original formulation of the AIP is in the de Branges-Rovnyak transcription of the functional model for a contraction operator and its unitary dilation. Our contribution is to express the AIP in the more intrinsically geometric, coordinate-free transcription of the model. We prove the main result on the AIP (the correspondence between solutions of the AIP and unitary colligation extensions of the partially defined isometric colligation constructed from the data of the problem) directly in this framework. Alternatively of course, one could appeal to the proof in the original formulation and then change coordinates to our setting, but we feel that a direct proof in the coordinate-free setting is instructive. Our approach also gives a scattering-theoretic interpretation of the AIP. Among all the possible applications which have already been discussed in the literature, we discuss one such in this framework, namely the commutant lifting theorem; this amounts to a coordinate-free transcription of the work in [Ku]. An alternative scattering-theoretic proof of the Commutant Lifting Theorem has appeared in the work of Arocena, Cotlar and Sadosky (see [ACS], [Ar1], [CS] and [Ar2]); we shall compare these various approaches in Section 4.

In order to better explain the coordinate-free formulation of the AIP, we have included a review of the connections between linear system theory, operator model theory and scattering theory. In all three theories a central object is an operator-valued function $\omega(z)$ with a representation of the form $\omega(z) = D + zC(I - zA)^{-1}B$ where

$$U = \begin{bmatrix} A & B \\ C & D \end{bmatrix} : \begin{bmatrix} \mathcal{H} \\ \mathcal{E}_1 \end{bmatrix} \longrightarrow \begin{bmatrix} \mathcal{H} \\ \mathcal{E}_2 \end{bmatrix}$$

is a unitary transformation. This object goes under the names of *transfer function*, *characteristic operator function* or *scattering function* depending on whether one is doing system theory, operator model theory or scattering theory, respectively. Of course these ideas are not original with us. For the system theory aspects, we refer to [An], for the operator model theory to [deBR] and [SzNF], and for scattering theory to [LP]. For the various connections among and elaborations on these theories, some good sources are [AdAr], [H], [BC] and [NV].

The paper is organized as follows. After this Introduction, Section 1 gives the precise formulation and main result concerning the solution set of the Abstract Interpolation Problem from [KaKhY, KhY] and sketches how a classical Nevanlinna-Pick interpolation problem fits into the AIP framework. Section 2 provides a review of the different roles played by the characteristic function of a unitary colligation in system theory, scattering theory and operator model theory, and gives the coordinate-free version of the AIP as a problem involving an abstract scattering system. Section 3 gives a direct proof of the main result on the AIP in the coordinate-free context, and finally, Section 4 shows how the AIP can be specialized to pick up the Commutant Lifting Theorem.

The first author would like to thank Sasha Kheifets and Victor Vinnikov for numerous conversations on the topic of this paper. Both authors would like to express our thanks to the Mathematical Sciences Research Institute in Berkeley, California for support in bringing us all together for the *Holomorphic Spaces* program in the Fall of 1995.

1 The Abstract Interpolation Problem: Statement and Main Result

The precise statement of the Abstract Interpolation Problem (AIP) requires some preliminaries. Define $\mathcal{S}(\mathcal{E}_1, \mathcal{E}_2)$ to be the space of Schur-class functions with input space $\mathcal{E}_1$ and output space equal to $\mathcal{E}_2$, i.e. $\mathcal{S}(\mathcal{E}_1, \mathcal{E}_2)$ consists of functions $\omega(z)$ which are analytic on the open unit disk $\mathbf{D}$ with values in the space $\mathcal{L}(\mathcal{E}_1, \mathcal{E}_2)$ of bounded linear operators from $\mathcal{E}_1$ to $\mathcal{E}_2$ and such that the operator-norm $\|\omega(z)\|$ is at most 1 for all $z \in \mathbf{D}$. Associated with any Schur-class function $\omega \in \mathcal{S}(\mathcal{E}_1, \mathcal{E}_2)$ is the de Branges-Rovnyak model space H^{ω} (see [deBR] but [NV] for this formulation) defined by

$$\mathcal{H}^{\omega} = \left(\begin{bmatrix} I & \omega \\ \omega^* & I \end{bmatrix}^{1/2} \begin{bmatrix} L_2(\mathcal{E}_2) \\ L_2(\mathcal{E}_1) \end{bmatrix} \right) \cap \begin{bmatrix} H_2(\mathcal{E}_2) \\ H_2(\mathcal{E}_1)^{\perp} \end{bmatrix} \tag{1.1}$$

with norm equal to the so-called lifted norm

$$\left\| \begin{bmatrix} I & \omega \\ \omega^* & I \end{bmatrix}^{1/2} \begin{bmatrix} f \\ g \end{bmatrix} \right\|^2_{\mathcal{H}^{\omega}} = \left\| Q \begin{bmatrix} f \\ g \end{bmatrix} \right\|^2_{L_2(\mathcal{E}_2) \oplus L_2(\mathcal{E}_1)}$$

where Q is the orthogonal projection onto the orthogonal complement of the kernel of $\begin{bmatrix} I & \omega \\ \omega^* & I \end{bmatrix}$ (considered as a multiplication operator). Here, for any Hilbert space $\mathcal{E}$, $L_2(\mathcal{E})$ consists of functions $f(z)$ defined almost everywhere on the unit circle $\mathbf{T} = \partial \mathbf{D}$ with values in $\mathcal{E}$ and with Fourier series $f(z) \equiv \sum_{n=-\infty}^{\infty} f_n z^n$ having coefficients $f_n \in \mathcal{E}$ square-summable in norm ($\|f\|^2_{L_2(\mathcal{E})} = \sum_{n=-\infty}^{\infty} \|f_j\|^2_{\mathcal{E}} < \infty$), $H_2(\mathcal{E})$ is the subspace of $L_2(\mathcal{E})$ consisting of those functions $f(z)$ with Fourier coefficients f_n vanishing for $n < 0$ (so $f(z) \sim \sum_{n=0}^{\infty} f_n z^n$), and the orthogonal complement $H_2(\mathcal{E})^{\perp}$ consists of those functions $f(z)$ in $L_2(\mathcal{E})$ of the form $f(z) \sim \sum_{n=-\infty}^{-1} f_j z^j$. Note also that elements $f(z) \sim \sum_{n=0}^{\infty} f_n z^n$ of $H_2(\mathcal{E})$ have analytic extensions to the disk $\mathbf{D}$ given by $f(z) = \sum_{n=0}^{\infty} f_n z^n$ while elements $f(z) \sim \sum_{n=-\infty}^{-1} f_n z^n$ of $H_2(\mathcal{E})^{\perp}$ have conjugate-analytic continuations to $\mathbf{D}$ given by $f(z) = \sum_{n=1}^{\infty} f_{-n} \bar{z}^n$.

The AIP then is: *find and characterize all pairs (F, ω) such that (i) F is a contractive linear operator from X_D into the de Branges-Rovnyak space $\mathcal{H}^{\omega}$ associated with ω and (ii) the identity*

$$[F(T_1 x)](z) = [F(z T_2 x)](z) - \begin{bmatrix} I & \omega(z) \\ \omega(z)^* & I \end{bmatrix} \begin{bmatrix} -M_2 \\ M_1 \end{bmatrix} x \tag{1.2}$$

holds for all $x \in \mathcal{X}$ and for almost all $z \in \mathbf{T}$.

More explicitly, as F is required to map $\mathcal{X}_D$ into H^{ω} and elements of $H^{\omega} \subset \begin{bmatrix} H_2(\mathcal{E}_2) \\ H_2(\mathcal{E}_1)^{\perp} \end{bmatrix}$ consist of two components, we may write $F(x) = \begin{bmatrix} F_1(x) \\ F_2(x) \end{bmatrix}$ with $F_1(x) \in H_2(\mathcal{E}_2)$ and $F_2(x) \in H_2(\mathcal{E}_1)^{\perp}$. Note that then $F_1(x)$ has analytic continuation to $\mathbf{D}$ while $F_2(x)$ has conjugate analytic continuation to $\mathbf{D}$. In more detail (1.2) may be written as

$$\begin{aligned} [F_1(T_1 x)](z) &= [F_1(z \cdot T_2 x)](z) + M_2 x - \omega(z) M_1 x \\ [F_2(T_1 x)](z) &= [F_2(\bar{z}^{-1} \cdot T_2 x)](z) + \omega(z)^* M_2 x - M_1 x \end{aligned} \tag{1.3}$$

for $z \in \mathbf{D}$.

As a particular example, consider the right tangential Nevanlinna-Pick interpolation problem: *Given n nonzero $p \times 1$ column vectors $u_1, \ldots, u_n$, n $q \times 1$*

column vectors $v_1, \dots, v_n$ and distinct points $w_1, \dots, w_n$ in the unit disk $\mathbf{D}$, find and characterize the set of $q \times p$ matrix Schur-class functions $\omega \in \mathcal{S}(\mathbf{C}^p, \mathbf{C}^q)$ which satisfy the interpolation conditions

$$\text{(INT)} \qquad \omega(w_j)u_j = v_i \text{ for } j = 1, \dots, n.$$

It is a classical result that a necessary and sufficient condition for existence of solutions is that the associated Pick matrix $\Lambda = \left[\frac{u_i^* u_j - v_i^* v_j}{1 - \overline{w}_i w_j}\right]$ be positive-semidefinite. We sketch the proof of the sufficiency of this condition in the framework of the AIP.

Sketch of sufficiency via AIP. We set $\mathcal{X} = \mathbf{C}^n$ with Hermitian form D induced by Λ ($D(x, y) = y^* \Lambda x$), and $\mathcal{E}_1 = \mathbf{C}^p$, $\mathcal{E}_2 = \mathbf{C}^q$ with the standard Euclidean inner products. Note that by our hypothesis on the Pick matrix Λ we have that D is positive semi-definite. We take $T_1 = W$ where W is the diagonal matrix with diagonal entries equal to $w_1, \dots, w_n$ and $T_2 = I_n$ as operators on $\mathbf{C}^n$ and set $M_1 = M_U := [\, u_1 \ \cdots \ u_n \,]$ and $M_2 = M_V := [\, v_1 \ \cdots \ v_n \,]$ as operators from $\mathcal{X}$ to $\mathcal{E}_1$ and from $\mathcal{X}$ to $\mathcal{E}_2$ respectively. One can check the identity

$$(1.4) \qquad W^* \Lambda W + U^* U = \Lambda + V^* V$$

and hence these objects form an admissible data set for an AIP. The AIP with this data set is: *Find all pairs (F, ω) with $\omega \in \mathcal{S}(\mathbf{C}^p, \mathbf{C}^q)$ such that $F = \begin{bmatrix} F_1 \\ F_2 \end{bmatrix}$ is a linear contraction operator from $\mathcal{X}_\Lambda$ to $\mathcal{H}^\omega$ for which the identities*

$$(1.5a) \qquad [F_1(w_j e_j)](z) = z[F_1(e_j)](z) + v_j - \omega(z)u_j$$

$$(1.5b) \qquad [F_2(w_j e_j)](z) = \overline{z}^{-1}[F_2(e_j)](z) + \omega(z)^* v_j - u_j$$

for $j = 1, \dots, n$ where e_j is the jth standard basis vector for $\mathbf{C}^n$.

Note that we can solve (1.5a) for F_1 in terms of $\omega(z)$:

$$(1.6a) \qquad [F_1(e_j)](z) = \frac{\omega(z)u_j - v_j}{z - w_j}.$$

As the left-hand side of (1.6a) must be in $H^2(\mathbf{C}^q)$ we see that $\omega(z)$ must satisfy the interpolation conditions (INT). Similarly we can solve for F_2 in (1.5b) to get

$$(1.6b) \qquad [F_2(e_j)](z) = \frac{\omega(z)^* v_j - u_j}{w_j - \overline{z}^{-1}} = -\overline{z}\frac{\omega(z)^* v_j - u_j}{1 - w_j \overline{z}_j}.$$

We see that the right-hand side of (1.6b) is automatically in $H^2(\mathbf{C}^p)^\perp$ and so no interpolation conditions appear as was the case for (1.6a). It turns out that if ω is Schur class and satisfies the interpolation conditions (INT), then it is automatic that

$$\begin{bmatrix} \dfrac{\omega(z)u_j - v_j}{z - w_j} \\ -\overline{z}\dfrac{\omega(z)^* w_j - u_j}{1 - w_j \overline{z}} \end{bmatrix} \in \mathcal{H}^\omega.$$

We may then use (1.5a) and (1.5b) to define $F = \begin{bmatrix} F_1 \\ F_2 \end{bmatrix}$. Then in addition F is automatically contractive (in fact isometric) from $\mathcal{X}_\Lambda$ to $\mathcal{H}^{\omega}$ and (ω, F) is a solution of the AIP for this data set arising from the right tangential interpolation problem. We have thus established a one-to-one correspondence between solutions of the AIP with this data set on the one hand, and Schur-class solutions of the interpolation conditions (INT) on the other. We can now appeal to the general theory of the AIP (see [KaKhY] and [KhY]) to see that positive-semidefiniteness of Λ is sufficient for the existence of Schur-class solutions of the interpolation conditions (INT). Of course the AIP theory does much more: it also provides a simple linear-fractional parametrization for the set of all solutions, including in degenerate cases.

Remark: For this special case of right tangential interpolation with finitely many nodes, we see that F is in fact an isometry and is uniquely determined by ω whenever (F, ω) is a solution of the AIP. This is not the case for other examples (such as boundary Nevanlinna-Pick interpolation or the transform of the Hamburger moment problem from the line to the circle (see [Kh4] for details) where it can happen that F is not isometric, or in the most general case of commutant lifting (see [K]) where it can happen that ω does not uniquely determine F.

To suggest the form of the solution of the AIP for the general case, we continue with our example of right tangential Nevanlinna-Pick interpolation. Note that the identity (1.4) implies that the formula

$$V : \begin{bmatrix} W \\ M_U \end{bmatrix} x \longrightarrow \begin{bmatrix} I \\ M_V \end{bmatrix} x \tag{1.7}$$

is a partially defined, well-defined isometric transformation with domain $\mathcal{D}_V = \operatorname{im} \begin{bmatrix} W \\ M_U \end{bmatrix} \subset \begin{bmatrix} \mathcal{X}_\Lambda \\ \mathbf{C}^p \end{bmatrix}$ and with range equal to $\mathcal{R}_V = \operatorname{im} \begin{bmatrix} I \\ V \end{bmatrix} \subset \begin{bmatrix} \mathcal{X}_\Lambda \\ \mathbf{C}^q \end{bmatrix}$. Suppose that $U : \begin{bmatrix} \mathcal{H} \\ \mathbf{C}^p \end{bmatrix} \to \begin{bmatrix} \mathcal{H} \\ \mathbf{C}^q \end{bmatrix}$ (where $\mathcal{H}$ is a Hilbert space containing $\mathcal{X}_\Lambda$) is any unitary extension of V of the indicated form, and write U in block matrix form as

$$U = \begin{bmatrix} A & B \\ C & D \end{bmatrix} : \begin{bmatrix} \mathcal{H} \\ \mathbf{C}^p \end{bmatrix} \longrightarrow \begin{bmatrix} \mathcal{H} \\ \mathbf{C}^q \end{bmatrix}.$$

Since by assumption U extends the partially defined V given by (1.7), we get

$$\begin{bmatrix} A & B \\ C & D \end{bmatrix} \begin{bmatrix} W \\ M_U \end{bmatrix} x = \begin{bmatrix} I \\ M_V \end{bmatrix} x \tag{1.8}$$

for all vectors x in $\mathcal{X}_\Lambda = \mathbf{C}^n$. In particular, taking x equal to the jth standard basis vector e_j for $\mathbf{C}^n$ leads to

$$\begin{bmatrix} A & B \\ C & D \end{bmatrix} \begin{bmatrix} w_j e_j \\ u_j \end{bmatrix} = \begin{bmatrix} e_j \\ v_j \end{bmatrix}. \tag{1.9}$$

From the first block row of (1.9) we see that

$$e_j = (I - w_j A)^{-1} B u_j.$$

Substitution of this into the second block row of (1.9) gives

$$[D + w_j C(I - w_j A)^{-1} B] u_j = v_j. \tag{1.10}$$

But since $\begin{bmatrix} A & B \\ C & D \end{bmatrix}$ is unitary, the general theory of unitary colligations (see e.g. [ADRdS], where the more general Krein space case is discussed) tells us that its characteristic function $\omega(z) = D + zC(I - zA)^{-1}B$ is in the Schur class $\mathcal{S}(\mathbf{C}^p, \mathbf{C}^q)$. From (1.10) we see that such an ω solves the interpolation conditions (INT).

By the preceding discussion, we see that in this way we have also produced a solution (ω, F) of the AIP for the data set associated with the right tangential interpolation problem. One can also reverse the process. If ω interpolates, one can realize ω as the characteristic function $\omega(z) = D + zC(I - zA)^{-1}B$ of a unitary colligation $U = \begin{bmatrix} A & B \\ C & D \end{bmatrix}$ which satisfies (1.10), and hence is a unitary extension of V. In this way we have established a one-to-one correspondence between unitary extensions U of V and Schur-class interpolants ω satisfying (INT). For details of this approach, we refer to [BT] and [AgMc] where a more general several-variable framework is considered.

Now we return to the general setting of the AIP. As mentioned in the Introduction, whenever $(T_1, T_2, M_1, M_2, D(\cdot,\cdot), \mathcal{X}, \mathcal{E}_1, \mathcal{E}_2)$ is an admissible AIP data set, so the Potapov identity (0.1) is satisfied, and then (0.2) gives a well-defined partial isometry with domain

$$\mathcal{D}_V = \operatorname{im} \begin{bmatrix} T_1 \\ M_1 \end{bmatrix} \subset \begin{bmatrix} \mathcal{X}_D \\ \mathcal{E}_1 \end{bmatrix}$$

and with range space equal to

$$\mathcal{R}_V = \operatorname{im} \begin{bmatrix} T_2 \\ M_2 \end{bmatrix} \subset \begin{bmatrix} \mathcal{X}_D \\ \mathcal{E}_2 \end{bmatrix}.$$

As in the Introduction, we let $\mathcal{X}_D$ be the completion of $\mathcal{X}$ (with vectors with zero D-self-inner-product identified with 0) in the D inner product, and we use the same symbols T_1 and T_2 for the operators from $\mathcal{X}$ into $\mathcal{X}_D$ given by T_1 and T_2 respectively followed by the canonical quotient map. As also mentioned in the Introduction, solutions of the AIP can be characterized in terms of unitary colligation extensions U of the partial isometric colligation V. More precisely, by a *unitary colligation* we mean any unitary operator U from a Hilbert space $\mathcal{K}$ to a Hilbert space $\mathcal{K}'$ where the domain Hilbert space has the direct sum form $\mathcal{K} = \mathcal{H} \oplus \mathcal{E}_1$ and the image Hilbert space $\mathcal{K}'$ has the direct sum form $\mathcal{H} \oplus \mathcal{E}_2$.

The salient feature is that the first component space of $\mathcal{K}'$ must be the same as the first component of $\mathcal{K}$ but the second component spaces may be different; in applications, the second component spaces $\mathcal{E}_1$ (the *input space*) and $\mathcal{E}_2$ (the *output space*) are fixed throughout, and the first component space $\mathcal{H}$ (the *state space*) is variable in the course of various manipulations. We may view V given by (0.2) as a partially defined isometric colligation. By a unitary colligation extension U we mean an extension $U = \begin{bmatrix} A & B \\ C & D \end{bmatrix} : \begin{bmatrix} \mathcal{H} \\ \mathcal{E}_1 \end{bmatrix} \to \begin{bmatrix} \mathcal{H} \\ \mathcal{E}_2 \end{bmatrix}$ where $\mathcal{H} \supset \mathcal{X}_D$.

The following is the main result on the AIP.

Theorem 1.1 *Let $(T_1, T_2, D(\cdot,\cdot), M_1, M_2, \mathcal{D}, \mathcal{E}_1, \mathcal{E}_2)$ be an admissible interpolation data set for an AIP. Then solutions exist and are in one-to-one correspondence with unitary colligation extensions*

$$U = \begin{bmatrix} A & B \\ C & D \end{bmatrix} : \begin{bmatrix} \mathcal{H} \\ \mathcal{E}_1 \end{bmatrix} \longrightarrow \begin{bmatrix} \mathcal{H} \\ \mathcal{E}_2 \end{bmatrix}$$

of the partial isometric colligation V defined by (0.2). More explicitly, if $U = \begin{bmatrix} A & B \\ C & D \end{bmatrix}$ is a unitary colligation extension of V, and $(F, \omega(z))$ is defined by

$$[Fx](z) = \begin{bmatrix} F_1 x \\ F_2 x \end{bmatrix}(z) = \begin{bmatrix} C(I - zA)^{-1}x \\ \bar{z}B^*(I - \bar{z}A^*)^{-1}x \end{bmatrix}$$
$$\omega(z) = D + zC(I - zA)^{-1}B,$$

then (F, ω) solves the AIP and every solution of the AIP arises in this way.

Actually the theory on the AIP from [KaKhY] and [KhY] goes somewhat farther than that presented in Theorem 1.1. Specifically, by using a technique of Arov and Grossman (see [ArG]), it is possible to obtain an explicit linear-fractional parametrization of all the unitary colligation extensions U of V, and of the associated solutions (F, ω) of the AIP, in terms of a free parameter Schur-class function $s(z) \in \mathcal{S}(\mathcal{N}_1, \mathcal{N}_2)$ where $\mathcal{N}_1 = \begin{bmatrix} \mathcal{X}_D \\ \mathcal{E}_1 \end{bmatrix} \ominus \mathcal{D}_V$ and $\mathcal{N}_2 = \begin{bmatrix} \mathcal{X}_D \\ \mathcal{E}_2 \end{bmatrix} \ominus \mathcal{R}_V$. However we will not discuss this aspect here.

2 Unitary Colligations and Abstract Scattering Systems

Let

$$U = \begin{bmatrix} A & B \\ C & D \end{bmatrix} : \begin{bmatrix} \mathcal{H} \\ \mathcal{E}_1 \end{bmatrix} \longrightarrow \begin{bmatrix} \mathcal{H} \\ \mathcal{E}_\in \end{bmatrix}$$

be a unitary colligation, i.e. U is unitary as an operator between the spaces of the indicated form. Associated with any such unitary colligation is its characteristic function

$$\omega_U(z) = D + zC(I - zA)^{-1}B. \tag{2.1}$$

For the history and general theory of unitary colligations, even in the general Krein space framework, we refer to [ADRdS]. This characteristic function $\omega_U(z)$ has proved to be fundamental in a number of guises: (1) as the transfer function of a discrete-time linear conservative system, (2) as the scattering function for an abstract scattering system, and (3) as the characteristic operator function in the functional model of a completely nonunitary contraction operator. We give a brief sketch of these ideas.

2.1 *Unitary Systems*

Consider the discrete-time, linear system

$$\begin{aligned} x(n+1) &= Ax(n) + Bu(n) \\ y(n) &= Cx(n) + Du(n). \end{aligned} \tag{2.2}$$

where we assume that $U = \begin{bmatrix} A & B \\ C & D \end{bmatrix}$ is unitary.

The system (2.1) is *conservative* in the sense that

$$\|x(n+1)\|^2 - \|x(n)\|^2 = \|u(n)\|^2 - \|y(n)\|^2$$

which can be iterated to

$$\|x(N_2+1)\|^2 - \|x(N_1)\|^2 = \sum_{n=N_1}^{N_2} \{\|u(n)\|^2 - \|y(n)\|^2\}. \tag{2.2a}$$

If we impose the initial condition $x(N) = 0$ and suppose that $u \in \ell_2(\mathcal{E}_1)$ with $u(n) = 0$ for $n < N$, then the recursion in (2.2) gives us a unique solution $x(n)$ (where we set $x(n) = 0$ for $n < N$) and the inequality (2.2a) (with $N_1 = N$ so $x(N) = 0$ and letting N_2 tend to infinity) gives us that the resulting $\{y(n)\}_{n=-\infty}^{\infty}$ (where we set $y(n) = 0$ for $n < N$) satisfies $\|y\|_{\ell_2}^2 \leq \|u\|_{\ell_2}^2$. If we take the limiting case of this situation as $N \to -\infty$, we find that, for a general element $u \in \ell_2(\mathcal{E}_1)$, there is a unique choice of sequence $\{x(n)\}_{n=-\infty}^{\infty}$ with $\lim_{n\to-\infty} x(n) = 0$ resulting in a unique $y \in \ell_2(\mathbf{Z}, \mathcal{E}_2)$. If we introduce the Fourier transform $\{v(n)\}_{n=-\infty}^{\infty} \to \widehat{v}(z) = \sum_{n=-\infty}^{\infty} v(n)z^n$ mapping $\ell_2(\mathcal{K})$ to $L_2(\mathcal{K})$ for any Hilbert space $\mathcal{K}$, then (2.1) transforms to

$$\begin{aligned} z^{-1}\widehat{x}(z) &= A\widehat{x}(z) + B\widehat{u}(z) \\ \widehat{y}(z) &= C\widehat{x}(z) + \widehat{u}(z). \end{aligned}$$

Solving for $\widehat{y}$ in terms of $\widehat{u}$ yields

$$\widehat{y}(z) = \omega_U(z)\widehat{u}(z)$$

where

$$\omega_U(z) = D + zC(I - zA)^{-1}B$$

is the *characteristic function* of the colligation U.

Since $\|x(N_1)\| \to 0$ as $N_1 \to -\infty$, we see from (2.2a) that

$$\sum_{n=-\infty}^{\infty} \|u(n)\|^2 - \sum_{n=-\infty}^{\infty} \|y(n)\|^2 = \lim_{N_2 \longrightarrow \infty} \|x(N_2)\|^2 \geq 0.$$

If it happens that $x(N) \to 0$ as $N \to \infty$ as well over all state trajectories, then $\|y\|^2_{\ell_2} = \|u\|^2_{\ell_2}$, $\|\widehat{y}\|^2_{L_2(\mathcal{E}_2)} = \|\widehat{u}\|^2_{L_2(\mathcal{E}_1)}$ and ω is an inner function (the C_{00} case in the operator model theory setting).

2.2 Abstract Scattering Systems and Operator Model Theory

Following Lax and Phillips (see [LP] and [AdAr]), we define an Abstract Scattering System (ASS) as a unitary operator $\mathcal{U}$ on a Hilbert space $\mathcal{K}$ together with two distinguished subspaces $\mathcal{G}_1$ and $\mathcal{G}_2$ such that

(i) $\mathcal{U}\mathcal{G}_1 \subset \mathcal{G}_1$, $\cap_{n=0}^{\infty}\mathcal{U}^n\mathcal{G}_1 = \{0\}$,

(ii) $\mathcal{U}^*\mathcal{G}_2 \subset \mathcal{G}_2$, $\cap_{n=0}^{\infty}\mathcal{U}^{*n}\mathcal{G}_2 = \{0\}$, and

(iii) $\mathcal{G}_1$ is orthogonal to $\mathcal{G}_2$.

The ASS is said to be *minimal* if $\widetilde{\mathcal{G}}_1 + \widetilde{\mathcal{G}}_2$ is dense in $\mathcal{K}$, where $\widetilde{\mathcal{G}}_1 = \text{clos.} \cup_{n=0}^{\infty} \mathcal{U}^{*n}\mathcal{G}_1$ and $\widetilde{\mathcal{G}}_2 = \text{clos.} \cup_{n=0}^{\infty}\mathcal{U}^n\mathcal{G}_2$. Given any ASS $(\mathcal{U}, \mathcal{K}, \mathcal{G}_1, \mathcal{G}_2)$, the *wandering subspaces* $\mathcal{E}_1$ and $\mathcal{E}_2$ are defined by

$$\mathcal{E}_1 = \mathcal{G}_1 \ominus \mathcal{U}\mathcal{G}_1, \quad \mathcal{E}_2 = \mathcal{U}\mathcal{G}_2 \ominus \mathcal{G}_2.$$

Then necessarily we have the internal orthogonal sum decompositions

$$\mathcal{G}_1 = \oplus_{n=0}^{\infty}\mathcal{U}^n\mathcal{E}_1, \quad \mathcal{G}_2 = \oplus_{n=-\infty}^{-1}\mathcal{U}^{*n}\mathcal{E}_2.$$

Introduce the Fourier representations determined by $\mathcal{E}_1$ and $\mathcal{E}_2$

$$\Phi_{\mathcal{E}_1}k = \{P_{\mathcal{E}_1}\mathcal{U}^{*n}k\}_{n=-\infty}^{\infty}, \quad \Phi_{\mathcal{E}_2}k = \{P_{\mathcal{E}_2}\mathcal{U}^{*n}k\}_{n=-\infty}^{\infty}.$$

Then $\Phi_{\mathcal{E}_1}$ is a partial isometry from $\mathcal{K}$ into $\ell_2(\mathcal{E}_1)$ with initial space equal to $\widetilde{\mathcal{G}}_1$ and $\Phi_{\mathcal{E}_1}$ maps $\mathcal{G}_1$ isometrically onto $\ell_2(\mathbf{N}, \mathcal{E}_1)$. Similarly, $\Phi_{\mathcal{E}_2}$ is a partial isometry from $\mathcal{K}$ onto $\ell_2(\mathcal{E}_2)$ with initial space equal to $\widetilde{\mathcal{G}}_2$ and $\Phi_{\mathcal{E}_2}$ maps $\mathcal{G}_2$ isometrically onto $\ell_2(\mathbf{N}_-, \mathcal{E}_2)$. If we define the *scattering space* $\mathcal{H}$ to be

$$\mathcal{H} = \mathcal{K} \ominus [\mathcal{G}_1 + \mathcal{G}_2],$$

then the operator

$$\tau = \begin{bmatrix} \Phi_{\mathcal{E}_2}|_{\mathcal{G}_2} \\ P_{\mathcal{H}} \\ \Phi_{\mathcal{E}_1}|_{\mathcal{G}_1} \end{bmatrix}$$

can be viewed as a unitary identification map between $\mathcal{K}$ and $\ell_2(\mathbf{N}_-, \mathcal{G}_2) \oplus \mathcal{H} \oplus \ell_2(\mathbf{N}, \mathcal{G}_1)$. The image $\mathcal{U}' = \tau\mathcal{U}\tau^*$ of $\mathcal{U}$ under this identification then is an operator on $\ell_2(\mathbf{N}_-, \mathcal{G}_2) \oplus \mathcal{H} \oplus \ell_2(\mathbf{N}, \mathcal{G}_1)$ with matrix representation necessarily of the form

$$\mathcal{U}' = \begin{bmatrix} \ddots &&&&&&& \\ & I_{\mathcal{E}_2} &&&&&& \\ && I_{\mathcal{E}_2} &&&&& \\ &&& \beta & \boxed{\alpha} &&& \\ &&& \delta & \gamma &&& \\ &&&&& I_{\mathcal{E}_1} && \\ &&&&&& I_{\mathcal{E}_1} & \\ &&&&&&& \ddots \end{bmatrix} \tag{2.3}$$

(where the boxed entry α is the $\mathcal{H}$ to $\mathcal{H}$ block matrix entry and where all unspecified entries are equal to 0), where the operator

$$\begin{bmatrix} \alpha & \beta \\ \gamma & \delta \end{bmatrix} : \begin{bmatrix} \mathcal{H} \\ \mathcal{E}_2 \end{bmatrix} \longrightarrow \begin{bmatrix} \mathcal{H} \\ \mathcal{E}_1 \end{bmatrix}$$

is unitary. We shall also need the matrix representation for the adjoint $\mathcal{U}^*$ of $\mathcal{U}$, namely

$$\mathcal{U}'^* = \tau\mathcal{U}^*\tau^* = \begin{bmatrix} \ddots &&&&&&& \\ & I_{\mathcal{E}_2} &&&&&& \\ && I_{\mathcal{E}_2} &&&&& \\ &&& C & D &&& \\ &&& \boxed{A} & B &&& \\ &&&&& I_{\mathcal{E}_1} && \\ &&&&&& I_{\mathcal{E}_1} & \\ &&&&&&& \ddots \end{bmatrix} \tag{2.4}$$

where

$$\begin{bmatrix} A & B \\ C & D \end{bmatrix} := \begin{bmatrix} \alpha^* & \gamma^* \\ \beta^* & \delta^* \end{bmatrix} : \begin{bmatrix} \mathcal{H} \\ \mathcal{E}_1 \end{bmatrix} \longrightarrow \begin{bmatrix} \mathcal{H} \\ \mathcal{E}_2 \end{bmatrix}$$

is unitary, and is of the form of a unitary colligation as considered in Section 2.1. Conversely, if we start with a unitary colligation $U = \begin{bmatrix} A & B \\ C & D \end{bmatrix} : \begin{bmatrix} \mathcal{H} \\ \mathcal{E}_1 \end{bmatrix} \rightarrow \begin{bmatrix} \mathcal{H} \\ \mathcal{E}_2 \end{bmatrix}$, let $\vec{x} = \{x(n)\}_{n=-\infty}^{\infty}$, $\vec{u} = \{u(n)\}_{n=-\infty}^{\infty}$ and $\vec{y} = \{y(n)\}_{n=-\infty}^{\infty}$ be a trajectory of the unitary system (2.2) with $\vec{u} \in \ell_2(\mathcal{E}_1)$, $\lim_{n\to-\infty} x(n) = 0$ and $\vec{y} \in \ell_2(\mathcal{E}_2)$. In general, if $\mathcal{E}$ is a Hilbert space, $\ell_2(\mathcal{E})$ denotes the space of $\mathcal{E}$-valued sequences $\{e_n\}$

indexed by the integers $n = \cdots -1, 0, 1, \dots$, $\ell_2(\mathbf{N}_-, \mathcal{E})$ denotes the space of such sequences $\{e_n\}$ indexed by the negative integers $\mathbf{N}_- = \{n : n = \dots, -2, -1\}$ and $\ell_2(\mathbf{N}, \mathcal{E})$ is the space of such sequences $\{e_n\}$ indexed by the nonnegative integers $\mathbf{N} = \{n : n = 0, 1, 2, \dots\}$. Define

$$\mathcal{U}' : \begin{bmatrix} \ell_2(\mathbf{N}_-, \mathcal{E}_2) \\ \mathcal{H} \\ \ell_2(\mathbf{N}, \mathcal{E}_1) \end{bmatrix} \longrightarrow \begin{bmatrix} \ell_2(\mathbf{N}_-, \mathcal{E}_2) \\ \mathcal{H} \\ \ell_2(\mathbf{N}, \mathcal{E}_1) \end{bmatrix}$$

by

$$\mathcal{U}'^* : \begin{bmatrix} P_{\ell_2(\mathbf{N}_-, \mathcal{E}_2)}\vec{y} \\ P_{[\mathcal{H}]_0}\vec{x} \\ P_{\ell_2(\mathbf{N}, \mathcal{E}_1)}\vec{u} \end{bmatrix} \longrightarrow \begin{bmatrix} P_{\ell_2(\mathbf{N}_-, \mathcal{E}_2)}\mathbf{U}^*\vec{y} \\ P_{[\mathcal{H}]_0}\mathbf{U}^*\vec{x} \\ P_{\ell_2(\mathbf{N}, \mathcal{E}_1)}\mathbf{U}^*\vec{u} \end{bmatrix}$$

where $\mathbf{U}^*$ is the backward shift operator

$$\mathbf{U}^* : \{v(n)\}_{n=-\infty}^{\infty} \longrightarrow \{v(n+1)\}_{n=-\infty}^{\infty}$$

on $\ell_2(\mathcal{E}_2)$, $\ell_\infty(\mathcal{H})$ or $\ell_2(\mathcal{E}_1)$. Physically, at each point in time, one takes a snapshot of the system trajectory, namely the past output string, the current state and the current and future input string. The definition of $\mathcal{U}'^*$ then is to map the current snapshot at time $t = 0$ to the next snapshot at time $t = 1$. In any case, $\mathcal{U}'^*$ is well-defined and has the form (2.4); in this way we embed the unitary colligation $U = \begin{bmatrix} A & B \\ C & D \end{bmatrix}$ into the ASS

$$\left(\mathcal{U}', \begin{bmatrix} \ell_2(\mathbf{N}_-, \mathcal{E}_2) \\ \mathcal{H} \\ \ell_2(\mathbf{N}, \mathcal{E}_1) \end{bmatrix}, \begin{bmatrix} 0 \\ 0 \\ \ell_2(\mathbf{N}, \mathcal{E}_1) \end{bmatrix}, \begin{bmatrix} \ell_2(\mathbf{N}_-, \mathcal{E}_2) \\ 0 \\ 0 \end{bmatrix}\right). \tag{2.4a}$$

In this way we have a one-to-one correspondence between ASS's and unitary colligations. For more on this connection between operator model theory and scattering, we refer to the paper of Helton [H].

The point of view of operator model theory is to arrive at this same structure but from a different starting point. Namely, one begins with a contraction operator T on a Hilbert space $\mathcal{H}$ and seeks to construct its unitary dilation $\mathcal{U}$, namely a unitary operator $\mathcal{U}$ on a Hilbert space $\mathcal{K} \supset \mathcal{H}$ such that $P_{\mathcal{H}}\mathcal{U}^n|_{\mathcal{H}} = T^n$ for $n = 1, 2, 3, \dots$. By a lemma of Sarason [S1], necessarily $\mathcal{K} = \mathcal{G}_2 \oplus \mathcal{H} \oplus \mathcal{G}_1$ where $\mathcal{U}\mathcal{G}_1 \subset \mathcal{G}_1$, and $\mathcal{U}^*\mathcal{G}_2 \subset \mathcal{G}_2$. If $\mathcal{U}$ is minimal as a unitary dilation of T, then the axioms (i), (ii) and (iii) for an ASS necessarily hold for the system $(\mathcal{K}, \mathcal{U}, \mathcal{G}_1, \mathcal{G}_2)$, and we arrive at (2.3) as the form for the minimal unitary dilation of T, where $\alpha = T$ and $\begin{bmatrix} T & \beta \\ \gamma & \delta \end{bmatrix} : \begin{bmatrix} \mathcal{H} \\ \mathcal{E}_2 \end{bmatrix} \rightarrow \begin{bmatrix} \mathcal{H} \\ \mathcal{E}_1 \end{bmatrix}$ is unitary. A particular choice of β, γ, δ is

$$\begin{bmatrix} T & \beta \\ \gamma & \delta \end{bmatrix} = \begin{bmatrix} T & D_{T^*} \\ D_T & -T^* \end{bmatrix} : \begin{bmatrix} \mathcal{H} \\ \mathcal{D}_{T^*} \end{bmatrix} \longrightarrow \begin{bmatrix} \mathcal{H} \\ \mathcal{D}_T \end{bmatrix}$$

where D_T and D_{T^*} are the defect operators

$$D_T = (I - T^*T)^{\frac{1}{2}}, \quad D_{T^*} = (I - TT^*)^{\frac{1}{2}}$$

and $\mathcal{D}_T$ and $\mathcal{D}_{T^*}$ are the defect spaces

$$\mathcal{D}_T = \text{clos. im } D_T, \quad \mathcal{D}_{T^*} = \text{clos. im } D_{T^*}.$$

This particular choice is sometimes called the Halmos dilation of T.

Given an Abstract Scattering System $(\mathcal{U}, \mathcal{K}, \mathcal{G}_1, \mathcal{G}_2)$, the scattering operator Ω is defined as the operator from $\ell_2(\mathcal{E}_1)$ into $\ell_2(\mathcal{E}_2)$ given by

$$\Omega = \Phi_{\mathcal{E}_2} P_{\widetilde{\mathcal{G}}_2} \Phi_{\mathcal{E}_1}^* = \Phi_{\mathcal{E}_2} \Phi_{\mathcal{E}_1}^*.$$

We may use the Fourier transform

$$\{v(n)\}_{n=-\infty}^{\infty} \longrightarrow \sum_{n=-\infty}^{\infty} v(n) z^n =: \widehat{v}(z)$$

to identify $\ell_2(\mathcal{E}_1)$ with $L_2(\mathcal{E}_1)$ and $\ell_2(\mathcal{E}_2)$ with $L_2(\mathcal{E}_2)$. Thus we may define $\widehat{\Phi}_{\mathcal{E}_i}$ $(i = 1, 2)$ to be the composition of $\Phi_{\mathcal{E}_i}$ with the Fourier transform. Then $\widehat{\Phi}_{\mathcal{E}_1}$ is a partial isometry from $\mathcal{K}$ into $L_2(\mathcal{E}_1)$ with initial space $\widetilde{\mathcal{G}}_1$ which maps $\mathcal{G}_1$ isometrically onto $H_2(\mathcal{E}_1)$ and $\widehat{\Phi}_{\mathcal{E}_2}$ is a partial isometry from $\mathcal{K}$ into $L_2(\mathcal{E}_2)$ with initial space $\widetilde{\mathcal{G}}_2$ which maps $\mathcal{G}_2$ isometrically onto $H_2(\mathcal{E}_2)^{\perp}$. The corresponding version $\widehat{\Omega} = \widehat{\Phi}_{\mathcal{E}_2} \widehat{\Phi}_{\mathcal{E}_1}^*$ of Ω is a contraction operator which intertwines the operator M_z of multiplication by the coordinate function z on the spaces $L_2(\mathcal{E}_1)$ and $L_2(\mathcal{E}_2)$, and, as a consequence of the assumed orthogonality between $\mathcal{G}_1$ and $\mathcal{G}_2$, maps $H_2(\mathcal{E}_1)$ into $H_2(\mathcal{E}_2)$. Hence $\widehat{\Omega}$ is necessarily equal to the operator M_ω of multiplication by a function $\omega(z) \in \mathcal{S}(\mathcal{E}_1, \mathcal{E}_2)$. In the context of operator model theory, $\omega(z)$ is the characteristic operator function of the contraction operator $T = \alpha$ which leads to the construction of the functional model for T (see [SzNF]).

It is well known (see e.g. Proposition 2.2 on page 246 of [SzNF] or Lemma 11 on page 413 of [NV]) that $\omega(z) = D + zC(I - zA)^{-1}B$, i.e. that the scattering function $\omega(z)$ of the ASS $(\mathcal{U}, \mathcal{K}, \mathcal{G}_1, \mathcal{G}_2)$ is equal to the characteristic function $\omega_U(z)$ of the unitary colligation $U = \begin{bmatrix} A & B \\ C & D \end{bmatrix}$ embedded in the matrix representation for $\mathcal{U}^*$ as in (2.4). We include a proof of this fact here to emphasize the connections between the system theory and scattering theory ideas.

Proposition 2.1 *Let $\omega(z)$ be the scattering function (so $M_\omega = \widehat{\Phi}_{\mathcal{E}_2} \widehat{\Phi}_{\mathcal{E}_1}^*$) for the ASS $(\mathcal{U}, \mathcal{K}, \mathcal{G}_1, \mathcal{G}_2)$ (with $\mathcal{E}_1 = \mathcal{G}_1 \ominus \mathcal{U}\mathcal{G}_1$ and $\mathcal{E}_2 = \mathcal{U}\mathcal{G}_2 \ominus \mathcal{G}_2$) and let the operator $\mathcal{U}^*$ have the matrix representation $\mathcal{U}'^*$ as given in (2.4). Then $\omega(z)$ is given by*

$$\omega(z) = D + zC(I - zA)^{-1}B.$$

Proof: By definition, $\omega(z)\widehat{u}(z) = \widehat{y}(z)$ where $\widehat{u}(z) = \sum_{n=0}^{\infty} u(n)z^n$ and $\widehat{y}(z) = \sum_{n=0}^{\infty} y(n)z^n$ means that

$$\begin{aligned} y(n) &= P_{\mathcal{E}_2}\mathcal{U}^{*n}\left(\sum_{k=0}^{\infty}\mathcal{U}^k u(k)\right) \\ &= P_{\mathcal{U}^*\mathcal{E}_2}\mathcal{U}^{*n+1}\left(\sum_{k=0}^{\infty}\mathcal{U}^k u(k)\right). \end{aligned}$$

Define $x(0) = 0$ and in general

$$x(n) = P_{\mathcal{H}}\mathcal{U}'^{*n}\begin{bmatrix} 0 \\ 0 \\ \{u(k)\}_{k\geq 0} \end{bmatrix}.$$

From the upper triangular form of $\mathcal{U}'^*$ in (2.4) we see that then

$$\begin{aligned} x(n+1) &= Ax(n) + Bu(n) \\ y(n) &= Cx(n) + Du(n) \end{aligned}$$

and hence the input-output map of the system (2.2) transforms $\{u(n)\}_{n=0}^{\infty}$ to $\{y(n)\}_{n=0}^{\infty}$. Upon applying Fourier transform we see that $\omega_U(z) = \omega(z)$ as required. □

We next address the question of to what extent does the scattering function $\omega(z)$ determine the ASS $(\mathcal{U}, \mathcal{K}, \mathcal{G}_1, \mathcal{G}_2)$. Let us say that two ASS's $(\mathcal{K}, \mathcal{U}, \mathcal{G}_1, \mathcal{G}_2)$ and $(\mathcal{K}', \mathcal{U}', \mathcal{G}_1', \mathcal{G}_2')$ with the same wandering subspaces $\mathcal{G}_1 \ominus \mathcal{U}\mathcal{G}_1 = \mathcal{E}_1 = \mathcal{G}_1' \ominus \mathcal{U}\mathcal{G}_2'$ and $\mathcal{U}\mathcal{G}_2 \ominus \mathcal{G}_2 = \mathcal{E}_2 = \mathcal{U}\mathcal{G}_2' \ominus \mathcal{G}_2'$ are *unitarily equivalent* if there is a unitary operator $\mathcal{V} : \mathcal{K} \to \mathcal{K}'$ such that $\mathcal{V}\mathcal{U} = \mathcal{U}'\mathcal{V}$ and $\mathcal{V}|_{\mathcal{E}_1} = I_{\mathcal{E}_1}$ and $\mathcal{V}|_{\mathcal{E}_2} = I_{\mathcal{E}_2}$. The following result establishes the essential one-to-one correspondence between minimal Abstract Scattering Systems and their associated scattering functions.

Theorem 2.2 *Suppose ω is a Schur-class function in $\mathcal{S}(\mathcal{E}_1, \mathcal{E}_2)$. Then there exists an Abstract Scattering System $(\mathcal{U}, \mathcal{K}, \mathcal{G}_1, \mathcal{G}_2)$ having ω as its scattering function. Any two minimal ASS's with the same scattering function ω are unitarily equivalent.*

Proof: Given a Schur-class function $\omega(z) \in \mathcal{S}(\mathcal{E}_1, \mathcal{E}_2)$, we can construct an ASS $(\mathcal{U}, \mathcal{K}, \mathcal{G}_1, \mathcal{G}_2)$ with $\mathcal{G}_1 \ominus \mathcal{U}\mathcal{G}_1 = \mathcal{E}_1$ and $\mathcal{U}\mathcal{G}_2 \ominus \mathcal{G}_2 = \mathcal{E}_2$ and having $\omega(z)$ as its scattering function as follows. It is known (see e.g. [ADRdS]) that there exists a unitary colligation $U = \begin{bmatrix} A & B \\ C & D \end{bmatrix} : \begin{bmatrix} \mathcal{H} \\ \mathcal{E}_1 \end{bmatrix} \to \begin{bmatrix} \mathcal{H} \\ \mathcal{E}_2 \end{bmatrix}$ having $\omega(z)$ as its characteristic function

(2.5) $$\omega(z) = D + zC(I - zA)^{-1}B.$$

(More concretely, one could use the Pavlov or de Branges-Rovnyak model ASS for a given ω given below to realize explicitly a particular choice of A, B, C, D.) Now embed U into the unitary operator $\mathcal{U}'^*$ on $\ell_2(\mathbf{N}, \mathcal{E}_2) \oplus \mathcal{H} \oplus \ell_2(\mathbf{N}, \mathcal{E}_1)$ according to the formula (2.4). Then the ASS given by (2.4a) scattering function equal to the preassigned function ω. One can also show that this ASS is minimal if and only if the unitary realization (2.5) is *closely connected* (i.e. the span of the images of $A^n B$ and of $A^{*k} C^*$ over $n, k = 0, 1, 2, \ldots$ is dense in $\mathcal{H}$).

Suppose now that $(\mathcal{U}, \mathcal{K}, \mathcal{G}_1, \mathcal{G}_2)$ is an ASS with scattering function ω. Define an operator $\mathcal{V}_P$ on $\widetilde{\mathcal{G}}_2 + \widetilde{\mathcal{G}}_1$ into $\begin{bmatrix} L_2(\mathcal{E}_2) \\ L_2(\mathcal{E}_1) \end{bmatrix}$ by

$$\mathcal{V}_P(g_2 + g_1) = \begin{bmatrix} \widehat{\Phi}_{\mathcal{E}_2}(g_2) \\ \widehat{\Phi}_{\mathcal{E}_1}(g_1) \end{bmatrix}.$$

Let us compute, for $g_2, g_2' \in \widetilde{\mathcal{G}}_2$ and $g_1, g_1' \in \widetilde{\mathcal{G}}_1$,

$$\begin{aligned} \langle g_2 + g_1, g_2' + g_1' \rangle &= \langle g_1, g_1' \rangle + \langle g_2, g_1' \rangle + \langle g_1, g_2' \rangle + \langle g_1, g_1' \rangle \\ &= \langle \widehat{\Phi}_{\mathcal{E}_2} g_2, \widehat{\Phi}_{\mathcal{E}_2} g_2' \rangle + \langle \omega^* \widehat{\Phi}_{\mathcal{E}_2} g_2, \widehat{\Phi}_{\mathcal{E}_1} g_1' \rangle \\ &\quad + \langle \omega \widehat{\Phi}_{\mathcal{E}_1} g_1, \widehat{\Phi}_{\mathcal{E}_2} g_2' \rangle + \langle \widehat{\Phi}_{\mathcal{E}_1} g_1, \widehat{\Phi}_{\mathcal{E}_1} g_1' \rangle \\ &= \left\langle \begin{bmatrix} I & \omega \\ \omega^* & I \end{bmatrix} \begin{bmatrix} \widehat{\Phi}_{\mathcal{E}_2} g_2 \\ \widehat{\Phi}_{\mathcal{E}_1} g_1 \end{bmatrix}, \begin{bmatrix} \widehat{\Phi}_{\mathcal{E}_2} g_2' \\ \widehat{\Phi}_{\mathcal{E}_1} g_1' \end{bmatrix} \right\rangle. \end{aligned} \tag{2.6}$$

Hence, if we let $\mathcal{K}_P^\omega$ be the completion of $\begin{bmatrix} L_2(\mathcal{E}_2) \\ L_2(\mathcal{E}_1) \end{bmatrix}$ in the inner product

$$\left\langle \begin{bmatrix} h_1 \\ h_2 \end{bmatrix}, \begin{bmatrix} h_1 \\ h_2 \end{bmatrix} \right\rangle_{\mathcal{K}_P^\omega} = \left\langle \begin{bmatrix} I & \omega \\ \omega^* & I \end{bmatrix} \begin{bmatrix} h_1 \\ h_2 \end{bmatrix}, \begin{bmatrix} h_1 \\ h_2 \end{bmatrix} \right\rangle_{L_2(\mathcal{E}_2) \oplus L_2(\mathcal{E}_1)}$$

(with any elements with zero self-inner-product identified to 0), then $\mathcal{V}_P$ extends uniquely to a well-defined unitary operator from $\mathcal{K}_0 :=$ clos. $(\widetilde{\mathcal{G}}_1 + \widetilde{\mathcal{G}}_2) \subset \mathcal{K}$ onto $\mathcal{K}_P^\omega$. Moreover if we set $\mathcal{U}' = M_z$ (multiplication by the coordinate function z on both components) on $\mathcal{K}_P^\omega$, $\mathcal{G}_1' = \begin{bmatrix} 0 \\ H_2(\mathcal{E}_1) \end{bmatrix} \subset \mathcal{K}_P^\omega$ and $\mathcal{G}_2' = \begin{bmatrix} H_2(\mathcal{E}_2)^\perp \\ 0 \end{bmatrix} \subset \mathcal{K}_P^\omega$, then $\mathcal{V}_P$ establishes a unitary equivalence between the ASS $(\mathcal{U}|_{\mathcal{K}_0}, \mathcal{K}_0, \mathcal{G}_1, \mathcal{G}_2)$ and the particular ASS $(\mathcal{K}_P^\omega, \mathcal{U}', \mathcal{G}_1', \mathcal{G}_2')$ with scattering function equal to $\omega(z)$. In particular, if the ASS $(\mathcal{U}, \mathcal{K}, \mathcal{G}_1, \mathcal{G}_2)$ is minimal, then $\mathcal{K} = \mathcal{K}_0$ and $(\mathcal{U}, \mathcal{K}, \mathcal{G}_1, \mathcal{G}_2)$ is unitarily equivalent to the particular ASS $(\mathcal{U}' = M_z, \mathcal{K}_P^\omega, \mathcal{G}_1', \mathcal{G}_2')$. We conclude that any two minimal ASS's with the same scattering function ω are unitarily equivalent as asserted. □

The particular model $\left(M_z, \mathcal{K}_P^\omega, \begin{bmatrix} 0 \\ H_2(\mathcal{E}_1) \end{bmatrix}, \begin{bmatrix} H_2(\mathcal{E}_2)^\perp \\ 0 \end{bmatrix}\right)$ for a minimal ASS with scattering function equal to ω appearing in the proof of Theorem 2.2 is called the *Pavlov transcription* in [NV]. The scattering or model space

$\mathcal{H}_P^{\omega} = \mathcal{K}_P^{\omega} \ominus \left\{ \begin{bmatrix} H_2(\mathcal{E}_2)^{\perp} \\ H_2(\mathcal{E}_1) \end{bmatrix} \right\}$ is then given at least formally by

$$\mathcal{H}_P^{\omega} = \begin{bmatrix} I & \omega \\ \omega^* & I \end{bmatrix}^{-1} \begin{bmatrix} H_2(\mathcal{E}_2) \\ H_2(\mathcal{E}_1)^{\perp} \end{bmatrix}.$$

One gets other concrete transcriptions of a minimal ASS with given scattering function ω by using a different spectral representation for the unitary operator $\mathcal{U}$. For a given $\omega \in \mathcal{S}(\mathcal{E}_1, \mathcal{E}_2)$, define

$$\text{(2.7)} \qquad \mathcal{K}_{BR}^{\omega} = \begin{bmatrix} I & \omega \\ \omega^* & I \end{bmatrix}^{\frac{1}{2}} \begin{bmatrix} L_2(\mathcal{E}_2) \\ L_2(\mathcal{E}_1) \end{bmatrix}$$

with inner product

$$\left\langle \begin{bmatrix} I & \omega \\ \omega^* & I \end{bmatrix}^{\frac{1}{2}} \begin{bmatrix} h \\ k \end{bmatrix}, \begin{bmatrix} I & \omega \\ \omega^* & I \end{bmatrix}^{\frac{1}{2}} \begin{bmatrix} h \\ k \end{bmatrix} \right\rangle_{\mathcal{K}_{BR}^{\omega}} = \left\langle Q \begin{bmatrix} h \\ k \end{bmatrix}, \begin{bmatrix} h \\ k \end{bmatrix} \right\rangle_{L_2(\mathcal{E}_2) \oplus L_2(\mathcal{E}_1)}$$

where Q is the orthogonal projection onto the orthogonal complement of the kernel of $\begin{bmatrix} I & \omega \\ \omega^* & I \end{bmatrix}$. Suppose as above that $(\mathcal{U}, \mathcal{K}, \mathcal{G}_1, \mathcal{G}_2)$ is a minimal ASS with scattering function $\omega \in \mathcal{S}(\mathcal{E}_1, \mathcal{E}_2)$, so $\widetilde{\mathcal{G}}_2 + \widetilde{\mathcal{G}}_1$ is dense in $\mathcal{K}$. Define $\mathcal{V}_{BR}$ on $\widetilde{\mathcal{G}}_2 + \widetilde{\mathcal{G}}_1$ into $L_2(\mathcal{E}_2) \oplus L_2(\mathcal{E}_1)$ by

$$\mathcal{V}_{BR}(g_2 + g_1) = \begin{bmatrix} I & \omega \\ \omega^* & I \end{bmatrix} \begin{bmatrix} \widehat{\Phi}_{\mathcal{E}_2} g_2 \\ \widehat{\Phi}_{\mathcal{E}_1} g_1 \end{bmatrix}.$$

From (2.6) we see that $\mathcal{V}_{BR}$ is a well-defined isometry from a dense subset $\widetilde{\mathcal{G}}_2 + \widetilde{\mathcal{G}}_1$ of $\mathcal{K}$ onto a dense subset $\begin{bmatrix} I & \omega \\ \omega^* & I \end{bmatrix} \begin{bmatrix} L_2(\mathcal{E}_2) \\ L_2(\mathcal{E}_1) \end{bmatrix}$ of $\mathcal{K}_{BR}^{\omega}$ and hence extends to a unitary identification map between $\mathcal{K}$ and $\mathcal{K}_{BR}^{\omega}$. Under this identification map, $\mathcal{U}$ transforms to $\mathcal{U}'' = \mathcal{V}_{BR} \mathcal{U} \mathcal{V}_{BR}^* = M_z$, $\mathcal{G}_1$ transforms to

$$\mathcal{G}_1'' = \mathcal{V}_{BR} \mathcal{G}_1 = \begin{bmatrix} \omega \\ I \end{bmatrix} H_2(\mathcal{E}_1),$$

and $\mathcal{G}_2$ transforms to

$$\mathcal{G}_2'' = \mathcal{V}_{BR} \mathcal{G}_2 = \begin{bmatrix} I \\ \omega^* \end{bmatrix} H_2(\mathcal{E}_2)^{\perp}.$$

A simple computation shows that the scattering/model space $\mathcal{H}$ transforms to

$$\begin{aligned} \mathcal{H}_{BR}^{\omega} &= \mathcal{K}_{BR}^{\omega} \ominus \left\{ \begin{bmatrix} I & \omega \\ \omega^* & I \end{bmatrix} \begin{bmatrix} H_2(\mathcal{E}_2)^{\perp} \\ H_2(\mathcal{E}_1) \end{bmatrix} \right\} \\ &= \begin{bmatrix} I & \omega \\ \omega^* & I \end{bmatrix}^{\frac{1}{2}} \begin{bmatrix} L_2(\mathcal{E}_2) \\ L_2(\mathcal{E}_1) \end{bmatrix} \bigcap \begin{bmatrix} H_2(\mathcal{E}_2) \\ H_2(\mathcal{E}_1)^{\perp} \end{bmatrix}, \end{aligned}$$

i.e., $\mathcal{H}^{\omega}_{BR}$ is equal to the de Branges-Rovnyak space $\mathcal{H}^{\omega}$ as defined in (1.1) in Section 1. This particular transcription

$$\left(M_z,\ \mathcal{K}^{\omega}_{BR},\ \begin{bmatrix} \omega \\ I \end{bmatrix} H_2(\mathcal{E}_1),\ \begin{bmatrix} I \\ \omega^* \end{bmatrix} H_2(\mathcal{E}_2)^{\perp}\right) \tag{2.8}$$

of a minimal ASS with scattering function equal to ω is the *de Branges-Rovnyak transcription* in the terminology of [NV]. Note that the scattering function of the de Branges-Rovnyak model ASS (2.8) is actually $\widetilde{\omega}(z) = i_2\omega(z)i_1^{-1}$ where

$$i_1 : \mathcal{E}_1 \longrightarrow \begin{bmatrix} I \\ \omega \end{bmatrix} \mathcal{E}_1 \subset \mathcal{K}^{\omega}_{BR},$$

$$i_2 : \mathcal{E}_2 \longrightarrow \begin{bmatrix} I \\ \omega^* \end{bmatrix} \mathcal{E}_2 \subset \mathcal{E}^{\omega}_{BR}$$

are unitary identification maps

$$i_1 : e_1 \longrightarrow \begin{bmatrix} I \\ \omega \end{bmatrix} e_1,\quad i_2 : e_2 \longrightarrow \begin{bmatrix} I \\ \omega^* \end{bmatrix} e_2. \tag{2.9}$$

Two Schur-class functions $\omega \in \mathcal{S}(\mathcal{E}_1, \mathcal{E}_2)$ and $\widetilde{\omega} \in \mathcal{S}(\widetilde{\mathcal{E}}_1, \widetilde{\mathcal{E}}_2)$ differing in such a trivial way are said to *coincide*.

Other possible transcriptions are the Sz.-Nagy-Foias transcription and the transcription associated with a regular factorization of the scattering function ω, but we shall not need those here.

2.3 The Abstract Interpolation Problem in Coordinate-free Form

Suppose now that we are given an admissible abstract interpolation data set

$$(T_1, T_2, D(\cdot,\cdot), M_1, M_2, \mathcal{X}, \mathcal{E}_1, \mathcal{E}_2)$$

and an ASS $(\mathcal{U}, \mathcal{K}, \mathcal{G}_1, \mathcal{G}_2)$ such that the wandering subspaces $\mathcal{G}_1 \ominus \mathcal{U}\mathcal{G}_1$ and $\mathcal{U}\mathcal{G}_2 \ominus \mathcal{G}_2$ are identified with $\mathcal{E}_1$ and $\mathcal{E}_2$ respectively in the AIP interpolation data set. In this way we are free to identify im M_1 and im M_2 as subspaces of the Hilbert space $\mathcal{K}$ in the ASS.

We are now ready to present the Coordinate-Free version of the Abstract Interpolation Problem (CF-AIP): *Given an admissible AIP data set, find a minimal ASS* $(\mathcal{U}, \mathcal{K}, \mathcal{G}_1, \mathcal{G}_2)$ *with wandering subspaces* $\mathcal{G}_1 \ominus \mathcal{U}\mathcal{G}_1 = \mathcal{E}_1$ *and* $\mathcal{U}\mathcal{G}_2 \ominus \mathcal{G}_2 = \mathcal{E}_2$ *prescribed as in the AIP data set, together with a contractive linear operator*

$$\mathbf{F} : \mathcal{X}_D \longrightarrow \mathcal{H} := \mathcal{K} \ominus \{\mathcal{G}_1 + \mathcal{G}_2\}$$

such that

$$\mathbf{F}T_1 + M_1 = \mathcal{U}\mathbf{F}T_2 + M_2. \tag{2.10}$$

The next order of business is to demonstrate that the CF-AIP is just the coordinate-free transcription of the AIP. Suppose that (ω, F) solves the AIP. Consider ω as the scattering function for a minimal ASS and consider the de Branges-Rovnyak transcription (2.8) associated with ω. We can view this ASS more abstractly in coordinate-free terms as $(\mathcal{U}, \mathcal{K}, \mathcal{G}_1, \mathcal{G}_2)$ where $\mathcal{E}_1 = \mathcal{G}_1 \ominus \mathcal{U}\mathcal{G}_1$, $\mathcal{E}_2 = \mathcal{U}\mathcal{G}_2 \ominus \mathcal{G}_2$ with $M_\omega : L_2(\mathcal{E}_1) \to L_2(\mathcal{E}_2)$ equal to $\widehat{\Phi}_{\mathcal{E}_2}\widehat{\Phi}^*_{\mathcal{E}_1}$. Viewed in these abstract terms, the contraction operator $F : \mathcal{X}_D \to \mathcal{H}^\omega_{BR}$ becomes a contraction operator $\mathbf{F} : \mathcal{X}_D \to \mathcal{H} = \mathcal{K} \ominus \{\mathcal{G}_1 + \mathcal{G}_2\}$. The identity (1.2) can be rewritten as

$$[F(T_1x)](z) = z[F(T_2x)](z) + \begin{bmatrix} I \\ \omega(z)^* \end{bmatrix} M_2x - \begin{bmatrix} \omega(z) \\ I \end{bmatrix} M_1x. \tag{2.11}$$

When moving from the de Branges-Rovnyak transcription $\mathcal{K}^\omega_{BR}$ to the coordinate-free setting $\mathcal{K}$, we use the identification maps i_1^{-1} and i_2^{-1} given by (2.9) to identify $\begin{bmatrix} \omega(z) \\ I \end{bmatrix} \mathcal{E}_1 \subset \mathcal{K}^\omega_{DB}$ with $\mathcal{E}_1 \subset \mathcal{K}$ and $\begin{bmatrix} I \\ \omega(x)^* \end{bmatrix} \mathcal{E}_2 \subset \mathcal{K}^\omega_{DB}$ with $\mathcal{E}_2 \subset \mathcal{K}$. Hence the equation (2.11) translates to

$$\mathbf{F}T_1x = \mathcal{U}\mathbf{F}T_2x + M_2x - M_1x$$

for all x in $\mathcal{X}_D$, and we see that $((\mathcal{U}, \mathcal{K}, \mathcal{G}_1, \mathcal{G}_2), \mathbf{F})$ is a solution of the CF-AIP.

Conversely, if $((\mathcal{U}, \mathcal{K}, \mathcal{G}_1, \mathcal{G}_2), \mathbf{F})$ is a solution of the CF-AIP, define ω to be the scattering function of $(\mathcal{U}, \mathcal{K}, \mathcal{G}_1, \mathcal{G}_2)$ and set $F = \mathcal{V}_{BR}\mathbf{F}$. Then it is easily checked that (ω, F) is a solution of the AIP.

3 Main Theorem on Coordinate-free Absract Interpolation Problem

The following is the coordinate-free version of Theorem 1.1 above.

Theorem 3.1 *Let $(T_1, T_2, D(\cdot,\cdot), M_1, M_2, \mathcal{X}, \mathcal{E}_1, \mathcal{E}_2)$ be an admissible data set for an AIP, and let $V : \mathcal{D}_V \to \mathcal{R}_V$ be the partial isometric colligation defined by (0.2). Then the solutions of the CF-AIP are in bijective correspondence (up to unitary equivalence) with unitary colligation extensions U of the partial isometric colligation V.*

More explicitly, if

$$U = \begin{bmatrix} A & B \\ C & D \end{bmatrix} : \begin{bmatrix} \mathcal{H} \\ \mathcal{E}_1 \end{bmatrix} \longrightarrow \begin{bmatrix} \mathcal{H} \\ \mathcal{E}_2 \end{bmatrix} \quad \text{where } \mathcal{H} \supset \mathcal{X}_D$$

is such a unitary colligation extension of V, embed U in an ASS $(\mathcal{U}, \mathcal{K}, \mathcal{G}_1, \mathcal{G}_2)$ such that $\mathcal{U}^*$ has the matrix representation $\mathcal{U}'^*$ as in (2.4). Let $\mathcal{K}_0 = clos.\{\widetilde{\mathcal{G}}_1 + \widetilde{\mathcal{G}}_2\}$ and let $(\mathcal{U}_0, \mathcal{K}_0, \mathcal{G}_1, \mathcal{G}_2)$ (where $\mathcal{U}_0 = \mathcal{U}|_{\mathcal{K}_0}$) be the minimal ASS contained in $(\mathcal{U}, \mathcal{K}, \mathcal{G}_1, \mathcal{G}_2)$. Finally set

$$\mathcal{H}_0 = clos.[\widetilde{\mathcal{G}}_1 + \widetilde{\mathcal{G}}_2] \ominus [\mathcal{G}_1 + \mathcal{G}_2]$$

and set $\mathbf{F} = P_{\mathcal{H}_0}|_{\mathcal{X}_D}$. Then $((\mathcal{U}_0, \mathcal{K}_0, \mathcal{G}_1, \mathcal{G}_2), \mathbf{F})$ is a solution of the CF-AIP.

Conversely, suppose that $((\mathcal{U}_0, \mathcal{K}_0, \mathcal{G}_1, \mathcal{G}_2), \mathbf{F})$ *is a solution of the CF-AIP and set* $\mathcal{H}_0 = \mathcal{K}_0 \ominus [\mathcal{G}_1 + \mathcal{G}_2]$. *Associated with this solution of the CF-AIP is a Hilbert space* $\mathcal{H}_u$ *(which is trivial exactly when* $\mathbf{F}$ *is isometric), a unitary operator* U_u *on* $\mathcal{H}_u$, *and an embedding* i *of* $\mathcal{X}_D$ *into* $\mathcal{H} := \begin{bmatrix} \mathcal{H}_u \\ \mathcal{H}_0 \end{bmatrix}$. *Then* $(U_u \oplus \mathcal{U}, \mathcal{H}_u \oplus \mathcal{K}, 0 \oplus \mathcal{G}_1, 0 \oplus \mathcal{G}_2)$ *is a (possibly nonminimal) ASS. Let*

$$U = \begin{bmatrix} A & B \\ C & D \end{bmatrix} : \begin{bmatrix} \mathcal{H} \\ \mathcal{E}_1 \end{bmatrix} \longrightarrow \begin{bmatrix} \mathcal{H} \\ \mathcal{E}_2 \end{bmatrix}$$

(where $\mathcal{H} = \mathcal{H}_u \oplus \mathcal{H}_0$*) be the unitary colligation embedded in the matrix representation* $\mathcal{U}'^*$ *in (2.4) associated with this latter ASS. Then* U *is a unitary colligation extension of the partial isometric colligation*

$$\widetilde{V} : \begin{bmatrix} i & 0 \\ 0 & I_{\mathcal{E}_1} \end{bmatrix} \mathcal{D}_V \longrightarrow \begin{bmatrix} i & 0 \\ 0 & I_{\mathcal{E}_2} \end{bmatrix} \mathcal{R}_V$$

defined by

$$\widetilde{V} : \begin{bmatrix} iT_1 \\ M_1 \end{bmatrix} x \longrightarrow \begin{bmatrix} iT_2 \\ M_2 \end{bmatrix} x. \tag{3.1}$$

Moreover we recover $\mathbf{F}$ *as* $\begin{bmatrix} 0 \\ \mathbf{F} \end{bmatrix} = P_{0 \oplus \mathcal{H}_0} i$.

Proof: Suppose that $U = \begin{bmatrix} A & B \\ C & D \end{bmatrix}$ is a unitary colligation extension of V. Thus U satisfies the operator identity

$$\begin{bmatrix} A & B \\ C & D \end{bmatrix} \begin{bmatrix} T_1 \\ M_1 \end{bmatrix} = \begin{bmatrix} T_2 \\ M_2 \end{bmatrix}$$

and hence $\begin{bmatrix} \alpha & \beta \\ \gamma & \delta \end{bmatrix} := \begin{bmatrix} A & B \\ C & D \end{bmatrix}^*$ satisfies the operator identity

$$\begin{bmatrix} \alpha & \beta \\ \gamma & \delta \end{bmatrix} \begin{bmatrix} T_2 \\ M_2 \end{bmatrix} = \begin{bmatrix} T_1 \\ M_1 \end{bmatrix}.$$

Embed U in an ASS $(\mathcal{U}, \mathcal{K}, \mathcal{G}_1, \mathcal{G}_2)$ so that $\mathcal{U}^*$ has the matrix representation $\mathcal{U}'^*$ as in (2.4). From the matrix representation (2.3) for $\mathcal{U}'$ we deduce that

$$\mathcal{U}' : \begin{bmatrix} \vdots \\ 0 \\ M_2 x \\ T_2 x \\ 0 \\ 0 \\ \vdots \end{bmatrix} \longrightarrow \begin{bmatrix} \vdots \\ 0 \\ 0 \\ T_1 x \\ M_1 x \\ 0 \\ \vdots \end{bmatrix}. \tag{3.2}$$

Set $\mathcal{K}_0 = \widetilde{\mathcal{G}}_1 + \widetilde{\mathcal{G}}_2$ and

$$\mathbf{F} = P_{\mathcal{H}_0}|_{\mathcal{X}_D} = P_{\mathcal{K}_0}|_{\mathcal{X}_D}$$

where $\mathcal{H}_0 = \text{clos.}[\widetilde{\mathcal{G}}_1 + \widetilde{\mathcal{G}}_2] \ominus [\mathcal{G}_1 + \mathcal{G}_2] \subset \mathcal{H}$. As $\mathcal{K}_0$ is reducing for $\mathcal{U}$, it follows that $\mathcal{U}$ commutes with $P_{\mathcal{K}_0}$. Hence from formula (3.2) we see that

$$\mathcal{U}_0' : \begin{bmatrix} \vdots \\ 0 \\ M_2x \\ \mathbf{F}T_2x \\ 0 \\ 0 \\ \vdots \end{bmatrix} \longrightarrow \begin{bmatrix} \vdots \\ 0 \\ 0 \\ \mathbf{F}T_1x \\ M_1x \\ 0 \\ \vdots \end{bmatrix},$$

or, in coordinate-free notation,

$$\mathcal{U}_0(\mathcal{U}_0^* M_2 + \mathbf{F}T_2)x = (\mathbf{F}T_1 + M_1)x$$

for all $x \in \mathcal{X}$. We conclude that $((\mathcal{U}_0, \mathcal{K}_0, \mathcal{G}_1, \mathcal{G}_2), \mathbf{F})$ is a solution of the CF-AIP.

Conversely, suppose that $((\mathcal{U}_0, \mathcal{K}_0, \mathcal{G}_1, \mathcal{G}_2), \mathbf{F})$ is a solution of the CF-AIP. Then $\mathbf{F} : \mathcal{X}_D \to \mathcal{H}_0$ where $\mathcal{H}_0 = \mathcal{K}_0 \ominus [\mathcal{G}_1 + \mathcal{G}_2]$. Let $U_0 = \begin{bmatrix} A_0 & B_0 \\ C_0 & D_0 \end{bmatrix}$ be the unitary colligation embedded in the matrix representation $\mathcal{U}_0'^*$ for $\mathcal{U}_0^*$ as in (2.4). The identity (2.10) satisfied by any solution of the CF-AIP then unravels to the operator equation

$$\begin{bmatrix} A_0 & B_0 \\ C_0 & D_0 \end{bmatrix} \begin{bmatrix} \mathbf{F}T_1 \\ M_1 \end{bmatrix} = \begin{bmatrix} \mathbf{F}T_2 \\ M_2 \end{bmatrix}. \tag{3.3}$$

Let $\mathcal{D}_\mathbf{F} = \text{im } D_\mathbf{F}$ (where $D_\mathbf{F} = (I - \mathbf{F}^*\mathbf{F})^{\frac{1}{2}}$) be the defect space of $\mathbf{F}$. We next check that the formula

$$V_0 : D_\mathbf{F}T_1x \longrightarrow D_\mathbf{F}T_2x \tag{3.4}$$

can be used to give a well-defined isometry from $\mathcal{D}_{V_0} := \text{clos. im } D_\mathbf{F}T_1$ onto $\mathcal{R}_{V_0} := \text{clos. im } D_\mathbf{F}T_2$. Recall from our assumption (0.1) for an admissible interpolation data set that

$$\|T_1x\|^2 - \|T_2x\|^2 = \|M_2x\|^2 - \|M_1x\|^2 \tag{3.5}$$

(where $\|T_1x\|^2$ and $\|T_2x\|^2$ are computed in $\mathcal{X}_D$). On the other hand, from (2.10) we see that

$$\mathbf{F}T_1x + M_1x = \mathcal{U}_0(\mathbf{F}T_2x + \mathcal{U}_0^* M_2x)$$

and hence

$$
\begin{aligned}
\|\mathbf{F}T_1x\|^2+\|M_1x\|^2 &= \|\mathcal{U}_0(\mathbf{F}T_2x+\mathcal{U}_0^*M_2x)\|^2\\
&= \|\mathbf{F}T_2x+\mathcal{U}_0^*M_2x\|^2\\
&= \|\mathbf{F}T_2x\|^2+\|\mathcal{U}_0^*M_2x\|^2\\
&= \|\mathbf{F}T_2x\|^2+\|M_2x\|^2.
\end{aligned}
\tag{3.6}
$$

Combining (3.5) and (3.6) gives

$$
\begin{aligned}
\|D_{\mathbf{F}}T_1x\|^2 &= \|T_1x\|^2-\|\mathbf{F}T_1x\|^2\\
&= [\|T_2x\|^2+\|M_2x\|^2-\|M_1x\|^2]\\
&\quad -[\|\mathbf{F}T_2x\|^2+\|M_2x\|^2-\|M_1x\|^2]\\
&= \|T_2x\|^2-\|\mathbf{F}T_2x\|^2\\
&= \|D_{\mathbf{F}}T_2X\|^2
\end{aligned}
\tag{3.7}
$$

and hence (3.4) gives a well-defined isometry from $\mathcal{D}_{V_0}$ onto $\mathcal{R}_{V_0}$ as claimed.
Let V_u on $\mathcal{H}_u \supset \mathcal{D}_{\mathbf{F}}$ be any unitary extension of V_0, and set

$$
U=\begin{bmatrix} V_u & 0 & 0\\ 0 & A_0 & B_0\\ 0 & C_0 & D_0\end{bmatrix}:\begin{bmatrix}\mathcal{H}_u\\ \mathcal{H}_0\\ \mathcal{E}_1\end{bmatrix}\longrightarrow\begin{bmatrix}\mathcal{H}_u\\ \mathcal{H}_0\\ \mathcal{E}_2\end{bmatrix}.
$$

Then we may consider U to be a unitary colligation

$$
U=\begin{bmatrix} A & B\\ C & D\end{bmatrix}:\begin{bmatrix}\mathcal{H}\\ \mathcal{E}_1\end{bmatrix}\longrightarrow\begin{bmatrix}\mathcal{H}\\ \mathcal{E}_2\end{bmatrix}
$$

with $\mathcal{H}=\begin{bmatrix}\mathcal{H}_u\\ \mathcal{H}_0\end{bmatrix}$. Then (3.3) and (3.4) combine to give us the operator identity

$$
U\begin{bmatrix} i & 0\\ 0 & I_{\mathcal{E}_1}\end{bmatrix}\begin{bmatrix}T_1\\ M_1\end{bmatrix}\longrightarrow\begin{bmatrix} i & 0\\ 0 & I_{\mathcal{E}_2}\end{bmatrix}\begin{bmatrix}T_2\\ M_2\end{bmatrix}
\tag{3.8}
$$

where $i=\begin{bmatrix}D_{\mathbf{F}}\\ \mathbf{F}\end{bmatrix}$ is an isometric embedding of $\mathcal{X}_D$ into $\mathcal{H}=\begin{bmatrix}\mathcal{H}_u\\ \mathcal{H}_0\end{bmatrix}$. Then by (3.8) we see that U is a unitary colligation which extends the partial isometric colligation $\widetilde{V}$ as in (3.1). Finally, from the form $i=\begin{bmatrix}D_{\mathbf{F}}\\ \mathbf{F}\end{bmatrix}$ of i, we see that we recover $\mathbf{F}$ as $\begin{bmatrix}0\\ \mathbf{F}\end{bmatrix}=P_{0\oplus\mathcal{H}_0}i$. The theorem follows. □

Remark: Suppose that

$$
U=\begin{bmatrix} A & B\\ C & D\end{bmatrix}:\begin{bmatrix}\mathcal{H}\\ \mathcal{E}_1\end{bmatrix}\longrightarrow\begin{bmatrix}\mathcal{H}\\ \mathcal{E}_2\end{bmatrix}
$$

is a unitary colligation (not necessarily minimal, so $\mathcal{H} = \mathcal{H}_u \oplus \mathcal{H}_0$ in the notation above) with characteristic function equal to $\omega(z) = D + zC(I - zA)^{-1}B$. Then it can be shown (see Section 5 of [BC]) that the map

$$x \longrightarrow \begin{bmatrix} C(I - zA)^{-1}x \\ \bar{z}B^*(I - \bar{z}A^*)^{-1}x \end{bmatrix} \tag{3.9}$$

is a partial isometry from $\mathcal{H}$ onto $\mathcal{H}^{\omega}_{BR}$ with initial space equal to $\mathcal{H}_0$. Furthermore, this map provides a unitary equivlence between the minimal part U_0 of the unitary colligation U

$$U_0 = \begin{bmatrix} P_{\mathcal{H}_0} & 0 \\ 0 & I \end{bmatrix} \begin{bmatrix} A & B \\ C & D \end{bmatrix}\Big|_{\mathcal{H}_0 \oplus \mathcal{E}_1}.$$

and the de Branges-Rovnyak model unitary colligation

$$U_{BR} = \begin{bmatrix} A_{BR} & B_{BR} \\ C_{BR} & D_{BR} \end{bmatrix} : \begin{bmatrix} \mathcal{H}^{\omega}_{BR} \\ \mathcal{E}_1 \end{bmatrix} \longrightarrow \begin{bmatrix} \mathcal{H}^{\omega}_{BR} \\ \mathcal{E}_2 \end{bmatrix}$$

where

$$A_0 : \begin{bmatrix} f(z) \\ g(z) \end{bmatrix} \longrightarrow \begin{bmatrix} [f(z) - f(0)]/z \\ \bar{z}[g(z) - \omega(z)^* f(0)] \end{bmatrix}$$

$$B_0 : e_1 \longrightarrow \begin{bmatrix} I - \omega(z)\omega(0)^* \\ \omega(z)^* - \omega(0)^* \end{bmatrix} e_1$$

$$C_0 : \begin{bmatrix} f(z) \\ g(z) \end{bmatrix} \longrightarrow f(0)$$

$$D_0 : e_1 \longrightarrow \omega(0)e_1$$

In deriving these formulas from those in [BC], one must be aware that here we are using Nikolski-Vasyunin coordinates $(f_{NV}(z), g_{NV}(z))$ for the elements of $\mathcal{H}^{\omega}_{BR}$ (with conjugate analytic second coordinate) rather than the original de Branges-Rovnyak coordinates $(f_{BR}(z), g_{BR}(z))$ (with analytic second coordinate); the correspondence between these two sets of coordinates is simply

$$\begin{bmatrix} f_{NK}(z) \\ g_{NK}(z) \end{bmatrix} = \begin{bmatrix} f_{BR}(z) \\ \bar{z}g_{BR}(\bar{z}) \end{bmatrix}.$$

In this way we see that the second component of the identification map $i = \begin{bmatrix} D_{\mathbf{F}} \\ \mathbf{F} \end{bmatrix}$ is just the map (3.9) and we recover the formula for F in Theorem 1.1.

4 The Commutant Lifting Theorem in Coordinate-free Form

The Commutant Lifting Theorem can be expressed as follows.

Theorem 4.1 (Commutant Lifting Theorem) *Suppose that we have two Abstract Scattering Systems*

$$(\mathcal{U}_{T'}, \mathcal{K}_{T'}, \mathcal{G}_{1T'}, \mathcal{G}_{2T'}), \quad (\mathcal{U}_T, \mathcal{K}_T, \mathcal{G}_{1T}, \mathcal{G}_{2T}).$$

We set

$$\mathcal{H}_{T'} = \mathcal{K}_{T'} \ominus [\mathcal{G}_{1T'} + \mathcal{G}_{2T'}], \quad \mathcal{H}_T = \mathcal{K}_T \ominus [\mathcal{G}_{1T} + \mathcal{G}_{2T}]$$

and define

$$T' = P_{\mathcal{H}_{T'}} \mathcal{U}_{T'}\big|_{\mathcal{H}_{T'}}, \quad T = P_{\mathcal{H}_T} \mathcal{U}_T\big|_{\mathcal{H}_T}$$

so that $\mathcal{U}_{T'}$ and $\mathcal{U}_T$ are the Sz.-Nagy unitary dilations of T' and T respectively. Suppose in addition that $X : \mathcal{H}_{T'} \to \mathcal{H}_T$ is such that

$$XT' = TX, \quad \|X\| \leq 1.$$

Then there is a $Y : \mathcal{K}_{T'} \to \mathcal{K}_T$ such that

(i) $Y : \mathcal{G}_{1T'} \to \mathcal{G}_{1T}$, $Y^* : \mathcal{G}_{2T} \to \mathcal{G}_{2T'}$, *and* $P_{H_T} Y\big|_{\mathcal{H}_{T'}} = X$,

(ii) $\|Y\| \leq 1$, *and*

(iii) $Y\mathcal{U}_{T'} = \mathcal{U}_T Y$.

We remark that there are a couple of alternative ways to express the conclusion of Theorem 4.1. Specifically one can formulate the problem in terms of the restriction $Y_r = Y|_{\mathcal{H}_{T'}+\mathcal{G}_{1T'}}$ of Y to the subspace $\mathcal{H}_{T'} + \mathcal{G}_{1T'}$ and demand

(i′) $Y_r : \mathcal{G}_{1T'} \to \mathcal{G}_{1T}$, $Y_r : (\mathcal{H}_{T'} + \mathcal{G}_{1T'}) \to (\mathcal{H}_T + \mathcal{G}_{1T})$ and $P_{\mathcal{H}_T} Y_r = X P_{\mathcal{H}_{T'}}|_{\mathcal{H}_{T'}+\mathcal{G}_{1T'}}$,

(ii′) $\|Y_r\| \leq 1$, and

(iii′) $Y_r\, \mathcal{U}_{T'}|_{\mathcal{H}_{T'}+\mathcal{G}_{1T'}} = \mathcal{U}_T|_{\mathcal{H}_T+\mathcal{G}_{1T}}\, Y_r$.

Alternatively, one can formulate the problem in terms of the compression

$$Y_{r*} = P_{\mathcal{G}_{2T}+\mathcal{H}_T}\, Y|_{\mathcal{G}_{2T'}+\mathcal{H}_{T'}}$$

of Y to the subspace $\mathcal{G}_{2T'} + \mathcal{H}_{T'}$ and demand that

(i″) $Y_{r*} : (\mathcal{G}_{2T'}+\mathcal{H}_{T'}) \to (\mathcal{G}_{2T}+\mathcal{H}_T)$, $Y_{r*} : \mathcal{H}_{T'} \to \mathcal{H}_T$ and $Y_{r*}|_{\mathcal{H}_{T'}} = X$.

(ii″) $\|Y_{r*}\| \leq 1$, and

(iii″) $Y_{r*} P_{\mathcal{G}_{2T'}+\mathcal{H}_{T'}}\, \mathcal{U}_{T'}|_{\mathcal{G}_{2T'}+\mathcal{H}_{T'}} = P_{\mathcal{G}_{2T}+\mathcal{H}_T} \mathcal{U}_T Y_{r*}$.

If Y satisfies (i), (ii) and (iii), then one easily checks that $Y_r = Y|_{\mathcal{H}_{T'}+\mathcal{G}_{1T'}}$ satisfies (i′), (ii′) and (iii′). Conversely, if Y_r is a solution of (i′), (ii′) and (iii′), then there is a unique extension of Y_r to an operator $Y : \mathcal{K}_{T'} \to \mathcal{K}_T$ which satisfies (i), (ii) and (iii). A similar story holds with respect to the conditions (i″), (ii″) and (iii″). We shall have occasion to use both these alternative formulations.

The purpose of this section is to relate this commutant lifting theorem to our coordinate-free Abstract Interpolation Problem, as has already been done by

Kupin [Ku] in the model setting. The first order of business is to identify the admissible AIP data set associated with a Commutant Lifting problem

Lemma 4.2 *Suppose that $(\mathcal{U}_{T'}, \mathcal{K}_{T'}, \mathcal{G}_{1T'}, \mathcal{G}_{2T'})$ and $(\mathcal{U}_T, \mathcal{K}_T, \mathcal{G}_{1T}, \mathcal{G}_{2T})$ are two ASS's, and that T' on $\mathcal{H}_{T'}$ and T on $\mathcal{H}_T$ and $X : \mathcal{H}_{T'} \to \mathcal{H}_T$ are operators as in the hypotheses of the Commutant Lifting Theorem (Theorem 4.1). Let $\mathcal{E}_{1T'} = \mathcal{G}_{1T'} \ominus \mathcal{U}_{T'}\mathcal{G}_{1T'}$ and $\mathcal{E}_{1T} = \mathcal{G}_{1T} \ominus \mathcal{U}_T\mathcal{G}_{1T}$ be the forward wandering subspaces in the unitary dilation spaces $\mathcal{K}_{T'}$ and $\mathcal{K}_T$ for T' and T respectively. Let*

$$\mathcal{X} = \mathcal{H}_{T'}, \quad \mathcal{E}_1 = \mathcal{E}_{1T'}, \quad \mathcal{E}_2 = \mathcal{E}_{1T},$$
$$T_1 = T', \quad T_2 = I_{\mathcal{H}_{T'}}, \quad D(\cdot,\cdot) = \langle (I - X^*X)\cdot, \cdot\rangle_{\mathcal{H}_{T'}},$$
$$M_1 = P_{\mathcal{E}_{1T'}}\mathcal{U}_{T'}, \quad M_2 = P_{\mathcal{E}_{1T}}\mathcal{U}_T X.$$

Then $(T_1, T_2, D(\cdot,\cdot), M_1, M_2, \mathcal{X}, \mathcal{E}_1, \mathcal{E}_2)$ is an admissible AIP data set.

Proof: The only nontrivial statement to be verified is the Potapov identity (0.1) for this data set. Note that (0.1) for our setting specializes to

$$\langle (I - X^*X)T'x, T'x\rangle + \|P_{\mathcal{E}_{1T'}}\mathcal{U}_{T'}x\|^2 = \langle (I - X^*X)x, x\rangle + \|P_{\mathcal{E}_{1T}}\mathcal{U}_T Xx\|^2$$

for all $x \in \mathcal{H}_{T'}$. To verify this, note that

$$\begin{aligned}
&\langle (I - X^*X)T'x, T'x\rangle + \|T_{\mathcal{E}_{1T'}}\mathcal{U}_{T'}x\|^2 \\
&= \langle T'^*(I - X^*X)T'x, x\rangle + \|P_{\mathcal{E}_{1T'}}\mathcal{U}_{T'}x\|^2 \\
&= \|T'x\|^2 - \|TXx\|^2 + \|P_{\mathcal{E}_{1T'}}\mathcal{U}_{T'}\|^2 \\
&= \|P_{\mathcal{H}_{T'}}\mathcal{U}_{T'}x\|^2 + \|P_{\mathcal{E}_{1T'}}\mathcal{U}_{T'}x\|^2 - \|TXx\|^2 = \|\mathcal{U}_{T'}x\|^2 - \|TXx\|^2 \\
&= \|x\|^2 - \left[\|\mathcal{U}_T Xx\|^2 - \|P_{\mathcal{E}_{1T}}\mathcal{U}_T Xx\|^2\right] \\
&= \|x\|^2 - \|Xx\|^2 + \|P_{\mathcal{E}_{1T}}\mathcal{U}_T Xx\|^2 \\
&= \langle (I - X^*X)x, x\rangle + \|P_{\mathcal{E}_{1T}}\mathcal{U}_T Xx\|^2.
\end{aligned}$$

The result follows. □

To avoid confusion when working with the correspondence between the data for the Commutant Lifting Theorem versus the data for an Abstract Interpolation Problem as in Lemma 4.2, it will be convenient to introduce identification maps

$$i_1 : \widetilde{\mathcal{G}}_1 \longrightarrow \widetilde{\mathcal{G}}_{1T'}, \quad i_2 : \widetilde{\mathcal{G}}_2 \longrightarrow \widetilde{\mathcal{G}}_{1T}$$

defined by

$$i_1 : \sum_{j=-\infty}^{\infty} \mathcal{U}^j e_j \longrightarrow \sum_{j=-\infty}^{\infty} \mathcal{U}_{T'}^j e_j, \quad e_j \in \mathcal{E}_1 = \mathcal{E}_{1T'}, \tag{4.1}$$

$$i_2 : \sum_{j=-\infty}^{\infty} \mathcal{U}^j f_j \longrightarrow \sum_{j=-\infty}^{\infty} \mathcal{U}_T^j f_j, \quad f_j \in \mathcal{E}_2 = \mathcal{E}_{1T}. \tag{4.2}$$

We shall use this convention in the discussion below.

The correspondence between solutions of the CF-AIP for the data set as in Lemma 4.2 and solutions of the Commutant Lifting problem is given by the following coordinate-free version of the result of Kupin [Ku].

Theorem 4.3 *Let $(\mathcal{U}_{T'}, \mathcal{K}_{T'}, \mathcal{G}_{1T'}, \mathcal{G}_{2T'})$ and $(\mathcal{U}_T, \mathcal{K}_T, \mathcal{G}_{1T}, \mathcal{G}_{2T})$ are two ASS's, let T' on $\mathcal{H}_{T'}$ and T on $\mathcal{H}_T$ and $X : \mathcal{H}_{T'} \to \mathcal{H}_T$ be as in the hypotheses of the Commutant Lifting Theorem (Theorem 4.1) and let $(T_1, T_2, D(\cdot,\cdot), M_1, M_2, \mathcal{X}, \mathcal{E}_1, \mathcal{E}_2)$ be the admissible AIP data set as in Lemma 4.2.*

(i) Suppose that $((\mathcal{U}, \mathcal{K}, \mathcal{G}_1, \mathcal{G}_2), \mathbf{F})$ is a solution of the CF-AIP for the data set given in Lemma 4.2. Define $Y_r : (\mathcal{H}_{T'} + \mathcal{G}_{1T'}) \to (\mathcal{H}_T + \mathcal{G}_{1T})$ by

$$Y_r|_{\mathcal{H}_{T'}} = X + i_2 P_{\widetilde{\mathcal{G}}_2}\mathbf{F}$$

and

$$Y_r|_{\mathcal{G}_{1T'}} = i_2 P_{\mathcal{G}_2} i_1^{-1}\Big|_{\mathcal{G}_{1T'}}.$$

Then Y_r satisfies conditions (i′), (ii′) and (iii′) in the Commutant Lifting Theorem.

Conversely, suppose that Y satisfies conditions (i), (ii) and (iii) in the Commutant Lifting Theorem. Let $(\mathcal{U}, \mathcal{K}, \mathcal{G}_1, \mathcal{G}_2)$ be any ASS with wandering subspaces $\mathcal{E}_1 := \mathcal{G}_1 \ominus \mathcal{U}\mathcal{G}_1 = \mathcal{E}_{1T'}$ and $\mathcal{E}_2 := \mathcal{U}\mathcal{G}_2 \ominus \mathcal{G}_2 = \mathcal{E}_{1T}$, and with scattering operator Ω equal to $i_2^{-1} Y i_1$. Let $\mathbf{F}$ be the unique operator $\mathbf{F} : \mathcal{H}_{T'}^{D_X} \to \mathcal{H}$ (where $\mathcal{H} = \mathcal{K} \ominus [\mathcal{G}_1 + \mathcal{G}_2]$ and where $\mathcal{H}_{T'}^{D_X}$ is the completion of $\mathcal{H}_{T'}$ in the inner product $\langle (I - X^*X)\cdot, \cdot\rangle$) defined weakly by

$$\langle \mathbf{F}x, g_2 + g_1\rangle = \langle i_2^{-1}(Y - X)x, g_2\rangle + \langle i_1^{-1}(-Y^*X + I)x, g_1\rangle \tag{4.3}$$

for $x \in \mathcal{H}_{T'}$, $g_2 \in \widetilde{\mathcal{G}}_2$ and $g_1 \in \widetilde{\mathcal{G}}_1$. Then $((\mathcal{U}, \mathcal{K}, \mathcal{G}_1, \mathcal{G}_2), \mathbf{F})$ is a solution of the CF-AIP associated with the AIP data given in Lemma 4.2.

Proof: Define Y_r as in part (i) in the statement of the Theorem. We verify condition (ii′) that Y_r is a contraction as follows. Let $k' = h' + g' \in \mathcal{H}_{T'} + \mathcal{G}_{1T'}$. Then

$$\begin{aligned}
\|Y_r k'\|^2 = \|Y_r(h' + g')\|^2 &= \|Xh' + i_2 P_{\mathcal{G}_2}\mathbf{F}h' + i_2 P_{\mathcal{G}_2} i_1^{-1} g'\|^2_{\mathcal{H}_T + \mathcal{G}_{1T}} \\
&= \|Xh'\|^2_{\mathcal{H}_{T'}} + \|i_2 P_{\mathcal{G}_2}\mathbf{F}h' + i_2 P_{\mathcal{G}_2} i_1^{-1} g'\|^2_{\mathcal{G}_{1T}} \\
&\le \|Xh'\|^2_{\mathcal{H}_{T'}} + \|\mathbf{F}h' + i_1^{-1} g'\|^2_{\mathcal{H}+\mathcal{G}_1} \\
&= \|Xh'\|^2_{\mathcal{H}_{T'}} + \|\mathbf{F}h'\|^2_{\mathcal{H}} + \|g'\|^2_{\mathcal{G}_{1T'}} \\
&\le \|h'\|^2_{\mathcal{H}_{T'}} + \|g'\|^2_{\mathcal{G}_{1T'}} = \|k'\|^2_{\mathcal{H}_{T'}+\mathcal{G}_{1T'}}
\end{aligned}$$

as required.

Next, set $S' = \mathcal{U}_{T'}|_{\mathcal{H}_{T'}+\mathcal{G}_{1T'}}$ and $S = \mathcal{U}_T|_{\mathcal{H}_T+\mathcal{G}_{1T}}$. Then we check the intertwining condition (ii′) as follows. With $k' = h' + g'$ as above, we have

$$\begin{aligned} Y_r(S'k') &= Y_r(S'(h'+g')) = Y_r(T'h' + i_1 M_1 h' + S'g') \\ &= [XT'h' + i_2 P_{\mathcal{G}_2}\mathbf{F}T'h'] + [i_2 P_{\mathcal{G}_2} M_1 h' + i_2 P_{\mathcal{G}_2} i_1^{-1} S'g'] \\ &= XT'h' + i_2 P_{\mathcal{G}_2}[S\mathbf{F}h' + M_2 h' - M_1 h'] \\ &\quad + [i_2 P_{\mathcal{G}_2} M_1 h' + i_2 P_{\mathcal{G}_2} i_1^{-1} S'g']. \end{aligned}$$

Write $XT'h' = TXh' = SXh' - i_2 M_2 h'$. Hence cancellation of terms in the above expression reduces us to

$$\begin{aligned} Y_r(S'(h''+g')) &= SXh' + i_2 P_{\mathcal{G}_2} S\mathbf{F}h' + i_2 P_{\mathcal{G}_2} i_1^{-1} S'g' \\ &= S(Xh' + i_2 P_{\mathcal{G}_2^+}\mathbf{F}h' + i_2 P_{\mathcal{G}_2} i_1^{-1} g') \\ &= SY_r k' \end{aligned}$$

as desired, and (ii′) is verified.

The conditions (i′) are obvious from the definition of Y_r and part (i) of the Theorem follows.

Next suppose that Y is a solution of the Commutant Lifting problem and define $((\mathcal{U}, \mathcal{K}, \mathcal{G}_1, \mathcal{G}_2), \mathbf{F})$ as in the statement of part (ii) of the Theorem. We must first check that $\mathbf{F}$ is well-defined contraction operator from $\mathcal{H}_{T'}^{D_X}$ into $\mathcal{K}$. By assumption $\widetilde{\mathcal{G}}_1 + \widetilde{\mathcal{G}}_2$ is dense in $\mathcal{K}$, so there is at most one $\mathbf{F}$ satisfying the conditions (4.3). To verify existence, it suffices to show that

$$\|\langle \mathbf{F}x, g_2 + g_1\rangle_{\mathcal{K}}\|^2 \le \langle (I - X^*X)x, x\rangle_{\mathcal{H}_{T'}} \cdot \|g_2 + g_1\|_{\mathcal{K}}^2 \tag{4.4}$$

and that

$$i_2^{-1}(Y - X)x \in \widetilde{\mathcal{G}}_2 \ominus \mathcal{G}_2, \quad i_1^{-1}(-Y^*X + I)x \in \widetilde{\mathcal{G}}_1 \ominus \mathcal{G}_1 \tag{4.5}$$

for all $x \in \mathcal{H}_{T'}$. Indeed, condition (4.4) guarantees the existence of a unique contraction mapping $F : \mathcal{H}_{T'} \to \mathcal{K}$, and then condition (4.5) guarantees that the image of this contraction $\mathbf{F}$ is in $\mathcal{H} = \mathcal{K} \ominus [\mathcal{G}_1 + \mathcal{G}_2]$. Verification of (4.5) is straightforward from the lifting property (i) of Y. To verify condition (4.4), note that, for $x \in \mathcal{H}_{T'}$, $g_2 \in \widetilde{\mathcal{G}}_2$ and $g_1 \in \widetilde{\mathcal{G}}_1$, we have

$$\begin{aligned} \|\langle \mathbf{F}x, g_2 + g_1\rangle\|^2 &= \|\langle i_2^{-1}(Y - X)x, g_2\rangle + \langle i_1^{-1}(-YX + I)x, g_1\rangle\|^2 \\ &= \|\langle (Y - X)x, i_2 g_2\rangle_{\widetilde{\mathcal{G}}_{1T}} + \langle (-Y^*X + I)x, i_1 g_1\rangle_{\widetilde{\mathcal{G}}_{1T'}}\|^2 \\ &= \left\langle \begin{bmatrix} I & Y \\ Y^* & I \end{bmatrix} \begin{bmatrix} -X \\ I \end{bmatrix} x, \begin{bmatrix} i_2 g_2 \\ i_1 g_1 \end{bmatrix} \right\rangle_{\widetilde{\mathcal{G}}_{1T} \oplus \widetilde{\mathcal{G}}_{1T'}} \\ &\le \left\langle \begin{bmatrix} I & Y \\ Y^* & I \end{bmatrix} \begin{bmatrix} -X \\ I \end{bmatrix} x, \begin{bmatrix} -X \\ I \end{bmatrix} x \right\rangle \end{aligned}$$

$$
\begin{aligned}
&\cdot \left\langle \begin{bmatrix} I & Y \\ Y^* & I \end{bmatrix} \begin{bmatrix} i_2 g_2 \\ i_1 g_1 \end{bmatrix}, \begin{bmatrix} i_2 g_2 \\ i_1 g_1 \end{bmatrix} \right\rangle \\
= &\ \{\langle(-X+Y)x, -Xx\rangle + \langle(-Y^*X+I)x, x\rangle\} \\
&\cdot \|g_2 + g_1\|_{\mathcal{K}}^2 \\
= &\ \langle(I - X^*X)x, x\rangle \cdot \|g_2 + g_1\|_{\mathcal{K}}^2
\end{aligned}
$$

as required. Here we used the Cauchy-Schwarz inequality for the inner product induced by the operator matrix $\begin{bmatrix} I & Y \\ Y^* & I \end{bmatrix}$ on $\widetilde{\mathcal{G}}_2 \oplus \widetilde{\mathcal{G}}_1$ and we used the fact that $i_2^{-1} Y i_1$ is the scattering matrix for the ASS $(\mathcal{U}, \mathcal{K}, \mathcal{G}_1, \mathcal{G}_2)$ to identify the norm of $g_2 + g_1$ in $\mathcal{K}$. We conclude that $\mathbf{F}$ is a well-defined contraction operator from $\mathcal{H}_{T'}$ into $\mathcal{H}$.

It remains to verify that $\mathbf{F}$ satisfies the commutation identity (2.10) for the AIP data set as in Lemma 4.2, i.e., the equation

$$
[\mathbf{F}T' + i_1^{-1} P_{\mathcal{E}_{1T'}} \mathcal{U}_{T'}]x = [\mathcal{U}\mathbf{F} + P_{\mathcal{E}_{1T}} \mathcal{U}_T X]x \tag{4.6}
$$

for $x \in \mathcal{H}_{T'}$. Projection onto $\widetilde{\mathcal{G}}_2$ followed by the inclusion map i_2 transforms (4.6) to

$$
[(Y - X)T' + Y P_{\mathcal{E}_{1T'}} \mathcal{U}_{T'}]x = [\mathcal{U}_T(Y - X) + P_{\mathcal{E}_{1T}} \mathcal{U}_T X]x \tag{4.7}
$$

while projection onto $\widetilde{\mathcal{G}}_1$ followed by the identification map i_1 transforms (4.6) to

$$
[(I - Y^*X)T' + P_{\mathcal{E}_{1T'}} \mathcal{U}_{T'}]x = [\mathcal{U}_{T'}(I - Y^*X) + Y^* P_{\mathcal{E}_{1T}} \mathcal{U}_T X]x \tag{4.8}
$$

for $x \in \mathcal{H}_{T'}$. Conversely, it is easy to see that simultaneous validity of (4.7) and (4.8) implies (4.6), since $\widetilde{\mathcal{G}}_1 + \widetilde{\mathcal{G}}_2$ is dense in $\mathcal{K}$ by assumption. To prove (4.7), use

$$
\begin{aligned}
P_{\mathcal{E}_{1T'}} \mathcal{U}_{T'} x &= (I - P_{\mathcal{H}_{T'}}) \mathcal{U}_{T'} x \\
&= \mathcal{U}_{T'} x - T' x
\end{aligned}
$$

and similarly,

$$
\begin{aligned}
P_{\mathcal{E}_{1T}} \mathcal{U}_T X &= \mathcal{U}_T X - T X \\
&= \mathcal{U}_T X - X T'
\end{aligned}
$$

for $x \in \mathcal{H}_{T'}$. Hence the left hand side of (4.7) becomes

$$
[YT' - XT' + Y(\mathcal{U}_{T'} - T')]x = [\mathcal{U}_T Y - XT']x
$$

while the right hand side becomes

$$
[(\mathcal{U}_T Y - \mathcal{U}_T X) + (\mathcal{U}_T X - XT')]x = [\mathcal{U}_T Y - XT']x
$$

and (4.7) follows. The identity (4.8) follows in a similar way. This completes the proof of Theorem 4.3 □

The Main Theorem on the CF-AIP (Theorem 3.1) gives the correspondence between solutions of the CF-AIP and unitary colligation extensions of the partial isometric colligation V (as defined in (0.2)) while Theorem 4.3 identifies the connection between the Commutant Lifting problem and the CF-AIP for an associated data set. Hence there is a one-to-one correspondence between solutions of the Commutant Lifting problem and unitary extensions of the isometry V associated with the Commutant Lifting AIP data set. As we did for the case of right tangential Nevanlinna-Pick interpolation in Section 1, we now show how to see this directly for the case of the Commutant Lifting Theorem. For simplicity we consider only the special case here where T' and T are $C_{0\cdot}$ contractions (so powers of T and of T' converge strongly to zero), so that $\mathcal{K}_{T'} = \widetilde{\mathcal{G}}_{1T'}$ and $\mathcal{K}_T = \widetilde{\mathcal{G}}_{1T}$.

Theorem 4.4 *Suppose that* $(\mathcal{U}_{T'}, \mathcal{K}_{T'}, \mathcal{G}_{1T'}, \mathcal{G}_{2T'})$ *and* $(\mathcal{U}_T, \mathcal{K}_T, \mathcal{G}_{1T}, \mathcal{G}_{2T})$ *are two ASS's as in the Commutant Lifting Theorem subject to the additional hypotheses that* $\mathcal{K}_{T'} = \widetilde{\mathcal{G}}_{1T'}$, $\mathcal{K}_T = \widetilde{\mathcal{G}}_{1T}$, *let* T' *on* $\mathcal{H}_{T'}$ *and* T *on* $\mathcal{H}_T$ *and* $X : \mathcal{H}_{T'} \to \mathcal{H}_T$ *be as in Theorem 4.1, and let* $(T_1, T_2, D(\cdot,\cdot), M_1, M_2, \mathcal{X}, \mathcal{E}_1, \mathcal{E}_2)$ *be the admissible AIP data set as in Lemma 4.2. Let* $V : \mathcal{D}_V \to \mathcal{R}_V$ *be the partial isometric colligation defined by*

$$\mathcal{D}_V = \begin{bmatrix} T' \\ P_{\mathcal{E}_{1T'}}\mathcal{U}_{T'} \end{bmatrix} \mathcal{H}_{T'} \subset \begin{bmatrix} \mathcal{H}_{T'}^{\mathcal{D}_X} \\ \mathcal{E}_{1T'} \end{bmatrix}, \quad \mathcal{R}_V = \begin{bmatrix} I \\ P_{\mathcal{E}_{1T}}\mathcal{U}_T X \end{bmatrix} \mathcal{H}_{T'} \subset \begin{bmatrix} \mathcal{H}_{T'}^{\mathcal{D}_X} \\ \mathcal{E}_{1T} \end{bmatrix}$$

$$V : \begin{bmatrix} T' \\ P_{\mathcal{E}_{1T'}}\mathcal{U}_{T'} \end{bmatrix} x \longrightarrow \begin{bmatrix} I \\ P_{\mathcal{E}_{1T}}\mathcal{U}_T X \end{bmatrix} x.$$

Explicitly, if

$$U = \begin{bmatrix} A & B \\ C & D \end{bmatrix} : \begin{bmatrix} \mathcal{H} \\ \mathcal{E}_{1T'} \end{bmatrix} \longrightarrow \begin{bmatrix} \mathcal{H} \\ \mathcal{E}_{1T} \end{bmatrix}$$

(where $\mathcal{H} \supset \mathcal{H}_{T'}^{\mathcal{D}_X}$*) is such a unitary colligation extension, let*

$$\omega(z) = D + zC(I - zA)^{-1}B \in \mathcal{S}(\mathcal{E}_{1T'}, \mathcal{E}_{1T})$$

be its characteristic function. Define $Y : \widetilde{\mathcal{G}}_{1T'} \to \widetilde{\mathcal{G}}_{1T}$ *in terms of its Fourier transform by*

$$\widehat{Y} = M_\omega : L_2(\mathcal{E}_{1T'}) \longrightarrow L_2(\mathcal{E}_{1T}).$$

Then Y *is a solution of conditions (i), (ii) and (iii) in the Commutant Lifting Theorem.*

Proof: Suppose that $U = \begin{bmatrix} A & B \\ C & D \end{bmatrix}$ is such a unitary colligation extension and let Y be defined as in the statement of the Theorem. Set

$$Y_{r*} = P_{\mathcal{G}_{2T}+\mathcal{H}_T} Y|_{\mathcal{G}_{2T'}+\mathcal{H}_{T'}}.$$

By assumption $\mathcal{K}_T = \widetilde{\mathcal{G}}_{1T}$ and hence

$$\mathcal{G}_{2T} + \mathcal{H}_T = \mathcal{K}_T \ominus \mathcal{G}_{1T} = \oplus_{n=-1}^{-\infty} \mathcal{U}_T^n \mathcal{E}_{1T} =: \mathcal{G}_{1T}^-$$

and similarly,

$$\mathcal{G}_{2T'} + \mathcal{H}_{T'} = \mathcal{G}_{1T'}^-.$$

Trivially, Y_{r*} satisfies conditions (ii″) and (iii″) in the third formulation of the conclusion of the Commutant Lifting Theorem 4.1. It remains only to check that

$$Y_{r*}|_{\mathcal{H}_{T'}} = X. \tag{4.9}$$

As im $X \subset \mathcal{H}_T \subset \mathcal{G}_{1T}^-$, (4.9) follows if we show that the $\mathcal{E}_{1T}$-Fourier coefficients for Yh of negative index match with the corresponding $\mathcal{E}_{1T}$-Fourier coefficient for Xh for each $h \in \mathcal{H}_{T'}$, i.e.,

$$P_{\mathcal{E}_{1T}} \mathcal{U}_T^n Y \big|_{\mathcal{H}_{T'}} = P_{\mathcal{E}_{1T}} \mathcal{U}_T^n X \text{ for } n = 1, 2, 3, \ldots. \tag{4.10}$$

We now use the hypothesis that $U = \begin{bmatrix} A & B \\ C & D \end{bmatrix}$ is a unitary colligation extension of V. Hence

$$\begin{aligned} [AT' + BP_{\mathcal{E}_{1T'}} \mathcal{U}_{T'}]h &= h \\ [CT' + P_{\mathcal{E}_{1T'}} \mathcal{U}_{T'}]h &= P_{\mathcal{E}_{1T}} \mathcal{U}_T Xh \end{aligned} \tag{4.11}$$

for all $h \in \mathcal{H}_{T'}$. From the first equation in (4.11) we get

$$AT'h + BP_{\mathcal{E}_{1T'}} \mathcal{U}_{T'} h = h. \tag{4.12}$$

Substitution of $T'h$ for h gives

$$AT'^2 h + BP_{\mathcal{E}_{1T'}} \mathcal{U}_{T'} T'h = T'h. \tag{4.13}$$

Substitution of this expression (4.13) for $T'h$ back into (4.12) gives

$$A[AT'^2 h + BP_{\mathcal{E}_{1T'}} \mathcal{U}_{T'} T'h] + BP_{\mathcal{E}_{1T'}} \mathcal{U}_{T'} h = h$$

or

$$A^2 T'^2 h + ABP_{\mathcal{E}_{1T'}} \mathcal{U}_{T'} T'h + BP_{\mathcal{E}_{1T'}} \mathcal{U}_{T'} h = h.$$

An inductive argument now gives

$$A^n T'^n h + \sum_{k=0}^{n-1} A^k BP_{\mathcal{E}_{1T'}} \mathcal{U}_{T'} T'^k h = h. \tag{4.14}$$

By assumption T'^n converges strongly to 0 as n tends to infinity. Taking n to infinity in (4.14) therefore gives

$$\sum_{k=0}^{\infty} A^k B P_{\mathcal{E}_{1T'}} \mathcal{U}_{T'} T'^k h = h$$

for all $h \in \mathcal{H}_{T'}$. Now use the second equation in (4.11) to get

$$\left[C \left(\sum_{k=0}^{\infty} A^k B P_{\mathcal{E}_{1T'}} \mathcal{U}_{T'} T'^{k+1} \right) + D P_{\mathcal{E}_{1T'}} \mathcal{U}_{T'} \right] h = P_{\mathcal{E}_{1T}} \mathcal{U}_T X h. \tag{4.15}$$

Note that

$$\begin{aligned} P_{\mathcal{E}_{1T'}} \mathcal{U}_{T'} T'^{k+1} &= P_{\mathcal{E}_{1T}} \mathcal{U}_{T'} P_{\mathcal{H}_{T'}} \mathcal{U}_{T'}^{k+1} \Big|_{\mathcal{H}_{T'}} \\ &= P_{\mathcal{E}_{1T'}} \mathcal{U}_{T'} P_{\mathcal{G}_{1T'}^-} \mathcal{U}_{T'}^{k+1} \Big|_{\mathcal{H}_{T'}} \\ &= P_{\mathcal{E}_{1T'}} \mathcal{U}_{T'} \mathcal{U}_{T'}^{k+1} \Big|_{\mathcal{H}_{T'}} \end{aligned}$$

and hence the left hand side of (4.15) is identifiable exactly as

$$P_{\mathcal{E}_{1T}} Y \mathcal{U}_{T'} h = P_{\mathcal{E}_{1T}} \mathcal{U}_T Y h.$$

We have thus established (4.10) for the case $n = 1$. The case for a general $n > 1$ follows from (4.15) by replacing h with $T'^{n-1}h$. The Theorem now follows.

□

As mentioned in the Introduction, Arocena in [Ar2] has given a somewhat different (perhaps more natural) scattering-theoretic proof of the Commutant Lifting Theorem. The idea goes back even earlier (see [Ar1] and [CS]) and originally was formulated without any scattering interpretation ([ACS]). For a lucid synopsis of this approach we also refer to [S2]. The formulation presented below is best suited for purposes of comparison with the main topic of this paper, namely, CF-AIP. We assume all the notation introduced above for the Commutant Lifting Theorem. Introduce a Hilbert space $\mathcal{K}_0$ by $\mathcal{K}_0 = \mathcal{G}_2 \dot{+} \mathcal{G}_1$ where $\mathcal{G}_2 = \mathcal{G}_{2T} \oplus \mathcal{H}_T$ and $\mathcal{G}_1 = \mathcal{H}_{T'} \oplus \mathcal{G}_{1T}$ with inner product on $\mathcal{K}$ equal to that induced by the operator block matrix

$$\begin{bmatrix} I_{\mathcal{G}_2} & X P_{\mathcal{H}_{T'}} \\ X^* P_{\mathcal{H}_T} & I_{\mathcal{G}_1} \end{bmatrix}.$$

Due to the intertwining relation $XT' = TX$, it works out that the operator

$$\mathcal{V}_0 : \mathcal{U}_T^* x_1 \dot{+} x_2 \longrightarrow x_1 \dot{+} \mathcal{U}_{T'} x_2$$

with domain equal to $\mathcal{U}_T^* \mathcal{G}_2 \dot{+} \mathcal{G}_1$ and range space equal to $\mathcal{G}_2 \dot{+} \mathcal{U}_{T'} \mathcal{G}_1$ is isometric. Note that $\mathcal{G}_2 \dot{+} \{0\}$ is simply invariant for $\mathcal{V}_0^*$ (and hence can be thought of as an *incoming subspace* for $\mathcal{V}_0$) and that $\{0\}\dot{+}\mathcal{G}_1$ is simply invariant for V_0 (and hence corresponds to an *outgoing subspace* for V_0). However, unlike the standing assumption in this paper, here we have that the incoming subspace $\mathcal{G}_2$ is not orthogonal to the outgoing subspace $\mathcal{G}_1$. Nevertheless, $(\mathcal{V}_0, \mathcal{K}_0, \mathcal{G}_1, \mathcal{G}_2)$ defines a *partial* (nonorthogonal) *abstract scattering system* (partial, since $\mathcal{V}_0$ is only partially defined isometric rather than fully defined unitary and nonorthogonal since $\mathcal{G}_1$ and $\mathcal{G}_2$ are not orthogonal). In this context it makes sense to look for embeddings of $(\mathcal{V}_0, \mathcal{K}_0, \mathcal{G}_1, \mathcal{G}_2)$ into a (nonorthogonal) abstract scattering system $(\mathcal{V}, \mathcal{K}, \mathcal{G}_1, \mathcal{G}_2)$, i.e. an embedding of $\mathcal{K}_0$ into a larger Hilbert space $\mathcal{K}$ on which is defined a unitary operator $\mathcal{V}$ which extends $\mathcal{V}_0$. It is shown in [Ar] that solutions Y of the Commutant Lifting Theorem correspond to the scattering operators ω associated with such abstract scattering system extensions $(\mathcal{V}, \mathcal{K}, \mathcal{G}_1, \mathcal{G}_2)$. The special case of the Commutant Lifting Theorem corresponding to Nehari's Theorem was worked out in a similar framework by Kheifets in [Kh5]. By way of comparison, we can view the CF-AIP framework as follows. We are given a partial abstract scattering system $(\mathcal{K}_0, \mathcal{V}_0, \mathcal{G}_1, \mathcal{G}_2)$ with incoming subspace $\mathcal{G}_2$ simply invariant for $\mathcal{V}_0^*$ and outgoing subspace $\mathcal{G}_1$ simply invariant for $\mathcal{V}_0$ and with $\mathcal{G}_2$ orthogonal to $\mathcal{G}_1$ in the $\mathcal{K}_0$-inner product and together spanning the whole space $\mathcal{K}_0$. Due to the orthogonality of $\mathcal{G}_2$ and $\mathcal{G}_1$, no extra intertwining condition is needed in order to guarantee that $\mathcal{V}_0 : \mathcal{V}_0^* g_2 \oplus g_1 \to g_2 \oplus \mathcal{V}_0 g_1$ is isometric from $\mathcal{V}_0^*\mathcal{G}_2 \oplus \mathcal{G}_1$ onto $\mathcal{G}_2 \oplus \mathcal{V}_0\mathcal{G}_1$. Just as in the Arocena setup, one seeks an embedding of the partial abstract scattering system $(\mathcal{V}_0, \mathcal{K}_0, \mathcal{G}_1, \mathcal{G}_2)$ into a full abstract scattering system $(\mathcal{V}, \mathcal{K}, \mathcal{G}_1, \mathcal{G}_2)$, but subject to the side condition (not present in the Arocena formalism) that this full abstract scattering system fit with the AIP data set satisfying the Potapov identity (the interpolation conditions). An interesting exercise would be to formulate a general framework which would contain the CF-AIP problem and the Arocena framework as extreme special cases. Ironically, both extreme special cases would contain the Commutant Lifting Theorem.

References

[AdAr] V.M. Adamyan and D.Z. Arov, On unitary coupling of semiunitary operators, *Dokl. Akad. Nauk. Arm. SSR* XLIII, **5** (1966), 257–263 [in Russian] = *Amer. Math. Soc. Transl.* **95** (1970), 75–129.

[AgMc] J. Agler and J. McCarthy, Nevanlinna-Pick interpolation on the bidisk, preprint, 1997.

[ADRdS] D. Alpay, A. Dijksman, J. Rovnyak and H. de Snoo, *Schur Functions, Operator Colligations, and Reproducing Kernel Pontryagin Spaces*, Operator Theory: Advances and Applications, OT96 Birkhauser-Verlag, 1997.

[An] A. Antoulas (Ed.), *Mathematical System Theory: The Influence of R.E. Kalman*, Springer-Verlag, Berlin-Heidlelberg-New York, 1991.

[Ar1] R. Arocena, Generalized Toeplitz kernels and dilations of intertwining operators, *Intgral Equations and Operator Theory* **6** (1983), 759–778.

[Ar2] R. Arocena, Unitary extensions of isometries and contractive intertwining diltions, in *The Gohberg Anniversary Collection: Vol. II: Topics in Analysis and Operator Theory* (Ed. H. Dym, S. Goldberg, M.A. Kaashoek and P. Lancaster), OT vol. **41**, Birkhauser-Verlag, Basel, 1989, pp. 13–23.

[ACS] R. Arocena, M. Cotlar, and C. Sadosky, Weighted inequalities in L^2 and lifting properties, in *Mathematical Analysis and Applications, Part A* (Ed. L. Nachbin), Adv. in Math. Suppl. Studies 7, Academic Press, New York and London, 1981, pp. 95–128.

[ArG] D.Z. Arov and L.Z. Grossman, Scattering matrices in the theory of unitary extensions of isometric operators, *Math. Nachrichten* **157** (1992), 105–123.

[BC] J.A. Ball and N. Cohen, De Branges-Rovnyak operator models and systems theory: a survey, in *Topics in Matrix and Operator Theory* (Ed. H. Bart, I. Gohberg and M.A. Kaashoek), Operator Theory: Advances and Applications vol. **50**, Birkhauser-Verlag, Basel, 1991, pp. 93–136.

[BT] J.A. Ball and T.T. Trent, Unitary colligations, reproducing kernel Hilbert spaces and Nevanlinna-Pick interpolation in several variables, *J. Functional Analysis*, 157 (1998), 1–61.

[deBR] L. de Branges and J.R. Rovnyak, Canonical models in quantum scattering theory, in *Perturbation Theory and its Applications in Quantum Mechanics* (Ed. C.H. Wilcox), Wiley, New York, 1966, pp. 295–392.

[CS] M. Cotlar and C. Sadosky, Prolongements des formes de Hankel généralisées en formes de Toeplitz, *C.R. Acad. Sci. Paris Sér. I Math.* **305** (1987), 167–170.

[FF] C. Foias and A.E. Frazho, *The Commutant Lifting Approach to Interpolation Problems*, Operator Theory: Advances and Applications vol. **44**, Birkhauser-Verlag, 1990.

[H] J.W. Helton, Discrete time systems, operator models, and scattering theory, *J. Functional Analysis* **16** (1974), 15–38.

[KaKhY] V.E. Katsnelson, A.Ya. Kheifets and P.M. Yuditskii, Abstract interpolation problem and isometric operators extension theory [in Russian], in *Operators in Functional Spaces and Questions of Function Theory: Collected Scientific Papers*, Kiev, Naukova Dumka, 1987, 83–96.

[Kh1] A.Ya. Kheifets, Parseval equality in abstract interpolation problem and coupling of open systems, *Teor. Funk., Funk. Anl. i ikh Prolozhen.* **49** (1988), 112–120; **50** (1988), 98–103 [in Russian] = *J. Soviet Math.* **49** (1990), 1114–1120; **49** (1990), 1307–1310.

[Kh2] A.Ya. Kheifets, Generalized bitangential Schur-Nevanlinna-Pick problem and the related Parseval equality, *Teor. Funk., Funk. Anal. i ikh Prolozhen.* **54** (1990), 89–96 [in Russian] = *J. Soviet Math.* **58** (1992), 358–364.

[Kh3] A.Ya. Kheifets, Nevanlinna-Adamjan-Arov-Krein theorem in semi-determinate case, *Teor. Funkt., Funkt. Anal. i ikh Prilozhen.* **56** (1991), 128–137 [in Russian] = *J. Math. Sciences* **76** (1995), 2542–2549.

[Kh4] A.Ya. Kheifets, Hamburger moment problem: Parseval equality and Arov-singularity, *J. Functional Analysis* **130** (1995), 310–333.

[Kh5] A.Ya. Kheifets, Parametrization of solutions of the Nehari problem and nonorthogonal dynamics, this volume.

[KhY] A.Ya. Kheifets and P.M. Yuditskii, An analysis and extension of V.P. Potapov's approach to interpolation problems with applications to the generalized bitangential Schur-Nevanlinna-Pick problem and j-inner-outer factorization, in

Matrix and Operator Valued Functions: The Vladimir Petrovich Potapov memorial Volume (Ed. I. Gohberg and L.A. Sakhnovich), Operator Theory: Advances and Applications vol. **72**, Birkhauser-Verlag, Basel, 1994, pp. 133–161.

[Ku] S. Kupin, Lifting theorem as a special case of abstract interpolation problem, *Zeitschrift fur Analysis und ihre Anwendungen* **15** (1996), 789–798.

[LP] P.D. Lax and R.S. Phillips, *Scattering Theory*, Academic Press, New York-London, 1967.

[NV] N.K. Nikolskii and V.I. Vasyunin, A unified approach to function models, and the transcription problem, in *The Gohberg Anniversary Collection Volume II: Topics in Analysis and Operator Theory* (Ed. H. Dym, S. Goldberg, M.A. Kaashoek and P. Lancaster), Operator Theory: Advances and Applications vol. **41**, Birkhauser-Verlag, Basel 1989, pp. 405–434.

[Sz.-NF] B. Sz.-Nagy and C. Foias, *Harmonic Analysis of Operators on Hilbert Space*, North Holland/American Elsevier, Amsterdam-New York, 1970.

[S1] D. Sarason, On spectral sets having connected complement, *Acta Aci. Math.* (Szeged) **26** (1965), 289–299.

[S2] D. Sarason, New Hilbert spaces from old, in *Paul Halmos: Celebrating 50 Years of Mathematics* (Ed. J. Ewing and F.W. Gehring), Springer, Berlin, 1991, pp. 195–204.

J.A. Ball
Department of Mathematics
Virginia Tech
Blacksburg, Virginia 24061 U.S.A.
ball@math.vt.edu

Tavan T. Trent
Department of Mathematics
University of Alabama
Tuscaloosa, Alabama 35406 U.S.A.
ttrent@gp.as.ua.edu

The first author was partially supported by the National Science Foundation grant DMS-9500912.

Operator Theory:
Advances and Applications, Vol. 115

On Classes of Realizable Operator-valued R-functions

S.V. Belyi and E.R. Tsekanovskii

In this paper we consider realization problems (see [5]–[7]) for operator-valued R-functions acting on a Hilbert space E ($\dim E < \infty$) as linear-fractional transformations of the transfer operator-valued functions (characteristic functions) of linear stationary conservative dynamic systems (Brodskiĭ-Livšic rigged operator colligations). We specialize three subclasses of the class of all realizable operator-valued R-functions [7]. We give complete proofs of direct and inverse realization theorems for each subclass announced in [5], [6].

1 Introduction

Realization theory of different classes of operator-valued (matrix-valued) functions as transfer operator-functions of linear systems plays an important role in modern operator and systems theory. Almost all realizations in the modern theory of non-selfadjoint operators and its applications deal with systems (operator colligations) in which the main operators are *bounded* linear operators [8], [10–16], [19], [23]. The realization with an *unbounded* operator as a main operator in a corresponding system has not been investigated thoroughly because of a number of essential difficulties usually related to unbounded non-selfadjoint operators.

This paper is the logical continuation of the results stated and proved in [7]. We consider realization problems for operator-valued R-functions acting on a finite dimensional Hilbert space E as linear-fractional transformations of the transfer operator-functions of linear stationary conservative dynamic systems (l.s.c.d.s.) θ of the form

$$\begin{cases} (\mathbb{A} - zI)x = KJ\varphi_- \\ \varphi_+ = \varphi_- - 2iK^*x \end{cases} \qquad \left(\operatorname{Im} \mathbb{A} = KJK^*\right),$$

or

$$\theta = \begin{pmatrix} \mathbb{A} & K & J \\ \mathfrak{H}_+ \subset \mathfrak{H} \subset \mathfrak{H}_- & & E \end{pmatrix}.$$

In the system θ above $\mathbb{A}$ is a bounded linear operator, acting from $\mathfrak{H}_+$ into $\mathfrak{H}_-$, where $\mathfrak{H}_+ \subset \mathfrak{H} \subset \mathfrak{H}_-$ is a rigged Hilbert space, $\mathbb{A} \supset T \supset A$, $\mathbb{A}^* \supset T^* \supset A$, A is a Hermitian operator in $\mathfrak{H}$, T is a non-Hermitian operator in $\mathfrak{H}$, K is a linear bounded operator from E into $\mathfrak{H}_-$, $J = J^* = J^{-1}$ is acting in E, $\varphi_\pm \in E$, φ_- is an input vector, φ_+ is an output vector, and $x \in \mathfrak{H}_+$ is a vector of the inner state of the system θ. The operator-valued function

$$W_\theta(z) = I - 2iK^*(\mathbb{A} - zI)^{-1}KJ \qquad (\varphi_+ = W_\theta(z)\varphi_-),$$

is the transfer operator-valued function of the system θ.

In [7] we established criteria for a given operator-valued R-function $V(z)$ to be realized in the form

$$V(z) = i[W_\theta(z) + I]^{-1}[W_\theta(z) - I]J.$$

It was shown that an operator-valued R-function

$$V(z) = Q + F \cdot z + \int_{-\infty}^{+\infty} \left(\frac{1}{t-z} - \frac{t}{1+t^2}\right) dG(t),$$

acting on a Hilbert space E ($\dim E < \infty$) with some invertibility condition can be realized if and only if

$$F = 0 \quad \text{and} \quad Qe = \int_{-\infty}^{+\infty} \frac{t}{1+t^2} dG(t)e,$$

for all $e \in E$ such that

$$\int_{-\infty}^{+\infty} (dG(t)e, e)_E < \infty.$$

In terms of realizable operator-valued R-functions we specialize in subclasses of the following types:

(1) a subclass for which $\overline{\mathfrak{D}(A)} = \mathfrak{H}$, $\mathfrak{D}(T) \neq \mathfrak{D}(T^*)$

(2) a subclass for which $\overline{\mathfrak{D}(A)} \neq \mathfrak{H}$, $\mathfrak{D}(T) = \mathfrak{D}(T^*)$

(3) a subclass for which $\overline{\mathfrak{D}(A)} \neq \mathfrak{H}$, $\mathfrak{D}(T) \neq \mathfrak{D}(T^*)$

To prove the direct and inverse realization theorems for operator-valued R-functions in each subclass we build a functional model which generally speaking is an unbounded version of the Brodskiĭ-Livšic model with diagonal real part. This model for bounded linear operators was constructed in [11]. In the recent paper [4] the realization problems for contractive operator-valued functions are considered in terms of systems of the special kind. However, as it follows from [5], [7] not every contractive in the lower half-plane function can be realized by the Brodskiĭ-Livšic rigged operator colligation.

2 Preliminaries

Let $\mathfrak{H}$ denote a Hilbert space with inner product (x, y) and let A be a closed linear Hermitian operator, i.e. $(Ax, y) = (x, Ay)$ ($\forall x, y \in \mathfrak{D}(A)$), acting on a Hilbert space $\mathfrak{H}$ with generally speaking, non-dense domain $\mathfrak{D}(A)$. Let $\mathfrak{H}_0 = \overline{\mathfrak{D}(A)}$ and A^* be the adjoint to the operator A (we consider A as acting from $\mathfrak{H}_0$ into $\mathfrak{H}$).

We denote $\mathfrak{H}_+ = \mathfrak{D}(A^*)$ $((\overline{\mathfrak{D}(A^*)} = \mathfrak{H})$ with inner product

$$(f, g)_+ = (f, g) + (A^* f, A^* g) \qquad (f, g \in \mathfrak{H}_+) \tag{1}$$

and then construct the *rigged* Hilbert space $\mathfrak{H}_+ \subset \mathfrak{H} \subset \mathfrak{H}_-$ [9], [7]. Here $\mathfrak{H}_-$ is the space of all linear functionals over $\mathfrak{H}_+$ that are continuous with respect to $\|\cdot\|_+$. The norms of these spaces are connected by the relations $\|x\| \leq \|x\|_+$ $(x \in \mathfrak{H}_+)$, $\|x\|_- \leq \|x\|$ $(x \in \mathfrak{H})$. The Riesz-Berezanskii operator (see [7]) $\mathcal{R}$ maps $\mathfrak{H}_-$ onto $\mathfrak{H}_+$ such that

$$\begin{aligned} &(x, y)_- = (x, \mathcal{R}y) = (\mathcal{R}x, y) = (\mathcal{R}x, \mathcal{R}y)_+ \qquad (x, y \in \mathfrak{H}_-) \\ &(u, v)_+ = (u, \mathcal{R}^{-1}v) = (\mathcal{R}^{-1}u, v) = (\mathcal{R}^{-1}x, \mathcal{R}^{-1}y)_- \qquad (u, v \in \mathfrak{H}_+) \end{aligned} \tag{2}$$

In what follows we use symbols $(+)$, $(\cdot)$, and $(-)$ to indicate the norms $\|\cdot\|_+$, $\|\cdot\|$, and $\|\cdot\|_-$ by which geometrical and topological concepts are defined in $\mathfrak{H}_+$, $\mathfrak{H}$, and $\mathfrak{H}_-$, respectively.

In the above settings $\mathfrak{D}(A) \subset \mathfrak{D}(A^*)(= \mathfrak{H}_+)$ and $A^* y = PAy$ $(\forall y \in \mathfrak{D}(A))$, where P is an orthogonal projection of $\mathfrak{H}$ onto $\mathfrak{H}_0$. Let

$$\mathfrak{L} := \mathfrak{H} \ominus \mathfrak{H}_0 \quad \mathfrak{M}_\lambda := (A - \lambda I)\mathfrak{D}(A) \quad \mathfrak{N}_\lambda := (\mathfrak{M}_{\bar\lambda})^\perp. \tag{3}$$

The subspace $\mathfrak{N}_\lambda$ is called a *defect subspace* of A for the point $\bar\lambda$. The cardinal number $\dim\mathfrak{N}_\lambda$ remains constant when λ is in the upper half-plane. Similarly, the number $\dim\mathfrak{N}_\lambda$ remains constant when λ is in the lower half-plane. The numbers $\dim\mathfrak{N}_\lambda$ and $\dim\mathfrak{N}_{\bar\lambda}$ $(\text{Im}\lambda < 0)$ are called the *defect numbers* or *deficiency indices* of operator A [8]. The subspace $\mathfrak{N}_\lambda$ which lies in $\mathfrak{H}_+$ is the set of solutions of the equation $A^* g = \lambda P g$.

Let now P_λ be the orthogonal projection onto $\mathfrak{N}_\lambda$, set

$$\mathfrak{B}_\lambda = P_\lambda \mathfrak{L}, \qquad \mathfrak{N}'_\lambda = \mathfrak{N}_\lambda \ominus \overline{\mathfrak{B}_\lambda} \tag{4}$$

It is easy to see that $\mathfrak{N}'_\lambda = \mathfrak{N}_\lambda \cap \mathfrak{H}_0$ and $\mathfrak{N}'_\lambda$ is the set of solutions of the equation $A^* g = \lambda g$ (see [27]), when $A^* : \mathfrak{H} \to \mathfrak{H}_0$ is the adjoint operator to A.

The subspace $\mathfrak{N}'_\lambda$ is the defect subspace of the densely defined Hermitian operator PA on $\mathfrak{H}_0$ (see [27]). The numbers $\dim\mathfrak{N}'_\lambda$ and $\dim\mathfrak{N}'_{\bar\lambda}$ $(\text{Im}\lambda < 0)$ are called *semi-defect numbers* or the *semi-deficiency indices* of the operator A [18]. The von Neumann formula

$$\mathfrak{H}_+ = \mathfrak{D}(A^*) = \mathfrak{D}(A) + \mathfrak{N}_\lambda + \mathfrak{N}_{\bar\lambda}, \qquad (\text{Im}\lambda \neq 0), \tag{5}$$

holds, but this decomposition is not direct for a non-densely defined operator A. There exists a generalization of von Neumann's formula [3], [27] to the case of a non-densely defined Hermitian operator (direct decomposition). We call an operator A *regular*, if PA is a closed operator in $\mathfrak{H}_0$. For a regular operator A we have

$$\mathfrak{H}_+ = \mathfrak{D}(A) + \mathfrak{N}'_\lambda + \mathfrak{N}'_{\bar{\lambda}} + \mathfrak{N}, \qquad (\mathrm{Im}\lambda \neq 0) \tag{6}$$

where $\mathfrak{N} := \mathcal{R}\mathfrak{L}$, $\mathcal{R}$ is the Riesz-Berezanskii operator. This is a generalization of von Neumann's formula. For $\lambda = \pm i$ we obtain the $(+)$-orthogonal decomposition

$$\mathfrak{H}_+ = \mathfrak{D}(A) \oplus \mathfrak{N}'_i \oplus \mathfrak{N}'_{-i} \oplus \mathfrak{N}. \tag{7}$$

Let $\tilde{A}$ be a closed Hermitian extension of the operator A. Then $\mathfrak{D}(\tilde{A}) \subset \mathfrak{H}_+$ and $P\tilde{A}x = A^*x$ $(\forall x \in \mathfrak{D}(\tilde{A}))$. According to [27] a closed Hermitian extension $\tilde{A}$ is said to be *regular* if $P\tilde{A}$ is closed. This implies that $\mathfrak{D}(\tilde{A})$ is $(+)$-closed. According to the theory of extensions of closed Hermitian operators A with non-dense domain [18], an operator U $(\mathfrak{D}(U) \subseteq \mathfrak{N}_i, \mathfrak{R}(U) \subseteq \mathfrak{N}_{-i})$ is called an *admissible operator* if $(U - I)f_i \in \mathfrak{D}(A)$ $(f_i \in \mathfrak{D}(U))$ only for $f_i = 0$. Then (see [3]) any symmetric extension $\tilde{A}$ of the non-densely defined closed Hermitian operator A, is defined by an isometric admissible operator U, $\mathfrak{D}(U) \subseteq \mathfrak{N}_i$, $\mathfrak{R}(U) \subseteq \mathfrak{N}_{-i}$ by the formula

$$\tilde{A}f_{\tilde{A}} = Af_A + (-if_i - iUf_i), \quad f_A \in \mathfrak{D}(A)$$

where $\mathfrak{D}(\tilde{A}) = \mathfrak{D}(A)\dot{+}(U - I)\mathfrak{D}(U)$. The operator $\tilde{A}$ is self-adjoint if and only if $\mathfrak{D}(U) = \mathfrak{N}_i$ and $\mathfrak{R}(U) = \mathfrak{N}_{-i}$.

Let us denote now by $P^+_{\mathfrak{N}}$ the orthogonal projection operator in $\mathfrak{H}_+$ onto $\mathfrak{N}$. We introduce a new inner product $(\cdot, \cdot)_1$ defined by

$$(f, g)_1 = (f, g)_+ + (P^+_{\mathfrak{N}} f, P^+_{\mathfrak{N}} g)_+ \tag{8}$$

for all $f, g \in \mathfrak{H}_+$. The obvious inequality

$$\|f\|^2_+ \leq \|f\|^2_1 \leq 2\|f\|^2_+$$

shows that the norms $\|\cdot\|_+$ and $\|\cdot\|_1$ are topologically equivalent. It is easy to see that the spaces $\mathfrak{D}(A)$, $\mathfrak{N}'_i$, $\mathfrak{N}'_{-i}$, $\mathfrak{N}$ are (1)-orthogonal. We write $\mathfrak{M}_1$ for the Hilbert space $\mathfrak{M} = \mathfrak{N}'_i \oplus \mathfrak{N}'_{-i} \oplus \mathfrak{N}$ with inner product $(f, g)_1$. We denote by $\mathfrak{H}_{+1}$ the space $\mathfrak{H}_+$ with norm $\|\cdot\|_1$, and by $\mathcal{R}_1$ the corresponding Riesz-Berezanskii operator related to the rigged Hilbert space $\mathfrak{H}_{+1} \subset \mathfrak{H} \subset \mathfrak{H}_{-1}$.

Denote by $[\mathfrak{H}_1, \mathfrak{H}_2]$ the set of all linear bounded operators acting from a Hilbert space $\mathfrak{H}_1$ into a Hilbert space $\mathfrak{H}_2$.

Definition: An operator $\mathbb{A} \in [\mathfrak{H}_+, \mathfrak{H}_-]$ is a *bi-extension* of A if both $\mathbb{A} \supset A$ and $\mathbb{A}^* \supset A$ hold.

If $\mathbb{A} = \mathbb{A}^*$, then $\mathbb{A}$ is called self-adjoint bi-extension of the operator A. It was mentioned in [7] that every self-adjoint bi-extension $\mathbb{A}$ of the regular Hermitian operator A is of the form:

$$\mathbb{A} = AP^+_{\mathfrak{D}(A)} + \left[A^* + \mathcal{R}_1^{-1}(S - \frac{i}{2}P^+_{\mathfrak{N}'_i} + \frac{i}{2}P^+_{\mathfrak{N}'_{-i}})\right] P^+_{\mathfrak{M}},$$

where S is an arbitrary (1)-self-adjoint operator in $[\mathfrak{M}_1, \mathfrak{M}_1]$. We write $\mathfrak{S}(A)$ for the class of bi-extensions of A. This class is closed in the weak topology and is invariant under taking adjoints (see [3], [27]).

Let $\mathbb{A}$ be a bi-extension of Hermitian operator A. The operator $\hat{A}f = \mathbb{A}f$, $\mathfrak{D}(\hat{A}) = \{f \in \mathfrak{H}, \mathbb{A}f \in \mathfrak{H}\}$ is called the *quasikernel* of $\mathbb{A}$. If $\mathbb{A} = \mathbb{A}^*$ and $\hat{A}$ is a quasi-kernel of $\mathbb{A}$ such that $A \neq \hat{A}$, $\hat{A}^* = \hat{A}$ then $\mathbb{A}$ is said to be a *strong* self-adjoint bi-extension of A.

Definition: We say that a closed densely defined linear operator T acting on a Hilbert space $\mathfrak{H}$ belongs to the class Ω_A if:

(1) $T \supset A, T^* \supset A$ where A is a closed Hermitian operator;

(2) $(-i)$ is a regular point of T.[1]

It was mentioned in [27] that lineals $\mathfrak{D}(T)$ and $\mathfrak{D}(T^*)$ are $(+)$-closed, the operators T and T^* are $(+, \cdot)$-bounded. The following theorem [27] is an analogue to von Neumann's formulae for the class Ω_A.

Theorem 1 *If an operator T belongs to the class Ω_A, then*

$$\begin{cases} \mathfrak{D}(T) = \mathfrak{D}(A)\dot{+}(I - \Phi)\mathfrak{N}_i \\ \mathfrak{D}(T^*) = \mathfrak{D}(A)\dot{+}(\Phi^* - I)\mathfrak{N}_{-i} \end{cases}$$

where Φ and Φ^* are admissible operators in $[\mathfrak{N}_i, \mathfrak{N}_{-i}]$ and $[\mathfrak{N}_{-i}, \mathfrak{N}_i]$ respectively.

There is a modification of the last theorem [27], [28].

Theorem 2 *I. For each operator of the class Ω_A there exists an operator M on the space $\mathfrak{M}_1$ with the following properties:*

(1) *$\mathfrak{D}(M) = \mathfrak{N}'_i \oplus \mathfrak{N}$ and $\mathfrak{R}(M) = \mathfrak{N}'_{-i} \oplus \mathfrak{N}$;*

(2) *$Mx + x = 0$ only for $x = 0$, and $M^*x + x = 0$ only for $x = 0$. Moreover, the following hold:*

$$\begin{cases} \mathfrak{D}(T) = \mathfrak{D}(A) \oplus (M + I)(\mathfrak{N}'_i \oplus \mathfrak{N}) \\ \mathfrak{D}(T^*) = \mathfrak{D}(A) \oplus (M^* + I)(\mathfrak{N}'_{-i} \oplus \mathfrak{N}) \end{cases} \tag{9}$$

II. Conversely, for each pair of (1)-adjoint operators M and M^* in $[\mathfrak{M}_1, \mathfrak{M}_1]$ with the properties (1) and (2) formulas (9) give a corresponding operator T in class Ω_A. Moreover, if $f = g + (M + I)\varphi$, $g \in \mathfrak{D}(A)$, and $\varphi \in \mathfrak{N}'_i \oplus \mathfrak{N}$, then

$$Tf = Ag + A^*(I + M)\varphi + i\mathcal{R}_1^{-1}P^+_{\mathfrak{N}}(I - M)\varphi \quad (f \in \mathfrak{D}(T)), \tag{10}$$

Similarly, if $f = g + (M^* + I)\psi$, $g \in \mathfrak{D}(A)$, and $\psi \in \mathfrak{N}'_{-i} \oplus \mathfrak{N}$, then

$$T^*f = Ag + A^*(I + M^*)\psi + i\mathcal{R}_1^{-1}P^+_{\mathfrak{N}}(M^* - I)\psi \quad (f \in \mathfrak{D}(T)), \tag{11}$$

The following theorems can be found in [27], [28].

---

[1] The condition, that $(-i)$ is a regular point in the definition of the class Ω_A is not essential. It is sufficient to require the existence of some regular point for T.

Theorem 3 *Let T be an operator of Ω_A class such that A is the maximal Hermitian part of T and T^*. Let M be the corresponding operator from the Theorem 2 with the properties* (1) *and* (2). *Then the operators $MM^* - I$ and $M^*M - I$ are invertible in $\mathfrak{M}$.*

Definition: A regular operator A is called *O-operator* if its semidefect numbers (defect numbers of an operator PA) are equal to zero.

Theorem 4 *Let T be an operator of the class Ω_A where A is a regular Hermitian operator. Then the following statements are valid:*

(1) *If A is an O-operator then*

$$\mathfrak{D}(T) = \mathfrak{D}(T^*) = \mathfrak{H}_+$$

and the operator $T - T^*$ is $(\cdot, \cdot)$-continuous.

(2) *If A is not an O-operator then either $\mathfrak{D}(T)$ or $\mathfrak{D}(T^*)$ does not coincide with $\mathfrak{H}_+$.*

Proof: Since T is an operator of the class Ω_A then $\mathfrak{D}(T)$ and $\mathfrak{D}(T^*)$ are subspaces of $\mathfrak{H}_+$. Let M and M^* be the operators defined in the Theorem 2. In this case $\mathfrak{D}(M) = \mathfrak{N}'_i \oplus \mathfrak{N}$, $\mathfrak{R}(M) \subseteq \mathfrak{N}'_{-i} \oplus \mathfrak{N}$, $\mathfrak{D}(M^*) = \mathfrak{N}'_{-i} \oplus \mathfrak{N}$, and $\mathfrak{R}(M^*) \subseteq \mathfrak{N}'_i \oplus \mathfrak{N}$. Formulas (9) imply that $\mathfrak{R}(M + I)$ and $\mathfrak{R}(M^* + I)$ are $(+)$- and (1)-subspaces as well. Consider the (1)-orthogonal complements

$$[\mathfrak{R}(M + I)]^{\perp} \quad \text{and} \quad [\mathfrak{R}(M^* + I)]^{\perp}.$$

Let us assume that A is not an O-operator. Then the semidefect numbers of A are not both zero. For any $y \in [\mathfrak{R}(M^* + I)]^{\perp}$ and for any $x \in \mathfrak{N}'_{-i} \oplus \mathfrak{N}$ we have

$$((M^* + I)x, y)_1 = 0.$$

Furthermore, using (1)-orthogonality relation one can show that

$$\begin{aligned}
((M^* + I)x, y)_1 &= \left((M^* + I)x,\, P^+_{\mathfrak{N}'_i} y + P^+_{\mathfrak{N}'_{-i}} y + P^+_{\mathfrak{N}} y\right)_1 \\
&= \left(M^* x,\, P^+_{\mathfrak{N}'_i} y + P^+_{\mathfrak{N}'_{-i}} y + P^+_{\mathfrak{N}} y\right)_1 \\
&\quad + \left(x,\, P^+_{\mathfrak{N}'_i} y + P^+_{\mathfrak{N}'_{-i}} y + P^+_{\mathfrak{N} y}\right)_1 \\
&= \left(M^* x,\, P^+_{\mathfrak{N}'_i} y + P^+_{\mathfrak{N}} y\right)_1 + \left(M^* x,\, P^+_{\mathfrak{N}'_{-i}} y\right)_1 \\
&\quad + \left(x,\, P^+_{\mathfrak{N}'_{-i}} y + P^+_{\mathfrak{N}} y\right)_1 + \left(x,\, P^+_{\mathfrak{N}'_i} y\right)_1 \\
&= \left(x,\, M(P^+_{\mathfrak{N}'_i} + P^+_{\mathfrak{N}})y\right)_1 + \left(x,\, (P^+_{\mathfrak{N}'_{-i}} + P^+_{\mathfrak{N}})y\right)_1 = 0.
\end{aligned}$$

Therefore, since M maps $\mathfrak{N}'_i \oplus \mathfrak{N}$ into $\mathfrak{N}'_{-i} \oplus \mathfrak{N}$ we have that

$$M(P^+_{\mathfrak{N}'_i} + P^+_{\mathfrak{N}})y = -(P^+_{\mathfrak{N}'_{-i}} + P^+_{\mathfrak{N}})y, \tag{12}$$

Let us denote $z = (P^+_{\mathfrak{N}'_i} + P^+_{\mathfrak{N}})y$. Then, obviously,

$$P^+_{\mathfrak{N}}(M + I)z = 0, \qquad (z \in \mathfrak{N}'_i \oplus \mathfrak{N}), \tag{13}$$

Hence, if $y \in \left[\mathfrak{R}(M^* + I)\right]^{\perp}$ then

$$z = (P^+_{\mathfrak{N}'_i} + P^+_{\mathfrak{N}})y \in \operatorname{Ker}\left[P^+_{\mathfrak{N}}(M + I)z\right] \quad \text{and} \quad y = z - P^+_{\mathfrak{N}'_{-i}} Mz.$$

Let now $z \in \operatorname{Ker}\left[P^+_{\mathfrak{N}}(M + I)\right]$. We show that the vector $y = z - P^+_{\mathfrak{N}'_{-i}} Mz$ belongs to $\left[\mathfrak{R}(M^* + I)\right]^{\perp}$. To do that it is sufficient to show that for indicated vector y the relation (12) holds. Indeed,

$$\begin{aligned} -(P^+_{\mathfrak{N}'_{-i}} + P^+_{\mathfrak{N}})y &= -P^+_{\mathfrak{N}} z + P^+_{\mathfrak{N}'_{-i}} Mz = P^+_{\mathfrak{N}} Mz + P^+_{\mathfrak{N}'_{-i}} Mz \\ &= Mz = M(P^+_{\mathfrak{N}'_i} + P^+_{\mathfrak{N}})y. \end{aligned}$$

Hence,

$$\left[\mathfrak{R}(M^* + I)\right]^{\perp} = (I - P^+_{\mathfrak{N}'_i} M)\{\operatorname{Ker}[P^+_{\mathfrak{N}}(M + I)]\}. \tag{14}$$

It can be shown similarly, that

$$[\mathfrak{R}(M + I)]^{\perp} = (I - P^+_{\mathfrak{N}'_i} M)\{\operatorname{Ker}[P^+_{\mathfrak{N}}(M^* + I)]\}. \tag{15}$$

Let us assume that $\left[\mathfrak{R}(M^* + I)\right]^{\perp} = 0$. It is easy to see that equality $(I - P^+_{\mathfrak{N}'_i} M)z = 0$ implies that if $z = 0$ then $\operatorname{Ker}[P^+_{\mathfrak{N}}(M^* + I)] = 0$. Then operator $(M^* + I)$ maps $\mathfrak{N}'_{-i} \oplus \mathfrak{N}$ onto $\mathfrak{M}$. Therefore, there exists vector $x \neq 0$, $x \in \mathfrak{N}'_{-i} \oplus \mathfrak{N}$ such that $P^+_{\mathfrak{N}}(M^* + I)x = 0$ and so $\operatorname{Ker}[P^+_{\mathfrak{N}}(M^* + I)] \neq 0$. Thus, $[\mathfrak{R}(M + I)]^{\perp} \neq 0$. Together with formulas (9) that proves the first part of the theorem.

Let now A be a regular O-operator, i.e. $\mathfrak{N}'_i = \mathfrak{N}'_{-i} = \{0\}$ and consequently $\mathfrak{M} = \mathfrak{N}$. Let us assume that x is $(+)$-orthogonal to $\mathfrak{D}(T)$. According to the formulas (9) x is $(+)$-orthogonal to $\mathfrak{D}(A)$ and therefore belongs to $\mathfrak{N}$. On the other hand (9) imply that x is $(+)$-orthogonal to $(M + I)\mathfrak{N}$. Hence, $(M^* + I)x = 0$. Using Theorem 1 we conclude that $x = 0$. Therefore, $\mathfrak{D}(T)$ is $(+)$-dense in $\mathfrak{H}_+$. In the same way one can prove that $\mathfrak{D}(T^*)$ is $(+)$-dense in $\mathfrak{H}_+$. □

Definition: An operator $\mathbb{A}$ in $[\mathfrak{H}_+, \mathfrak{H}_-]$ is called a $(*)$-extension of an operator T of the class Ω_A if both $\mathbb{A} \supset T$ and $\mathbb{A}^* \supset T^*$.

This $(*)$-extension is called *correct*, if an operator $\mathbb{A}_R = \frac{1}{2}(\mathbb{A} + \mathbb{A}^*)$ is a strong self-adjoint bi-extension of an operator A. It is easy to show that if $\mathbb{A}$ is a $(*)$-extension of T, the T and T^* are quasi-kernels of $\mathbb{A}$ and $\mathbb{A}^*$, respectively.

Definition: We say the operator T of the class Ω_A belongs to the class Λ_A if

(1) T admits a correct $(*)$-extension;

(2) A is a maximal common Hermitian part of T and T^*.

The following theorem can be found in [27].

Theorem 5 *Let an operator T belong to Ω_A and M be an operator in $[\mathfrak{M}, \mathfrak{M}]$ that is related to T by Theorem 2. Then T belongs to Λ_A if and only if there exists either* (1)*-isometric operator or* $(\cdot)$*-isometric operator U in $[\mathfrak{N}'_i, \mathfrak{N}'_{-i}]$ such that*

$$\begin{cases} (U+I)\mathfrak{N}'_i + (M+I)(\mathfrak{N}'_i \oplus \mathfrak{N}) = \mathfrak{M}, \\ (U+I)\mathfrak{N}'_i + (M+I)(\mathfrak{N}'_i \oplus \mathfrak{N}) = \mathfrak{M}. \end{cases} \tag{16}$$

Corollary 1 *If a closed Hermitian operator A has finite and equal defect indices then the class Ω_A coincides with the class Λ_A.*

Let A be a closed Hermitian operator on $\mathfrak{H}$ and $\mathfrak{h}$ be a Hilbert space such that $\mathfrak{H}$ is a subspace of $\mathfrak{h}$. Let $\tilde{A}$ be a self-adjoint extension of A on $\mathfrak{h}$, and $\tilde{E}(t)$ be the spectral function of $\tilde{A}$. An operator function $R_\lambda = P_{\mathfrak{H}}(\tilde{A} - \lambda I)^{-1}|_{\mathfrak{H}}$ is called a *generalized resolvent* of A, and $E(t) = P_{\mathfrak{H}}\tilde{E}(t)|_{\mathfrak{H}}$ is the corresponding *generalized spectral function.* Here

$$R_\lambda = \int_{-\infty}^{\infty} \frac{dE(t)}{t-\lambda} \quad (\mathrm{Im}\lambda \neq 0). \tag{17}$$

If $\mathfrak{h} = \mathfrak{H}$ then R_λ and $E(t)$ are called *canonical resolvent* and *canonical spectral function*, respectively. According to [21] we denote by $\hat{R}_\lambda$ the $(-, \cdot)$-continuous operator from $\mathfrak{H}_-$ into $\mathfrak{H}$ which is adjoint to $R_{\bar{\lambda}}$:

$$(\hat{R}_\lambda f, g) = (f, R_{\bar{\lambda}} g) \quad (f \in \mathfrak{H}_-,\ g \in \mathfrak{H}). \tag{18}$$

It follows that $\hat{R}_\lambda f = R_\lambda f$ for $f \in \mathfrak{H}$, so that $\hat{R}_\lambda$ is an extension of R_λ from $\mathfrak{H}$ to $\mathfrak{H}_-$ with respect to $(-, \cdot)$-continuity. The function $\hat{R}_\lambda$ of the parameter λ, $(\mathrm{Im}\lambda \neq 0)$ is called the *extended generalized (canonical) resolvent* of the operator A. We write $\aleph$ to denote the family of all finite intervals on the real axis. It is known [21] that if $\Delta \in \aleph$ then $E(\Delta)\mathfrak{H} \subset \mathfrak{H}_+$ and the operator $E(\Delta)$ is $(\cdot, +)$-continuous. We denote by $\hat{E}(\Delta)$ the $(-, \cdot)$-continuous operator from $\mathfrak{H}_-$ to $\mathfrak{H}$ that is adjoint to $E(\Delta) \in [\mathfrak{H}, \mathfrak{H}_+]$. Similarly,

$$(\hat{E}(\Delta) f, g) = (f, E(\Delta) g) \quad (f \in \mathfrak{H}_-,\ g \in \mathfrak{H}) \tag{19}$$

One can easily see that $\hat{E}(\Delta)f = E(\Delta)f, \forall f \in \mathfrak{H}$, so that $\hat{E}(\Delta)$ is the extension of $E(\Delta)$ by continuity. We say that $\hat{E}(\Delta)$, as a function of $\Delta \in \aleph$, is the *extended generalized (canonical) spectral function* of A corresponding to the self-adjoint extension $\tilde{A}$ (or to the original spectral function $E(\Delta)$). It is known [21] that $\hat{E}(\Delta) \in [\mathfrak{H}_-, \mathfrak{H}_+]$, $\forall \Delta \in \aleph$, and $(\hat{E}(\Delta)f, f) \geq 0$ for all $f \in \mathfrak{H}_-$. It is also known [21] that the complex scalar measure $(E(\Delta)f, g)$ is a complex function of bounded variation on the real axis. However, $(\hat{E}(\Delta)f, g)$ may be unbounded for $f, g \in \mathfrak{H}_-$.

Now let $\hat{R}_\lambda$ be an extended generalized (canonical) resolvent of a closed Hermitian operator A and let $\hat{E}(\Delta)$ be the corresponding extended generalized (canonical) spectral function. It was shown in [21] that for any $f, g \in \mathfrak{H}_-$,

(20) $$\int_{-\infty}^{+\infty} \frac{|d(\hat{E}(\Delta)f, g)|}{1+t^2} < \infty$$

and the following integral representation holds

(21) $$\hat{R}_\lambda - \frac{\hat{R}_i + \hat{R}_{-i}}{2} = \int_{-\infty}^{+\infty} \left(\frac{1}{t-\lambda} - \frac{t}{1+t^2}\right) d\hat{E}(t)$$

Lemma 6 ([1],[7]) *Let* $\mathbb{A} = AP^+_{\mathfrak{D}(A)} + [A^* + \mathcal{R}_1^{-1}(S - \frac{i}{2}P^+_{\mathfrak{N}'_i} + \frac{i}{2}P^+_{\mathfrak{N}'_{-i}})]P^+_{\mathfrak{M}}$ *be a strong self-adjoint bi-extension of a regular Hermitian operator A with the quasi-kernel* $\hat{A}$ *and let* $\hat{E}(\Delta)$ *be the extended generalized (canonical) spectral function of* $\hat{A}$*. Then for every* $f \in \mathfrak{H} \oplus L$*,* $f \neq 0$*, and for every* $g \in \mathfrak{H}_-$ *there is an integral representation*

(22) $$(\bar{R}_\lambda f, g) = \int_{-\infty}^{+\infty} \left(\frac{1}{t-\lambda} - \frac{t}{1+t^2}\right) d(\hat{E}(t)f, g) + \frac{1}{2}((\hat{R}_i + \hat{R}_{-i})f, g).$$

Here $L = \mathfrak{R}\Big[\mathcal{R}_1^{-1}(P^+_{\mathfrak{M}}S - \frac{i}{2}P^+_{\mathfrak{N}'_i} + \frac{i}{2}P^+_{\mathfrak{N}'_{-i}})\Big]$, $\bar{R}_\lambda = \overline{(\mathbb{A} - \lambda I)^{-1}}$.

Theorem 7 ([7]) *Let* $\mathbb{A} = AP^+_{\mathfrak{D}(A)} + [A^* + \mathcal{R}_1^{-1}(S - \frac{i}{2}P^+_{\mathfrak{N}'_i} + \frac{i}{2}P^+_{\mathfrak{N}'_{-i}})]P^+_{\mathfrak{M}}$ *be a strong self-adjoint bi-extension of a regular Hermitian operator A with the quasi-kernel* $\hat{A}$ *and let* $\hat{E}(\Delta)$ *be the generalized (canonical) spectral function of* $\hat{A}$*,* $F = \mathfrak{H}_+ \ominus \mathfrak{D}(\hat{A})$*,* $L = \mathcal{R}_1^{-1}(P^+_{\mathfrak{M}}S - \frac{i}{2}P^+_{\mathfrak{N}'_i} + \frac{i}{2}P^+_{\mathfrak{N}'_{-i}})F$*. Then for any* $f \in L \dot{+} \mathfrak{L}$*,* $f \neq 0$*,*

(23) $$\int_{-\infty}^{+\infty} d(\hat{E}(t)f, f) = \infty, \qquad \text{if } f \notin \mathfrak{L},$$

and

(24) $$\int_{-\infty}^{+\infty} d(\hat{E}(t)f, f) < \infty, \qquad \text{if } f \in \mathfrak{L}.$$

Moreover, there exist real constants b and c such that

$$c\|f\|_-^2 \le \int_{-\infty}^{+\infty} \frac{d(\hat{E}(t)f,f)}{1+t^2} \le b\|f\|_-^2 \tag{25}$$

for all $f \in L\dot{+}\mathfrak{L}$.

In another form Theorem 7 is in [1]. We briefly sketch the proof of this theorem.

Proof: Let us choose a point z with $\mathrm{Im} z \neq 0$ to be a regular point of the operator $\mathbb{A}$ and consider function $f(z)$ defined for all $f \in L\dot{+}\mathfrak{L}$ by the formula:

$$f(z) = \big((\mathbb{A} - zI)^{-1} f, f\big).$$

It can be seen that $f(z) = \overline{f(\bar{z})}$ and

$$\mathrm{Im}\, f(z) = \mathrm{Im}\, z\|(\mathbb{A} - \bar{z}I)^{-1} f\|^2,$$

which means that $f(z)$ is an analytic R-function (see [17]) and according to the Lemma 6 has the integral representation

$$f(z) = \int_{-\infty}^{+\infty} \left(\frac{1}{t-z} - \frac{t}{1+t^2}\right) d\big(\hat{E}(t)f, f\big) + \frac{1}{2}\big((\hat{R}_i + \hat{R}_{-i})f, f)\big).$$

This representation implies that

$$\lim_{\eta\to\infty} \frac{\mathrm{Im}\, f(i\eta)}{\eta} = 0,$$

and therefore (see [13], [17])

$$\sup_{\eta>0} \eta \mathrm{Im}\, f(i\eta) = \int_{-\infty}^{+\infty} d(\hat{E}(t)f, f).$$

Now, let us pick $f \in L\dot{+}\mathfrak{L}$ such that $f \notin \mathfrak{L}$ and show that in this case

$$\sup_{\eta>0} \eta \mathrm{Im}\, f(i\eta) = \infty.$$

It can be shown that for any $f \in L\dot{+}\mathfrak{L}$ there are $x_0(i\eta) \in \mathfrak{D}(\hat{A})$, $f_1 \in F$ and $f_2 \in \mathfrak{L}$ such that

$$(\mathbb{A} + i\eta I)^{-1} f = x_0(i\eta) + f_1,$$

$$x_0(i\eta) = (\hat{A} + i\eta I)^{-1}\big[f_2 - \mathcal{R}_1 P_{\mathfrak{N}} S f_1 - (\mathbb{A}^* + i\eta I) f_1\big],$$

or

$$i\eta(x_0(i\eta) + f_1) = -\hat{A}x_0(i\eta) - A^* f_1 + f_2 - \mathcal{R}_1 P_{\mathfrak{N}} S f_1.$$

The vectors $x_0(i\eta)$ and f_1 are $(+1)$-orthogonal and hence

$$\|(\mathbb{A}+i\eta I)^{-1}f\|_+^2 = \|x_0(i\eta)\|_+^2 + \|f_1\|_+^2.$$

If we assume that the number set $\{\|x_0(i\eta)\|_+^2\}$ is bounded, i.e.

$$\sup\{\|x_0(i\eta)\|_+\} = c < \infty,$$

then, due to the $(+,\cdot)$-continuity of the operator $\hat{A}$ (see [28]), there exists a constant $d > 0$, such that for all $x_0(i\eta) \in \mathfrak{D}(\hat{A})$

$$\|\hat{A}x_0(i\eta)\| \le d\|x_0(i\eta)\|_+ \le d\sqrt{c},$$

and

$$\begin{aligned}\|x_0(i\eta) + f_1\| &\le \frac{1}{\eta}\|f_2 - \hat{A}x_0(i\eta) - A^*f_1\| \\ &\le \frac{1}{\eta}\left(d\sqrt{c} + \|f_2\| + \|A^*f_1\|\right).\end{aligned}$$

This implies $\lim_{\eta\to\infty} x_0(i\eta) = -f_1$. The set $\{x_0(i\eta)\}$ is bounded in $\mathfrak{H}_+$ and therefore weakly compact. This means there exists such an element $x_0 \in \mathfrak{H}_+$ that

$$\lim_{\eta_n\to\infty}(x_0(i\eta_n),\varphi) = (x_0,\varphi), \quad \forall\varphi \in \mathfrak{H}_-,$$

where $\{x_0(i\eta_n)\}$ is a sequence of the elements of the set $\{x_0(i\eta)\}$ and $x_0 \in \mathfrak{H}_+$. Thus $x_0 = -f_1$. On the other hand

$$\mathfrak{D}(\hat{A}) = \mathfrak{D}(A) \oplus \operatorname{Ker}\left[P_{\mathfrak{M}}^+ S - \frac{i}{2}P_{\mathfrak{N}'_i} + \frac{i}{2}P_{\mathfrak{N}'_{-i}}\right],$$

is a subspace in $\mathfrak{H}_+$ and must be weakly closed providing $x_0 \in \mathfrak{D}(\hat{A})$. Considering the fact that $f_1 \in F$, $F = \mathfrak{H}_+ \ominus \mathfrak{D}(\hat{A})$, and $x_0 = -f_1$ we obtain a contradiction.

Hence for all $f \in L\dot{+}\mathfrak{L}$, $f \notin \mathfrak{L}$

$$\int_{-\infty}^{+\infty} d(\hat{E}(t)f,f) = \sup_{\eta>0}\eta\operatorname{Im} f(i\eta) = \infty.$$

To prove relation (23) we assume that $f \in \mathfrak{L}$. In this case $f_1 = 0$ and $(\mathbb{A}+i\eta I)^{-1}f = x_0(i\eta)$. The latter yields

$$\|(\hat{A}+i\eta I)x_0(i\eta)\|^2 = \|f\|^2.$$

Further it is not hard to get the inequality

$$\eta^2\|x_0)i\eta)\|^2 \le \|(\hat{A}+i\eta I)x_0(i\eta)\|^2 = \|f\|^2,$$

that implies

$$\eta\operatorname{Im} f(i\eta) = \eta^2\|(\mathbb{A}+i\eta I)^{-1}f\| \le \|f\|^2 < \infty.$$

The last inequality proves (24).

It can be shown that $(\mathbb{A} + iI)^{-1} \in \mathfrak{N}_{-i}$ for all $f \in L\dot{+}\mathfrak{L}$. The norms $\|\cdot\|$ and $\|\cdot\|_+$ are equivalent on $\mathfrak{N}_{\pm i}$ and so are the norms $\|\cdot\|$ and $\|\cdot\|_-$ (see [28]). Therefore

$$c\|f\|_-^2 \leq \operatorname{Im} f(i) \leq b\|f\|_-^2, \quad b > 0,\ c > 0 - \text{const}.$$

Combining this with

$$\operatorname{Im} f(i) = \frac{1}{2i}\big((\bar{R}_i - \bar{R}_{-i})f, f\big) = \int_{-\infty}^{+\infty} \frac{d(\hat{E}(t)f, f)}{1+t^2},$$

we obtain the relation (25). □

Corollary 2 *In the settings of Theorem 7 for all $f, g \in L\dot{+}\mathfrak{L}$*

$$\left|\left(\frac{\hat{R}_i + \hat{R}_{-i}}{2} f, g\right)\right| \leq a\sqrt{\int_{-\infty}^{+\infty} \frac{d(\hat{E}(t)f, f)}{1+t^2}} \cdot \sqrt{\int_{-\infty}^{+\infty} \frac{d(\hat{E}(t)g, g)}{1+t^2}}, \tag{26}$$

where $a > 0$ is a constant (*see* [1]).

3 Linear Stationary Conservative Dynamic Systems

In this section we consider linear stationary conservative dynamic systems (l. s. c. d. s.) θ of the form

$$\begin{cases} (\mathbb{A} - zI) = KJ\varphi_- \\ \varphi_+ = \varphi_- - 2iK^*x \end{cases} \qquad \big(\operatorname{Im} \mathbb{A} = KJK^*\big). \tag{27}$$

In a system θ of the form (27) $\mathbb{A}$, K and J are bounded linear operators in Hilbert spaces, φ_- is an input vector, φ_+ is an output vector, x is an inner state vector of the system θ. For our purposes we need the following more precise definition:

Definition: The array

$$\theta = \begin{pmatrix} \mathbb{A} & K & J \\ \mathfrak{H}_+ \subset \mathfrak{H} \subset \mathfrak{H}_- & & E \end{pmatrix} \tag{28}$$

is called a linear stationary conservative dynamic system (l.s.c.d.s.) or Brodskiĭ-Livšic rigged operator colligation if

(1) $\mathbb{A}$ is a correct $(*)$-extension of an operator T of the class Λ_A.

(2) $J = J^* = J^{-1} \in [E, E]$, $\dim E < \infty$.

(3) $\mathbb{A} - \mathbb{A}^* = 2iKJK^*$, where $K \in [E, \mathfrak{H}_-]$ $\quad (K^* \in [\mathfrak{H}_+, E])$.

In this case, the operator K is called a *channel operator* and J is called a *direction operator* [10], [20]. A system θ of the form (30) will be called a *scattering* system (*dissipative* operator colligation) if $J = I$. We will associate with the system θ an operator-valued function

$$W_\theta(z) = I - 2iK^*(\mathbb{A} - zI)^{-1}KJ \tag{29}$$

which is called *a transfer operator-valued function* of the system θ or a characteristic operator-valued function of Brodskiĭ-Livšic rigged operator colligations. It may be shown [10], that the transfer operator-function of the system θ of the form (28) has the following properties:

$$\begin{aligned} W_\theta^*(z)JW_\theta(z) - J &\geq 0 \quad (\operatorname{Im} z > 0, z \in \rho(T)) \\ W_\theta^*(z)JW_\theta(z) - J &= 0 \quad (\operatorname{Im} z = 0, z \in \rho(T)) \\ W_\theta^*(z)JW_\theta(z) - J &\leq 0 \quad (\operatorname{Im} z < 0, z \in \rho(T)) \end{aligned} \tag{30}$$

where $\rho(T)$ is the set of regular points of an operator T. Similar relations take place if we change $W_\theta(z)$ to $W_\theta^*(z)$ in (30). Thus, a transfer operator-valued function of the system θ of the form (28) is J-contractive in the lower half-plane on the set of regular points of an operator T and J-unitary on real regular points of an operator T.

Let θ be a l. s. c. d. s. of the form (28). We consider an operator-valued function

$$V_\theta(z) = K^*(\mathbb{A}_R - zI)^{-1}K. \tag{31}$$

The transfer operator-function $W_\theta(z)$ of the system θ and an operator-function $V_\theta(z)$ of the form (31) are connected by the relation

$$V_\theta(z) = i[W_\theta(z) + I]^{-1}[W_\theta(z) - I]J \tag{32}$$

As it is known [7] an operator-function $V(z) \in [E, E]$ is called an *operator-valued R-function* if it is holomorphic in the upper half-plane and $\operatorname{Im} V(z) \geq 0$ when $\operatorname{Im} z > 0$.

It is known [17], [22], [27] that an operator-valued R-function acting on a Hilbert space E ($\dim E < \infty$) has an integral representation

$$V(z) = Q + F \cdot z + \int_{-\infty}^{+\infty} \left(\frac{1}{t - z} - \frac{t}{1 + t^2} \right) dG(t), \tag{33}$$

where $Q = Q^*$, $F \geq 0$ in the Hilbert space E, $G(t)$ is non-decreasing operator-function on $(-\infty, +\infty)$ for which

$$\int_{-\infty}^{+\infty} \frac{dG(t)}{1 + t^2} \in [E, E].$$

Definition: We call an operator-valued R-function acting on a Hilbert space E ($\dim E < \infty$) realizable if in some neighborhood of the point $(-i)$, the function $V(z)$ can be represented in the form

$$V(z) = i[W_\theta(z) + I]^{-1}[W_\theta(z) - I]J \tag{34}$$

where $W_\theta(z)$ is a transfer operator-function of some l.s.c.d.s. θ with the direction operator J ($J = J^* = J^{-1} \in [E, E]$).

Definition: An operator-valued R-function $V(z) \in [E, E]$ ($\dim E < \infty$) will be said to be a member of the class $N(R)$ if in the representation (33) we have

$$\text{i)} \quad F = 0,$$

$$\text{ii)} \quad Qe = \int_{-\infty}^{+\infty} \frac{t}{1+t^2} dG(t)e$$

for all $e \in E$ such that

$$\int_{-\infty}^{+\infty} (dG(t)e, e)_E < \infty.$$

The next result is proved in [7].

Theorem 8 *Let θ be a l.s.c.d.s. of the form* (28), $\dim E < \infty$. *Then the operator-function $V_\theta(z)$ of the form* (31), (32) *belongs to the class $N(R)$.*

The following converse result was also established in [7].[2]

Theorem 9 *Suppose that the operator-valued function $V(z)$ is acting on a finite-dimensional Hilbert space E and belong to the class $N(R)$. Then $V(z)$ admits a realization by the system θ of the form* (28) *with a preassigned direction operator J for which $I + iV(-i)J$ is invertible.*

Remark: It was mentioned in [7] that when $J = I$ the invertibility condition for $I + iV(\lambda)J$ is satisfied automatically.

Now we are going to introduce three distinct subclasses of the class of realizable operator-valued functions $N(R)$.

Definition: An operator-valued R-function $V(z) \in [E, E]$ ($\dim E < \infty$) of the class $N(R)$ is said to be a member of the subclass $N_0(R)$ if in the representation (33)

$$\int_{-\infty}^{+\infty} (dG(t)e, e)_E = \infty, \qquad (e \in E, e \neq 0).$$

[2]The method of rigged Hilbert spaces for the solving of inverse problems of the theory of characteristic operator-valued functions was introduced in [26] and developed further in [1].

Consequently, the operator-function $V(z)$ of the class $N_0(R)$ has the representation

$$V(z) = Q + \int_{-\infty}^{+\infty} \left(\frac{1}{t - z} - \frac{t}{1 + t^2} \right) dG(t), \qquad (Q = Q^*). \tag{35}$$

Note, that the operator Q can be an arbitrary self-adjoint operator on the Hilbert space E.

Definition: An operator-valued R-function $V(z) \in [E, E]$ $(\dim E < \infty)$ of the class $N(R)$ is said to be a member of the subclass $N_1(R)$ if in the representation (33)

$$\int_{-\infty}^{+\infty} (dG(t)e, e)_E < \infty, \qquad (e \in E). \tag{36}$$

It is easy to see that the operator-valued function $V(z)$ of the class $N_1(R)$ has a representation

$$V(z) = \int_{-\infty}^{+\infty} \frac{1}{t - z} dG(t) \tag{37}$$

Definition: An operator-valued R-function $V(z) \in [E, E]$, $(\dim E < \infty)$ of the class $N(R)$ is said to be a member of the subclass $N_{01}(R)$ if the subspace

$$E_\infty = \left\{ e \in E : \int_{-\infty}^{+\infty} (dG(t)e, e)_E < \infty \right\}$$

possesses a property: $E_\infty \neq \emptyset$, $E_\infty \neq E$.

One may notice that $N(R)$ is a union of three distinct subclasses $N_0(R)$, $N_1(R)$ and $N_{01}(R)$. The following theorem is an analogue of the Theorem 8 for the class $N_0(R)$.

Theorem 10 *Let θ be a l. s. c. d. s. of the form* (28), $\dim E < \infty$ *where A is a linear closed Hermitian operator with dense domain and $\mathfrak{D}(T) \neq \mathfrak{D}(T^*)$. Then the operator-valued function $V_\theta(z)$ of the form* (31), (32) belongs to the class $N_0(R)$.

Proof: Relying on Theorem 8 an operator-valued function $V_\theta(z)$ of the system θ mentioned in the statement belongs to the class $N(R)$. Since $N_0(R)$ is a subclass of $N(R)$, it is sufficient to show that

$$\int_{-\infty}^{+\infty} (dG(t)e, e)_E = \infty, \qquad (e \in E, e \neq 0).$$

According to Theorem 7, if for some vector $f \in E$ we have that $Kf \notin \mathfrak{L}$ where $\mathfrak{L} = \mathfrak{H} \ominus \overline{\mathfrak{D}(A)}$, then

$$\int_{-\infty}^{+\infty} (dG(t)f, f)_E = \infty, \quad \text{where } G(t) = K^* E(t) K, \tag{38}$$

$E(t)$ is an extended generalized spectral function of the operator $\hat{A}$. Here $\hat{A}$ is the quasi-kernel of an operator

$$\mathbb{A}_R = \frac{\mathbb{A} + \mathbb{A}^*}{2}.$$

It is given that A is a closed Hermitian operator with dense domain ($\overline{\mathfrak{D}(A)} = \mathfrak{H}$), which implies that $\mathfrak{L} = \emptyset$. Thus, for any $f \in E$ such that $f \neq 0$ we have

$$Kf \notin \mathfrak{L},$$

and (38) holds. Therefore, $V_\theta(z)$ belongs to the class $N_0(R)$. □

Note that the condition (38) has also appeared in [14], [15]. Theorem 11 below is a version of the Theorem 9 for the class $N_0(R)$.

Theorem 11 *Let an operator-valued function $V(z)$ acting on a finite-dimensional Hilbert space E belong to the class $N_0(R)$. Then it admits a realization by the system θ of the form* (28) *with a preassigned directional operator J for which $I + iV(-i)J$ is invertible, densely defined closed Hermitian operator A, and $\mathfrak{D}(T) \neq \mathfrak{D}(T^*)$.*

Proof: Since $N_0(R)$ is a subclass of $N(R)$ then all conditions of Theorem 9 are satisfied and operator-valued function $V(z) \in N_0(R)$ is a realizable one. Thus, all we have to show is that $\overline{\mathfrak{D}(A)} = \mathfrak{H}$ and $\mathfrak{D}(T) \neq \mathfrak{D}(T^*)$.

We will briefly repeat the framework of the proof of Theorem 9.

Let $C_{00}(E, (-\infty, +\infty)$ be the set of continuous compactly supported vector-valued functions $f(t)$ $(-\infty < t < +\infty)$ with values in a finite dimensional Hilbert space E. We introduce an inner product

$$(f, g) = \int_{-\infty}^{+\infty} (G(dt) f(t), g(t))_E \tag{39}$$

for all $f, g \in C_{00}(E, (-\infty, +\infty))$. To construct a Hilbert space we identify with zero all the functions $f(t)$ such that $(f, f) = 0$, make a completion, and obtain a new Hilbert space $L^2_G(E)$.

Let $\mathfrak{D}_0$ be the set of the continuous vector-valued (with values in E) functions $f(t)$ such that not only

$$\int_{-\infty}^{+\infty} (dG(t) f(t), f(t))_E < \infty, \tag{40}$$

holds but also

(41) $$\int_{-\infty}^{+\infty} t^2 (dG(t) f(t), f(t))_E < \infty,$$

is true. We introduce an operator $\hat{A}$ on $\mathfrak{D}_0$ in the following way

(42) $$\hat{A} f(t) = t f(t).$$

Below we denote again by $\hat{A}$ the closure of Hermitian operator $\hat{A}$ (42). Moreover, $\hat{A}$ is self-adjoint in $L^2_G(E)$. Now let $\tilde{\mathfrak{H}}_+ = \mathfrak{D}(\hat{A})$ with an inner product

(43) $$(f, g)_{\tilde{\mathfrak{H}}_+} = (f, g) + (\hat{A} f, \hat{A} g)$$

for all $f, g \in \tilde{\mathfrak{H}}_+$. We equip the space $L^2_G(E)$ with spaces $\tilde{\mathfrak{H}}_+$ and $\tilde{\mathfrak{H}}_-$:

(44) $$\tilde{\mathfrak{H}}_+ \subset L^2_G(E) \subset \tilde{\mathfrak{H}}_-$$

and denote by $\tilde{\mathcal{R}}$ the corresponding Riesz-Berezanskii operator, $\tilde{\mathcal{R}} \in [\tilde{\mathfrak{H}}_-, \tilde{\mathfrak{H}}_+]$. After straightforward calculations on the vectors $e(t) = e$, $e \in E$ we obtain

(45) $$\tilde{\mathcal{R}} e = \frac{e}{1 + t^2}, \quad e \in E.$$

Let us now consider the set

(46) $$\mathfrak{D}(A) = \tilde{\mathfrak{H}}_+ \ominus \tilde{\mathcal{R}} E,$$

where by $\ominus$ we mean orthogonality in $\tilde{\mathfrak{H}}_+$. We define an operator A on $\mathfrak{D}(A)$ by the following expression

(47) $$A = \hat{A}\big|_{\mathfrak{D}(A)}$$

Obviously A is a closed Hermitian operator.

Since $V(z)$ is a member of the class $N_0(R)$ then (38) holds for all $e \in E$. Consequently, in the $(-)$-orthogonal decomposition

$$E = E_\infty \oplus F_\infty, \quad \text{where} \quad F_\infty = E_\infty^\perp$$

the first term $E_\infty = 0$. So that $E = F_\infty$ and (46) can be written as

$$\mathfrak{D}(A) = \tilde{\mathfrak{H}}_+ \ominus \tilde{\mathcal{R}} F_\infty.$$

Let us note again that in the formula above we are talking about $(+)$-orthogonal difference.

If we identify the space E with the space of functions $e(t) = e$, $e \in E$ we obtain

(48) $$L^2_G(E) \ominus \overline{\mathfrak{D}(A)} = E_\infty.$$

The right-hand side of (48) is zero in our case and we can conclude that

$$\overline{\mathfrak{D}(A)} = L^2_G(E) = \mathfrak{H}.$$

Let us now show that $\mathfrak{D}(T) \neq \mathfrak{D}(T^*)$. We already found out that our operator A is densely defined. This implies that its defect subspaces coincide with the semi-defect subspaces. In particular, $\mathfrak{N}_{\pm i} = \mathfrak{N}'_{\pm i}$. Using the same technique that we used in the proof of Theorem 9 (see [7]) we obtain

$$\mathfrak{N}'_{\pm i} = \mathfrak{N}_{\pm i} = \left\{ f(t) \in L^2_G(E),\ f(t) = \frac{e}{t \pm i},\quad e \in E \right\}. \tag{49}$$

For the pair of admissible operators $\Phi \in [\mathfrak{N}_i, \mathfrak{N}_{-i}]$ and $\Phi^* \in [\mathfrak{N}_{-i}, \mathfrak{N}_i]$ where

$$\Phi\left(\frac{e}{t-i}\right) = \frac{e}{t+i},\quad e \in E \tag{50}$$

we have that

$$\mathfrak{D}(T) = \mathfrak{D}(A) \dot{+} (I - \Phi)\mathfrak{N}_i,$$
$$\mathfrak{D}(T^*) = \mathfrak{D}(A) \dot{+} (I - \Phi^*)\mathfrak{N}_{-i}.$$

Direct calculations show that

$$(I - \Phi)\left(\frac{e}{t-i}\right) = \frac{e}{t-i} - \frac{e}{t+i} = \frac{2ie}{t^2+1},\quad e \in E,$$

and

$$(I - \Phi^*)\left(\frac{e}{t+i}\right) = \frac{e}{t+i} - \frac{e}{t-i} = -\frac{2ie}{t^2+1},\quad e \in E,$$

Therefore,

$$(I - \Phi)\mathfrak{N}_i = \left\{ \frac{2ie}{t^2+1},\quad e \in E \right\}, \tag{51}$$

and

$$(I - \Phi^*)\mathfrak{N}_{-i} = \left\{ -\frac{2ie}{t^2+1},\quad e \in E \right\}. \tag{52}$$

Applying Theorem 1 we conclude that $\mathfrak{D}(T) = \mathfrak{D}(T^*)$ if and only if $\mathfrak{N}_{\pm i} = 0$, which is not true. Therefore, the condition $\mathfrak{D}(T) \neq \mathfrak{D}(T^*)$ is satisfied and the proof of the theorem is complete. □

Similar results for the class $N_1(R)$ can be obtained in the following two theorems.

Theorem 12 *Let θ be a l. s. c. d. s. of the form* (28), dim $E < \infty$ *where A is a linear closed Hermitian O*-operator *and $\mathfrak{D}(T) = \mathfrak{D}(T^*)$. Then operator-valued function $V_\theta(\lambda)$ of the form* (31), (32) *belongs to the class $N_1(R)$.*

Proof: As in the Theorem 10 we already know that the operator-valued function $V_\theta(\lambda)$ belongs to the class $N(R)$. Therefore it is enough to show that

$$\int_{-\infty}^{+\infty} (dG(t)e, e)_E < \infty,$$

for all $e \in E$ and (37) holds.

Since it is given that A is closed Hermitian O-operator we can use Theorem 4 saying that for the system θ

$$\mathfrak{D}(T) = \mathfrak{D}(T^*) = \mathfrak{H}_+ = \mathfrak{D}(A^*).$$

This fact implies that the $(*)$-extension $\mathbb{A}$ coincides with operator T. Consequently, $\mathbb{A}^* = T^*$ and our system θ has a form

(53)
$$\theta = \begin{pmatrix} T & K & J \\ \mathfrak{H}_+ \subset \mathfrak{H} \subset \mathfrak{H}_- & & E \end{pmatrix},$$

where

$$\operatorname{Im} T = \frac{T - T^*}{2i} = KJK^*.$$

Taking into account that $\dim E < \infty$ and $K : E \to \mathfrak{H}_-$ we conclude that $\dim \mathfrak{R}(\operatorname{Im} T) < \infty$.

Let

$$T = T_R + i \operatorname{Im} T,$$
$$T^* = T_R - i \operatorname{Im} T,$$

where

$$T_R = \frac{T + T^*}{2}.$$

In our case the operator K is acting from the space E into the space $\mathfrak{H}$. Therefore $Ke = g$ belongs to $\mathfrak{H}$ for all $e \in E$. For the operator-valued function $V_\theta(\lambda)$ we can derive an integral representation for all $f \in E$

(54)
$$\big(V_\theta(\lambda) f, f\big)_E = \big(K^*(T_R - \lambda I)^{-1} K f, f\big)_E = \left(K^* \int_{-\infty}^{+\infty} \frac{dE(t)}{t - \lambda} K f, f\right)_E$$
$$= \int_{-\infty}^{+\infty} \frac{d\big(K^* E(t) K f, f\big)_E}{t - \lambda},$$

where $E(t)$ is the complete set of spectral orthoprojections of the operator T_R. Denote

$$G(t) = K^* E(t) K.$$

Then

$$\int_{-\infty}^{+\infty} d(G(t)e,e) = \int_{-\infty}^{+\infty} d(K^*E(t)Ke,e) = \int_{-\infty}^{+\infty} d(E(t)Ke,Ke)$$
$$= \int_{-\infty}^{+\infty} d(E(t)g,g) = (g,g)\int_{-\infty}^{+\infty} dE(t) = (g,g)$$
$$= (Ke,Ke) = (K^*Ke,e) = (\operatorname{Im} Te,e) < \infty,$$

for all $e \in E$. Using standard techniques we obtain the representation (37) from the representation (54). This completes the proof of the theorem. □

Theorem 13 *Suppose that an operator-valued function $V(z)$ is acting on a finite-dimensional Hilbert space E and belongs to the class $N_1(R)$. Then it admits a realization by the system θ of the form* (28) *with a preassigned directional operator J for which $I + iV(-i)J$ is invertible, a linear closed regular Hermitian O-operator A with a non-dense domain, and $\mathfrak{D}(T) = \mathfrak{D}(T^*)$.*

Proof: Similarly to Theorem 11 we can say that since $N_1(R)$ is a subclass of $N(R)$ then it is sufficient to show that operator A is a closed Hermitian O-operator with a non-dense domain and $\mathfrak{D}(T) = \mathfrak{D}(T^*)$.

Once again we introduce an operator $\hat{A}$ by the formula (42), an operator A by the formula (47) and note that

$$\mathfrak{D}(A) = \tilde{\mathfrak{H}}_+ \ominus \tilde{\mathcal{R}}E.$$

Let us recall, that since $V(z)$ belongs to the class $N_1(R)$ then

$$\int_{-\infty}^{+\infty} (dG(t)e,e)_E < \infty, \quad \forall e \in E.$$

That means that in the $(-)$-orthogonal decomposition

$$E = E_\infty \oplus F_\infty$$

the second term $F_\infty = 0$ and therefore $E = E_\infty$. Then

$$\mathfrak{D}(A) = \tilde{\mathfrak{H}}_+ \ominus \tilde{\mathcal{R}}E_\infty.$$

Combining this, formula (48), and the fact that $E_\infty \neq 0$ we obtain that $\overline{\mathfrak{D}(A)} \neq \mathfrak{H} = L^2_G(E)$. Relying on the proof of Theorem 9 (see [7]) we let

$$A_1 = \hat{A}\big|_{\mathfrak{D}(A_1)}, \qquad \mathfrak{D}(A_1) = \tilde{\mathfrak{H}}_+ \ominus \tilde{\mathcal{R}}E_\infty.$$

The following obvious inclusions hold: $A \subset A_1 \subset \hat{A}$. Moreover, a set

$$\mathfrak{D}(A_1) = \tilde{\mathfrak{H}}_+ \ominus \tilde{\mathcal{R}}E_\infty$$

in our case coincides with $\mathfrak{D}(A)$ and operator A_1 (defined on $\mathfrak{D}(A_1)$) with A. Now it is not difficult to see that

$$\mathfrak{D}(A^*) = \mathfrak{H}_+ = \tilde{\mathfrak{H}}_+,$$

the rigged Hilbert space $\tilde{\mathfrak{H}}_+ \subset \mathfrak{H} \subset \tilde{\mathfrak{H}}_-$ coincides with $\mathfrak{H}_+ \subset \mathfrak{H} \subset \mathfrak{H}_-$ and $\mathcal{R} = \tilde{\mathcal{R}}$. Indeed, $\tilde{\mathfrak{H}}_+ = \mathfrak{D}(\hat{A})$ by the definition, in [7] we have shown that $\mathfrak{D}(A_1^*) = \mathfrak{D}(\hat{A})$, and $D(A_1) = D(A)$ above. All together it yields $\mathfrak{H}_+ = \tilde{\mathfrak{H}}_+$.

Let $\mathfrak{N}'_{\pm i}$ be the semidefect subspaces of operator A and $\mathfrak{N}^0_{\pm i}$ be the defect subspaces of operator A_1, described in the second part of the proof of Theorem 9 (see [7]). It was shown that

$$\mathfrak{N}^0_{\pm i} = \left\{ f(t) \in L^2_G(E),\ f(t) = \frac{e}{t \pm i},\quad e \in E_\infty \right\}, \tag{55}$$

and

$$\mathfrak{N}'_{\pm i} = \mathfrak{N}_i \ominus \mathfrak{N}^0_{\pm i}, \tag{56}$$

where $\mathfrak{N}_i$ are defect spaces of the operator A. In our case $A = A_1$ therefore

$$\mathfrak{N}'_{\pm i} = 0.$$

This implies that the semidefect numbers of operator A are equal to zero. Hence, A is an O-operator.

Note that A is also a regular Hermitian operator. Thus, Theorem 4 is applicable and yields

$$\mathfrak{D}(T) = \mathfrak{D}(T^*).$$

This completes the proof of the theorem. □

The following two theorems will complete our framework by establishing direct and inverse realization results for the remaining subclass of realizable operator-valued R-functions $N_{01}(R)$.

Theorem 14 *Let θ be a l. s. c. d. s. of the form* (28), $\dim E < \infty$ *where A is a linear closed Hermitian operator with non-dense domain and $\mathfrak{D}(T) \neq \mathfrak{D}(T^*)$. Then the operator-valued function $V_\theta(z)$ of the form* (31), (32) *belongs to the class $N_{01}(R)$.*

Proof: We know that $V_\theta(z)$ belongs to the class $N(R)$. To prove the statement of the theorem we only have to show that in the $(-)$-orthogonal decomposition $E = E_\infty \oplus F_\infty$ both components are non-zero. In other words we have to show the existence of such vectors $e \in E$ that

$$\int_{-\infty}^{+\infty} d(G(t)e, e) = \infty, \tag{57}$$

and vectors $f \in E$, $f \neq 0$ that

$$\int_{-\infty}^{+\infty} d(G(t)f, f) < \infty. \tag{58}$$

Let $\mathfrak{H}_0 = \overline{\mathfrak{D}(A)}$ and $\mathfrak{L} = \mathfrak{H} \ominus \mathfrak{H}_0$. Since $\overline{\mathfrak{D}(A)} = \mathfrak{H}_0 \neq \mathfrak{H}$, $\mathfrak{L}$ is non-empty. $K^{-1}\mathfrak{L}$ is obviously a subset of E. Moreover, according to Theorem 7 for all $f \in K^{-1}\mathfrak{L}$ (58) holds. Thus, $K^{-1}\mathfrak{L}$ is a non-zero subset of E_∞.

Now we have to show that the vectors satisfying (57) make a non-zero subset of E as well. Indeed, the condition

$$\mathfrak{D}(T) \neq \mathfrak{D}(T^*)$$

implies that a certain part of $\mathfrak{R}(K) \subseteq \overline{\mathfrak{R}(\mathbb{A} - \mathbb{A}^*) + \mathfrak{L}} \subseteq L \dot{+} \mathfrak{L}$ where L was defined in Theorem 7 essentially belongs to L. Otherwise we could have re-traced our steps and show that $\mathfrak{D}(T) = \mathfrak{D}(T^*)$. Therefore, there exist $g \in \mathfrak{H}_-$, $g \notin \mathfrak{L}$, $f \in E$ such that $Kf = g \notin \mathfrak{L}$. Then according to Theorem 7 for this $f \in E$ (57) holds. The proof of the theorem is complete. □

Theorem 15 *Suppose that an operator-valued function $V(z)$ is acting on a finite-dimensional Hilbert space E and belongs to the class $N_{01}(R)$. Then it admits a realization by the system θ of the form* (28) *with a preassigned directional operator J for which $I + iV(-i)J$ is invertible, a linear closed regular Hermitian operator A with a non-dense domain, and $\mathfrak{D}(T) \neq \mathfrak{D}(T^*)$.*

Proof: Once again all we have to show is that $\overline{\mathfrak{D}(A)} \neq \mathfrak{H}$. We have already mentioned (48) that $L^2_G(E) \ominus \overline{\mathfrak{D}(A)} = E_\infty$. This implies that $\mathfrak{D}(A)$ is dense in $\mathfrak{H}$ if and only if $E_\infty = 0$. Since the class $N_{01}(R)$ assumes the existence of non-zero vectors $f \in E$ such that (58) is true we can conclude that $E_\infty \neq 0$ and therefore $\overline{\mathfrak{D}(A)} \neq \mathfrak{H}$.

In the proofs of Theorems 11 and 13 we have shown that $\mathfrak{D}(T) = \mathfrak{D}(T^*)$ in case when $F_\infty = 0$. If $F_\infty \neq 0$ then $\mathfrak{D}(T) \neq \mathfrak{D}(T^*)$. The definition of the class $N_{01}(R)$ implies that $F_\infty \neq 0$. Thus we have $\mathfrak{D}(T) \neq \mathfrak{D}(T^*)$. The proof is complete. □

Let us consider examples of the realization in the classes $N(R)$.

Example 1 This example is to illustrate the realization in $N_0(R)$ class. Let

$$Tx = \frac{1}{i}\frac{dx}{dt},$$

with

$$\mathfrak{D}(T) = \left\{x(t) \,\middle|\, x(t) - \text{ abs. continuous}, x'(t) \in L^2_{[0,l]}, x(0) = 0\right\},$$

be differential operator in $\mathfrak{H} = L^2_{[0,l]}$ $(l > 0)$. Obviously,

$$T^* x = \frac{1}{i}\frac{dx}{dt},$$

with

$$\mathfrak{D}(T^*) = \left\{ x(t) \,\middle|\, x(t) - \text{ abs. continuous}, x'(t) \in L^2_{[0,l]}, x(l) = 0 \right\},$$

is its adjoint. Consider a Hermitian operator A [27]

$$\begin{aligned} Ax &= \frac{1}{i}\frac{dx}{dt}, \\ \mathfrak{D}(A) &= \left\{ x(t) \,\middle|\, x(t) - \text{ abs. continuous}, x'(t) \in L^2_{[0,l]}, x(0) = x(l) = 0 \right\}, \end{aligned}$$

and its adjoint A^*

$$\begin{aligned} A^* x &= \frac{1}{i}\frac{dx}{dt}, \\ \mathfrak{D}(A^*) &= \left\{ x(t) \,\middle|\, x(t) - \text{ abs. continuous}, x'(t) \in L^2_{[0,l]} \right\}. \end{aligned}$$

Then $\mathfrak{H}_+ = \mathfrak{D}(A^*) = W_2^1$ is a Sobolev space with scalar product

$$(x, y)_+ = \int_0^l x(t)\overline{y(t)}\,dt + \int_0^l x'(t)\overline{y'(t)}\,dt.$$

Construct rigged Hilbert space [9]

$$W_2^1 \subset L^2_{[0,l]} \subset (W_2^1)_-,$$

and consider operators

$$\begin{aligned} \mathbb{A}x &= \frac{1}{i}\frac{dx}{dt} + ix(0)\left[\delta(x-l) - \delta(x)\right], \\ \mathbb{A}^* x &= \frac{1}{i}\frac{dx}{dt} + ix(l)\left[\delta(x-l) - \delta(x)\right], \end{aligned}$$

where $x(t) \in W_2^1$, $\delta(x)$, $\delta(x-l)$ are delta-functions in $(W_2^1)_-$. It is easy to see that

$$\mathbb{A} \supset T \supset A, \qquad \mathbb{A}^* \supset T^* \supset A,$$

and

$$\theta = \begin{pmatrix} \frac{1}{i}\frac{dx}{dt} + ix(0)[\delta(x-l) - \delta(x)] & K & -1 \\ W_1^2 \subset L^2_{[0,l]} \subset (W_2^1)_- & & \mathbb{C}^1 \end{pmatrix} \quad (J = -1)$$

is the Brodskiĭ-Livšic rigged operator colligation where

$$Kc = c \cdot \frac{1}{\sqrt{2}}[\delta(x-l) - \delta(x)], \quad (c \in \mathbb{C}^1)$$

$$K^*x = \left(x, \frac{1}{\sqrt{2}}[\delta(x-l) - \delta(x)]\right) = \frac{1}{\sqrt{2}}[x(l) - x(0)],$$

and $x(t) \in W_2^1$. Also

$$\frac{\mathbb{A} - \mathbb{A}^*}{2i} = -\left(\cdot, \frac{1}{\sqrt{2}}[\delta(x-l) - \delta(x)]\right) \frac{1}{\sqrt{2}}[\delta(x-l) - \delta(x)].$$

The characteristic function of this colligation can be found as follows

$$W_\theta(\lambda) = I - 2iK^*(\mathbb{A} - \lambda I)^{-1} K J = e^{i\lambda l}.$$

Consider the following R-function (hyperbolic tangent)

$$V(\lambda) = -i \tanh\left(\frac{i}{2}\lambda l\right).$$

Obviously this function can be realized as follows

$$\begin{aligned} V(\lambda) &= -i \tanh\left(\frac{i}{2}\lambda l\right) = -i \frac{e^{\frac{i}{2}\lambda l} - e^{-\frac{i}{2}\lambda l}}{e^{\frac{i}{2}\lambda l} + e^{-\frac{i}{2}\lambda l}} = -i \frac{e^{i\lambda l} - 1}{e^{i\lambda l} + 1} \\ &= i\,[W_\theta(\lambda) + I]^{-1}\,[W_\theta(\lambda) - I]\,J. \qquad (J = -1) \end{aligned}$$

The following simple example showing the realization for $N_1(R)$ class.

Example 2 Consider bounded linear operator in $\mathbb{C}^2$:

$$T = \begin{pmatrix} i & i \\ -i & 1 \end{pmatrix}$$

Let x be an element of $\mathbb{C}^2$ such that

$$x = \begin{pmatrix} x_1 \\ x_2 \end{pmatrix},$$

and φ be a row vector $\varphi = \begin{pmatrix} 1 & 0 \end{pmatrix}$ and let $J = 1$. Obviously,

$$T^* = \begin{pmatrix} -i & i \\ -i & 1 \end{pmatrix}.$$

It is clear that $\mathfrak{D}(T) = \mathfrak{D}(T^*)$. Now we can find

$$\frac{T - T^*}{2i} = \begin{pmatrix} 1 & 0 \\ 0 & 0 \end{pmatrix}$$

and show that φ above is the only channel vector such that

$$\frac{T-T^*}{2i}x = (x,\varphi)J\varphi.$$

Thus, operator T can be included in the system

$$\theta = \begin{pmatrix} T & K & J \\ \mathbb{C}^2 & & \mathbb{C}^1 \end{pmatrix},$$

with

$$Kc = \begin{pmatrix} c & 0 \end{pmatrix}, \quad c \in \mathbb{C}^1$$

$$K^* x = x_1, \quad x = \begin{pmatrix} x_1 \\ x_2 \end{pmatrix} \in \mathbb{C}^2,$$

Then $W_\theta(\lambda)$ is represented by the formula

$$W_\theta(\lambda) = \frac{\lambda^2 + (1-i)\lambda - 1 - 1}{\lambda^2 - (1+i)\lambda - 1 + i}.$$

Its linear-fractional transformation is a R-function and

$$V_\theta(\lambda) = \frac{1-\lambda}{\lambda^2 - \lambda - 1}$$

can therefore be realized as follows

$$V_\theta(\lambda) = i\,[W_\theta(\lambda) + I]^{-1}\,[W_\theta(\lambda) - I]\,J.$$

Example 3 In order to present the realization in $N_{01}(R)$ class we will use Examples 1 and 2.

Consider the system

$$\theta = \begin{pmatrix} & \mathbb{A} & K & J \\ W_1^2 \otimes \mathbb{C}^2 \subset L^2_{[0,l]} \otimes \mathbb{C}^2 \subset (W_2^1)_- \otimes \mathbb{C}^2 & & & \mathbb{C}^2 \end{pmatrix},$$

where $\mathbb{A}$ is a diagonal block-matrix

$$\mathbb{A} = \begin{pmatrix} \mathbb{A}_1 & 0 \\ 0 & T \end{pmatrix},$$

with

$$\mathbb{A}_1 = \frac{1}{i}\frac{dx}{dt} + ix(0)\,[\delta(x-l) - \delta(x)]$$

from Example 1, and

$$T = \begin{pmatrix} i & i \\ -i & 1 \end{pmatrix},$$

from Example 2. Operator K here is defined as a diagonal operator block-matrix

$$K = \begin{pmatrix} K_1 & 0 \\ 0 & K_2 \end{pmatrix},$$

with operators K_1 and K_2 from Examples 1 and 2, respectively,

$$J = \begin{pmatrix} -1 & 0 \\ 0 & 1 \end{pmatrix}.$$

It can be easily shown that

$$W_\theta(\lambda) = \begin{pmatrix} e^{i\lambda l} & 0 \\ 0 & \dfrac{\lambda^2 + (1-i)\lambda - 1 - 1}{\lambda^2 - (1+i)\lambda - 1 + i} \end{pmatrix},$$

and

$$V_\theta(\lambda) = \begin{pmatrix} -i \tanh\left(\frac{i}{2}\lambda l\right) & 0 \\ 0 & \dfrac{1-\lambda}{\lambda^2 - \lambda - 1} \end{pmatrix},$$

is an operator-valued function of class $N_{01}(R)$.

Acknowledgement

We would like to thank Frietz Gesztesy, Henk de Snoo, and referee for valuable discussions.

References

[1] Yu.M. Arlinskiĭ, *On inverse problem of the theory of characteristic functions of unbounded operator colligations*, Dopovidi Akad. Nauk Ukrain. RSR **2** (1976), no. Ser. A, 105–109.

[2] Yu.M. Arlinskiĭ and E.R. Tsekanovskiĭ, *Regular (∗)-extension of unbounded operators, characteristic operator-functions and realization problems of transfer mappings of linear systems*, Preprint, VINITI,-2867.-79 Dep. - 72 p.

[3] ———, *The method of equipped spaces in the theory of extensions of Hermitian operators with a nondense domain of definition*, Sibirsk. Mat. Zh. **15** (1974), 243–261.

[4] D.R. Arov and M.A. Nudelman, *Passive linear stationary dynamical scattering systems with continuous time*, Integr. Equat. Oper. Th. **24** (1996), 1–45.

[5] S.V. Belyi and E.R. Tsekanovskiĭ, *Classes of operator-valued R-functions and their realization by conservative systems*, Dokl. Akad. Nauk SSR **321** (1991), no. 3, 441–445.

[6] ———, *Realization and factorization problems for J-contractive operator-valued functions in half-plane and systems with unbounded operators*, Systems and Networks: Mathematical Theory and Applications **2** (1994), Akademie Verlag, 621–624.

[7] S.V. Belyi and E.R. Tsekanovskiĭ, *Realization theorems for operator-valued R-functions*, Operator theory: Advances and Applications **98** (1997), Birkhäuser Verlag Basel, 55–91.

[8] H. Bart, I. Gohberg and M.A. Kaashoek, *Minimal factorizations of matrix and operator-functions. Operator theory: Advances and Applications*, Birkhäuser Verlag Basel, 1979.

[9] Ju. M. Berezanskii *Expansion in eigenfunctions of self-adjoint operators*, vol. 17, Transl. Math. Monographs, AMS, Providence RI, 1968.

[10] M.S. Brodskiĭ, *Triangular and Jordan representations of linear operators*, Moscow, Nauka, 1969.

[11] M.S. Brodskiĭ and M.S. Livšic, *Spectral analysis of non-selfadjoint operators and intermediate systems*, Uspekhi Matem. Nauk **XIII** (1958), no. 1 (79), 3–84.

[12] J.A. Ball and Nir Cohen, *De Branges-Rovnyak operator models and systems theory: a survey*, In book: Operator Theory: Advances and Applications, Birkhäuser Verlag Basel **50** (1991), 93–136.

[13] F. Gesztesy and E.R. Tsekanovskiĭ, *On matrix-valued Herglotz functions*, preprint.

[14] S. Hassi, M. Kaltenbäck and H. de Snoo, *Triplets of Hilbert spaces and Friedrichs extensions associated with the subclass N_1 of Nevanlina functions*, Jour. of Operator Theory **37** (1997), 155–181.

[15] S. Hassi, M. Kaltenbäck and H. de Snoo, *On some subclasses of Nevalinna functions*, Jour. Annal. Appls. **15** (1997), 45–55.

[16] J.W. Helton, *Systems with infinite-dimensional state space: the Hilbert space approach*, Proc. IEEE **64** (1976), no. 1, 145–160.

[17] I.S. Kač and M.G. Krein, *The R-functions – analytic functions mapping the upper half-plane into itself*, Supplement I to the Russian edition of F.V. Atkinson, Discrete and continuous boundary problems, (1968), Mir, Moscow (Russian) (English translation: Amer. Math. Soc. Transl. (2) 103 (1974), 1–18).

[18] M.A. Krasnoselskii, *On self-adjoint extensions of Hermitian operators*, Ukrain. Mat. Zh. **1** (1949), 21–38.

[19] M.S. Livšic, *On spectral decomposition of linear non-selfadjoint operators*, Math. Sbornik **34** (1954), no. 76, 145–198.

[20] ———, *Operators, oscillations, waves*, Moscow Nauka, 1966.

[21] Ju.L. Šmuljan, *Extended resolvents and extended spectral functions of Hermitian operator*, Math. USSR Sbornick **13** (1971), no. 3, 435–450.

[22] ———, *On operator-valued R-functions*, Sibirsk. Mat. Zh. **12** (1971), no. 2, 442–452.

[23] B. Sz.-Nagy and C. Foias, *Harmonic analysis of operators on Hilbert space*, North-Holland Pub. Co., Amsterdam, 1970.

[24] E.R. Tsekanovskiĭ, *Generalized self-adjoint extensions of symmetric operators*, Dokl. Akad. Nauk SSR **178** (1968), 1267–1270.

[25] ———, *On the description and uniqueness of the generalized extensions of quasi-Hermitian operators*, Functional Anal. Appl. **3** (1969), 79–80.

[26] ———, *Analytical properties of the resolvent matrix-valued functions and inverse problem*, Abstracts of the All Union Conference on Complex Analysis, Kharkov, FTINT **3** (1971), 233–235.

[27] E.R. Tsekanovskiĭ and Ju.L. Šmuljan, *The Theory of bi-extensions of operators on rigged Hilbert spaces. Unbounded operator colligations and characteristic functions.*, Russian Math. Surveys **32** (1977), no. 5, 69–124.

[28] E.R. Tsekanovskiĭ and Ju.L. Šmuljan, *Method of generalized functions in the theory of extensions of unbounded linear operators*, Donetsk State University, Donetsk, 1973.

[29] V.E. Tsekanovskiĭ and E.R. Tsekanovskiĭ, *Stieltjes operator-functions with the gaps and their realization by conservative systems*, Proceedings of the International symposium MTNS-91 **1** (1992), 37–43.

[30] G. Weiss, *The representation of regular linear systems on Hilbert spaces*, International series of Numerical Mathematics **91** (1989), 401–415.

Department of Mathematics
Troy State University
Troy, AL 36082
sbelyi@trojan.troyst.edu

Department of Mathematics
P.O. Box 2044
Niagara University, NY 14109
tsekanov@niagara.edu

1991 Mathematics Subject Classification
Primary 47A10, 47B44; Secondary 46E20, 46F05

Operator Theory:
Advances and Applications, Vol. 115

Submodules of $L^2(\mathbb{T}^N)$

Douglas N. Clark and Sarah H. Ferguson

We compute $\mathrm{Ext}^1_{\mathrm{A}(\mathbb{D}^N)}$ groups with certain submodules of $L^2(\mathbb{T}^N)$ in either variable. The obstruction to the vanishing of these groups with the Hardy space in the first variable is formulated in terms of a lifting problem for a class of Hankel operators.

1 Introduction

The work of Douglas and Paulsen [DoPa], suggests that homological methods can become a powerful tool in multi-variable operator theory. One obstacle to the application of these tools is the lack of knowledge of projectives in the category of Hilbert modules over the polydisk algebra, $\mathrm{A}(\mathbb{D}^N)$. In [CC1], J. F. Carlson and the first author began the project of computing first order Ext groups of Hilbert modules over the disk algebra, $\mathrm{A}(\mathbb{D})$. The authors used the Yoneda realization of $\mathrm{Ext}^1_{\mathrm{A}(\mathbb{D})}(H, K)$ as short exact sequences of Hilbert $\mathrm{A}(\mathbb{D})$-modules beginning in H and ending in K modulo the usual equivalence. The group $\mathrm{Ext}^1_{\mathrm{A}(\mathbb{D})}(H, K)$ has a concrete realization as a space of operators from H into K and thus the problem of computing it is not only tractable but of interest independent of the cohomology.

In this note we demonstrate some of the techniques and results of our recent papers by computing first order Ext groups of Hilbert $\mathrm{A}(\mathbb{D}^N)$-modules. The modules we are interested in are certain distinguished submodules of $L^2(\mathbb{T}^N)$ which we regard as a Hilbert $\mathrm{A}(\mathbb{D}^N)$-module via pointwise multiplication by functions in $\mathrm{A}(\mathbb{D}^N)$. For notational convenience we will focus on the two-variable case but make note of the generalizations to the analogous submodules of $L^2(\mathbb{T}^N)$. The paper is divided into 4 sections with the homological preliminaries contained in section 2. Section 3 contains the computations of the first order groups considered here with the exception of the group $\mathrm{Ext}^1_{\mathrm{A}(\mathbb{D}^2)}(H^2, \mathcal{M})$, $\mathcal{M} \subseteq L^2(\mathbb{T}^2)$ a submodule, which is considered in section 4. The obstruction to the vanishing of this latter group is formulated in terms of a lifting problem for Hankel operators which map $H^2(\mathbb{T}^2)$ into the quotient $\mathcal{M}^\perp$. Section 4 also contains a discussion on what is known about the so-called *big* and *little* Hankel operators on $H^2(\mathbb{T}^2)$.

In more detail, let I^2 denote the right half plane of $L^2(\mathbb{T}^2)$ consisting of those functions $f \in L^2(\mathbb{T}^2)$ whose Fourier coefficients are supported in $\mathbb{Z}^+ \times \mathbb{Z}$ where

February 2, 1998 final draft.
The second author was partially supported by a grant from the National Science Foundation.

$\mathbb{Z}^+$ denotes the nonnegative integers. The upper half plane, denoted by J^2, consists of those functions $f \in L^2(\mathbb{T}^2)$ whose Fourier coefficients are supported in $\mathbb{Z}\times\mathbb{Z}^+$. Set $K^2 = I^2 + J^2$. In section 3 we show that the groups $\text{Ext}^1_{\text{A}(\mathbb{D}^2)}(\cdot,\cdot)$ with I^2 or J^2 in either variable vanish when restricted to submodules of $L^2(\mathbb{T}^2)$. We also characterize the groups $\text{Ext}^1_{\text{A}(\mathbb{D}^2)}(K^2,\mathcal{M})$ as a certain quotient of submodules of $L^\infty(\mathbb{T}^2)$ functions and use this together with results in [F3] on big Hankel operators to show directly that $\text{Ext}^1(K^2,K^2) \neq (0)$. Our computations yield the table of $\text{Ext}^1_{\text{A}(\mathbb{D}^2)}$ groups below. There is a question mark for the group $\text{Ext}^1_{\text{A}(\mathbb{D}^2)}(H^2,K^2)$ since we do not have conclusive evidence that this group vanishes. This is discussed in more detail in the last section.

$\text{Ext}^1_{\text{A}(\mathbb{D}^2)}(\cdot,\cdot)$	H^2	I^2	J^2	K^2
H^2	$\neq 0$	0	0	?
I^2	0	0	0	0
J^2	0	0	0	0
K^2	H^∞	0	0	$\neq 0$

2 Preliminaries

2.1 Derivations or 1-cocycles

If $\mathcal{A}$ is a function algebra then a Hilbert $\mathcal{A}$-module is a Hilbert space H together with a bounded homomorphism (i.e., a representation) $\pi_H : \mathcal{A} \to \mathcal{B}(H)$. Multiplication by $a \in \mathcal{A}$ is defined by $a \cdot h = \pi(a)h$, $h \in H$. Let (K, π_K) be another Hilbert $\mathcal{A}$-module. By an $\mathcal{A}$-module map from H into K we mean a bounded operator $T : H \to K$ satisfying $T\pi_H(a) = \pi_K(a)T$ for all $a \in \mathcal{A}$. We let $\text{Hom}_{\mathcal{A}}(H,K)$ denote the space of $\mathcal{A}$-module maps from H into K.

An $H - K$ derivation is a bounded linear map $\delta : \mathcal{A} \to \mathcal{B}(H,K)$ satisfying $\delta(ab) = \pi_K(a)\delta(b) + \delta(a)\pi_H(b)$ for all $a, b \in \mathcal{A}$. The derivation δ is inner if there is an operator $X \in \mathcal{B}(H,K)$ such that $\delta(a) = \pi_K(a)X - X\pi_H(a)$, $a \in \mathcal{A}$. Let $\text{Der}_{\mathcal{A}}(H,K)$ and $\text{Inn}_{\mathcal{A}}(H,K)$ denote the space of $H-K$ derivations and inner derivations, respectively. Recall from [CC1] that the group $\text{Ext}^1_{\mathcal{A}}(H,K)$ is defined using short exact sequences of Hilbert $\mathcal{A}$-modules and this definition yields groups which are isomorphic to a quotient of certain bilinear maps. These bilinear maps or co-cycles are readily identified with derivations which yields an isomorphism between $\text{Ext}^1_{\mathcal{A}}(H,K)$ and the first-order Hochschild group $H^1(\mathcal{A}, \mathcal{B}(H,K)) = \text{Der}_{\mathcal{A}}(H,K)/\text{Inn}_{\mathcal{A}}(H,K)$, see [F1].

The following proposition gives a necessary and sufficient condition for a linear map $\delta : \mathcal{A} \to \mathcal{B}(H,K)$ to determine an $H-K$ derivation. This is useful in computing Ext^1 groups over finitely generated algebras and in particular, yields a concrete characterization of Ext^1 over the polydisk algebra in terms of a joint similarity problem for a certain class of N-tuples of operators.

Proposition 2.1 *Let (H, π_H) and (K, π_K) be Hilbert modules over a unital Banach algebra $\mathcal{A}$. A linear map $\delta : \mathcal{A} \to \mathcal{B}(H, K)$ is an $H - K$ derivation if and only if $\pi : \mathcal{A} \to \mathcal{B}(H \oplus K)$ defined by*

$$\pi(a) = \begin{pmatrix} \pi_K(a) & \delta(a) \\ 0 & \pi_H(a) \end{pmatrix}$$

is a bounded representation of $\mathcal{A}$ on the Hilbert space direct sum $H \oplus K$. In this case, the derivation δ is inner if and only if there is an operator $X \in \mathcal{B}(H, K)$ such that

$$\begin{pmatrix} I_K & X \\ 0 & I_H \end{pmatrix} \begin{pmatrix} \pi_K(a) & \delta(a) \\ 0 & \pi_H(a) \end{pmatrix} \begin{pmatrix} I_K & -X \\ 0 & I_H \end{pmatrix} = \begin{pmatrix} \pi_K(a) & 0 \\ 0 & \pi_H(a) \end{pmatrix}$$

for all $a \in \mathcal{A}$.

Proof: Note that the map π defined above is bounded if and only if δ is bounded. A direct computation shows that π is a homomorphism if and only if δ is a derivation. If $X \in \mathcal{B}(H, K)$ then

$$\begin{pmatrix} I_K & X \\ 0 & I_H \end{pmatrix} \begin{pmatrix} \pi_K(a) & \delta(a) \\ 0 & \pi_H(a) \end{pmatrix} \begin{pmatrix} I_K & -X \\ 0 & I_H \end{pmatrix}$$
$$= \begin{pmatrix} \pi_K(a) & \delta(a) - (\pi_K(a)X - X\pi_H(a)) \\ 0 & \pi_H(a) \end{pmatrix}$$

for all $a \in \mathcal{A}$. The statement on inner derivations now follows. □

2.2 Ext^1 *Over the Polydisk Algebra*

Let $\mathrm{A}(\mathbb{D}^N)$ denote the polydisk algebra. That is, $\mathrm{A}(\mathbb{D}^N)$ is the uniform closure of the analytic polynomials in N variables over the polydisk, $\mathbb{D}^N$. Since $\mathrm{A}(\mathbb{D}^N)$ is generated by the coordinate functions z_i $i = 1, \dots, N$, every Hilbert $\mathrm{A}(\mathbb{D}^N)$-module H is uniquely determined by N commuting operators $T_1, \dots, T_N$ defined by $T_i h = z_i \cdot h$, $h \in H$. In other words, for each polynomial $p \in \mathbb{C}[z_1, \dots, z_n]$, $p \cdot h = p(T_1, \dots, T_n)h$. Note that in our definition of a Hilbert module we require that $p \mapsto p(T_1, \dots, T_n)$ extends to a bounded representation of $\mathrm{A}(\mathbb{D}^N)$ on H. Thus the N-tuple $(T_1, \dots, T_N)$ must be *jointly polynomially bounded*. In other words, there is a constant $C > 0$ such that

(1) $$\|p(T_1, \dots, T_N)\| \le C \sup_{z \in \mathbb{D}^N} |p(z_1, \dots, z_N)| \quad p \in \mathbb{C}[z_1, \dots, z_N].$$

If the inequality above is satisfied with $C = 1$ then the $\mathrm{A}(\mathbb{D}^N)$-module H is said to be contractive. H is said to be *cramped* if H is similar to a contractive

$A(\mathbb{D}^N)$-module. In other words, H is cramped if there exists an invertible operator R such that the N-tuple $(R^{-1}T_1R, \ldots, R^{-1}T_NR)$ satisfies the inequality above with $C = 1$.

Let K be another Hilbert $A(\mathbb{D}^N)$-module with $V_ik = z_i \cdot k$, $i = 1, \ldots, N$. If $\delta : \mathcal{A} \to \mathcal{B}(H, K)$ is a linear map and $X_i = \delta(z_i)$, $i = 1, \ldots, N$, then by Proposition 2.1, δ is an $H - K$ derivation if and only if the operators $R_i = \begin{pmatrix} V_i & X_i \\ 0 & T_i \end{pmatrix}$ $i = 1, \ldots, N$ mutually commute and are jointly polynomially bounded. Furthermore, if δ is an $H - K$ derivation then δ is inner if and only if there is an operator $Y \in \mathcal{B}(H, K)$ such that $X_i = V_iY - YT_i$ $i = 1, \ldots, N$.

3 Submodules $L^2(\mathbb{T}^N)$

Let $L^2(\mathbb{T}^N)$ denote the space of Lebesgue measurable, square-integrable functions on the N-torus, $\mathbb{T}^N$. Since the distinguished boundary of the polydisk is $\mathbb{T}^N$, the polydisk algebra $A(\mathbb{D}^N)$ is isometrically isomorphic to the algebra $A(\mathbb{T}^N)$ defined to be the closure in $C(\mathbb{T}^N)$ of the analytic trigonometric polynomials. The space $L^2(\mathbb{T}^N)$ will be viewed as a Hilbert module over the polydisk algebra via pointwise multiplication by functions in $A(\mathbb{T}^N)$. Throughout, U_i $i = 1, \ldots, N$ will denote the bilateral shifts on $L^2(\mathbb{T}^N)$ defined by $U_if(\zeta_1, \ldots, \zeta_N) = \zeta_i f(\zeta_1, \ldots, \zeta_N)$.

By a submodule of $L^2(\mathbb{T}^N)$ we mean a norm closed subspace which is invariant under U_i $i = 1, \ldots, N$. A distinguished submodule of $L^2(\mathbb{T}^N)$ is the Hardy space $H^2(\mathbb{T}^N)$ consisting of those functions whose Fourier transform is supported on $\mathbb{Z}^{+N} = \mathbb{Z}^+ \times \cdots \times \mathbb{Z}^+$. Throughout we denote the unilateral shifts on $H^2(\mathbb{T}^N)$ by $S_i = U_i|_{H^2}$ $i = 1, \ldots, N$.

Let $\mathcal{M}$ be a submodule of $L^2(\mathbb{T}^N)$. We make $\mathcal{M}^\perp = L^2(\mathbb{T}^N) \ominus \mathcal{M}$ into a Hilbert $A(\mathbb{D}^N)$-module by defining the action of $f \in A(\mathbb{T}^N)$ by $f \cdot g = P_{\mathcal{M}}^\perp(fg)$, $g \in \mathcal{M}^\perp$. As a module, $\mathcal{M}^\perp$ is isomorphic to the quotient module $L^2(\mathbb{T}^N)/\mathcal{M}$. Because of this we will use the notation $L^2(\mathbb{T}^N)/\mathcal{M}$ to denote the Hilbert $A(\mathbb{D}^N)$-module $\mathcal{M}^\perp$ just defined.

In [CC2] it was shown that $L^2(\mathbb{T}^N)$ is projective (and hence injective) in the category of Hilbert $A(\mathbb{D}^N)$-modules. A consequence of this which will be used in the sequel is the following observation. For $\psi \in L^\infty(\mathbb{T}^N)$, we let $M_\psi : L^2(\mathbb{T}^N) \to L^2(\mathbb{T}^N)$ denote the multiplication operator $M_\psi f = \psi f$. If $\mathcal{M}, \mathcal{N} \subseteq L^2(\mathbb{T}^N)$ are submodules then $\mathrm{Hom}_{A(\mathbb{D}^N)}(\mathcal{N}, \mathcal{M}) = \{M_\psi|_\mathcal{N} \mid \psi\mathcal{N} \subseteq \mathcal{M}\}$. In particular, $\mathrm{Hom}_{A(\mathbb{D}^N)}(\mathcal{M}, L^2) = \{M_\psi|_\mathcal{M} \mid \psi \in L^\infty(\mathbb{T}^N)\}$ and $\mathrm{Hom}_{A(\mathbb{D}^N)}(H^2, \mathcal{M}) = \mathcal{M} \cap L^\infty(\mathbb{T}^N)$.

Now consider the short exact sequence

$$0 \longrightarrow \mathcal{M} \longrightarrow L^2(\mathbb{T}^N) \longrightarrow L^2(\mathbb{T}^N)/\mathcal{M} \longrightarrow 0$$

where the first map is the inclusion and the second map is the orthogonal projection onto $\mathcal{M}^\perp$. Since $L^2(\mathbb{T}^N)$ is injective, the long exact sequence (constructed in

[CC1] and extended through the second order groups in [F2]) yields the following short exact sequence

$$\mathrm{Hom}_{A(\mathbb{D}^N)}(\mathcal{N}, \mathcal{M}) \longrightarrow \mathrm{Hom}_{A(\mathbb{D}^N)}(\mathcal{N}, L^2) \longrightarrow \mathrm{Hom}_{A(\mathbb{D}^N)}(\mathcal{N}, L^2/\mathcal{M})$$
$$\xrightarrow{\pi} \mathrm{Ext}^1_{A(\mathbb{D}^N)}(\mathcal{N}, \mathcal{M}) \longrightarrow 0.$$

The map π sends a module map $\mathcal{A} \in \mathrm{Hom}(\mathcal{N}, L^2/\mathcal{M})$ to the equivalence class of the $\mathcal{N} - \mathcal{M}$ derivation $\delta_A(f) = P_{\mathcal{M}} M_f A$, $f \in A(\mathbb{T}^N)$, and so every $\mathcal{N} - \mathcal{M}$ derivation up to equivalence is determined by an operator in $\mathrm{Hom}_{A(\mathbb{D}^N)}(\mathcal{N}, L^2/\mathcal{M})$. By exactness we can characterize the obstruction to the vanishing of the groups $\mathrm{Ext}^1_{A(\mathbb{D}^N)}(\mathcal{M}, \mathcal{N})$ in terms of a joint commutant lifting problem. We record this only for future reference.

Proposition 3.1 $\mathrm{Ext}^1(\mathcal{N}, \mathcal{M}) = (0)$ *if and only if*

$$\mathrm{Hom}_{A(\mathbb{D}^N)}(\mathcal{N}, L^2/\mathcal{M}) = \{P^{\perp}_{\mathcal{M}} M_{\psi}|_{\mathcal{N}} \mid \psi \in L^{\infty}(\mathbb{T}^N)\}.$$

We will restrict our attention to certain distinguished submodules of $L^2(\mathbb{T}^2)$. The results presented here easily generalize to the analogous submodules of $L^2(\mathbb{T}^N)$ but for notational convenience and clarity we do only the two-variable case. Let I^2 denote the right half plane of $L^2(\mathbb{T}^2)$ consisting of those functions $f \in L^2(\mathbb{T}^2)$ whose Fourier transform is supported on $\mathbb{Z}^+ \times \mathbb{Z}$ where $\mathbb{Z}^+$ denotes the nonnegative integers. Note that I^2 is a submodule of $L^2(\mathbb{T}^2)$ and that $U_1|_{I^2}$ is an isometry on I^2 and $U_2|_{I^2}$ is unitary on I^2. Indeed, as a Hilbert space $I^2 \cong H^2 \otimes L^2$, $U_1|_{I^2}$ is just the shift of infinite mutiplicity $S \otimes I$ and $U_2|_{I^2}$ is unitarily equivalent to $I \otimes U$ where U is the bilateral shift on L^2. Similarly, let J^2 denote the upper half plane $L^2 \otimes H^2$ and note that in this case, $U_1|_{J^2} \cong U \otimes I$ is a unitary operator on J^2 and $U_2|_{J^2} \cong I \otimes S$ is an isometry on J^2.

Theorem 3.2 *If* $\mathcal{M} \subseteq L^2(\mathbb{T}^2)$ *is a submodule then the groups* $\mathrm{Ext}^1_{A(\mathbb{D}^2)}(\mathcal{M}, I^2)$, $\mathrm{Ext}^1_{A(\mathbb{D}^2)}(\mathcal{M}, J^2)$, $\mathrm{Ext}^1_{A(\mathbb{D}^2)}(I^2, \mathcal{M})$ *and* $\mathrm{Ext}^1_{A(\mathbb{D}^2)}(J^2, \mathcal{M})$ *all vanish.*

Proof: We first settle the case with I^2 in the first variable. Since $U_1|_{I^2}$ is an isometry on I^2 and $U_2|_{I^2}$ is a unitary on I^2, I^2 is projective in the cramped category of Hilbert $A(\mathbb{D}^2)$-modules, [CC2]. This implies, in particular, that every module map from I^2 into the quotient $L^2/\mathcal{M}$ has a lifting to a module map from I^2 into $L^2(\mathbb{T}^2)$. Since $\mathrm{Hom}_{A(\mathbb{D}^2)}(I^2, L^2) = \{M_{\psi}|_{I^2} \mid \psi \in L^{\infty}(\mathbb{T}^2)\}$, it follows that $\mathrm{Hom}_{A(\mathbb{D}^2)}(I^2, L^2/\mathcal{M}) = \{P^{\perp}_{\mathcal{M}} M_{\psi}|_{I^2} \mid \psi \in L^{\infty}(\mathbb{T}^2)\}$. By Proposition 3.1, $\mathrm{Ext}^1_{A(\mathbb{D}^2)}(I^2, \mathcal{M}) = (0)$. The same argument shows that $\mathrm{Ext}^1_{A(\mathbb{D}^2)}(J^2, \mathcal{M}) = (0)$.

To see that $\mathrm{Ext}^1_{A(\mathbb{D}^2)}(\mathcal{M}, I^2) = (0)$ note that the compression $P^{\perp}_{I^2} U_1|_{I^{2\perp}}$ is a coisometry on $I^{2\perp}$ and $P^{\perp}_{I^2} U_2|_{I^{2\perp}}$ is a unitary on $I^{2\perp}$. Thus the quotient L^2/I^2 is injective in the cramped category of Hilbert $A(\mathbb{D}^2)$-modules, [CC2].

This implies that every $A(\mathbb{D}^2)$-module map from $\mathcal{M}$ into $L^2(\mathbb{T}^2)/I^2$ extends to a module map from $L^2(\mathbb{T}^2)$ into $L^2(\mathbb{T}^2)/I^2$. Since $L^2(\mathbb{T}^2)$ is projective [CC2], $\mathrm{Ext}^1_{A(\mathbb{D}^2)}(L^2, I^2) = (0)$ and thus by Proposition 3.1, $\mathrm{Hom}_{A(\mathbb{D}^2)}(L^2, L^2/I^2) = \{P^{\perp}_{I^2}M_\psi \mid \psi \in L^\infty(\mathbb{T}^2)\}$. It follows that $\mathrm{Hom}_{A(\mathbb{D}^2)}(\mathcal{M}, L^2/I^2) = \{P^{\perp}_{I^2}M_\psi|_{\mathcal{M}} \mid \psi \in L^\infty(\mathbb{T}^2)\}$. Therefore, by Proposition 3.1 $\mathrm{Ext}^1_{A(\mathbb{D}^2)}(\mathcal{M}, I^2) = (0)$. The same argument shows that $\mathrm{Ext}^1_{A(\mathbb{D}^2)}(\mathcal{M}, J^2) = (0)$. □

Define the submodule $K^2 \subseteq L^2(\mathbb{T}^2)$ to be the set of functions whose Fourier transform is supported on the 1st, 2nd and 4th quadrant of $L^2(\mathbb{T}^2)$. Alternatively, $K^2 = I^2 + J^2$. We can view the Hilbert space direct sum $I^2 \oplus J^2$ as a Hilbert $A(\mathbb{D}^2)$-module with the module multiplication given by the direct sum of the actions on I^2 and J^2. Since $H^2(\mathbb{T}^2) = I^2 \cap J^2$ we have a short exact sequence of Hilbert $A(\mathbb{D}^2)$-modules

$$0 \longrightarrow H^2(\mathbb{T}^2) \longrightarrow I^2 \oplus J^2 \longrightarrow K^2 \longrightarrow 0$$

where the first map sends $f \in H^2(\mathbb{T}^2)$ to the pair $(f, -f)$ and the second map sends the pair $(f, g) \in I^2 \oplus J^2$ to the sum $f + g \in K^2$.

We observe that for any Hilbert $A(\mathbb{D}^2)$-module H, $\mathrm{Ext}^1_{A(\mathbb{D}^2)}(H, I^2 \oplus J^2) \cong \mathrm{Ext}^1_{A(\mathbb{D}^2)}(H, I^2) \oplus \mathrm{Ext}^1_{A(\mathbb{D}^2)}(H, J^2)$. In particular, this holds for a submodule $\mathcal{M} \subseteq L^2(\mathbb{T}^2)$. By the long exact sequence together with Theorem 3.2, we have the exact sequence

$$\begin{aligned}\mathrm{Hom}_{A(\mathbb{D}^2)}(K^2, \mathcal{M}) &\longrightarrow \mathrm{Hom}_{A(\mathbb{D}^2)}(I^2 \oplus J^2, \mathcal{M}) \longrightarrow \mathrm{Hom}_{A(\mathbb{D}^2)}(H^2, \mathcal{M}) \\ &\longrightarrow \mathrm{Ext}^1_{A(\mathbb{D}^2)}(K^2, \mathcal{M}) \longrightarrow 0.\end{aligned}$$

Now the map from $\mathrm{Hom}_{A(\mathbb{D}^2)}(I^2 \oplus J^2, \mathcal{M}) \cong \mathrm{Hom}_{A(\mathbb{D}^2)}(I^2, \mathcal{M}) \oplus \mathrm{Hom}_{A(\mathbb{D}^2)}(J^2, \mathcal{M})$ into $\mathrm{Hom}_{A(\mathbb{D}^2)}(H^2, \mathcal{M})$ above sends a pair (f_1, f_2) with $f_1 I^2 \subseteq \mathcal{M}$ and $f_2 J^2 \subseteq \mathcal{M}$ to the sum $f_1 + f_2$ which multiplies K^2 into $\mathcal{M}$. Since $f H^2(\mathbb{T}^2) \subseteq \mathcal{M}$ if and only if $f \in \mathcal{M} \cap L^\infty(\mathbb{T}^2)$ the following characterization of the obstruction to the vanishing of the groups $\mathrm{Ext}^1_{A(\mathbb{D}^2)}(K^2, \mathcal{M})$ follows from exactness of the sequence above.

Proposition 3.3 *Let $\mathcal{M} \subseteq L^2(\mathbb{T}^2)$. Then $\mathrm{Ext}^1_{A(\mathbb{D}^2)}(K^2, \mathcal{M}) = (0)$ if and only if for every $f \in \mathcal{M} \cap L^\infty(\mathbb{T}^2)$ there exists $f_1, f_2 \in \mathcal{M} \cap L^\infty(\mathbb{T}^2)$ satisfying $f_1 I^2 \subseteq \mathcal{M}$, $f_2 J^2 \subseteq \mathcal{M}$ and $f = f_1 + f_2$.*

Unlike the one variable case, there is no Beurling type characterization of the submodules of $H^2(\mathbb{T}^2)$. In particular, there are submodules $H^2(\mathbb{T}^2)$ which do not contain any bounded functions other than the zero function, [Ru]. The following corollary implies that a submodule $\mathcal{M} \subseteq H^2(\mathbb{T}^2)$ contains a nonzero bounded function if and only if $\mathrm{Ext}^1_{A(\mathbb{D}^2)}(K^2, \mathcal{M}) \neq (0)$.

Corollary 3.4 *If $\mathcal{M} \subseteq H^2(\mathbb{T}^2)$ is a submodule then* $\mathrm{Ext}^1_{A(\mathbb{D}^2)}(K^2, \mathcal{M}) \cong \mathcal{M} \cap H^\infty(\mathbb{T}^2)$. *In particular,* $\mathrm{Ext}^1_{A(\mathbb{D}^2)}(K^2, H^2) \cong H^\infty(\mathbb{T}^2)$.

Proof: By the exact sequence above, $\mathrm{Ext}^1_{A(\mathbb{D}^2)}(K^2, \mathcal{M})$ can be identified algebraically with a quotient of $\mathcal{M} \cap L^\infty(\mathbb{T}^2)$. Specifically, $\mathrm{Ext}^1_{A(\mathbb{D}^2)}(K^2, \mathcal{M})$ is isomorphic to $\mathcal{M} \cap L^\infty(\mathbb{T}^2)$ modulo $\{f \in \mathcal{M} \cap L^\infty(\mathbb{T}^2) \mid fI^2 \subseteq \mathcal{M} \text{ and } fJ^2 \subseteq \mathcal{M}\}$. We note that the only function which multiplies I^2 into $H^2(\mathbb{T}^2)$ is the zero function. The same holds true for J^2. Thus if $\mathcal{M} \subseteq H^2(\mathbb{T}^2)$ then $\mathrm{Ext}^1_{A(\mathbb{D}^2)}(K^2, \mathcal{M}) \cong \mathcal{M} \cap L^\infty(\mathbb{T}^2) = \mathcal{M} \cap H^\infty(\mathbb{T}^2)$. □

Note that $I^2 \cap L^\infty(\mathbb{T}^2)$ is just the subspace of $L^\infty(\mathbb{T}^2)$ consisting of those functions which are analytic in the first variable. We will denote this subspace by $H^\infty \otimes L^\infty$. It is easy to see that a function $f \in L^\infty(\mathbb{T}^2)$ multiplies I^2 into K^2 if and only if $f \in H^\infty \otimes L^\infty$. Similarly, $L^\infty \otimes H^\infty := J^2 \cap L^\infty(\mathbb{T}^2)$ is the subspace of functions in $L^\infty(\mathbb{T}^2)$ which are analytic in the second variable and can be characterized as those functions which multiply J^2 into K^2. Thus by Proposition 3.3, $\mathrm{Ext}^1_{A(\mathbb{D}^2)}(K^2, K^2) = (0)$ if and only if $K^2 \cap L^\infty(\mathbb{T}^2) = H^\infty \otimes L^\infty + L^\infty \otimes H^\infty$. It turns out that this is not the case, in fact one can show that $\mathrm{Ext}^1_{A(\mathbb{D}^2)}(K^2, K^2) \cong \mathrm{Ext}^1_{A(\mathbb{D}^2)}(H^2, H^2)$ and the group $\mathrm{Ext}^1_{A(\mathbb{D}^2)}(H^2, H^2)$ is known to be nonzero [CC2]. This is related to the so-called *big* Hankel problem which we will say more about below.

Alternatively, we give a concrete proof using Proposition 3.3 together with a result in [F3] to show that $\mathrm{Ext}^1_{A(\mathbb{D}^2)}(K^2, K^2) \neq (0)$.

Theorem 3.5 *The function $\phi(\zeta, \lambda) = \log(1-\bar{\zeta}\lambda) - \log(1-\zeta\bar{\lambda})$ is in $K^2 \cap L^\infty(\mathbb{T}^2)$ but ϕ is not in $H^\infty \otimes L^\infty + L^\infty \otimes H^\infty$.*

Proof: The function ϕ is certainly bounded and supported on K^2. Let $\phi_1(\zeta, \lambda) = \log(1 - \bar{\zeta}\lambda)$ and $\phi_2(\zeta, \lambda) = \log(1 - \zeta\bar{\lambda})$. Let $P^\perp : L^2(\mathbb{T}^2) \longrightarrow H^2(\mathbb{T}^2)^\perp$ be the orthogonal projection onto $H^2(\mathbb{T}^2)^\perp$ and consider the densely defined Hankel operators $\Gamma_{\phi_i} f = P^\perp(\phi_i f)$ $f \in A(\mathbb{T}^2)$, $i = 1, 2$. Since ϕ_1 lies in the second quadrant of $L^2(\mathbb{T}^2)$, Γ_{ϕ_1} maps into the second quadrant of $L^2(\mathbb{T}^2)$ and so can be realized as the operator on the infinite direct sum $\oplus_{i=1}^\infty H^2$ with operator matrix $\left(\frac{I_{H^2}}{i+j+1}\right)$. In particular, Γ_{ϕ_1} extends to a bounded operator on $H^2(\mathbb{T}^2)$ and similar reasoning shows that Γ_{ϕ_2} extends to a bounded operator on $H^2(\mathbb{T}^2)$. Thus the Hankel operator $\Gamma_\phi = \Gamma_{\phi_1} - \Gamma_{\phi_2}$ is bounded on $H^2(\mathbb{T}^2)$. A result in [F3] implies that neither Γ_{ϕ_1} nor Γ_{ϕ_2} has an $L^\infty(\mathbb{T}^2)$ symbol. In other words, there does not exist $\psi \in L^\infty(\mathbb{T}^2)$ such that $\Gamma_{\phi_1} = \Gamma_\psi$ or $\Gamma_{\phi_2} = \Gamma_\psi$. Now if ϕ is the sum of two functions f_1 and f_2 with $f_1 \in H^\infty \otimes L^\infty$ and $f_2 \in L^\infty \otimes H^\infty$ then

$$\Gamma_{\phi_1} - \Gamma_{\phi_2} = \Gamma_{f_1} + \Gamma_{f_2}. \tag{2}$$

Since Γ_{ϕ_1} and Γ_{f_1} both map $H^2(\mathbb{T}^2)$ into the second quadrant and the operators Γ_{ϕ_2} and Γ_{f_2} both map $H^2(\mathbb{T}^2)$ into the 4th quadrant the equation in (2) implies that $\Gamma_{\phi_1} = \Gamma_{f_1}$ which is impossible since f_1 is bounded. $\square$

4 Hankel Operators

In two variables there is more than one Hankel operator defined on $H^2(\mathbb{T}^2)$. There are the so called *big* Hankel operators which map into $H^2(\mathbb{T}^2)^\perp$ and the *little* Hankel operators which map into the space of complex conjugates of functions in $z_1z_2H^2(\mathbb{T}^2)$. In both cases the Hankel operators map into quotients of $L^2(\mathbb{T}^2)$. A problem of much interest is determining if the Hankel operators (big or little) are determined by bounded functions. Known results on this problem are stated below. A more general problem is that of determining when the Hankel operators defined on $H^2(\mathbb{T}^2)$ and mapping into a fixed quotient of $L^2(\mathbb{T}^2)$ have bounded symbols. By Proposition 3.1 this problem can be formulated in terms of the vanishing of $\mathrm{Ext}^1_{A(\mathbb{D}^2)}(H^2, \mathcal{M})$.

More explicitly, by a Hankel operator from $H^2(\mathbb{T}^2)$ into a quotient $L^2(\mathbb{T}^2)/\mathcal{M}$ we mean an operator $A : H^2(\mathbb{T}^2) \to \mathcal{M}^\perp$ satisfying $AS_i = P^\perp_{\mathcal{M}} U_i A$ $i = 1, 2$. In other words, a Hankel operator mapping into $\mathcal{M}^\perp$ is just an operator in $\mathrm{Hom}_{A(\mathbb{D}^2)}(H^2, L^2/\mathcal{M})$. If we set $\phi = A(1)$ where "1" denotes the constant function 1 then for $f \in A(\mathbb{T}^2)$, $Af = P^\perp_{\mathcal{M}}(\phi f)$. We write $\Gamma^{\mathcal{M}}_\phi$ for the operator A and call ϕ the symbol of the Hankel operator.

Note that any $\phi \in L^2(\mathbb{T}^2)$ satisfying $\|P^\perp_{\mathcal{M}}(\phi f)\| \le C\|f\|$ for some $C > 0$ and for all $f \in A(\mathbb{T}^2)$ extends to a bounded Hankel operator $\Gamma^{\mathcal{M}}_\phi$. In particular, if $\psi \in L^\infty(\mathbb{T}^2)$ then $\Gamma^{\mathcal{M}}_\psi$ is a part of multiplication by ψ namely, $P^\perp_{\mathcal{M}} M_\psi|_{H^2}$ and thus $\|\Gamma^{\mathcal{M}}_\psi\| \le \|\psi\|_\infty$. The Hankel operator $\Gamma^{\mathcal{M}}_\phi : H^2(\mathbb{T}^2) \to \mathcal{M}^\perp$ has a bounded symbol if there is a $\psi \in L^\infty(\mathbb{T}^2)$ such that $\Gamma^{\mathcal{M}}_\phi = \Gamma^{\mathcal{M}}_\psi = P^\perp_{\mathcal{M}} M_\psi|_{H^2}$. Equivalently, Γ_ϕ has a bounded symbol if $P^\perp_{\mathcal{M}}\phi = P^\perp_{\mathcal{M}}\psi$ for some $\psi \in L^\infty(\mathbb{T}^2)$. Thus every Hankel operator from $H^2(\mathbb{T}^2)$ into $\mathcal{M}^\perp$ has a bounded symbol if and only if $\mathrm{Hom}_{A(\mathbb{D}^2)}(H^2, L^2/\mathcal{M}) = \{P^\perp_{\mathcal{M}} M_\psi|_{H^2} \mid \psi \in L^\infty(\mathbb{T}^2)\}$. The following is just a reformulation of Proposition 3.1, for the modules H^2 and $\mathcal{M} \subseteq L^2(\mathbb{T}^2)$ in terms of Hankel operators mapping into $\mathcal{M}^\perp$.

Proposition 4.1 *Let $\mathcal{M} \subseteq L^2(\mathbb{T}^2)$ be a submodule. The following are equivalent.*

(1) $\mathrm{Ext}^1_{A(\mathbb{D}^2)}(H^2, \mathcal{M}) = (0)$.

(2) *If $\phi \in L^2(\mathbb{T}^2)$ and $C > 0$ such that $\|P^\perp_{\mathcal{M}}(\phi f)\| \le C\|f\|$ for all $f \in A(\mathbb{T}^2)$ then there exists a function $\psi \in L^\infty(\mathbb{T}^2)$ with $P^\perp_{\mathcal{M}}\phi = P^\perp_{\mathcal{M}}\psi$.*

For the submodule $\mathcal{M} = H^2(\mathbb{T}^2)$, it was first observed in [CC2] that the vanishing of the group $\mathrm{Ext}^1_{A(\mathbb{D}^2)}(H^2, H^2)$ is equivalent to the so-called *big* Hankel problem i.e., the problem of determining if the big Hankel operators $\Gamma_\phi : H^2 \to H^2(\mathbb{T}^2)^\perp$ have bounded symbols. In [CoSa], the authors proved the existence of a big Hankel operator with no bounded symbol thus implying that $\mathrm{Ext}^1_{A(\mathbb{D})2}(H^2, H^2) \neq (0)$, see [CC2]. Cotlar and Sadosky proved only the existence of such a Hankel operator using a duality argument and a characterization of the symbols in terms of their *restricted* BMO class. In [F3] concrete examples of symbols which did not lift to $L^\infty(\mathbb{T}^2)$ were constructed using techniques inspired by the cohomology and which yielded, in particular, *completely bounded* derivations of the bidisk algebra into $\mathcal{B}(H^2(\mathbb{T}^2))$ which are not inner. Such a derivation corresponds to a bounded big Hankel operator with no bounded symbol. This is in great contrast to derivations of C^*-algebras (mapping into $\mathcal{B}(H)$ with H a Hilbert space) which are necessarily inner if they are completely bounded, [Ch].

We should point out that Bakonyi and Timotin in [BT] used alternative methods to show that the function $\phi(\zeta, \lambda) = \log(1 - \overline{\zeta}\lambda)$ determines a bounded big Hankel operator on $H^2(\mathbb{T}^2)$ with no bounded symbol and thus answered (in the negative) a question posed in [CoSa] concerning the Nehari distance formula for big Hankel operators. Their question was also answered in [F3] and the non-lifting of ϕ to $L^\infty(\mathbb{T}^2)$ is a special case of the main result in [F3]. In fact, it can be shown [F4] that for any submodule $\mathcal{M} \subseteq L^2(\mathbb{T}^2)$, $\mathrm{Ext}^1_{A(\mathbb{D}^2)}(H^2, \mathcal{M}) = (0)$ if and only if the Nehari distance formula holds:

$$\mathrm{dist}_\infty(\psi, \mathcal{M} \cap L^\infty(\mathbb{T}^2)) \leq C \| P^\perp_{\mathcal{M}} M_\psi |_{H^2} \| \quad \text{for all } \psi \in L^\infty(\mathbb{T}^2).$$

For the submodule $\mathcal{M} = K^2$, the corresponding Hankel operators map $H^2(\mathbb{T}^2)$ into $K^{2\perp} = \overline{z_1 z_2 H^2(\mathbb{T}^2)}$. Note that such a Hankel operator can be identified, via the change of variable, $z_1, z_2 \to \overline{Z_1}, \overline{Z_2}$ as an operator $\gamma_\phi : H^2(\mathbb{T}^2) \to H^2(\mathbb{T}^2)$ which is densely defined by $\gamma_\phi(f) = P_+(\phi f(e^{-it}))$. These are the so-called *little* Hankel operators. The little Hankel operator γ_ϕ has a bounded symbol if and only if $P_+\phi = P_+\psi$ for some $\psi \in L^\infty(\mathbb{T}^2)$. V.V. Peller has informed us that a criterion for boundedness of a little Hankel operator can be deduced from a result in [CPL]. The Hankel operators considered there are regarded as paracommutators on $L^2(\mathbb{R} \times \mathbb{R})$ and in this setting the criterion for boundedness is that the symbol is in $BMO(\mathbb{R} \times \mathbb{R})$. We believe that by identifying *analytic* $H^1(\mathbb{T}^2)$ as a subspace of *harmonic* $H^1(\mathbb{T}^2)$, this characterization together with the duality results in [ChFe] implies that every little Hankel operator has a bounded symbol.

It would be nice to have an algebraic proof of the lifting (or non-lifting) of little Hankel operators by determining if the group $\mathrm{Ext}^1_{A(\mathbb{D}^2)}(H^2, K^2)$ vanishes. Partial results on this problem suggest that the group may not vanish.

References

[BT] M. Bakonyi and D. Timotin, *On a conjecture of Cotlar and Sadosky on multidimensional Hankel operators*, Preprint (1997).

[CC1] J.F. Carlson and D.N. Clark, *Cohomology and extensions of Hilbert modules*, J. Funct. Anal. **128** (1995), 278–306.

[CC2] J.F. Carlson and D.N. Clark, *Projectivity and extensions of Hilbert modules over* $A(\mathbb{D}^N)$, Mich. Math. J. **44** (1997), 365–373.

[CCFW] J.F. Carlson, D.N. Clark, C. Foias and J.P. Williams, *Projective Hilbert* $A(\mathbb{D})$*-Modules*, New York J. Math. **1** (1994), 26–38.

[ChFe] S.-Y. A. Chang and R. Fefferman, *Some recent developments in Fourier analysis and* H^p *theory on product spaces*, Bull. Amer. Math. Soc. **12** (1985), 1–43.

[Ch] E. Christensen, *Extensions of derivations II*, Math. Scand. **50** (1982), 111–122.

[CPL] L. Chun, L. Peng and D. Lizhong, *Paracommutators on product spaces*, Institute of Mathematics, Peking University, Beijing, Research Report **No. 14** (1989).

[CoSa] M. Cotlar and C. Sadosky, *Two distinguished subspaces of product BMO and Nehari-AAK theory for Hankel operators on the torus*, Int. Eqs. and Op. Th. **26** (1996), 273–304.

[DoPa] R.G. Douglas and V.I. Paulsen, *Hilbert Modules over Function Algebras*, Longman Scientific & Technical, New York, 1989.

[F1] S.H. Ferguson, *Polynomially bounded operators and Ext groups*, Proc. Amer. Math. Soc. **124** (1996) no. 9, 2779–2785.

[F2] S.H. Ferguson, *Ext, Analytic Kernels and Operator Ranges*, Dissertation, University of Houston, 1996.

[F3] S.H. Ferguson, *The Nehari Problem for the Hardy space of the torus*, to appear J. Operator Theory.

[F4] S.H. Ferguson, *Quotients of* $L^\infty(\mathbb{T}^N)$, preprint.

[Ru] W. Rudin, *Function Theory in Polydiscs*, W. A. Benjamin, Inc., New York, 1969.

Department of Mathematics
U. Georgia, Athens, GA
30602-7403
dnc@math.uga.edu

Department of Mathematics
Purdue University, West Lafayette
IN 47907-1395
sfergus@math.purdue.edu

Operator Theory:
Advances and Applications, Vol. 115

Kolmogorov Decompositions and the Realization of Time Dependent Systems

Tiberiu Constantinescu and Aurelian Gheondea

We formulate a generalized realization problem for time dependent systems on spaces with indefinite inner product. We use the Kolmogorov decomposition of Hermitian kernels as a main tool for the solution of this problem. The uniqueness property of the minimal realization is characterized. Applications are given to the realization of time independent systems, including the connection with the Azizov realization theorem and with de Branges-Rovnyak models. We give an example of a Toeplitz transfer map with a time dependent unitary realization but with no time independent unitary realization and an example of a function in a generalized Schur class for which one cannot run a Schur type algorithm.

1 Introduction

Many physical systems can be described by linear equations of the form

$$\Sigma \begin{cases} x_{n+1} = Ax_n + Bu_n \\ y_n \quad = Cx_n + Du_n, \quad n \in \mathbf{Z}. \end{cases}$$

For the engineering applications it is important to have realizations of this type with A, B, C, and D finite dimensional operators (see the classical references R.E. Kalman, P.L. Falb, and B.M. Arbib [18], and T. Kailath [17]), while the property that the block matrix $\begin{bmatrix} A & B \\ C & D \end{bmatrix}$ is unitary on some Hilbert space appears to be of interest in applications and developments as those presented in the references M.S. Livsic [21], M.S. Brodskii and M.S. Livsic [8], B. Sz.-Nagy and C. Foiaş [24], L. de Branges [5], L. de Branges and J. Rovnayk [7], P. Lax and R.S. Phillips [20]. There are also examples for which one has to consider more general systems with operator coefficients depending on n, see e.g. [10]. In other direction, it appeared to be of interest to consider realizations on indefinite inner product spaces, such as in [3], [6].

The main notion involved in the study of these systems is the so-called characteristic map and a key problem concerns the realization of various classes of maps as characteristic maps of unitary systems (see B. Sz.-Nagy and C. Foiaş [24], L. de Branges and J. Rovnayk [7], T.Ya. Azizov [3], and [6]). To be more precise, we will consider the general case of time dependent state-space models of discrete time linear systems given by the following equations:

$$\Sigma \begin{cases} x_{n+1} = A(n)x_n + B(n)u_n \\ y_n \quad = C(n)x_n + D(n)u_n, \quad n \in \mathbf{Z}, \end{cases}$$

where $A(n) \in \mathcal{L}(\mathcal{H}_n, \mathcal{H}_{n+1})$, $B(n) \in \mathcal{L}(\mathcal{U}_n, \mathcal{H}_{n+1})$, $C(n) \in \mathcal{L}(\mathcal{H}_n, \mathcal{Y}_n)$, and $D(n) \in \mathcal{L}(\mathcal{U}_n, \mathcal{Y}_n)$ are bounded linear operators between Kreĭn spaces. We recall that a Kreĭn space $\mathcal{H}$ is a complex linear space endowed with an inner product $[\cdot, \cdot]$ with the property that there exists a bounded linear operator J on $\mathcal{H}$ such that $J = J^* = J^{-1}$ and the new inner product on $\mathcal{H}$,

$$\langle h, g \rangle_J = [Jh, g], \quad h, g \in \mathcal{H},$$

is a positive definite inner product with respect to which $\mathcal{H}$ becomes a Hilbert space. An operator J as above will be referred to as a *fundamental symmetry* on $\mathcal{H}$. For two Kreĭn spaces $\mathcal{H}$ and $\mathcal{K}$ we denote by $\mathcal{H} \oplus \mathcal{K}$ the orthogonal direct sum Kreĭn space of $\mathcal{H}$ and $\mathcal{K}$. If $\mathcal{K}$ is a *Kreĭn subspace* of $\mathcal{H}$ then $\mathcal{H} \ominus \mathcal{K}$ denotes the orthogonal complement of $\mathcal{K}$.

$\mathcal{L}(\mathcal{H}, \mathcal{K})$ denotes the set of bounded linear operators from $\mathcal{H}$ into $\mathcal{K}$. For an operator $T \in \mathcal{L}(\mathcal{H}, \mathcal{K})$ we use the notation $T^\sharp \in \mathcal{L}(\mathcal{K}, \mathcal{H})$ to denote the *adjoint* of T, that is, $[Th, g] = [h, T^\sharp g]$ for $h, g \in \mathcal{H}$. An operator $U \in \mathcal{L}(\mathcal{H}, \mathcal{K})$ is *unitary* if $U^\sharp = U^{-1}$.

The system Σ is *unitary* provided that the *system matrix* at time n,

$$\begin{bmatrix} A(n) & B(n) \\ C(n) & D(n) \end{bmatrix},$$

is unitary for every $n \in \mathbf{Z}$. The evolution of the system is described by the *transfer map* $T_\Sigma = [T_{ij}]_{i,j \in \mathbf{Z}}$ of Σ, given by the formula:

$$T_{ij} = \begin{cases} D(i), & j = i, \\ C(i)B(j), & j = i - 1, \\ C(i)\tau_A(i, j+1)B(j), & \text{otherwise}, \end{cases}$$

where τ_A is called the *transition map* and is defined by

$$\tau_A(k, l) = \begin{cases} A(k-1)A(k-2)\dots A(l+1)A(l), & k > l \\ I_{\mathcal{H}_l}, & k = l \\ 0, & k < l. \end{cases}$$

If we look at τ_A as an infinite matrix with operator entries then it is lower triangular. The main realization problem that will be addressed in this paper concerns the characterization of those lower triangular block-operator matrices $T = [T_{ij}]_{i,j \in \mathbf{Z}}$ that appear as transfer maps of unitary systems.

We notice that if Σ is time independent, that is, $A(n) = A$, $B(n) = B$, $C(n) = C$, $D(n) = D$ for all $n \in \mathbf{Z}$, then T_Σ is Toeplitz lower triangular, that is, $T_{ij} = T_{i-j}$ for all $i, j \in \mathbf{Z}$. As we shall see in Section 3.A, the natural class of transfer maps for time independent unitary systems is identified with the class of operator valued functions that are holomorphic in a neighbourhood of zero. In the noncontractive setting, the realization of this class of functions was obtained by T.Ya. Azizov [3], L. de Branges [6], A. Dijksma, H. Langer and H. de Snoo [14], and [15]).

In Hilbert spaces, a natural minimality condition is associated to unitary systems and it is shown that minimal unitary systems with the same transfer map are unitarily equivalent (see J.A. Ball and N. Cohen [4] for a recent discussion). For unitary systems on indefinite metric spaces, the situation is completely different (cf. L. de Branges [6]), due to some problems emphasized already in the classical paper of L. Schwartz [22].

A system is called *minimal* if

$$\bigvee_{k\leq n-1} \tau_A(n, k+1)B(k)\mathcal{U}_k \vee \bigvee_{k\geq n} \tau_A(k, n)^\sharp C(k)^\sharp \mathcal{Y}_k = \mathcal{H}_n \tag{1.1}$$

for all integers n. Two systems Σ and Σ' are unitarily equivalent provided that there are unitary operators $\phi_n \in \mathcal{L}(\mathcal{H}_n, \mathcal{H}'_n)$ such that $A'(n)\phi_n = \phi_{n+1}A(n)$, $C'(n)^\sharp = \phi_n C(n)^\sharp$, $B'(n) = \phi_{n+1}B(n)$ and $D'(n) = D(n)$ for all $n \in \mathbf{Z}$. We will be interested in characterizing those transfer maps with the property that every two of their minimal unitary realizations are unitarily equivalent.

The purpose of this paper is to address the above general realization problem for time dependent systems on indefinite metric spaces and to emphasize the role played by a general construction referred to as Kolmogorov decomposition. A main consequence of our results will be the fact that, unlike the Hilbert space case, a Toeplitz transfer map may have a time dependent unitary realization but no unitary realization that is independent of time. Another issue that is completely different from the Hilbert space case concerns the uniqueness of the realization. We obtain a result that gives a characterization of those transfer maps with unique minimal unitary realization.

Spaces with indefinite inner products appear naturally in the modeling of linear systems as substitutes of Hilbert spaces. In this paper we allow also the input-output spaces to have indefinite inner products, since this makes the theory more symmetric. Occasionally this hypothesis may be restricted, with no loss of generality, to the Hilbert space case.

The paper is organized as follows. The main results of the paper, on the realization and the uniqueness of the realization of time varying systems, are presented in Section 2. Several applications are presented in the last section where we discuss time invariant systems, the connection with the realization theorem of Azizov, the connection with the de Branges-Rovnyak model and we find a function in a generalized Schur class for which one cannot run a Schur type algorithm.

2 Realization of Lower Triangular Families

In this section we indicate solutions of the problems mentioned in the previous section for lower triangular families of operators. In fact, these solutions will be consequences of some of our previous results on Kolmogorov decompositions of Hermitian kernels and their multiplicative decomposition obtained in [11].

We briefly recall the terminology used in connection with Hermitian kernels. Let $\mathbf{U} = \{\mathcal{U}_i\}_{i\in\mathbf{Z}}$ be a family of Kreĭn spaces with (indefinite) inner products denoted by $[\cdot,\cdot]_{\mathcal{U}_i}$. A mapping H defined on $\mathbf{Z}\times\mathbf{Z}$ such that $H(i,j)\in\mathcal{L}(\mathcal{U}_j,\mathcal{U}_i)$ for all $i,j\in\mathbf{Z}$ is called a *Hermitian kernel* provided that $H(i,j)=H(j,i)^\sharp$ for all $i,j\in\mathbf{Z}$. Letting $\mathcal{F}_0(\mathbf{U})$ denote the vector space of sequences $f=(f_k)_{k\in\mathbf{Z}}$ with finite support and such that $f_k\in\mathcal{U}_k$ for all $k\in\mathbf{Z}$ we can define an indefinite inner product on $\mathcal{F}_0(\mathbf{U})$, denoted also by $[\cdot,\cdot]$,

$$[f,g]=\sum_{k\in\mathbf{Z}}[f_k,g_k]_{\mathcal{U}_k},\quad f=(f_k)_{k\in\mathbf{Z}},\ g=(g_k)_{k\in\mathbf{Z}}\in\mathcal{F}_0(\mathbf{U}). \tag{2.1}$$

In addition, to any Hermitian kernel H there corresponds an indefinite inner product $[\cdot,\cdot]_H$ on $\mathcal{F}_0(\mathbf{U})$

$$[f,g]_H=\sum_{k,l\in\mathbf{Z}}[H(l,k)f_k,g_l]_{\mathcal{U}_k},\quad f=(f_k)_{k\in\mathbf{Z}},\ g=(g_k)_{k\in\mathbf{Z}}\in\mathcal{F}_0(\mathbf{U}). \tag{2.2}$$

With this definition, for two Hermitian kernels H and K, we denote $H\leq K$ whenever

$$[f,f]_H\leq[f,f]_K,\quad f\in\mathcal{F}_0(\mathbf{U}). \tag{2.3}$$

This definition is consistent with the definition of positive semidefinite kernels in the sense that a Hermitian kernel L is positive semidefinite if and only if $L\geq 0$, where 0 is the 0-kernel.

By definition, a *Kolmogorov decomposition* of H is a pair $(V;\mathcal{K})$, where $(\mathcal{K},[\cdot,\cdot])$ is a Kreĭn space and $V=\{V_i\}_{i\in\mathbf{Z}}$ is a family of linear operators with the following properties:

$$\text{(a) } V_i\in\mathcal{L}(\mathcal{U}_i,\mathcal{K}),\quad\text{for all } i\in\mathbf{Z}.$$
$$\text{(b) } \mathcal{K}=\bigvee_{i\in\mathbf{Z}}V_i\mathcal{U}_i.$$
$$\text{(c) } H(i,j)=V_i^\sharp V_j,\quad\text{for all } i,j\in\mathbf{Z}.$$

It was proved in [11], Theorem 3.1, that H admits a Kolmogorov decomposition if and only if there exists L a positive semidefinite kernel such that $-L\leq H\leq L$. It is known from [22] that the latter condition is equivalent to the possibility of writing H as a difference of two positive semidefinite kernels. It is also shown in [11], Theorem 3.3, that if we suppose $H(n,n) = I_{\mathcal{U}_n}$ for all $n \in \mathbf{Z}$ and it is known that H admits a Kolmogorov decomposition, then there exists a family $\{\mathcal{K}_n\}_{n\in\mathbf{Z}}$ of Kreĭn spaces and a family $\{W_n\}_{n\in\mathbf{Z}}$ of unitary operators such that $W_n \in \mathcal{L}(\mathcal{K}_{n+1}, \mathcal{K}_n)$ for all $n \in \mathbf{Z}$ and the pair $(\{V_n^0\}_{n\in\mathbf{Z}}; \mathcal{K}_0)$, where

$$V_n^0 = \begin{cases} W_{-1}^\sharp W_{-2}^\sharp \dots W_n^\sharp | \mathcal{U}_n, & n < 0 \\ P_{\mathcal{U}_0}^{\mathcal{K}_0} | \mathcal{U}_0, & n = 0 \\ W_0 W_1 \dots W_{n-1} | \mathcal{U}_n, & n > 0 \end{cases} \tag{2.4}$$

is a Kolmogorov decomposition of the kernel H.

We now relate the realization problem with the existence of a Kolmogorov decomposition for a certain Hermitian kernel. This relation can be viewed as a version of the well-known Herglotz representation for the functions in the Carathéodory class. First, we notice that we can assume without loss of generality that $\mathcal{U}_{i+1} = \mathcal{Y}_i$ for all $i \in \mathbf{Z}$, and let $T = [T_{ij}]_{i,j\in\mathbf{Z}}$, $T_{ij} \in \mathcal{L}(\mathcal{U}_j, \mathcal{U}_{i+1})$, be a lower triangular family. Then, for $i \leq j$, we introduce the block matrix

$$T^{[ij]} = [T_{lm}]_{i\leq l,m\leq j}.$$

The matrix

$$\tilde{T}^{[ij]} = \begin{bmatrix} 0 & 0 \\ T^{[ij]} & 0 \end{bmatrix} \in \mathcal{L}\Big(\bigoplus_{k=i}^{j+1} \mathcal{U}_k\Big)$$

is lower triangular with zero diagonal, therefore

$$G_{T^{[ij]}} = (I + \tilde{T}^{[ij]})(I - \tilde{T}^{[ij]})^{-1}$$

is well-defined. Finally,

$$H_{T^{[ij]}} = \frac{1}{2}(G_{T^{[ij]}} + G_{T^{[ij]}}^\sharp) \tag{2.5}$$

is a $(j-i+2) \times (j-i+2)$ Hermitian block matrix. The mapping $T^{[ij]} \to H_{T^{[ij]}}$ is a one-to-one mapping from the set of $(j-i+1) \times (j-i+1)$ lower triangular block matrices onto the set of $(j-i+2) \times (j-i+2)$ Hermitian matrices with identity on the main diagonal. The inverse operation is easily described as follows: let $H^{[ij]} = [H_{lm}]_{i\leq l,m\leq j+1}$ be a $(j-i+2) \times (j-i+2)$ Hermitian block matrix with $H_{ii} = I_{\mathcal{U}_i}$ for $k = i, i+1, \dots, j+1$. Define

$$G_{H^{[ij]}} = I + 2(\text{the strictly lower part of } H^{[ij]})$$

and then introduce

$$\tilde{T}^{[ij]} = (G_{H^{[ij]}} + I)^{-1}(G_{H^{[ij]}} - I).$$

Since the diagonal of the $G_{H^{[ij]}}$ is the identity, it follows that $\tilde{T}^{[ij]} = \begin{bmatrix} 0 & 0 \\ T_{H^{[ij]}} & 0 \end{bmatrix}$ for some lower triangular $(j-i+1) \times (j-i+1)$ block matrix $T_{H^{[ij]}}$. Finally, we see that the mapping $H^{[ij]} \to T_{H^{[ij]}}$ is the inverse of the mapping $T^{[ij]} \to H_{T^{[ij]}}$.

In addition, we see that we can repeat the same operation for any pair (i, j) and associate to the lower triangular family T a uniquely determined Hermitian kernel H_T with $H_T(i,i) = I_{\mathcal{U}_i}$ for all $i \in \mathbf{Z}$ and the correspondence

$$T \longleftrightarrow H_T \tag{2.6}$$

is one-to-one and onto.

The following result provides a solution to the realization problem for lower triangular families.

Theorem 2.1 *The lower triangular family $T = [T_{ij}]_{i,j\in\mathbf{Z}}$ of operators $T_{ij} \in \mathcal{L}(\mathcal{U}_i)$ admits a realization by a unitary system Σ if and only if there exists a positive definite kernel L such that*

$$-L \leq H_T \leq L.$$

Moreover, if this holds, then the realization of T by a unitary system Σ can be chosen minimal.

Proof: Suppose there exists a positive definite kernel L such that $-L \leq H_T \leq L$. By the above mentioned Theorem 3.1 in [11], the kernel H_T admits a Kolmogorov decomposition and since $H_T(i,i) = I_{\mathcal{U}_i}$ for all $i \in \mathbf{Z}$, we can use Theorem 3.3 in [11] in order to obtain the multiplicative structure described by (2.4) of this Kolmogorov decomposition of H_T. With respect to the decompositions $\mathcal{K}_n = \mathcal{H}_n \oplus \mathcal{U}_n$, each unitary operator W_n, $n \in \mathbf{Z}$, admits the representation

$$W_n = \begin{bmatrix} A_n & B_n \\ C_n & D_n \end{bmatrix}. \tag{2.7}$$

We claim that the transfer map of the unitary system

$$\Sigma \begin{cases} x_{n+1} = A_n^{\sharp} x_n + C_n^{\sharp} u_n \\ y_n \;\;\;\, = B_n^{\sharp} x_n + D_n^{\sharp} u_n \end{cases}, \quad n \in \mathbf{Z}, \tag{2.8}$$

where A_n, B_n, C_n, D_n are given by (2.7), is exactly T. To that end, we notice that for $i \leq j$,

$$H_{T^{[ij]}} = \left(\bigoplus_{k=i}^{j+1} P_{\mathcal{U}_k} \right) \begin{bmatrix} I_{\mathcal{U}_i} & W_i & & \dots & W_i W_{i+1} \dots W_j \\ W_i^{\sharp} & I_{\mathcal{U}_{i+1}} & W_{i+1} & \dots & \\ \vdots & & \ddots & & \\ & & & \ddots & \\ W_j^{\sharp} & \dots & & W_i^{\sharp} & I_{\mathcal{U}_{j+1}} \end{bmatrix} \Bigg| \bigoplus_{k=i}^{j+1} \mathcal{U}_k$$

and a direct computation shows that

$$G_{H_{T^{[ij]}}} = \left(\bigoplus_{k=i}^{j+1} P_{\mathcal{U}_k}\right)(I - W^{[ij]})^{-1}(I + W^{[ij]})\Big|\bigoplus_{k=i}^{j+1} \mathcal{U}_k,$$

where

$$W^{[ij]} = \begin{bmatrix} 0 & 0 & \dots & 0 & \\ W_i^\sharp & 0 & \dots & & \\ 0 & W_{i+1}^\sharp & 0 & & \\ \vdots & & \ddots & & \\ 0 & 0 & \dots & W_j^\sharp & 0 \end{bmatrix}.$$

Consequently,

$$\begin{aligned} \tilde{T}^{[ij]} &= \left(\left(\bigoplus_{k=i}^{j+1} P_{\mathcal{U}_k}\right)(I - W^{[ij]})^{-1}(I + W^{[ij]})\Big|\bigoplus_{k=i}^{j+1} \mathcal{U}_k + I\right)^{-1} \\ &\quad \times \left(\left(\bigoplus_{k=i}^{j+1} P_{\mathcal{U}_k}\right)(I - W^{[ij]})^{-1}(I + W^{[ij]})\Big|\bigoplus_{k=i}^{j+1} \mathcal{U}_k - I\right) \\ &= \left(\left(\bigoplus_{k=i}^{j+1} P_{\mathcal{U}_k}\right)(I - W^{[ij]})^{-1}\Big|\bigoplus_{k=i}^{j+1} \mathcal{U}_k\right)^{-1} \\ &\quad \times \left(\left(\bigoplus_{k=i}^{j+1} P_{\mathcal{U}_k}\right)(I - W^{[ij]})^{-1}W^{[ij]}\Big|\bigoplus_{k=i}^{j+1} \mathcal{U}_k\right). \end{aligned}$$

After a reordering of spaces we can write, with respect to an obvious orthogonal decomposition, that

$$W^{[ij]} = \begin{bmatrix} A^{[ij]} & C^{[ij]} \\ B^{[ij]} & D^{[ij]} \end{bmatrix},$$

so we can use a well-known inversion formula for 2×2 block-matrices and obtain

$$\begin{aligned} \tilde{T}^{[ij]} &= \left(\left(\bigoplus_{k=i}^{j+1} P_{\mathcal{U}_k}\right)\begin{bmatrix} I - A^{[ij]} & C^{[ij]} \\ -B^{[ij]} & I - D^{[ij]} \end{bmatrix}^{-1}\Big|\bigoplus_{k=i}^{j+1} \mathcal{U}_k\right)^{-1} \\ &\quad \times \left(\left(\bigoplus_{k=i}^{j+1} P_{\mathcal{U}_k}\right)\begin{bmatrix} I - A^{[ij]} & C^{[ij]} \\ -B^{[ij]} & I - D^{[ij]} \end{bmatrix}^{-1}\begin{bmatrix} A^{[ij]} & C^{[ij]} \\ B^{[ij]} & D^{[ij]} \end{bmatrix}\Big|\bigoplus_{k=i}^{j+1} \mathcal{U}_k\right) \\ &= K\left(\left(\bigoplus_{k=i}^{j+1} P_{\mathcal{U}_k}\right)\begin{bmatrix} * & * \\ -K^{-1}B^{[ij]}(I - A^{[ij]})^{-1} & K^{-1} \end{bmatrix}\right. \\ &\quad \left.\begin{bmatrix} A^{[ij]} & C^{[ij]} \\ B^{[ij]} & D^{[ij]} \end{bmatrix}\Big|\bigoplus_{k=i}^{j+1} \mathcal{U}_k\right) \\ &= K(K^{-1}(D^{[ij]} + B^{[ij]}(I - A^{[ij]})^{-1}C^{[ij]})) = D^{[ij]} \\ &\quad + B^{[ij]}(I - A^{[ij]})^{-1}C^{[ij]}, \end{aligned}$$

where $K = I - D^{[ij]} - B^{[ij]}(I - A^{[ij]})^{-1}C^{[ij]}$. The previous formula for $\tilde{T}^{[ij]}$ shows that $T = T_\Sigma$, where Σ is given by (2.8).

Conversely, if T admits a unitary realization, then the same kind of computations as above shows how to obtain a Kolmogorov decomposition of the kernel H_T. We can omit the details.

It remains to show that a Kolmogorov decomposition $(V^0; \mathcal{K}_0)$ of H_T given by (2.4) is minimal if and only if (1.1) holds. Each unitary operator W_n involved in formula (2.4) of V_n^0 was decomposed as in (2.7). Let $g \in \mathcal{K}_0$ be chosen such that $g \perp \bigvee_{n \in \mathbf{Z}} V_n^0 \mathcal{U}_n$. First, it follows that $g \in \mathcal{H}_0$. Then, since $g \perp W_0 \mathcal{U}_1$, it follows that

$$\left[\begin{bmatrix} g \\ 0 \end{bmatrix}, \begin{bmatrix} B_0 u_1 \\ D_0 u_1 \end{bmatrix}\right]_{\mathcal{K}_0} = 0, \quad \text{for all } u_1 \in \mathcal{U}_1,$$

so $[g, B_0 u_1]_{\mathcal{U}_0} = 0$ for all $u_1 \in \mathcal{U}_1$. Similarly, $[g, C_0^\sharp u_{-1}]_{\mathcal{U}_0} = 0$. We show by induction that

$$[g, A_0 A_1 \dots A_{k-1} B_k u_{k+1}]_{\mathcal{U}_0} = 0, \quad \text{for all } u_{k+1} \in \mathcal{U}_{k+1},\ k \geq 0,$$
$$[g, A_{-1}^\sharp A_{-2}^\sharp \dots A_{k-1}^\sharp C_k^\sharp u_k]_{\mathcal{U}_0} = 0, \quad \text{for all } u_k \in \mathcal{U}_k,\ k < 0,$$

and these equalities show that, indeed, $(V^0; \mathcal{K}_0)$ is minimal if and only if the corresponding realization of T is minimal in the sense of (1.1). □

As we have already mentioned in Introduction, it is known that the uniqueness property for minimal unitary realizations is no longer true in Kreĭn spaces (see [22], [6]). However, a characterization of the uniqueness property can be obtained in terms of a so-called Gram operator associated to H_T with respect to a positive semidefinite kernel L such that $-L \leq H_T \leq L$. Even though such a characterization is not intrinsic, i.e. is not in terms of the transfer map itself, it is still easy to use it in order to deduce most of the known uniqueness criteria.

We indicate now how to define the Gram operator of a positive semidefinite kernel L with the property that $-L \leq H \leq L$, where H is a given Hermitian kernel. We keep the assumption that $\mathcal{U}_{i+1} = \mathcal{Y}_i$ for all $i \in \mathbf{Z}$. Let $\mathcal{F}_0(\mathbf{U})$ denote the set of all families $f = \{f_i\}_{i \in \mathbf{Z}}$ with finite support, $f_i \in \mathcal{U}_n$ for all $i \in \mathbf{Z}$. Since H is a Hermitian kernel, we can introduce on $\mathcal{F}_0(\mathbf{U})$ an indefinite inner product $[\cdot, \cdot]_H$ by the formula

$$[f, g]_H = \sum_{i,j \in \mathbf{Z}} [H(i, j) f(j), g(i)]_{\mathcal{U}_i}, \quad f, g \in \mathcal{F}_0(\mathbf{U}).$$

Denote by $\mathcal{K}_L$ the quotient-completion of $(\mathcal{F}_0(\mathbf{U}), [\cdot, \cdot]_L)$ to a Hilbert space. More precisely, let $\mathcal{N}_L = \{f \in \mathcal{F}_0(\mathbf{U}) \mid [f, f]_L = 0\}$ denote the isotropic subspace of the positive semidefinite inner product space $(\mathcal{F}_0(\mathbf{U}), [\cdot, \cdot]_L)$. Then, consider the quotient $\mathcal{F}_0(\mathbf{U})/\mathcal{N}_L$ and complete it to a Hilbert space $\mathcal{K}_L$. Since $-L \leq H \leq L$, we obtain that $\mathcal{N}_L$ is contained into the isotropic subspace $\mathcal{N}_H$ of the inner product $(\mathcal{F}_0(\mathbf{U}), [\cdot, \cdot]_H)$. Therefore, $[\cdot, \cdot]_H$ uniquely induces an inner product on $\mathcal{K}_L$, also denoted by $[\cdot, \cdot]_H$, such that

$$|[f, g]_H| \leq [f, f]_L^{1/2} [g, g]_L^{1/2}, \quad f, g \in \mathcal{K}_L.$$

Using the Riesz representation theorem we get a selfadjoint and contractive operator $A_L \in \mathcal{L}(\mathcal{K}_L)$ such that

$$[f, g]_H = [A_L f, g]_L, \quad f, g \in \mathcal{K}_L. \tag{2.9}$$

A_L is the *Gram operator* associated to the kernel H with respect to the nonnegative kernel L. For an arbitrary bounded operator T, $\rho(T)$ denotes the resolvent set of T, namely the set of complex numbers λ with the property that $\lambda I - T$ is boundedly invertible.

Theorem 2.2 *Let $T = [T_{ij}]_{i,j \in \mathbf{Z}}$ be a lower triangular family of operators $T_{ij} \in \mathcal{L}(\mathcal{U}_i)$ admitting unitary realizations. Then any two minimal unitary realizations of T are unitarily equivalent if and only if the following property holds: for every positive semidefinite kernel L such that $-L \leq H_T \leq L$, there exists $\epsilon > 0$ such that either $(0, \epsilon) \subset \rho(A_L)$ or $(-\epsilon, 0) \subset \rho(A_L)$, where A_L denotes the Gram operator of H_T with respect to L.*

Proof: We have seen in the proof of Theorem 2.1 that a Kolmogorov decomposition $(V^0; \mathcal{K}_0)$ of H_T given by (2.4) is minimal if and only if (1.1) holds.

Next, we show that two Kolmogorov decompositions $(V; \mathcal{K}_0)$ and $(V'; \mathcal{K}_0')$ of H_T are unitarily equivalent if and only if their corresponding realizations are unitarily equivalent. The fact that $(V; \mathcal{K}_0)$ and $(V'; \mathcal{K}_0')$ of H_T are unitarily equivalent means that there exists a unitary operator $\Phi_0 : \mathcal{K}_0 \to \mathcal{K}_0'$ such that $V_n' = \Phi_0 V_n$ for all $n \in \mathbf{Z}$. In particular, for $n = 0$,

$$P_{\mathcal{U}_0}^{\mathcal{K}_0'} | \mathcal{U}_0 = \Phi_0 P_{\mathcal{U}_0}^{\mathcal{K}_0} | \mathcal{U}_0.$$

This shows that $\mathcal{U}_0$ is a reducing space of Φ_0 and $\Phi_0 = I_{\mathcal{U}_0} \oplus \phi_0$ for some unitary operator $\phi_0 \in \mathcal{L}(\mathcal{H}_0, \mathcal{H}_0')$. Then, we can inductively define the unitary operators $\Phi_n = I_{\mathcal{U}_n} \oplus \phi_n \in \mathcal{L}(\mathcal{K}_n, \mathcal{K}_n')$ such that $W_n' \Phi_{n+1} = \Phi_n W_n$ for all $n \in \mathbf{Z}$. It follows that the operators ϕ_n, $n \in \mathbf{Z}$, give a unitary equivalence of the realizations of T associated to $(V; \mathcal{K}_0)$ and $(V'; \mathcal{K}_0')$. An application of Theorem 4.1 in [11] concludes the proof. □

3 Applications

We discuss here a few applications of the realizations of time dependent systems. Thus, we deal with the realization problem for time independent systems and we show the connection with the de Branges-Rovnyak model. We also give an example of a Toeplitz lower triangular family of operators without a time invariant realization but admitting a time dependent realization and an example of a function in a generalized Schur class for which we cannot run the Schur algorithm.

A. Time Independent Systems. For a positive semidefinite Toeplitz kernel, the Kolmogorov decomposition gives rise to the Naĭmark dilation. However, for

Hermitian kernels the situation is more complicated. This is explained in [11] (the problem of existence of the Naĭmark dilation was previously delt with by M. Cotlar and C. Sadoski [12] and F. Pelaez [23]).

We briefly describe the setting in [11]. Let $\mathcal{U}$ be a Kreĭn space and consider the complex vector space $\mathcal{F}_0(\mathcal{U})$ of all functions $h\colon \mathbf{Z} \to \mathcal{U}$ with finite support ($\mathcal{F}_0(\mathcal{U}) = \mathcal{F}_0(\mathbf{U})$, where $\mathbf{U} = \{\mathcal{U}_i\}_{i\in\mathbf{Z}}$ such that $\mathcal{U}_i = \mathcal{U}$ for all $i \in \mathbf{Z}$).

On $\mathcal{F}_0(\mathcal{U})$ we consider two operators, the *forward shift* S_+ defined by $(S_+h)(n) = h(n-1)$, for all $h \in \mathcal{F}_0(\mathcal{U})$ and $n \in \mathbf{Z}$, and the *backward shift* S_- defined by $(S_-h)(n) = h(n+1)$, for all $h \in \mathcal{F}_0(\mathcal{U})$ and all $n \in \mathbf{Z}$. A positive semidefinite kernel L is called of *foreward shift bounded type* if S_+ is bounded with respect to the seminorm $\mathcal{F}_0(\mathcal{U}) \ni h \mapsto [h,h]_L^{1/2}$, that is,

$$[S_+h, S_+h]_L \leq C[h,h]_L, \quad h \in \mathcal{F}_0(\mathcal{U}),$$

for some constant $C \geq 0$. We denote by $\mathcal{R}_+^+(\mathcal{U})$ the class of all kernels L of foreward shift bounded type. Similarly, a positive semidefinite kernel L is called of *backward shift bounded type* if S_- is bounded with respect to the seminorm $\mathcal{F}_0(\mathcal{U}) \ni h \mapsto [h,h]_L^{1/2}$ and denote by $\mathcal{R}_-^+(\mathcal{U})$ the corresponding class of kernels. Also, let $\mathcal{R}_0^+(\mathcal{U}) = \mathcal{R}_+^+(\mathcal{U}) \cap \mathcal{R}_-^+(\mathcal{U})$.

We established in [11] that for a Toeplitz Hermitian kernel H there exists $L \in \mathcal{R}_0^+(\mathcal{U})$ such that $-L \leq H \leq L$, if and only if H admits a Naĭmark dilation $(U, Q; \mathcal{K})$, where $\mathcal{K}$ is a Kreĭn space, $U \in \mathcal{L}(\mathcal{K})$ is a unitary operator and $Q \in \mathcal{L}(\mathcal{U}, \mathcal{K})$, such that

$$\mathcal{K} = \bigvee_{n\in\mathbf{Z}} U^n Q\mathcal{U},$$

$$H(i,j) = Q^\sharp U^{i-j} Q, \quad i, j \in \mathbf{Z}.$$

Let $T = [T_{ij}]_{i,j\in\mathbf{Z}}$ be a lower triangular Toeplitz family of operators $T_{ij} \in \mathcal{L}(\mathcal{U}, \mathcal{Y})$. Without loss of generality, we can assume $\mathcal{U} = \mathcal{Y}$, and we see that the kernel H_T associated to T by the formula (2.5) is a Toeplitz Hermitian kernel.

The implication (iii) $\Rightarrow$(iv) in the next result is a theorem of T.Ya. Azizov (cf. Theorem 5.3.10 in [3]) for which we give an indirect proof, but connecting the logical path it turns out that our proof has a number of similarities with his original proof in that it uses the existence of a Naĭmark dilation (unitary dilation, in [3]). However, we emphasize the equivalence of (ii) and (iii) in the next theorem which shows that a certain majorization of the associated Hermitian kernel implies the convergence of a power series, at least in a neighbourhood of 0. In addition, this result enables us to relate the Cauchy-Hadamard convergence test of power series with the realization theory and Naĭmark dilations.

Theorem 3.1 *Let $T = [T_{ij}]_{i,j\in\mathbf{Z}}$ be a Toeplitz lower triangular family. Set $T_{i-j} = T_{ij}$ for $i \geq j$ and consider the formal power series*

$$\Theta(z) = \sum_{k\geq 0} T_k z^k.$$

The following assertions are equivalent:

(i) *T admits a time independent unitary realization.*

(ii) *There exists $L \in \mathcal{R}_0^+(\mathcal{U})$ such that $-L \leq H_T \leq L$.*

(iii) *The formal power series Θ converges in a neighbourhood of* 0.

(iv) *There exists a unitary operator $U = \begin{bmatrix} A & B \\ C & D \end{bmatrix}$ and $r > 0$ such that*

$$\Theta(z) = D - zC(I - zA)^{-1}B, \quad |z| < r.$$

In addition, if these assertions hold, then the realizations as in (i) *and* (iv) *can be chosen minimal.*

Proof: (ii)⇒(i) This is similar to the proof of Theorem 2.1. By Theorem 6.12 in [11], there exists a Naĭmark dilation of the kernel H_T. Since $H_T(0,0) = I_{\mathcal{U}}$, Q is a bounded isometric operator from $\mathcal{U}$ to $\mathcal{K}$ and so, the space $\mathcal{U}$ can be identified with the subspace $Q\mathcal{U}$ of $\mathcal{K}$. Therefore,

$$H_T(i,j) = P_{\mathcal{U}} U^{i-j}|\mathcal{U}, \quad i, j \in \mathbf{Z},$$

and

$$\mathcal{K} = \bigvee_{n \in \mathbf{Z}} U^n \mathcal{U}.$$

With respect to the decomposition $\mathcal{K} = \mathcal{H} \oplus \mathcal{U}$, the operator U admits the 2×2 block-matrix representation $\begin{bmatrix} A & B \\ C & D \end{bmatrix}$. Then the transfer map of the unitary system

$$\Sigma \begin{cases} x_{n+1} = A^\sharp x_n + C^\sharp u_n \\ y_n \quad = B^\sharp x_n + D^\sharp u_n, \end{cases} \tag{3.1}$$

is exactly Θ. Since the computations are similar to those in the proof of Theorem 2.1 we omit the details.

(i)⇒(ii) We notice that in the correspondence (2.6), H_T is a Toeplitz Hermitian kernel provided that Θ is the transfer map of a time independent system.

(i)⇒(iii) Since T admits a minimal unitary realization

$$\Sigma \begin{cases} x_{n+1} = Ax_n + Bu_n \\ y_n \quad = Cx_n + Du_n, \end{cases} \tag{3.2}$$

it follows that

$$T_k = T_{k0} = \begin{cases} D, & k = 0 \\ CA^kB, & k \geq 1 \end{cases} \tag{3.3}$$

and the Cauchy-Hadamard formula shows that the series Θ converges for $|z| < \|A\|^{-1}$.

(i)⇒(iv) As before we have (3.2) and (3.3) from which it readily follows that for some $r > 0$ the following representation holds:

$$\Theta(z) = D - zC(I - zA)^{-1}B, \quad |z| < r.$$

(iv)⇒(i) This is obvious.

(iii)⇒(ii) For a certain simplicity, and without any loss of generality, we work only with respect to the Hilbert structure of $\mathcal{U}$. Denote by $\langle\cdot,\cdot\rangle$ and $\|\cdot\|$ the Hilbertian inner product and, respectively, the corresponding norm on $\mathcal{U}$. Let Θ be a holomorphic function on a neighbourhood of zero,

$$\Theta(z) = \sum_{k\geq 0} T_k z^k, \tag{3.4}$$

and let $T = [T_{ij}]_{i,j\in\mathbf{Z}}$ be the associated Toeplitz lower triangular family, $T_{ij} = T_{i-j}$. Then, there is $r > 0$ such that Θ is holomorphic for $|z| < r$, and for r', $0 < r' < r$, there is $\alpha > 0$ such that

$$\|\Theta(z)\| \leq \alpha, \quad |z| < r'.$$

For $|z| < \min\{1, r', 1/\alpha\} = r''$, we can define the function

$$G_\Theta(z) = (I - z\Theta(z))^{-1}(I + z\Theta(z)).$$

We also have that there is $\rho > 0$ such that $\|G_\Theta(z)\| \leq C$ for $|z| \leq \rho$. Define

$$G_{\Theta,\rho}(z) = G_\Theta(\rho z), \quad |z| \leq 1,$$

and

$$H_{T,\rho}(\zeta) = \frac{G_{\Theta,\rho}(\zeta) + G^*_{\Theta,\rho}(\zeta)}{2}, \quad |\zeta| = 1.$$

Then, $\|H_{T,\rho}(\zeta)\| \leq C$ for $|\zeta| = 1$, so $H_{T,\rho} \in L^\infty(\mathcal{U})$. By a well-known result (see, for instance, Theorem IX.1.1 in [16]), it follows that the operator of multiplication by $H_{T,\rho}$, denoted also by $H_{T,\rho}$, is bounded on $L^2(\mathcal{U})$, equivalently, for some $\beta > 0$ we have

$$-\beta I \leq H_{T,\rho} \leq \beta I$$

where I denotes the identity kernel, that is, $I(n,n) = I_{\mathcal{U}}$ and $I(n,k) = 0$ for all $n \neq k$, $n, k \in \mathbf{Z}$.

Let the inner product $[\cdot,\cdot]$ on $\mathcal{F}_0(\mathcal{U})$ be defined as in (2.1). Notice that the shift operators $S_\pm$ are isometric with respect to the inner product $[\cdot,\cdot]$, in particular they are bounded with respect to the inner product $[\cdot,\cdot]_{\beta I}$ and hence we can use Theorem 6.8 in [11] and conclude that there is a Naĭmark dilation $(\mathcal{K}; U)$ of $H_{T,\rho}$, that is, $\mathcal{K}$ is a Kreĭn space and $U \in \mathcal{L}(\mathcal{K})$ a unitary operator such that

$$\mathcal{K} = \bigvee_{n\in\mathbf{Z}} U^n \mathcal{U}$$

and

$$\rho^j A_j = P_{\mathcal{U}} U^j | \mathcal{U}, \quad j \in \mathbf{Z}.$$

It follows that

$$A_j = P_{\mathcal{U}} (\rho^{-1} U)^j | \mathcal{U}, \quad j \in \mathbf{Z}.$$

If we consider the Taylor expansion of G_Θ in a neighbourhood of 0,

$$G_\Theta(z) = I + 2 \sum_{k \geq 1} z^k A_k,$$

then the block matrix representation of the kernel H_T defined by (2.6) is

$$(3.5) \qquad H_T \sim \begin{bmatrix} \ddots & & & & \\ & I & A_1 & A_2 & \\ & A_1^* & I & A_1 & \\ & A_2^* & A_1^* & I & \\ & & & & \ddots \end{bmatrix}.$$

Since the operator U is unitary, the operator $\rho^{-1} U$ is bounded with bounded inverse. Using Example 6.5 and the proof of Theorem 3.1 ((3) $\Rightarrow$ (1)) in [11], it follows that $-L \leq H_T \leq L$ for some $L \in \mathcal{R}_0^+(\mathcal{U})$. □

On the ground of the simple relation connecting the coefficients in the Taylor expansion (3.4) of Θ and the entries of the matrix of H_T in (3.5), Theorem 3.1 enables us to obtain a more tractable characterization of the Hermitian Toeplitz kernels admitting Naĭmark dilations (compare with [12], [23], and Theorem 6.8 in [11]).

Corollary 3.2 *Let H be a Hermitian Toeplitz kernel H and denote $A_{i-j} = H(i, j)$ for $i, j \in \mathbf{Z}$. Define $T_0 = A_1$ and then recursively*

$$T_n = A_{n+1} - A_1 T_{n-1} - \cdots A_n T_0, \quad n \geq 1.$$

A necessary and sufficient condition for the existence of a Naĭmark dilation of H is that $\lim \sup_{n \to \infty} \|T_n\|^{1/n} < \infty$.

Proof: This is an application of Theorem 3.1 and Cauchy-Hadamard formula. □

B. de Branges-Rovnyak models. We show a direct deduction of the de Branges-Rovnyak model from the realization based on the Naĭmark dilation. Let Θ be a holomorphic function in a neighbourhood of zero,

$$\Theta(z) = \sum_{k \geq 0} T_k z^k,$$

and let $T = [T_{ij}]_{i,j\in\mathbf{Z}}$ be the associated Toeplitz lower triangular family, $T_{ij} = T_{i-j}$. Then, there is $r > 0$ such that Θ is holomorphic for $|z| < r$, and for r', $0 < r' < r$, there is $M > 0$ such that

$$\|\Theta(z)\| \leq M, \quad |z| < r'.$$

For $|z| < \min\{1, r', 1/M\} = r''$, we can define the function

$$G_\Theta(z) = (I - z\Theta(z))^{-1}(I + z\Theta(z)),$$

which is also holomorphic for $|z| < r''$. For a function S on $|z| < r''$, we use the notation $\tilde{S}(z) = S^\sharp(\bar{z})$. Using this notation and the function G_Θ, we introduce a new kernel on $\{|z| < r''\} \times \{|z| < r''\}$ by the formula

$$\mathcal{F}_\Theta(z,w) = \frac{1}{2}\begin{bmatrix} \dfrac{G_\Theta(z) + G_\Theta^\sharp(w)}{1 - z\bar{w}} & \dfrac{G_\Theta(z) - G_\Theta(\bar{w})}{z - \bar{w}} \\ \dfrac{\tilde{G}_\Theta(z) - \tilde{G}_\Theta(\bar{w})}{z - \bar{w}} & \dfrac{\tilde{G}_\Theta(z) + \tilde{G}_\Theta^\sharp(w)}{1 - z\bar{w}} \end{bmatrix}.$$

This kernel can be viewed as a "frequency representation" of the Toeplitz kernel H_T since, formally, for all $|z|, |w| < r''$ we have

$$\mathcal{F}_\Theta(z,w) = \begin{bmatrix} \cdots & z^2 & z & 1 & 0 & 0 & 0 & \cdots \\ \cdots & 0 & 0 & 0 & 1 & z & z^2 & \cdots \end{bmatrix} H_T \begin{bmatrix} \vdots & \vdots \\ \bar{w}^2 & 0 \\ \bar{w} & 0 \\ 1 & 0 \\ 0 & 1 \\ 0 & \bar{w} \\ 0 & \bar{w}^2 \\ \vdots & \vdots \end{bmatrix}, \tag{3.6}$$

where H_T is the kernel associated to T by the formula (2.6). It is expected that the existence of a Naĭmark dilation for H_T will imply the existence of a Kolmogorov decomposition for $\mathcal{F}_\Theta$. In fact, the relation is quite simple. If $U \in \mathcal{L}(\mathcal{K})$ is a Naĭmark dilation of H_T, then we obtain by direct computation the following Herglotz type formula:

$$G_\Theta(z) = P_{\mathcal{U}}(I - zU)^{-1}(I + zU)|\mathcal{U}, \quad |z| < \rho.$$

We deduce that

$$\mathcal{F}_\Theta(z,w) = A^\sharp(z)A(w),$$

where

$$A(w) = [\, I_{\mathcal{K}} \quad U \,]\begin{bmatrix} (I - \bar{w}U^\sharp)^{-1}|\mathcal{U} & 0 \\ 0 & (I - \bar{w}U)^{-1}|\mathcal{U} \end{bmatrix}.$$

A simple argument shows that

$$\bigvee_{|w|<\rho} A(w)(\mathcal{U} \oplus \mathcal{U}) = \mathcal{K},$$

and we obtain that the state-space of a realization of Θ can be simply described as $\mathcal{H} = \mathcal{K} \ominus \mathcal{U}$.

Now, a legitimate question is whether a state-space can be directly obtained as the space of a Kolmogorov decomposition of a certain kernel associated to Θ. This question can be simply answered as follows. We can compute $A(w)$ using the matrix entries of U. Thus, we obtain

$$\begin{aligned} A(w) &= [I_{\mathcal{K}} \quad U]\begin{bmatrix} (I - \bar{w}U^{\sharp})^{-1} & 0 \\ 0 & (I - \bar{w}U)^{-1} \end{bmatrix}\Big|(\mathcal{U} \oplus \mathcal{U}) \\ &= [(I - \bar{w}U^{\sharp})^{-1} \quad U(I - \bar{w}U)^{-1}]\,|(\mathcal{U} \oplus \mathcal{U}) \\ &= \begin{bmatrix} \bar{w}(I - \bar{w}A^{\sharp})^{-1}C^{\sharp} & (I - \bar{w}A)^{-1}B \\ I_{\mathcal{U}} & \Theta(\bar{w}) \end{bmatrix}\begin{bmatrix} \tilde{S}^{-1}(\bar{w}) & 0 \\ 0 & S^{-1}(\bar{w}) \end{bmatrix}, \end{aligned}$$

where $S(z) = I - z\Theta(z)$. We see that we can write

$$A(w) = \left(\begin{bmatrix} 0 & 0 \\ I_{\mathcal{U}} & \tilde{\Theta}^{\sharp}(w) \end{bmatrix} + \begin{bmatrix} \bar{w}(I - \bar{w}A^{\sharp})^{-1}C^{\sharp} & (I - \bar{w}A)^{-1}B \\ 0 & 0 \end{bmatrix}\right)$$
$$\begin{bmatrix} \tilde{S}^{-1}(\bar{w}) & 0 \\ 0 & S^{-1}(\bar{w}) \end{bmatrix},$$

and define the functions

$$\begin{aligned} A_1(w) &= \begin{bmatrix} 0 & 0 \\ I_{\mathcal{U}} & \tilde{\Theta}^{\sharp}(w) \end{bmatrix}\begin{bmatrix} \tilde{S}^{-1}(\bar{w}) & 0 \\ 0 & S^{-1}(\bar{w}) \end{bmatrix}, \\ A_2(w) &= \begin{bmatrix} (I - \bar{w}A^{\sharp})^{-1}C^{\sharp} & (I - \bar{w}A)^{-1}B \\ 0 & 0 \end{bmatrix}\begin{bmatrix} \bar{w}\tilde{S}^{-1}(\bar{w}) & 0 \\ 0 & S^{-1}(\bar{w}) \end{bmatrix}. \end{aligned}$$

Using the minimality of U we deduce that

$$\begin{aligned} \bigvee_{|w|<\rho} A_1(w)(\mathcal{U} \oplus \mathcal{U}) &= \mathcal{U}, \\ \bigvee_{|w|<\rho} A_2(w)(\mathcal{U} \oplus \mathcal{U}) &= \mathcal{K} \ominus \mathcal{U}. \end{aligned}$$

The last relation implies that

$$\bigvee_{|w|<\rho} \begin{bmatrix} (I - \bar{w}A^{\sharp})^{-1}C^{\sharp} & (I - \bar{w}A)^{-1}B \\ 0 & 0 \end{bmatrix}(\mathcal{U} \oplus \mathcal{U}) = \mathcal{K} \ominus \mathcal{U},$$

and this shows that the state-space of the realization of Θ associated to U is obtained from the Kolmogorov decomposition of the kernel

$$\begin{aligned} B_\Theta(z,w) &= \begin{bmatrix} C(I-zA)^{-1} \\ B^\sharp(I-zA^\sharp)^{-1} \end{bmatrix} [\,(I-\bar{w}A^\sharp)^{-1}C^\sharp \quad (I-\bar{w}A)^{-1}B\,] \\ &= \begin{bmatrix} \dfrac{I-\Theta(z)\Theta^\sharp(w)}{1-z\bar{w}} & \dfrac{\Theta(z)-\Theta(\bar{w})}{z-\bar{w}} \\ \dfrac{\tilde{\Theta}(z)-\tilde{\Theta}(\bar{w})}{z-\bar{w}} & \dfrac{I-\tilde{\Theta}(z)\tilde{\Theta}^\sharp(w)}{1-z\bar{w}} \end{bmatrix}. \end{aligned}$$

B_Θ is the well-known de Branges-Rovnyak kernel (see [6], [1]).

C. Schur classes. Starting with the work of Kreĭn and Langer, [19], it appeared to be of interest to generalize the Schur class of functions to an indefinite setting. One possible extension (see [3]) is to say that a function holomorphic in a neighbourhood of zero belongs to $\mathcal{S}'_\kappa(\mathcal{U},\mathcal{Y})$, $\kappa \geq 0$, provided that it admits a minimal unitary realization with state-space a Pontryagin space of negative index κ. Another possible class (as defined in [19] and also discussed in [1]), is denoted by $\mathcal{S}_\kappa(\mathcal{U},\mathcal{Y})$ and consists of those functions holomorphic in a neighbourhood of zero with the property that the kernels

$$\frac{I-\Theta(z)\Theta^\sharp(w)}{1-z\bar{w}}, \quad B_\Theta, \quad B_{\tilde{\Theta}}, \quad \frac{I-\tilde{\Theta}(z)\tilde{\Theta}^\sharp(w)}{1-z\bar{w}},$$

have the same number κ of negative squares (see [1] for a detailed discussion). A first reason that these classes are more tractable is that their elements have the uniqueness property. An interesting question is whether a Schur-type algorithm can be developed for these classes of functions (see [13], [1]). A possible approach to this question can be based on the following result (see [16] for more details in the Hilbert space case).

Let $\Theta \in \mathcal{S}_\kappa(\mathcal{U},\mathcal{Y})$. It follows that $I-\Theta(0)\Theta^\sharp(0)$ and $I-\Theta^\sharp(0)\Theta(0)$ both have finite negative indices $\leq \kappa$. We suppose in addition that

$$\kappa_-(I-\Theta(0)\Theta^\sharp(0)) = \kappa_-(I-\Theta^\sharp(0)\Theta(0)) = \kappa_0 \leq \kappa. \tag{3.7}$$

We recall the concept of elementary rotation. For two fundamental symmetries J_1 and J_2 on $\mathcal{H}_1$ and, respectively, $\mathcal{H}_2$ and T an operator in $\mathcal{L}(\mathcal{H}_1,\mathcal{H}_2)$, we define the *defect operator* $D_T = |J_1 - T^*J_2T|^{1/2}$ and the *defect space* $\mathcal{D}_T = \overline{D_T\mathcal{H}_1}$. The defect space is viewed as a Kreĭn space with the indefinite inner product determined by the symmetry $J_T = \operatorname{sgn}(J_1 - T^*J_2T)$. It is known (see [2]) that there exists a uniquely determined unitary operator $R(T)$ from $\mathcal{H}_1 \oplus \mathcal{D}_{T^*}$ onto $\mathcal{H}_2 \oplus \mathcal{D}_T$ such that

$$R(T) = \begin{bmatrix} T & D_{T^*} \\ D_T & -L_{T^*}J_{T^*} \end{bmatrix}, \tag{3.8}$$

where $L_{T^*}\colon \mathcal{D}_{T^*} \to \mathcal{D}_T$ is the so-called *link operator* (see [2]). For example, the link operator is uniquely determined by the equation $L_{T^*}D_{T^*} = D_TJ_1T^*$ and it also satisfies the relation $(J_{T^*} - D_{T^*}J_1D_{T^*})|\mathcal{D}_{T^*} = L_{T^*}^*J_TL_{T^*}$.

The operator $R(T)$ is called *the elementary rotation* of T and it may be viewed as the unique solution of the problem of finding a unitary operator extending both the operator row matrix $[\,T \quad D_{T^*}\,]$ and the operator column matrix $\begin{bmatrix} T \\ D_T \end{bmatrix}$. We use this terminology in order to distinguish $R(T)$ from the so-called Julia operators as in [1]. Thus, according to [1], a *Julia operator* for T is any unitary operator

$$\begin{bmatrix} T & D_* \\ D & -L^\# \end{bmatrix}$$

in $\mathcal{L}(\mathcal{H}_1 \oplus \mathcal{D}_*, \mathcal{H}_2 \oplus \mathcal{D})$ such that $D^\# \in \mathcal{L}(\mathcal{D}, \mathcal{H}_1)$ and $D_* \in \mathcal{L}(\mathcal{D}_*, \mathcal{H}_2)$ have zero kernels. Obviously, $R(T)$ is a Julia operator. Under the condition (3.7), the elementary rotation is essentially unique.

Theorem 3.3 *Suppose that* $\kappa^-[\mathcal{U}] = \kappa^-[\mathcal{Y}] < \infty$ *and that* (3.7) *holds. Consider the function* $\Theta \in \mathcal{S}_\kappa(\mathcal{U}, \mathcal{Y})$ *and let*

$$\Theta(z) = D + zC(I - zA)^{-1}B$$

be a minimal unitary realization of Θ. *Let*

$$R(D) = \begin{bmatrix} D & \beta \\ \gamma & \delta \end{bmatrix} : \mathcal{U} \oplus \mathcal{D} \longrightarrow \mathcal{Y} \oplus \tilde{\mathcal{D}}$$

be the elementary rotation of D. *Then, there is* $\Theta_1 \in \mathcal{S}_{\kappa-\kappa_0}(\tilde{\mathcal{D}}, \mathcal{D})$ *such that*

$$\Theta(z) = D + z\beta\Theta_1(z)(I - z\delta\Theta_1(z))^{-1}\gamma, \tag{3.9}$$

if and only if

$$W = \begin{bmatrix} D & C \\ B & A \end{bmatrix} = (I_{\mathcal{Y}} \oplus \Phi^\sharp)(R(D) \oplus I)(I \oplus W_1)(I_{\mathcal{U}} \oplus \Phi), \tag{3.10}$$

where $W_1 = \begin{bmatrix} D_1 & C_1 \\ B_1 & A_1 \end{bmatrix}$, $\Theta_1(z) = D_1 + zC_1(I - zA_1)^{-1}B_1$ *is a minimal unitary realization of* Θ_1, *and* Φ *is a bounded unitary operator.*

Proof: Suppose (3.9) holds. Define

$$\begin{aligned} W' &= (R(D) \oplus I)(I \oplus W_1) \\ &= \begin{bmatrix} D & \beta D_1 & \beta C_1 \\ \gamma & \delta D_1 & \delta C_1 \\ 0 & B_1 & A_1 \end{bmatrix} : \mathcal{U} \oplus \tilde{\mathcal{D}} \oplus \mathcal{H}_1 \longrightarrow \mathcal{Y} \oplus \tilde{\mathcal{D}} \oplus \mathcal{H}_1, \end{aligned}$$

where $\mathcal{H}_1$ is the state-space of the given realization of Θ_1, so $\kappa_-(\mathcal{H}_1) = \kappa - \kappa_0$. It is easily checked that W' provides a minimal unitary realization for Θ, and since

$\kappa_-(\tilde{\mathcal{D}} \oplus \mathcal{H}_1) = \kappa$, it follows that a unitary operator $\Phi : \mathcal{H} \to \tilde{\mathcal{D}} \oplus \mathcal{H}_1$ exists such that

$$W = (I_{\mathcal{Y}} \oplus \Phi^\sharp)W'(I_{\mathcal{U}} \oplus \Phi) = (I_{\mathcal{Y}} \oplus \Phi^\sharp)(R(D) \oplus I)(I \oplus W_1)(I_{\mathcal{U}} \oplus \Phi).$$

The converse implication is similar and we can omit the details. □

As a conclusion of the previous result, if we want to develop a Schur-type algorithm, then we are faced with the problem of factoring a unitary operator $W = \begin{bmatrix} D & C \\ B & A \end{bmatrix}$ as in (3.10). This is reminiscent of Euler's description of $SO(3)$.

We now give an example for which this factorization cannot be realized. Set $\mathcal{U} = \mathbf{C}^3$, with the Euclidean structure, and consider $\mathcal{H}$ the Pontryagin space of negative index 1 obtained on $\mathbf{C}^3$ with respect to the fundamental symmetry

$$\begin{bmatrix} 1 & 0 & 0 \\ 0 & 0 & 1 \\ 0 & 1 & 0 \end{bmatrix}.$$

Then, define the operators A, B, C, D as given by the matrices:

$$D = \begin{bmatrix} 0 & 0 & 0 \\ 0 & 1 & 0 \\ 0 & 0 & 1 \end{bmatrix}, \quad C = \begin{bmatrix} 1 & 0 & 0 \\ 0 & 0 & 1 \\ 0 & 0 & 1 \end{bmatrix},$$

$$B = \begin{bmatrix} 1 & 0 & 0 \\ 0 & 0 & 0 \\ 0 & 1 & 1 \end{bmatrix}, \quad A = \begin{bmatrix} 0 & 0 & 0 \\ 0 & 0 & -1 \\ 0 & -1 & 1 \end{bmatrix}.$$

By direct computation it follows that the block matrix $\begin{bmatrix} D & C \\ B & A \end{bmatrix}$ is unitary. Moreover, we simply obtain that the elementary rotation of D is

$$R(D) = \begin{bmatrix} 0 & 0 & 0 & 1 \\ 0 & 1 & 0 & 0 \\ 0 & 0 & 1 & 0 \\ 1 & 0 & 0 & 0 \end{bmatrix} : \mathcal{U} \oplus \mathbf{C} \longrightarrow \mathcal{U} \oplus \mathbf{C},$$

so $\mathcal{D} = \mathbf{C}$, $\tilde{\mathcal{D}} = \mathbf{C}$, with the Euclidean structure. We also check that

$$\begin{aligned} B\mathcal{U} \vee AB\mathcal{U} \vee A^2B\mathcal{U} \ldots \vee &= \mathcal{H}, \\ C^\sharp\mathcal{U} \vee A^\sharp C^\sharp\mathcal{U} \vee A^{\sharp 2}C^\sharp\mathcal{U} \vee \ldots &= \mathcal{H}. \end{aligned}$$

We define the function

$$\Theta(z) = D + zC(I - zA)^{-1}B$$

and since $\begin{bmatrix} D & C \\ B & A \end{bmatrix}$ gives a minimal unitary realization of this function, it follows that the Hermitian kernel

$$\frac{I - \Theta(z)\Theta^{\sharp}(w)}{1 - z\bar{w}},$$

has one negative square. By a result in [1], it follows that $\Theta \in \mathcal{S}_1(\mathcal{U})$. We see that

$$\kappa_-(I - \Theta(0)\Theta^{\sharp}(0)) = \kappa_-(I - \Theta^{\sharp}(0)\Theta(0)) = 0.$$

Assuming that a function $\Theta_1 \in \mathcal{S}_1(\tilde{\mathcal{D}}, \mathcal{D})$ does exist subject to the condition (3.9), then it follows from Theorem 3.4 that (3.10) holds. This implies that there exists a matrix $[\, c_1 \quad c_2 \quad c_3 \,]$ such that

$$C = \beta\, [\, c_1 \quad c_2 \quad c_3 \,],$$

or

$$\begin{bmatrix} 1 & 0 & 0 \\ 0 & 0 & 1 \\ 0 & 0 & 1 \end{bmatrix} = \begin{bmatrix} 1 \\ 0 \\ 0 \end{bmatrix} [\, c_1 \quad c_2 \quad c_3 \,],$$

which is not possible. That is, the Schur algorithm does not work for Θ. This provides a negative answer to a question in [13] (see also [1], p. 186).

D. A Toeplitz Map without Time Invariant Realizations. We conclude this section with an example of a Toeplitz transfer map admitting a time variant unitary realization but no time invariant unitary realization. This contrasts with the Hilbert space case.

Theorem 3.4 *There exists a Toeplitz lower triangular family T with the property that there is no system*

$$\Sigma \begin{cases} x_0 &= 0 \\ x_{n+1} &= Ax_n + Bu_n \qquad n \geq 0, \\ y_n &= Cx_n + Du_n, \end{cases}$$

such that $T_\Sigma = T$, but there is a unitary time dependent system

$$\Sigma_0 \begin{cases} x_0 &= 0 \\ x_{n+1} &= A(n)x_n + B(n)u_n \quad , \quad n \geq 0 \\ y_n &= C(n)x_n + D(n)u_n, \end{cases}$$

such that $T_{\Sigma_0} = T$.

Proof: We construct a sequence of numbers $s_0 = 1, s_1, \ldots, s_n, \ldots,$ such that there is no (unitary) bounded operator U such that

$$s_n = P_{\mathbf{C}} U^n | \mathbf{C}, \quad n \geq 0. \tag{3.11}$$

To that end, we need some results concerning the structure of invertible Hermitian matrices. Using Theorem 2.4 in [9], we define $s_1, s_2, \dots, s_n, \dots$, such that

$$s_n = f(\gamma_1, \gamma_2, \dots, \gamma_{n-1}) + d_1^2 d_2^2 \dots d_{n-1}^2 \gamma_n, \tag{3.12}$$

where f is a certain function of parameters $\gamma_1, \gamma_2, \dots, \gamma_{n-1}$ which are complex numbers of absolute value different from 1 and $d_k = |1 - |\gamma_k|^2|^{1/2}$, $k \geq 1$. Suppose we found $\gamma_1, \gamma_2, \dots, \gamma_{n-1}$ strictly positive numbers such that $s_1 > 1$, $s_2 > 2^2$, $s_{n-1} > (n-1)^{n-1}$. Then formula (3.12) gives

$$s_n = k_{n-1} + l_{n-1}\gamma_n$$

with $k_{n-1} > 0$ and $l_{n-1} > 0$, so we can choose γ_n such that

$$\gamma_n > \frac{n^n - k_{n-1}}{l_{n-1}}$$

and then $s_n > n^n$. This shows that there is no (unitary) bounded operator U such that (3.11) holds. According to Theorem 3.1, the Toeplitz triangular family given by

$$T_0 = s_1, \ \dots, \ T_n = s_{n+1} - s_1 T_{n-1} - \dots - s_n T_0, \ \dots$$

admits no unitary time invariant realization. On the other hand, following [9] we can define the operators

$$U_1 = \begin{bmatrix} \ddots & & & & & \\ & 1 & 0 & & & \\ & 0 & d_1 & -\bar{\gamma}_1 & 0 & \dots \\ & & \gamma_1 & d_1 & 0 & \dots \\ & & 0 & 0 & 1 & \dots \\ & & & & & \ddots \end{bmatrix},$$

$$U_2 = \begin{bmatrix} \ddots & & & & & & \\ & 1 & 0 & & & & \\ & 0 & d_2 d_1 & * & * & & \dots \\ & & \gamma_1 & d_1 & 0 & & \dots \\ & & \gamma_2 d_1 & -\bar{\gamma}_1\gamma_2 & d_2 & & \\ & & 0 & 0 & 1 & \dots & \\ & & & & & & \ddots \end{bmatrix},$$

$$U_3 = \begin{bmatrix} \ddots & & & & & & \\ & 1 & 0 & & & & \\ & 0 & d_3 d_2 d_1 & * & * & * & \\ & & \gamma_1 & d_1 & 0 & 0 & \dots \\ & & \gamma_2 d_1 & -\gamma_2\bar{\gamma}_1 & d_2 & 0 & \dots \\ & & \gamma_3 d_2 d_1 & -\gamma_3 d_2 \bar{\gamma}_1 & -\gamma_3\bar{\gamma}_2 & d_3 & \dots \\ & & 0 & 0 & 0 & 1 & \ddots \end{bmatrix}, \dots,$$

and deduce from Theorem 2.4 in [9] that

$$P_{\mathbf{C}}U_1U_2\dots U_n \mid \mathbf{C} = s_n, \quad n \geq 1,$$

and U_n, $n \geq 1$, are unitary operators with respect to appropriate Kreĭn spaces. By Theorem 2.1, these operators give rise to a time dependent unitary realization of T. □

References

[1] D. Alpay, A. Dijksma, J. Rovnyak and H.S.V. de Snoo: *Schur Functions, Operator Colligations, and Reproducing Kernel Pontryagin Spaces*, Birkhäuser, Basel-Boston, 1997.

[2] Gr. Arsene, T. Constantinescu and A. Gheondea: Lifting of operators and prescribed numbers of negative squares, *Michigan J. Math.*, **39** (1987), 201–216.

[3] T.Ya. Azizov and I.S. Iokhvidov: *Linear Operators in Spaces with an Indefinite Metric*, Wiley&Sons, 1989.

[4] J.A. Ball and N. Cohen: de Branges-Rovnyak operator models and system theory: a survey, in *Oper. Theory: Adv. Appl.*, vol. 50, Birkhäuser, Basel-Boston, 1991, pp. 93–136.

[5] L. de Branges: *Hilbert Spaces of Entire Functions*, Prentice-Hall, Englewood Cliffs, N.J., 1968.

[6] L. de Branges: A construction of Krein spaces of analytic functions, *J. Functional Analysis* **98** (1991), 1–41.

[7] L. de Branges and J. Rovnyak: Canonical models in quantum mechanics, in *Perturbation Theory and its Applications in Quantum Mechanics*, Proc. Adv. Sem. Math. Res. Center., Madison WS 1965, pp. 295–392.

[8] M.S. Brodskii and M.S. Livsic: Spectral analysis of non-self-adjoint operators and intermediate systems, *Uspekhi Mat. Nauk* **13**(1958), 3–85.

[9] T. Constantinescu: Schur analysis for invertible matrices, *Rev. Roumaine Math. Pures Appl* **35** (1990), 509–521.

[10] T. Constantinescu: *Schur Parameters, Factorization and Dilation Problems*, Birkhäuser Verlag, Basel-Boston-Berlin, 1996.

[11] T. Constantinescu and A. Gheondea: Representations of Hermitian kernels by means of Kreĭn spaces, *Publ. RIMS*, Kyoto Univ. 33 (1997), 917–951.

[12] M. Cotlar and C. Sadoski: Toeplitz liftings of Hankel forms bounded by non-Toeplitz norms, *Integral Equations and Operator Theory* **14** (1991), 501–532.

[13] G. Christner and J. Rovnyak: Julia operators and the Schur algorithm, in *Harmonic Analysis and Operator Theory*, Contemporary Mathematics **189** (1995), pp. 135–160.

[14] A. Dijksma, H. Langer and H. S. de Snoo: Unitary colligations in Kreĭn spaces and their role in extension theory of isometries and symmetric linear relations in Hilbert spaces, in *Functional Analysis*. II, Lecture Notes in Math. no. **1242**, Springer Verlag, Berlin, 1987, pp. 1–42.

[15] A. Dijksma, H. Langer and H.S. de Snoo: Representations of holomorphic operator functions by means of resolvents of unitary or selfadjoint operators in Kreĭn spaces, in Oper. Theory: Adv. Appl., vol. 24, Birkhäuser, Basel-Boston, 1987, pp. 123–143.

[16] C. Foiaş and A. Frazho: *The Commutant Lifting Approach to Interpolation Problems*, Oper. Theory Adv. Appl., vol. 44, Birkhäuser, Basel and Boston, 1990.

[17] T. Kailath: *Linear Systems*, Prentice–Hall Englewood Cliffs, New Jersey, 1980.

[18] R.E. Kalman, P.L. Falb and B.M. Arbib: *Topics in Mathematical System Theory*, McGraw–Hill, New York, 1969.

[19] M.G. Krein and H. Langer: Uber einige Fortsetzungsprobleme, die eng mit det Theorie hermitescher Operatoren im Raume Π_κ zusammenhangen. I. Einige Funcktionenklassen und ihre Darstellungen, *Math. Nachr.* **77** (1977), 187–236; II. Verallgemeinerte Resolventen, u-Resolventen und ganze Operatoren, *J. Functional Analysis* **30**(1978), 390–447.

[20] P. Lax and R.S. Phillips: *Scattering Theory*, New York, 1967.

[21] M.S. Livsic: *Operators, oscillations, waves. Open systems*, Transl. Math. Monographs, vol. 34, Amer. Math. Soc., Providence, R.I., 1973.

[22] L. Schwartz: Sous espace Hilbertiens d'espaces vectoriel topologiques et noyaux associés (noyaux reproduisants), *J. Analyse Math.* **13** (1964), 115–256.

[23] F. Pelaez: An extension of Naĭmark's dilation theorem to Kreĭn spaces, in *Harmonic Analysis and Operator Theory*, Contemporary Mathematics, vol. 189, 1995.

[24] B. Sz.-Nagy and C. Foiaş: *Harmonic Analysis of Operators on Hilbert Space*, North Holland, New York, 1970.

T. Constantinescu
Department of Mathematics
University of Texas at Dallas
Richardson, TX 75083-0688
USA
tiberiu@utdallas.edu

A. Gheondea
Institutul de Matematică
al Academiei Române
C.P. 1-764, 70700 Bucureşti
România
gheondea@imar.ro

AMS Subject Classification: 47A48, 47N70, 47B50.

Operator Theory:
Advances and Applications, Vol. 115

Hankel Forms and Operators in Hardy Spaces with Two Szegő Weights

Mischa Cotlar and Cora Sadosky

The Nehari and Adamyan–Arov–Krein theory was extended in [CS93] to a Hankel operator acting in two weighted Hardy spaces with different measures. We now show that the theory can be developed further in the case the measures have Szegő weights. A notion of two-weight Carleson measures is introduced, and we prove a corresponding Carleson imbedding theorem in terms of vector-valued Hankel operators.

Introduction

In [CS93] we extended the theorems of Nehari and of Adamyan, Arov and Krein (Nehari–AAK) to the case of Hankel operators in two Hardy spaces with different measures on the circle. This article, though a continuation of that paper, is written in self-contained form.

The interest of considering the Nehari–AAK theorem in weighted Hardy spaces stems from several considerations. Even in the case of weighted spaces with respect to only *one general measure*, such a result includes not only the original Nehari–AAK, but also the Herglotz–Bochner representation theorem, the Helson–Szegő prediction theorem, the Fefferman H^1–BMO duality, and more. Moreover, as we learned after this paper was written, this theory plays an important role in the approximation of transfer functions of linear systems and other applications (see [LO98] and its references).

The first indication that it is advantageous to allow Hardy spaces with *two different measures* is that the Nehari theorem then has the Sz.Nagy–Foias commutant lifting as a corollary. A second advantage may be naturality: when we work not with Hankel forms but with their natural extension, abstract invariant forms in scattering structures, there is no reason why the forms two arguments should have values in the same space (see [Sad98] and its references).

While an isometric isomorphism exists between the classical Hardy spaces H^2 and the weighted $H^2(\mu)$, it only allows the translation from the unweighted to the weighted case of results based on the "Hilbert-space structure," while those results based on the usual identification of $H^2(\mathbb{T})$ and $H^2(\mathbb{D})$ do not translate. The notion of reproducing kernel in $H^2(\mathbb{D})$, the space of analytic functions inside the disk, is not part of the "Hilbert-space structure," and here we work with the kernels corresponding to the spaces $H^2(\mathbb{D};\mu)$, for the case of Szegő measures.

Sadosky was partially supported by NSF grant DMS-9742750, and by DOE grant FG02-97ER25341.

The singular numbers of a weighted Hankel operator or form are also expressed in terms of the weights.

We circumscribe our study to the case of Szegő measures to ensure that the Hankel operators of finite rank in the corresponding weighted spaces are not zero [CS93]. The Nehari–AAK theorem extends fully in this case, and the formulas for the distance from a bounded function to the classes of bounded analytic or bounded meromorphic functions can again be expressed in terms of the norms and the singular numbers of Hankel operators (see Section 3).

Carleson measures play an important role in the classical case, and Treil established in [Tre89], through a result of Bonsall [Bon84], a close relationship between Hankel operators and Carleson measures, providing a novel approach to the Carleson imbedding theorem. Again, an extension of Treil's result cannot be expected for the case of Hardy spaces with two general measures, but we obtain such an extension for Szegő measures.

This paper is organized in four sections. Section 1 includes notation and basic definitions, and the main result referred from [CS93] appears as Theorem 0.

Section 2 details the basic properties of Szegő measures, the corresponding model subspaces, and their reproducing kernels. The explicit expression of the projections involved in the translation between the unweighted and the weighted Hardy spaces are given there.

Hankel forms and operators acting in weighted Hardy spaces, as well as weighted Carleson measures, are considered in Section 3. The main results, valid for Szegő measures, are a weighted Bonsall theorem (both in scalar and vector-valued versions) and a weighted Carleson imbedding theorem. The extensions of those results to two different Szegő measures are given in Section 4.

The extensions given here should lead to new applications to free interpolation, including those obtained by Nikolskii and Vasyunin (see [Nik86]), as well as to the interpolation and approximation scheme of Leblond and collaborators. This approach also leads to new questions in prediction theory for stochastic processes. These, as well as the extension of the results of this paper to weighted Hardy spaces in $\mathbb{R}$ and in $\mathbb{T}^d$, and to de Branges–Rovnyak spaces will be considered elsewhere.

1 Notation and Basic Result

Let $\mathbb{T} = \{\xi \in \mathbb{C} : |\xi| = 1\}$ denote the unit circle and $\mathbb{D} = \{z \in \mathbb{C} : |z| < 1\}$, the unit disk. Let $\mathcal{P}$ be the vector space of all trigonometric polynomials in $\mathbb{T}$, $f \in \mathcal{P}$ if $f = \sum_{\text{finite}} a_n \xi^n$. Then $\mathcal{P} = \mathcal{P}_+ \dot{+} \mathcal{P}_-$, for $\mathcal{P}_+$ the linear span of the set $\{\xi^n : n \geq 0\}$, the subspace of analytic polynomials, and $\mathcal{P}_-$ the linear span of the set $\{\xi^n : n < 0\}$, the subspace of antianalytic polynomials.

For a positive finite measure μ defined on $\mathbb{T}$, $L^2(\mu) = L^2(\mathbb{T}; d\mu)$ is a Hilbert space under the scalar product $\langle f, g\rangle_\mu := \int_{\mathbb{T}} f\bar{g}\, d\mu$, and $H^2(\mu)$ and $H^2_-(\mu)$ are defined as the closures in $L^2(\mu)$ of $\mathcal{P}_+$ and $\mathcal{P}_-$, respectively. When $\mu = m$, the normalized Lebesgue measure on $\mathbb{T}$, we write $L^2(m) = L^2$, $H^2(m) = H^2$, and $H^2_-(m) = H^2_-$.

The shift operator $\mathcal{S}$ is defined in $\mathcal{P}$ by $\mathcal{S}f(\xi) = \xi f(\xi)$, and satisfies $\mathcal{S}\mathcal{P}_+ \subset \mathcal{P}_+$ and $\mathcal{S}^{-1}\mathcal{P}_- \subset \mathcal{P}_-$. It follows that $\mathcal{S}$ is a unitary operator in $L^2(\mu)$, and that $S_\mu := \mathcal{S}|H^2(\mu)$, its restriction to $H^2(\mu)$, is an isometry, while $\mathcal{S}^{-1}H^2_-(\mu) \subset H^2_-(\mu)$.

A sesquilinear form $B : \mathcal{P}_+ \times \mathcal{P}_- \to \mathbb{C}$ is called a *Hankel form* if

$$(1.1) \qquad B(\mathcal{S}f, g) = B(f, \mathcal{S}^{-1}g), \quad \text{for all}(f, g) \in \mathcal{P}_+ \times \mathcal{P}_-.$$

Let μ_1 and μ_2 be two given positive finite measures on $\mathbb{T}$. If a Hankel form $B : \mathcal{P}_+ \times \mathcal{P}_- \to C$ satisfies

$$(1.2) \qquad |B(f, g)| \leq c\, \|f\|_{\mu_1} \|g\|_{\mu_2}, \quad \text{for all}(f, g) \in \mathcal{P}_+ \times \mathcal{P}_-,$$

and some constant $c > 0$, it extends to a bounded form (also denoted by B) with domain $H^2(\mu_1) \times H^2_-(\mu_2)$. Thus B satisfies (1.2) if and only if there exists a bounded operator $\Gamma : H^2(\mu_1) \to H^2_-(\mu_2)$, associated with B through

$$(1.3) \qquad B(f, g) = \langle \Gamma f, g \rangle_{\mu_2}$$

with norm $\|\Gamma\| = \|B\|$. Such Γ is called a *Hankel operator*.

If Γ is a bounded operator in a Hilbert space and B is its associated form (1.3), their *singular numbers* are defined by

$$s_n(B) = s_n(\Gamma) := \inf\{\|\Gamma - \Gamma_n\| : \operatorname{rank} \Gamma_n \leq n\}, \quad n \geq 0,$$

so that $s_0(B) = s_0(\Gamma) = \|\Gamma\| = \|B\|$.

Theorem 0 (Weighted Nehari–AAK theorem [CS93]). *Given a bounded Hankel form $B : H^2(\mu_1) \times H^2_-(\mu_2) \to \mathbb{C}$, for μ_1 and μ_2 positive finite measures on $\mathbb{T}$, for every integer $n \geq 0$, there exist a complex measure ν_n on $\mathbb{T}$, and a subspace $M_n \subset H^2(\mu_1)$,* codim $M_n \leq n$, *such that*

$$B(f, g) = \int_{\mathbb{T}} f \bar{g}\, d\nu_n, \ \textit{for}\ (f, g) \in M_n \times H^2_-(\mu_2),$$

while

$$(1.4) \qquad s_n(B) = \sup\{\nu_n(\Delta)/(\mu_1(\Delta)\mu_2(\Delta))^{-\frac{1}{2}} : \textit{Borel set}\ \Delta \subset \mathbb{T}\}$$

and

$$s_n(B) = s_n(\Gamma) = \inf\{\|\Gamma - \Gamma_n\| : \operatorname{rank} \Gamma_n \leq n \ \textit{and}\ \Gamma_n \textit{Hankel}\},$$

where B and Γ are associated by (1.3).

When $\mu_1 = \mu_2 = m$, Theorem 0 is the classical Adamyan–Arov–Krein theorem [AAK71], and if, moreover, $n = 0$, we recover the classical Nehari theorem [Neh57]. If $\mu_1 = \mu_2 = \mu$ and $\epsilon > 0$, Theorem 0 applied to $B(f, g) := (1 + \epsilon) \int_{\mathbb{T}} f \bar{g}\, d\mu$ gives the Helson–Szegő theorem in prediction theories, with refinements.

Every $\varphi \in L^\infty$ defines a Hankel form B_φ by $B_\varphi(f,g) := \int_{\mathbb{T}} f\bar{g}\varphi\, dm$, and B_φ satisfies (1.1) and (1.2) for $\mu_1 = \mu_2 = m$. Then the theorems of Nehari and Adamyan–Arov–Krein can be restated as

$$\|B_\varphi\| = \operatorname{dist}(\varphi, H^\infty) \tag{1.5}$$

and

$$s_n(B_\varphi) = \operatorname{dist}(\varphi, H^\infty + R_n), \quad n \geq 0 \tag{1.6}$$

where H^∞ is the class of bounded analytic functions, and R_n that of rational functions with at most n poles in $\mathbb{D}$.

There is no natural extension of (1.5) and (1.6) for arbitrary measures μ_1, μ_2. Furthermore, if μ is a deterministic measure, i.e., if $H^2(\mu) = L^2(\mu)$, all Hankel forms of finite rank in $H^2(\mu) \times H^2_-(\mu)$ vanish [CS93]. In order to make precise the results of Theorem 0 we consider the class of Szegő measures.

Throughout this paper $A \cong B$ stands for $\frac{1}{c}B \leq A \leq cB$ for some $c > 1$.

2 Szegö Measures and Model Subspaces

A positive finite measure μ defined on $\mathbb{T}$ is called a *Szegő measure* if $d\mu = w\, dm$, for some *Szegő weight* w, satisfying

$$0 \leq w \in L^1, \quad \log w \in L^1. \tag{2.1}$$

In what follows it is assumed that the measure μ satisfies

$$H^2(\mu) \neq L^2(\mu) \quad \text{and} \quad \operatorname{supp}\mu = \text{infinite set}. \tag{2.2}$$

By a result of Szegő (see [GS58]), a Szegő weight w admits a unique factorization (up to a unimodular constant)

$$w(\xi) = \theta(\xi)\bar{\theta}(\xi), \quad \theta \in H^2 \text{an outer function}, \ \xi \in \mathbb{T}. \tag{2.3}$$

Thus, there exists an isometric isomorphism between $H^2(\mu)$ and H^2, defined by

$$H^2(\mu) \ni f \longmapsto h = \theta f \in H^2 \tag{2.4}$$

with

$$\int_{\mathbb{T}} |f|^2\, d\mu = \int_{\mathbb{T}} \frac{|h|^2}{|\theta|^2} w\, dm = \int_{\mathbb{T}} |h|^2\, dm.$$

Similarly, $H^2_-(\mu) \ni f_- \mapsto h_- = \bar{\theta} f \in H^2$ establishes an isometric isomorphism between $H^2_-(\mu)$ and H^2_-. This implies that every $f = h/\theta \in H^2(\mu)$ extends to $f(z)$, an holomorphic function in the disk $\mathbb{D}$, and $f(z)$ tends in a natural sense to $f(\xi)$ when $z \in \mathbb{D}$ tends radially to $\xi \in \mathbb{T}$. Moreover,

$$\int_{\mathbb{T}} |f(r\xi)\theta(r\xi) - f(\xi)\theta(\xi)|^2\, dm \longrightarrow 0 \quad \text{when } r \longrightarrow 1.$$

Thus, $H^2(w) = H^2(\mu)$ can be considered as a Hilbert space of holomorphic functions in $\mathbb{D}$, as H^2 is. As such, it has a *reproducing kernel* $K^\mu(\zeta, z) = K_z^\mu(\zeta)$, as H^2 has the reproducing kernel

(2.5) $$K(\zeta, z) = \frac{1}{1 - \bar{z}\zeta}, \quad z \in \mathbb{D}, \ \zeta \in \overline{\mathbb{D}}.$$

More precisely, for every $z \in \mathbb{D}$, there is a function $K_z^\mu \in H^2(\mu)$ such that

(2.6) $$f(z) = \langle f, K_z^\mu \rangle_\mu, \quad \text{for all} f \in H^2(\mu) \text{ and all } z \in \mathbb{D}.$$

Here

(2.7) $$K_z^\mu = \frac{K_z(\xi)}{\theta(\xi)\bar{\theta}(z)} = \frac{1}{1 - \bar{z}\xi} \frac{1}{\theta(\xi)} \frac{1}{\bar{\theta}(z)}, \quad z \in \mathbb{D}, \ \xi \in \mathbb{T}.$$

Moreover, the functions $K_z^\mu \in H^2(\mu)$ are eigenfunctions of S_μ^* with eigenvalues $\bar{z}$, where S_μ is the shift operator acting in $H^2(\mu)$ by multiplication by z. In fact, for all $f \in H^2(\mu)$, $z \in \mathbb{D}$,

(2.8) $$\langle S_\mu^* K_z^\mu, f \rangle_\mu = \langle K_z^\mu, S_\mu f \rangle_\mu = \langle K_z^\mu, zf \rangle_\mu = \overline{zf}(z) = \bar{z}\langle K_z^\mu, f \rangle_\mu.$$

On the other hand, while H_-^2 is the orthogonal complement of H^2, $H_-^2(\mu)$ needs not be orthogonal to $H^2(\mu)$ when $w \not\equiv 1$, and $\{\xi^k : k \geq 0\}$ is not an orthogonal basis for $H^2(\mu)$. Nevertheless, under the assumption that the support of μ is infinite, $\{\xi^k : k \geq 0\}$ can be transformed by the Gramm–Schmidt process in the orthogonal basis $\{\mathcal{E}_k : k \geq 0\}$ of $H^2(\mu)$ consisting of the analytic polynomials $\mathcal{E}_k$ of degree k, called the *Szegő polynomials*. From $\mathcal{E}_k(z) = \langle \mathcal{E}_k, K_z^\mu \rangle_\mu$, $\mathcal{E}_k(\zeta) = \langle \mathcal{E}_k, K_\zeta^\mu \rangle_\mu$ for $z \in \mathbb{D}$, $\zeta \in \mathbb{D}$, follows

(2.9) $$K^\mu(\zeta, z) =: K_z^\mu(\zeta) = \sum_{k=0}^{\infty} \mathcal{E}_k(\zeta)\overline{\mathcal{E}_k(z)}$$

and

$$K_z^\mu(\xi) = \sum_{k=0}^{\infty} \mathcal{E}_k(\xi)\overline{\mathcal{E}_k(z)}, \quad \text{for } \xi \in \mathbb{T}.$$

Since by the initial assumption on μ we have $H^2(\mu) \subsetneq L^2(\mu)$, it follows that $\xi H^2(\mu) \subsetneq H^2(\mu)$. By a theorem of Szegő, the distance ρ of the function $1 \in H^2(\mu)$ to the subspace $\xi H^2(\mu)$ is given by

(2.10) $$\rho = exp\left(\frac{1}{2}\int \log w \, dm\right).$$

Moreover, if π is the orthoprojection of 1 onto $\xi H^2(\mu)$, we have $1 - \pi \perp \xi H^2$ and $\rho = \|1 - \pi\|_\mu$. Then it is not hard to check that the function $|1 - \pi|^2 w$

is a constant, and thus, equal to ρ^2 (see [Hof62], Ch. 4, [CS93], p. 179). Since $w = |\theta|^2$, it holds that

$$1 - \pi(\xi) = \frac{\rho}{\theta(\xi)}. \tag{2.11}$$

Together (2.10) and (2.11) provide an explicit expression for $\theta(\xi)$.

For each $z \in \mathbb{D}$, let $E_z^\mu = \{\lambda K_z^\mu : \lambda \in \mathbb{D}\}$ be the one-dimensional subspace spanned by K_z^μ, and let $(E_z^\mu)^\perp = \{f \in H^2(\mu) : f \perp K_z^\mu\}$. E_z^μ is called a (one-dimensional) *model subspace* of $H^2(\mu)$.

In what follows, a Blaschke factor β_z is defined, for $z \in \mathbb{D}\backslash\{0\}$, $\zeta \in \overline{\mathbb{D}}$, by $\beta_z(\zeta) = \frac{|z|}{z}\frac{z-\zeta}{1-\bar{z}\zeta}$, and β_m denotes a Blaschke product of m Blaschke factors.

Lemma 2.1

$$(E_z^\mu)^\perp = \beta_z H^2(\mu), \quad \textit{for } \beta_z \textit{ a Blaschke factor.} \tag{2.12}$$

Proof: By the isometry between $H^2(\mu)$ and H^2, $f \in H^2(\mu)$ is orthogonal to K_z^μ in $H^2(\mu)$ if and only if $\theta f \in H^2$ is orthogonal to $\theta K_z^\mu = K_z/\bar{\theta}(z)$ in H^2. Therefore, if $E_z = \{\lambda K_z/\bar{\theta}(z) : \lambda \in \mathbb{C}\}$, then $E_z^\perp = \theta(E_z^\mu)^\perp$ and $(E_z^\mu)^\perp = \theta^{-1}E_z^\perp$. But $E_z^\perp$ is S^*-invariant and of codimension one, hence, by Beurling's theorem, $E_z^\perp = \beta_z H^2$, for β_z a one-factor Blaschke product, and the thesis follows. □

Let P_- be the orthoprojector of L^2 onto H_-^2, P_-^μ be the orthoprojector of $L^2(\mu)$ onto $H_-^2(\mu)$, and P_z^μ be the orthoprojector of $H^2(\mu)$ onto E_z^μ. Consider the normalized functions on $\mathbb{T}$,

$$\varphi_z(\xi) = \frac{K_z(\xi)}{\|K_z\|} = (1 - |z|^2)^{\frac{1}{2}} K_z(\xi) \tag{2.13}$$

and

$$\phi_z^\mu(\xi) = \frac{K_z^\mu(\xi)}{\|K_z^\mu\|_\mu}. \tag{2.14}$$

Since (2.7) implies

$$\|K_z^\mu\|_\mu^2 = (|\theta(z)|^2(1 - |z|^2))^{-1} \tag{2.15}$$

it follows that

$$\phi_z^\mu(\xi) = \frac{|\theta(z)|}{\bar{\theta}(z)}\frac{\varphi_z(\xi)}{\theta(\xi)}. \tag{2.16}$$

Lemma 2.2 *The following relations hold for all $f \in H^2(\mu)$:*

(a) $P_-^\mu f = \bar{\theta}^{-1}P_-(\bar{\theta} f)$.

(b) $P_z^\mu f = \dfrac{f(z)}{\|K_z^\mu\|_\mu} \cdot \dfrac{K_z^\mu}{\|K_z^\mu\|_\mu}$.

(c) $\|P_z^\mu f\|_\mu^2 = |f(z)|^2(1-|z|^2)|\theta(z)|^2$.

Proof: Since $H_-^2(\mu) = \bar{\xi}\,\overline{H^2(\mu)} = \{\bar{\xi}\bar{h}/\bar{\theta} : h \in H^2\}$, given $f \in L^2(\mu)$, let $P_-^\mu f = \bar{\xi}\bar{h}_0/\bar{\theta}$ for some $h_0 \in H^2$. Then $\langle f, g\rangle_\mu = \langle P_-^\mu f, g\rangle_\mu$ for all $g \in H_-^2(\mu)$, is equivalent to $\langle f, \bar{\xi}\bar{h}/\bar{\theta}\rangle_\mu = \langle \bar{\xi}\bar{h}_0/\bar{\theta}, \bar{\xi}\bar{h}/\bar{\theta}\rangle_\mu$ for all $h \in H^2$. Hence, since the weight w of μ is $w = \theta\bar{\theta}$, for all $h \in H^2$ we have

$$\langle f, \bar{\xi}\bar{h}/\bar{\theta}\rangle_\mu = \int f\theta\xi h = \int \bar{\xi}\bar{h}_o\xi h = \int \bar{\theta}P_-^\mu f\xi h,$$

or $P_-(\bar{\theta}f) = \bar{\theta}P_-^\mu f$, which is (a).

Given $f \in H^2(\mu)$, there is a $\lambda \in \mathbb{C}$ such that $P_z^\mu f = \lambda K_z^\mu$. Thus, $\langle f, K_z^\mu\rangle_\mu = \langle \lambda K_z^\mu, K_z^\mu\rangle_\mu$, and, solving for λ, $\lambda = \langle f, K_z^\mu\rangle_\mu / \langle K_z^\mu, K_z^\mu\rangle_\mu$. By (2.6) and (2.7) we obtain (b).

From (b), $\|P_z^\mu f\|_\mu^2 = |f(z)|^2\|K_z^\mu\|_\mu^{-2}$, which, by (2.15), is (c). □

The relation between the orthoprojectors is given next.

Proposition 2.3 *For $z \in \mathbb{D}$ and $\xi \in \mathbb{T}$, and β_z a Blaschke factor, define*

$$\beta_z^\mu(\xi) := \frac{\bar{\theta}(\xi)}{\theta(\xi)}\beta_z(\xi), \tag{2.17}$$

so that $|\beta_z^\mu| \equiv 1$ a.e. on $\mathbb{T}$. Then, for all $f \in H^2(\mu)$,

$$P_z^\mu f = \beta_z^\mu P_-^\mu \overline{\beta_z^\mu} f. \tag{2.18}$$

Proof: Given $f \in H^2(\mu)$, there is $g \in H^2$ such that $f = g/\theta$. Then, by the previous lemma, (2.7) and (2.15),

$$\begin{aligned}
(P_z^\mu f)(\xi) &= \big(\langle f, K_z^\mu\rangle_\mu/\|K_z^\mu\|^2\big)K_z^\mu(\xi)\\
&= \frac{K_z^\mu(\xi)}{\|K_z^\mu\|_\mu^2}\int_{\mathbb{T}} f(\xi)\overline{K_z^\mu(\xi)}\,d\mu\\
&= |\theta(z)|^2(1-|z|^2)\frac{K_z(\xi)}{\overline{\theta(z)}\theta(\xi)}\int_{\mathbb{T}} g(\xi)\overline{K_z(\xi)}\,dm(\xi)\cdot\frac{1}{\theta(z)}\\
&= (1-|z|^2)\langle g, K_z\rangle K_z(\xi)/\theta(\xi)\\
&= (P_z g/\theta)(\xi) = (1/\theta)\beta_z P_- \overline{\beta_z} g,
\end{aligned}$$

since K_z is the reproducing kernel of H^2, with $\|K_z\|^2 = 1-|z|^2$. Writing

$$g(\xi) = \theta(\xi)f(\xi) = (\theta/\bar{\theta})(\xi)(\bar{\theta}f)(\xi),$$

by Lemma 2.2(a) we have

$$P_z^\mu f = (1/\theta)\beta_z P_-^\mu \bar{\beta}_z(\theta/\bar{\theta})\bar{\theta} f,$$

which, by (2.17), reduces to

$$P_z^\mu f = \beta_z^\mu P_-^\mu \bar{\beta}_z^\mu f,$$

the thesis. □

For each $\psi \in L^\infty$, define the operators $\Gamma_\psi : H^2 \to H_-^2$, and $\Gamma_\psi^\mu : H^2(\mu) \to H_-^2(\mu)$, by

$$\Gamma_\psi g := P_-(\psi g), g \in H^2 \quad \text{and} \quad \Gamma_\psi^\mu f := P_-^\mu(\psi f), f \in H^2(\mu). \tag{2.19}$$

In what follows, to avoid complicated subscripts we will write $\Gamma_\psi = \Gamma[\psi]$ and $\Gamma_\psi^\mu = \Gamma^\mu[\psi]$.

Proposition 2.3 implies

$$\|P_z^\mu f\|_\mu^2 = \|\Gamma^\mu[\bar{\beta}_z^\mu]f\|_\mu^2, \quad f \in H^2(\mu). \tag{2.20}$$

Since by Lemma 2.2(a), $\Gamma^\mu[\psi]f = P_-^\mu(\psi f) = (1/\bar{\theta})\ P_-(\bar{\theta}\psi f) = (1/\bar{\theta})$ $P_-(\psi\bar{\theta}/\theta)(\theta f)$, we have

$$\Gamma^\mu[\psi]f = (1/\bar{\theta})\Gamma[\psi\bar{\theta}/\theta](\theta f). \tag{2.21}$$

Proposition 2.4 (a) *If $E \subset H^2(\mu)$ is an invariant subspace under S_μ, of finite codimension m, then $E = \beta_m H^2(\mu)$, for β_m an m-Blaschke product.*
(b) *$\Gamma^\mu[\psi]$ has finite rank m if $\psi = (1/\bar{\theta})\bar{\beta}_m h$, for $h \in H^2(\mu)$ and β_m an m-Blaschke product.*

Proof: (a) If $E \subset H^2(\mu)$ is S_μ-invariant, then $\theta E \subset H^2$ is S-invariant, since $\xi\theta E = \theta\xi E \subset \theta E$, and θE is also of codimension m. Then $\theta E = \beta_m H^2$, and $E = (\beta_m/\theta)H^2 = \beta_m H^2(\mu)$.

(b) By (2.21), $\Gamma^\mu[\psi](f) = 0$ if and only if $\Gamma[\psi\bar{\theta}/\theta](\theta f) = 0$. Since $f = \theta g$ is an isometric isomorphism between $H^2(\mu)$ and H^2, the kernels of $\Gamma^\mu[\psi]$ and $\Gamma[\psi\bar{\theta}/\theta]$ have the same codimension m, and, since $\ker\Gamma[\psi\bar{\theta}/\theta]$ is also S-invariant, by the Kronecker theorem, $\psi\bar{\theta}/\theta = \bar{\beta}_m\eta$, for some $\eta \in H^2$, and β_m an m-Blaschke product, which is (b) with $h = \theta\eta \in H^2$. □

3 Hankel Forms and Carleson Measures in $H^2(\mu)$

We start with a formal definition of Hankel forms in a weighted H^2 space. Given a positive finite measure μ on the circle, then every linear operator (not necessarily bounded) $\Gamma : \mathcal{P}_+ \to H_-^2(\mu)$ gives rise to a sesquilinear form $B_\Gamma : \text{P}_+ \times$

$\mathcal{P}_- \to \mathbb{C}$, defined by $B_\Gamma(f, g) := \langle \Gamma f, g\rangle_\mu$. The operator Γ is called *μ-Hankel* if its associated form B_Γ is Hankel, that is, if $B(\mathcal{S}f, g) = B(f, \mathcal{S}^{-1}g)$ for all $f \in \mathcal{P}_+, g \in \mathcal{P}_-$ (see Section 1). A Hankel form B is called *μ-Hankel* if $B = B_\Gamma$ for some μ-Hankel operator Γ (such Γ would be unique).

Every $\phi \in L^2(\mu)$ defines a μ-Hankel operator $\Gamma^\mu_\phi = \Gamma^\mu[\phi]$, by $\Gamma^\mu_\phi := P^\mu_-(\phi f)$, so that

$$B^\mu_\phi(f, g) := \int f\bar{g}\phi \, d\mu \tag{3.1}$$

is its associated μ-Hankel form. Conversely, every μ-Hankel operator Γ is of the form $\Gamma = \Gamma^\mu_\phi$ for some $\phi \in L^2(\mu)$, and every such ϕ is called a *μ-symbol* of Γ and of B_Γ, as in (3.1). In particular, $\phi_- = \Gamma 1$ is a μ-symbol for any μ-Hankel operator Γ.

Two $L^2(\mu)$ functions ϕ and ψ are μ-symbols of Γ if and only if

$$\psi \, d\mu = \phi \, d\mu + h \, dm \quad \text{for some } h \in H^2. \tag{3.2}$$

For every Hankel form $B : \mathcal{P}_+ \times \mathcal{P}_- \to \mathbb{C}$ the following conditions are equivalent: (i) B is μ-Hankel; (ii) the linear form $l : \mathcal{P} \to \mathbb{C}$, defined by $l(g) := \overline{B(1, g)}$ is continuous; (iii) B is the associated form of the μ-Hankel operator Γ^μ_ϕ, $\phi \in L^2(\mu)$.

A Hankel form B is called *μ-bounded* if there is a constant C such that

$$|B(f, g)| \le C \, \|f\|_\mu \, \|g\|_\mu, \quad \text{for all } (f, g) \in \mathcal{P}_+ \times \mathcal{P}_-, \tag{3.3}$$

and, in this case, B is μ-Hankel, and its associated μ-Hankel operator $\Gamma : \mathcal{P}_+ \to H^2_-(\mu)$ is also μ-bounded,

$$\|\Gamma f\|_\mu \le C \, \|f\|_\mu \quad \text{and} \quad \|\Gamma\|_\mu = \|B\|_\mu. \tag{3.4}$$

If $\varphi \in L^\infty$, then Γ^μ_φ and B^μ_φ are μ-bounded, with

$$\|\Gamma^\mu_\varphi\|_\mu \le \|\varphi\|_\infty, \tag{3.5}$$

since $|\int f\bar{g}\varphi \, d\mu| \le \|\varphi\|_\infty \, \|f\|_\mu \, \|g\|_\mu$. Thus, there is a one-to-one correspondence $\Gamma^\mu \leftrightarrow B^\mu$ between μ-bounded Hankel operators and μ-bounded Hankel forms.

We restrict ourselves to the case when the given measure μ is a Szegő measure, $d\mu = w \, dm$, $w = \theta\bar{\theta}$, for $\theta \in H^2$ an outer function, and write both $H^2(w) = H^2(\mu)$ and "w-bounded" for "μ-bounded." In this case Theorem 0 (for $\mu_1 = \mu_2 = \mu$, $d\mu = w \, dm$, and for $n = 0$), asserts that B is μ-bounded if and only if there exists a complex measure ν satisfying

$$\left|\int_D d\nu\right| \le \|B\|_\mu \int_D w \, dm \quad \text{for all Borel sets } D \subset \mathbb{T},$$

which implies $d\nu = w_0\,dm$, $|w_0(\xi)| \le \|B\|_\mu w(\xi)$, or $w_0(\xi) = \varphi(\xi)w(\xi)$ for $\xi \in \mathbb{T}$, with $\|\varphi\|_\infty \le \|B\|_\mu$. Thus, if the Hankel form B is μ-bounded, then $B = B_\varphi$, with

$$B_\varphi(f,g) = \int f\bar{g}\varphi\, w\,dm, \quad \|\varphi\|_\infty \le \|B_\varphi\|_\mu. \tag{3.6}$$

Corollary 3.1 *If μ is a Szegő measure with weight w, and B is a μ-Hankel form, then B is μ-bounded if and only if*

$$\Gamma 1 = P_-^\mu(h/w + \varphi), \quad \text{for } \varphi \in L^\infty \text{ and } h \in H^2, \tag{3.7}$$

where Γ is the associated μ-Hankel operator of B.

Remark: For $w \equiv 1$, (3.7) reduces to the well-known relation

$$\Gamma 1 = P_-\varphi \in P_-L^\infty = \mathrm{BMO}.$$

Proof: For $dm = w\,dm$, if B is μ-bounded, then $B = B_\varphi$ by (3.6), and (3.7) is an immediate consequence of (3.2). Conversely, if (3.7) holds, then by Lemma 2.2(a),

$$\Gamma 1 = \bar{\theta}^{-1}P_-(h/w + \varphi)\bar{\theta} = h/w + \varphi - \bar{\theta}^{-1}P(h/w+\varphi)\bar{\theta} = \varphi + \psi,$$

for $\psi\,dm = h\,dm - \theta P(h/w+\varphi)\bar{\theta}\,dm$, and $h - \theta P(h/w+\varphi)\bar{\theta} \in H^1$. Thus, by (3.2), $\Gamma_\psi = 0$ and $\Gamma = \Gamma_\varphi$, for $\varphi \in L^\infty$, which implies that Γ, and B, are μ-bounded. □

For $\varphi \in L^\infty$, B_φ is μ-bounded by (3.5), and by (3.6) and (3.2), there is a $\psi \in L^\infty$ satisfying $\|\psi\|_\infty = \|B_\varphi\|_\mu$ and $\varphi - \psi = h/w = \eta/\bar{\theta}$ for $\eta \in H^2(w)$. Then $\|\psi\|_\infty = \|\varphi - \eta/\bar{\theta}\|_\infty$ with $\eta \in H^2(w) \cap \bar{\theta}L^\infty$. This can be rewritten as

Corollary 3.2 *Given a Szegő measure $d\mu = w\,dm$, $w = \theta\bar{\theta}$, and $\varphi \in L^\infty$, the norm of the corresponding μ-Hankel form is*

$$\|B_\varphi\|_\mu = \mathrm{dist}\,(\varphi, \bar{\theta}^{-1}H^2(\mu) \cap L^\infty). \tag{3.8}$$

Remark: For $w \equiv 1$, (3.8) reduces to the statement of the Nehari theorem, $\|B_\varphi\| = \|\Gamma_\varphi\| = \mathrm{dist}\,(\varphi, H^\infty)$.

Similarly, the particular case of Theorem 0, for $d\mu_1 = d\mu_2 = d\mu = w\,dm$, a Szegő measure, can be restated for $n > 0$, as follows.

Corollary 3.3 *Given a Szegő measure $d\mu = w\,dm$, $w = \theta\bar{\theta}$, and $\varphi \in L^\infty$, the singular numbers of the corresponding μ-Hankel form are, for each $n > 0$,*

$$\begin{aligned} s_n(B_\varphi) &= \inf\{\|\beta_n\varphi - \eta/\bar{\theta}\|_\infty : \eta \in H^2(\mu) \cap \bar{\theta}L^\infty\} \\ &= \inf\{\|\varphi - (\bar{\beta}_n/\bar{\theta})\eta\|_\infty : \eta \in H^2(\mu) \cap \bar{\theta}L^\infty\} \\ &= \inf\{\|\varphi - \chi/\bar{\theta}\|_\infty : \chi \in H^2(\mu) + R_n(\mu)\}, \end{aligned}$$

where β_n is an n-Blaschke product, and $R_n(\mu) = \bar{\theta}^{-1}R_n$, R_n: the class of rational functions with n poles.

Remark: For $w \equiv 1$, the statement above reduces to that of the Adamyan–Arov–Krein theorem, $s_n(B_\varphi) = \text{dist}\,(\varphi, H^\infty + R_n)$.

In particular, given a Szegő measure μ, with weight $w = \theta\bar{\theta}$, a μ-bounded Hankel form B has finite rank if and only if $B = B_\varphi$, for $\varphi = (\theta/\bar{\theta})\beta h$, $h \in H^2$ and β a finite Blaschke product.

Remark: An explicit expression for the antianalytic symbol $\Gamma 1$ of a finite Hankel operator Γ of finite rank is given in [CS93] in terms of the function $1 - \pi$, given by (2.11), and expressible through the reproducing kernel of $H^2(\mu)$, or the Szegő polynomials (see Section 2).

The following result extends [Bon84] to weighted $H^2(\mu)$, for a Szegő measure μ. (See [Nik86], [Tre89].)

Theorem 3.4 (Bonsall's Theorem for Szegő measures). *Given a Szegő measure μ, with weight w, if $B^\mu : H^2(\mu) \times H^2_-(\mu) \to \mathbb{C}$ is a μ-Hankel form, then*

$$\begin{aligned} \|B^\mu\|_\mu &:= \sup\left\{\frac{|B^\mu(f,g)|}{\|f\|_\mu\,\|g\|_\mu} : (f,g) \in \mathcal{P}_+ \times \mathcal{P}_-\right\} \\ &\cong \sup\left\{\frac{|B^\mu(\phi_z^\mu, g)|}{\|g\|_\mu} : z \in \mathbb{D},\ g \in H^2_-(\mu)\right\}. \end{aligned} \tag{3.10}$$

Proof: Let $B : H^2 \times H^2_- \to \mathbb{C}$ be defined by $B(h_1, h_2) := B^\mu(h_1/\theta, h_2/\bar{\theta})$, for $(h_1, h_2) \in H^2 \times H^2_-$. Then B is a bounded Hankel form, with $\|B\| = \|B^\mu\|_\mu$. By Bonsall's Theorem,

$$\begin{aligned} \|B^\mu\|_\mu = \|B\| \cong \sup&\left\{\frac{|B(\varphi_z, h_2)|}{\|h_2\|} : z \in \mathbb{D}, h_2 \in H^2_-\right\} \\ = \sup&\left\{\|g\|_\mu^{-1}\left|B\left(\frac{\bar{\theta}(z)}{|\theta(z)|}\theta\phi_z^\mu, \bar{\theta}g\right)\right| : z \in \mathbb{D}, g \in H^2_-(\mu)\right\} \\ = \sup&\left\{\frac{|B^\mu(\phi_z^\mu, g)|}{\|g\|_\mu} : z \in \mathbb{D}, g \in H^2_-(\mu)\right\}, \end{aligned}$$

by (2.16). □

To each Szegő measure μ we have associated the class of μ-bounded operators $\{\Gamma^\mu\}$, or forms $\{B^\mu\}$, with $\Gamma^\mu = \Gamma^\mu_\varphi = P_-\varphi$ for $\varphi \in L^\infty$, and their μ-norms are given in terms of $\{\Gamma^\mu\phi_z^\mu\}$, $z \in \mathbb{D}$. Now we associate to each Szegő measure μ a class of μ-Carleson measures in the unit disk $\mathbb{D}$.

Definition 3.1 A positive measure σ defined in $\mathbb{D}$ is a μ-Carleson measure, for μ a given Szegő measure with weight w, if and only if there is a constant $C_{\mu,\sigma}$, such that for all $f \in P_+$,

$$\int\int_{\mathbb{D}} |f(z)|^2 \, d\sigma \leq C_{\mu,\sigma} \int_{\mathbb{T}} |f|^2 \, d\mu. \tag{3.11}$$

By Lemma 2.2(c) and (2.20), (3.11) can be rewritten as

$$\int\int_{\mathbb{D}} \|\Gamma^\mu[\bar{\beta}_z^\mu]f\|_\mu^2 |\theta(z)|^{-2}(1-|z|^2)^{-1} \, d\sigma \leq C_{\mu,\sigma} \int_{\mathbb{T}} |f|^2 w \, dm, \tag{3.12}$$

where $\theta(z)$ is the analytic extension of θ, the outer function satisfying $|\theta(\xi)|^2 \, dm = d\mu$. Observe that for $w \equiv 1$, the μ-Carleson measures are the classical Carleson measures in $\mathbb{D}$.

To each measure σ in $\mathbb{D}$ corresponds an operator $\Gamma : f \mapsto \Gamma f$, where f is a scalar-valued function in $H^2(\mu)$ and Γf is a vector-valued function, which assigns to each $z \in \mathbb{D}$ the function

$$(\Gamma f)(z) := \Gamma^\mu[\bar{\beta}_z^\mu]f \in H_-^2(\mu). \tag{3.13}$$

Thus, if σ is a μ-Carleson, and we define

$$d\tilde{\sigma}_\mu(z) := (1-|z|^2)^{-1}|\theta(z)|^{-2} d\sigma(z) \tag{3.14}$$

then $\Gamma f \in L^2(\mathbb{D}; \tilde{\sigma}_\mu; H_-^2(\mu))$, the space of square integrable functions with respect to $\tilde{\sigma}_\mu$ with values in $H_-^2(\mu)$. By the Fubini theorem, $L^2(\mathbb{D}; \tilde{\sigma}_\mu; H_-^2(\mu))$ coincides, as a function space, with $H_-^2(\mathbb{T}; \mu; L^2(\tilde{\sigma}_\mu))$, and then we have $\Gamma : H^2(\mu) \to H_-^2(\mu; L^2(\tilde{\sigma}_\mu))$, where the range is the subspace of antianalytic $L^2(\tilde{\sigma}_\mu)$-valued functions.

Theorem 3.5 (Carleson imbedding theorem for a Szegő weight). *Given a Szegő measure μ, let σ be a μ-Carleson measure in $\mathbb{D}$, and define $\tilde{\sigma}_\mu$ by* (3.14). *Then*

$$C_{\mu,\sigma} \cong \sup\left\{\int\int_{\mathbb{D}} \|\Gamma^\mu[\bar{\beta}_z^\mu]\phi_u^\mu\|_\mu d\tilde{\sigma}_\mu(z) : u \in \mathbb{D}\right\}. \tag{3.15}$$

Proof: Let us verify that the operator defined by (3.13) is Hankel. In fact, let $g \in H_-^2(\mu; L^2(\tilde{\sigma}_\mu))$, so that for each $\xi \in \mathbb{T}, g(\xi) \in L^2(\tilde{\sigma}_\mu)$, $g(\xi) = g(\xi, z)$, $z \in \mathbb{D}$. Then, for $f \in H^2(\mu)$,

$$\begin{aligned}\langle \Gamma\xi f, g(\xi,z)\rangle_{L^2(\mu;L^2(\tilde{\sigma}_\mu))} &= \langle \Gamma^\mu[\bar{\beta}_z^\mu]\xi f, g(\xi,z)\rangle \\ &= \int_{\mathbb{T}}\int\int_{\mathbb{D}} \Gamma^\mu[\bar{\beta}_z^\mu](\xi f)\overline{g(\xi,z)}\, d\tilde{\sigma}_\mu(z) w(\xi)\, dm(\xi) \\ &= \int\int_{\mathbb{D}}\int_{\mathbb{T}} \Gamma^\mu[\bar{\beta}_z^\mu](\xi f)\overline{g(\xi,z)} w(\xi)\, dm(\xi)\, d\tilde{\sigma}_\mu(z)\end{aligned}$$

$$= \int\int_{\mathbb{D}}\int_{\mathbb{T}} \Gamma^{\mu}[\bar{\beta}_z^{\mu}](f)\overline{\bar{\xi}g(\xi, z)}w(\xi)\,dm(\xi)\,d\tilde{\sigma}_{\mu}(z)$$
$$= \langle \Gamma f, \bar{\xi} g\rangle.$$

Assuming that Theorem 3.4 holds for μ-Hankel operators with vector-valued range, we have then

$$C_{\mu,\sigma} \cong \|\Gamma\| \cong \sup\{\|\Gamma\phi_u^{\mu}\|_{\mu} : u \in \mathbb{D}\}$$
$$= \sup\left\{\int\int_{\mathbb{D}} \|\Gamma^{\mu}[\bar{\beta}_z^{\mu}]\phi_u^{\mu}\|_{\mu}\,d\tilde{\sigma}_{\mu}(z) : \phi_u^{\mu}, u \in \mathbb{D}\right\}.$$

□

To complete the proof we need

Theorem 3.4a *Theorem 3.4 extends to vector-valued μ-Hankel operators Γ^{μ} : $H^2(\mu) \to H^2_-(\mu; \mathcal{H})$, for $\mathcal{H}$ a Hilbert space.*

Proof: The space $L^2(\mu; \mathcal{H})$ has elements $\mathbf{f} : \mathbb{T} \to \mathcal{H}$, and its scalar product is

$$\langle \mathbf{f}_1, \mathbf{f}_2\rangle_{\mu} = \int \langle \mathbf{f}_1(\xi), \mathbf{f}_2(\xi)\rangle_{\mathcal{H}}\, w(\xi)\,dm,$$

while $L^2(\mathcal{H})$ is the space of functions $\mathbf{g} : \mathbb{T} \to \mathcal{H}$ with

$$\langle \mathbf{g}_1, \mathbf{g}_2\rangle = \int \langle \mathbf{g}_1(\xi), \mathbf{g}_2(\xi)\rangle_{\mathcal{H}}\,dm.$$

Let $\mathcal{P}(\mathcal{H}) = \{f(\xi) = \sum_{\text{finite}} \mathbf{v}_k\xi^k : \mathbf{v}_k \in \mathcal{H}\}$ be the class of $\mathcal{H}$-valued trigonometric polynomials, so that $\mathcal{P}(\mathcal{H}) \subset \mathcal{L}^2(\mathcal{H}) \cap \mathcal{L}^2(\mu; \mathcal{H})$, $H^2(\mu; \mathcal{H})$ is the closure of $\mathcal{P}_+(\mathcal{H})$ in $L^2(\mu; \mathcal{H})$, and $H^2(\mathcal{H})$ that of $\mathcal{P}_+(\mathcal{H})$ in $L^2(\mathcal{H})$. The correspondence $H^2(\mu; \mathcal{H}) \ni \mathbf{f} \mapsto \theta\mathbf{f} = \mathbf{g} \in H^2(\mathcal{H})$ is an isometry from $H^2(\mu; \mathcal{H})$ into $H^2(\mathcal{H})$. Let us see that it is onto, i.e., that $\mathbf{g} \in H^2(\mathcal{H})$ implies $\mathbf{g}/\theta \in H^2(\mu; \mathcal{H})$, and to prove this it is enough to do it for $\mathbf{g} \in \mathcal{P}_+(\mathcal{H})$.

Since $1 \in H^2$ implies $1/\theta \in H^2(\mu)$, and $1/\theta$ is the L^2-limit of a sequence of polynomials $g_n \in \mathcal{P}_+$, it is sufficient to show that $g_n\mathbf{g} \in H^2(\mu; \mathcal{H})$ for $\mathbf{g} \in \mathcal{P}_+(\mathcal{H})$, and we may take $\mathbf{g} = \mathbf{v}\xi^k$, for $\mathbf{v} \in \mathcal{H}$, $k \geq 0$. But $g_n(\xi)\mathbf{v}\xi^k = \sum_r c_r\xi^r\mathbf{v}\xi^k \in \mathcal{P}_+(\mathcal{H}) \subset H^2(\mu; \mathcal{H})$.

Thus $\mathbf{f} \leftrightarrow \theta\mathbf{f}$ is an isometry of $H^2(\mu; \mathcal{H})$ onto $H^2(\mathcal{H})$. Now, it is known that the Bonsall's Theorem holds for Hankel operation acting from H^2 to $H^2_-(\mathcal{H})$ [Tre89]. Therefore, the proof of Theorem 3.4 based on the isometric isomorphism $f \leftrightarrow \theta f$, and on the relation between scalar-valued functions ϕ_z^{μ} and φ_z, applies to μ-Hankel bounded operators $\Gamma : H^2(\mu) \to H^2_-(\mu; \mathcal{H})$. □

4 Generalizations for Pairs of Szegő Weights

Given two finite positive measures on $\mathbb{T}$, μ_1 and μ_2, and a μ_2-Hankel operator $\Gamma : \mathcal{P}_+ \to H^2_-(\mu_2)$, we say that Γ is (μ_1, μ_2)-*bounded*, with norm $\leq c$ if, for all $(f, g) \in \mathcal{P}_+ \times \mathcal{P}_-$,

$$\text{(4.1)} \qquad \|\Gamma f\|_{\mu_2} \leq c\,\|f\|_{\mu_1} \Leftrightarrow |B(f,g)| \leq c\,\|f\|_{\mu_1}\|g\|_{\mu_2},$$

where B is the associated μ_2-Hankel form of Γ.

Set $\mu := \mu_1 + \mu_2$, so that μ_1 and μ_2 are absolutely continuous with respect to μ, $\mu_1 \ll \mu$, $\mu_2 \ll \mu$, and write

$$\text{(4.2)} \quad d\mu_1 = w_1\,d\mu,\ d\mu_2 = w_2\,d\mu,\ d\mu' := w'\,d\mu \quad \text{for } w' = \sqrt{w_1 w_2}.$$

Proposition 4.1 *Let μ_1 and μ_2 be two finite positive measures on $\mathbb{T}$, and $\mu = \mu_1 + \mu_2$, so that $\mu_1 \ll \mu$ and $\mu_2 \ll \mu$. Define $\mu' \ll \mu$ by* (4.2). *Then for every Hankel form $B : \mathcal{P}_+ \times \mathcal{P}_- \to \mathbb{C}$, the following conditions are equivalent:*

(a) $|B(f,g)| \leq \|f\|_{\mu_1}\|g\|_{\mu_2}$, *for all* $(f,g) \in \mathcal{P}_+ \times \mathcal{P}_-$;

(b) $|B(f,g)| \leq \|f\|_{\mu'}\|g\|_{\mu'}$, *for all* $(f,g) \in \mathcal{P}_+ \times \mathcal{P}_-$.

Remark: Hankel forms satisfying condition (a) are called (μ_1, μ_2)-bounded, with norm ≤ 1.

Proof: Assume first that (b) holds. Then, by Theorem 0, there exists a complex measure μ_0 on $\mathbb{T}$, satisfying

$$|\mu_0(\Delta)| \leq \mu'(\Delta) \quad \text{for all Borel sets } \Delta \subset \mathbb{T},$$

and such that

$$B(f,g) = \int f\bar{g}\,d\mu_0 \quad \text{for all } (f,g) \in \mathcal{P}_+ \times \mathcal{P}_-.$$

Since $\mu_0 \ll \mu' \ll \mu$, we write $d\mu_0 = w_0\,d\mu$, with $|w_0| \leq w' = \sqrt{w_1 w_2}$ a.e., and

$$|B(f,g)| \leq \int |f\bar{g}|w'\,d\mu \leq \left(\int |f|^2 w_1 d\mu\right)^{\frac{1}{2}} \left(\int |g|^2 w_2\,d\mu\right)^{\frac{1}{2}},$$

so that (a) holds. Conversely, assuming (a), by Theorem 0, there exists a complex measure ν on $\mathbb{T}$, satisfying

$$|\nu(\Delta)| \leq \sqrt{\mu_1(\Delta)\mu_2(\Delta)} \quad \text{for all Borel sets } \Delta \subset \mathbb{T},$$

and such that

$$B(f,g) = \int f\bar{g}\,d\nu, \quad \text{for all } (f,g) \in \mathcal{P}_+ \times \mathcal{P}_-.$$

Again we have $\nu \ll \mu$, and write $d\nu = v d\mu$, with $|v| \le \sqrt{w_1 w_2} = w'$ a.e. Thus

$$|B(f,g)| \le \int |f\bar{g}|\,|v|\,d\mu \le \int |f\bar{g}| w'\,d\mu \le \left(\int |f|^2 d\mu'\right)^{\frac{1}{2}} \left(\int |g|^2\,d\mu'\right),$$

so that (b) holds. □

Proposition 4.1 entails that the properties concerning (μ_1, μ_2)-*boundedness* of a Hankel form can be reduced to properties concerning μ'-boundedness, for the single measure $\mu' = w' d\mu$ defined by (4.2).

However, Proposition 4.1 does not ensure that properties concerning the *singular numbers* of (μ_1, μ_2)-bounded Hankel forms can be reduced to those corresponding to μ'-bounded forms. Neither does it ensure that the properties of μ'-Carleson measures can be translated to Carleson measures defined for pairs of measures μ_1, μ_2. Each of these questions has to be checked separately, for the appropriate pair-wise analog, to be stated explicitly below.

In what follows, the measures μ_1 and μ_2 are assumed to be Szegő measures with weights $w_1 = \theta_1 \bar{\theta}_1$ and $w_2 = \theta_2 \bar{\theta}_2$, respectively. Observe that is this case, $w' = \sqrt{w_1 w_2}$ is also a Szegő weight, since then $\frac{1}{2}\log w_1 + \frac{1}{2}\log w_2 = \log w'$ $\in L^1$.

From Theorem 0 follows that every (μ_1, μ_2)-bounded Hankel operator Γ is of the form $\Gamma = \Gamma^{\mu_1 \mu_2}[\phi]$, defined by

$$\Gamma^{\mu_1\mu_2}[\phi] f := P_-^{\mu_2} \phi \sqrt{\tfrac{w_1}{w_2}} f, \quad \text{for } f \in H^2(\mu_1), \tag{4.3}$$

and, by Lemma 2.2(a) and (2.21), for each $f \in H^2(w_1)$, we have

$$\Gamma^{\mu_1\mu_2}[\phi] f = \bar{\theta}_2^{-1} \Gamma \left[\phi \sqrt{\bar{\theta}_1/\theta_1} \sqrt{\bar{\theta}_2/\theta_2} \right] (\theta_1 f). \tag{4.4}$$

Theorem 4.2 (Bonsall's Theorem for two Szegő weights). *Given two Szegő measures μ_1 and μ_2, let $\Gamma^{\mu_1\mu_2} : H^2(\mu_1) \to H^2_-(\mu_2)$ be a (μ_1, μ_2)-bounded Hankel operator. Then*

$$\|\Gamma^{\mu_1\mu_2}\|_{\mu_1,\mu_2} \cong \sup\{\|\Gamma^{\mu_1\mu_2} \phi_z^{\mu_1}\|_{\mu_2} : z \in \mathbb{D}\}. \tag{4.5}$$

Proof: By Proposition 4.1, $\Gamma = \Gamma^{\mu_1\mu_2}$ is μ'-bounded, for $d\mu' = w' d\mu$, for $w' = \sqrt{w_1 w_2}$, where $d\mu_i = w_i d\mu$, for $i = 1, 2$. By Theorem 3.4, we have the norm equivalences

$$\|\Gamma^{\mu_1\mu_2}\|_{\mu_1,\mu_2} = \|\Gamma^{\mu'}\|_{\mu'} \cong \sup\{\|\Gamma^{\mu'} \phi_z^{\mu'}\|_{\mu'} : z \in \mathbb{D}\}$$

or, in terms of the associated Hankel form B,

$$\|B\|_{\mu_1,\mu_2} = \|B\|_{\mu'} \cong \sup\{|B(\phi_z^{\mu'}, g)|/\|g\|_{\mu'} : z \in \mathbb{D}, g \in \mathcal{P}_-\}.$$

From the relations (2.16) between $\phi_z^{\mu_1}$ and $\phi_z^{\mu'}$ with φ_z, and from

$$\|g\|_{\mu'}^2 = \int |g|^2 \sqrt{w_1 w_2}\, dm, \quad \|g\|_{\mu_2}^2 = \int |g|^2 w_2\, dm,$$

it follows that

$$\sup\{|B(\phi_z^{\mu'}, g)|/\|g\|_{\mu'} : z \in \mathbb{D}, g \in \mathcal{P}_-\}$$
$$= \sup\{|B(\phi_z^{\mu_1}, g)|/\|g\|_{\mu_2} : z \in \mathbb{D}, g \in \mathcal{P}_-\},$$

which is the thesis. □

Theorem 4.2a (Extension for Hankel operators with vector-valued range). *Given a Hilbert space $\mathcal{H}$, two Szegő measures μ_1 and μ_2, and a (μ_1, μ_2)-bounded Hankel operator $\Gamma : H^2(\mu_1) \to H_-^2(\mu_2; \mathcal{H})$, the result corresponding to* (4.5) *holds.*

Proof: The Nehari Theorem for two measures (Theorem 0 for $n = 0$) applies for Hankel forms $B : \mathcal{P}_+ \times \mathcal{P}_-(\mathcal{H}) \to \mathbb{C}$ where the elements of $\mathcal{P}_-(\mathcal{H})$ are $\mathcal{H}$-valued antianalytic polynomials. For the particular case when μ_1 and μ_2 are Szegő measures, this says that if such a form B is (μ_1, μ_2)-bounded, there exists an $\mathcal{H}$-valued measure ν on $\mathbb{T}$, such that

$$B(f, \mathbf{g}) = \int f \langle d\nu, \mathbf{g}\rangle_{\mathrm{H}}, \quad \text{for} f \in \mathcal{P}_+, \mathbf{g} \in \mathcal{P}_-(\mathcal{H}),$$

with

$$|\langle \nu(\Delta), \eta\rangle_{\mathrm{H}}| \le \|B\| \mu_1(\Delta)^{\frac{1}{2}} \mu_2(\Delta)^{\frac{1}{2}} \|\eta\|_{\mathrm{H}},$$

for $\eta \in \mathcal{H}$, and every Borel set $\Delta \subset \mathbb{T}$.

Setting $d\mu' = w'\, dm$ as in (4.2), it follows from the proof of Proposition 4.1 that the equivalence (a) ⇔ (b) holds for $B : \mathcal{P}_+ \times \mathcal{P}_-(\mathcal{H}) \to \mathbb{C}$, and the thesis follows. □

Definition 4.1 A measure $\sigma \ge 0$ in $\mathbb{D}$ is a (μ_1, μ_2)-Carleson measure for μ_1, μ_2, Szegő measures, if and only if

$$C_{\mu_1, \mu_2, \sigma} := \sup\{\|f\|_{\mu_1}^{-2} \int\int_{\mathbb{D}} \|\Gamma^{\mu_1 \mu_2}[\bar{\beta}_z^{\mu_1}] f\|_{\mu_2}^2\, d\tilde{\sigma}_{\mu_2}(z) : f \in H^2(\mu_1)\} < \infty,$$

where $d\tilde{\sigma}_{\mu_2}(z) = (1 - |z|^2)^{-1} |\theta_2(z)|^{-2} d\sigma(z)$, for $|\theta_2(\xi)|^2\, dm = d\mu_2$.

With a (μ_1, μ_2)-Carleson measure σ we associate the operator

$$\Gamma : H^2(\mu_1) \longrightarrow H_-^2(\mu_2; L^2(\tilde{\sigma}_{\mu_2})),$$

defined by

$$\text{(4.7)} \quad \Gamma f := \Gamma^{\mu_1 \mu_2}[\bar{\beta}_z^{\mu_1}] f \in L^2(\mathbb{D}; \tilde{\sigma}_{\mu_2}; H_-^2(\mu_2)) \equiv H_-^2(\mathbb{T}; \mu_2; L^2(\tilde{\sigma}_{\mu_2})),$$

for which $\|\Gamma\| \cong C_{\mu_1, \mu_2, \sigma}$.

Theorem 4.3 *If σ is a (μ_1, μ_2)-Carleson measure in $\mathbb{D}$, then*

$$C_{\mu_1,\mu_2,\sigma} \cong \sup\left\{\int\int_{\mathbb{D}} \|\Gamma^{\mu_1\mu_2}[\bar{\beta}_z^{\mu_1}]\phi_u^{\mu_1}\|_{\mu_2}^2\, d\tilde{\sigma}_{\mu_2}(z) : u \in \mathbb{D}\right\}.$$

Proof: The argument in the proof of Theorem 3.5 applies here, and, by Theorem 4.2a, it remains only to show that the operator Γ associated to σ through (4.7) is, indeed, Hankel. In fact, this follows from (2.21),

$$\begin{aligned}
\langle\Gamma\xi f, g(\xi,z)\rangle_{\mu_2} &= \int_{\mathbb{T}}\int\int_{\mathbb{D}} \Gamma^{\mu_1\mu_2}[\bar{\beta}_z^{\mu_1}](\xi f)\overline{g(\xi,z)}\, d\tilde{\sigma}_{\mu_2}(z)\, w_2(\xi)\, dm \\
&= \int_{\mathbb{T}}\int\int_{\mathbb{D}} \bar{\theta}_2^{-1}\Gamma\left[\bar{\beta}_z^{\mu_1}\sqrt{\frac{\bar{\theta}_1\bar{\theta}_2}{\theta_1\theta_2}}\right](\theta_1\xi f)\overline{g(\xi,z)}\, d\tilde{\sigma}_{\mu_2}(z)\, w_2(\xi)\, dm \\
&= \int\int_{\mathbb{D}}\int_{\mathbb{T}} \bar{\theta}_2^{-1} P_-\bar{\beta}_z^{\mu_1}\sqrt{\frac{\bar{\theta}_1\bar{\theta}_2}{\theta_1\theta_2}}(\xi\theta_1 f)\overline{g(\xi,z)} w_2(\xi)\, dm\, d\tilde{\sigma}_{\mu_2}(z) \\
&= \int_{\mathbb{T}}\int\int_{\mathbb{D}} \bar{\theta}_2^{-1}\Gamma\left[\bar{\beta}_z^{\mu_1}\sqrt{\frac{\bar{\theta}_1\bar{\theta}_2}{\theta_1\theta_2}}\right](\theta_1 f)\overline{\bar{\xi} g(\xi,z)}\, d\tilde{\sigma}_{\mu_2}(z) w_2(\xi)\, dm \\
&= \langle\Gamma f, \bar{\xi} g\rangle_{\mu_2}.
\end{aligned}$$

□

Now we can formulate an AAK-type result for (μ_1, μ_2)-Carleson measures.

Definition 4.2 Let μ_1, μ_2 be two Szegő measures, and let σ be a (μ_1, μ_2)-Carleson measure in $\mathbb{D}$. For each positive integer n define

$$s_{\mu_1,\mu_2,\sigma}(n) := \inf\left\{ \sup\{\|\beta_n f\|_{\mu_1}^{-2} \int\int_{\mathbb{D}} \|\Gamma^{\mu_1\mu_2}[\bar{\beta}_z^{\mu_1}]\beta_n f\|_{\mu_2}^2\, d\tilde{\sigma}_{\mu_2}(z) : f \in H^2(\mu_1)\} : \beta_n \text{ an } n\text{-Blaschke product}\right\}.$$

If Γ is the operator associated with σ through (4.7), it follows from Definition 4.2 that

$$s_{\mu_1,\mu_2,\sigma}(n) = \inf\{\|\Gamma\beta_n\| : \beta_n \text{ an } n\text{-Blaschke product}\},$$

and the argument in the proof of Theorem 4.3 yields

Theorem 4.4 *If σ is a (μ_1, μ_2)-Carleson measure in $\mathbb{D}$, then*

$$s_{\mu_1,\mu_2,\sigma}(n) = \inf\left\{ \sup_{u\in\mathbb{D}} \int\int_{\mathbb{D}} \|\Gamma^{\mu_1\mu_2}[\bar{\beta}_z^{\mu_1}]\beta_n\phi_u^{\mu_1}\|_{\mu_2}^2 \, d\tilde{\sigma}_{\mu_2}(z) : \beta_n \text{an } n\text{-Blaschke product} \right\}.$$

References

[AAK71] V.M. Adamyan, V.Z. Arov and M.G. Kreĭn, Analytic properties of Schmidt pairs of a Hankel operator and generalized Schur–Takagi problem. *Mat. Sbornik* **86** (1971), 33–73, In Russian.

[Bon84] F.F. Bonsall, Boundedness of Hankel matrices, *J. London Math. Soc.* (2) **29**(2) (1984), 289–300.

[CS93] Mischa Cotlar and Cora Sadosky, Abstract, weighted, and multidimensional Adamjan–Arov–Kreĭn theorems, and the singular numbers of Sarason commutants. *Integral Equations Operator Theory* **17** (1993), 169–201.

[GS58] Ulf Grenander and Gabor Szegö, *Toeplitz forms and their applications* California Monographs in Mathematical Sciences. University of California Press, Berkeley, 1958.

[Hof62] Kenneth Hoffman, *Banach Spaces of Analytic Functions*. Prentice-Hall Series in Modern Analysis. Prentice-Hall Inc., Englewood Cliffs, N. J., 1962.

[LO98] Juliette Leblond and Martine Olivi, Weighted H^2 approximation of transfer functions, *Mathematics of Control, Signals and Systems* **11** (1998), 28–39.

[Neh57] Zeev Nehari, On bounded bilinear forms. *Ann. of Math.* (2) **65** (1957), 153–162.

[Nik86] N.K. Nikol'skiĭ, *Treatise on the shift operator*. Number 273 in Grundlehren der mathematischen Wissenschaften. Springer, Berlin, 1986. With an appendix by S. V. Hruščev and V. V. Peller.

[Sad98] Cora Sadosky, Liftings of kernels shift-invariant in scattering systems. In Sheldon Axler, John McCarthy, and Donald Sararon, editors, *Holomorphic Spaces*, number 33 in Mathematical Sciences Research Institute Publications. Cambridge Univ. Press, New York, 1998.

[Tre89] S.R. Treil, Hankel operators, embedding theorems and bases of co-invariant subspaces of the multiple shift operator. *Algebra i Analiz* **1**(6) (1989), 200–234, translation in Leningrad Math. J. 1 (1990), no. 6, 1515–1548.

Mischa Cotlar
Facultad de Ciencias
Universidad Central de Venezuela
Caracas 1040, Venezuela
mcotlar@euler.ciens.ucv.ve

Cora Sadosky
Department of Mathematics
Howard University
Washington, DC 20059, USA
cs@scs.howard.edu

1991 Mathematics Subject Classification. 42B30, 47B35.

Operator Theory:
Advances and Applications, Vol. 115

The Structured Norm of a Hilbert Space Operator with respect to a Given Algebra of Operators

Avraham Feintuch and Alexander Markus

We consider the notion of structural singular value introduced to study robustness properties of systems for Hilbert space operators. We consider the cases where the underlying algebras are $\mathcal{L}(\mathcal{H})$, the algebra of all bounded linear operators on a Hilbert space $\mathcal{H}$, and nest algebras. We point out the relationship between these results and those of Yu. Shmul'yan on mappings on the unit ball of operators by fractional linear transformations and extend his results to nest algebras.

1 Introduction

It is an elementary fact that if $\mathcal{R}$ is a Banach algebra with identity and B_1 is its open unit ball, $B_1 = \{f \in \mathcal{R} : \|f\| < 1\}$, then $1 - gf$ is invertible in $\mathcal{R}$ for given $g \in \mathcal{R}$ and all $f \in B_1$ if $\|g\| \leq 1$. Of course there is nothing special about balls of radius one. For B_r, the open ball centered at zero with radius r, the required condition on g is $\|g\| \leq \frac{1}{r}$. In general this condition is not necessary. However, if $\mathcal{H}$ is a complex Hilbert space and $\mathcal{R} = \mathcal{L}(\mathcal{H})$, the algebra of all bounded linear operators on $\mathcal{H}$, the condition given above is also necessary. One way of formulating this fact is to use a notion introduced by Doyle [5] and Safonov [16] in the context of a finite dimensional Hilbert space. The *structured norm of* $A \in \mathcal{L}(\mathcal{H})$ relative to a given subalgebra $\mathcal{R} \subseteq \mathcal{L}(\mathcal{H})$ which contains the identity is the number

$$\mu_{\mathcal{R}}(A) = \frac{1}{\inf\{\|T\| : T \in \mathcal{R},\ 1 \in \sigma(TA)\}}.$$

Here $\sigma(TA)$ is the spectrum of the operator TA. Doyle called this number the *structured singular value*, a notion appropriate to the finite dimensional context, and this number has proved to be a powerful tool in robust system analysis and design. $\mu_{\mathcal{R}}$ gives a measure of robust stability with respect to certain types of perturbations which measure certain types of uncertainties in modelling of physical systems.

It seems appropriate in the infinite dimensional setting to refer to $\mu_{\mathcal{R}}(A)$ as the *structured norm of* A with respect to $\mathcal{R}$. Using this terminology we can reformulate our opening paragraph as the formula $\mu_{\mathcal{L}(\mathcal{H})}(A) = \|A\|$, the standard operator norm of A. We will give a slightly more general form of this later on.

In this paper we will mainly consider two types of algebras to play the role of $\mathcal{R}$, both of which have been considered in various contexts in the theory of robust control of linear systems. We first describe an abstract framework for what follows. For a given subalgebra (all our subalgebras will contain the identity) $\mathcal{R}$ of $\mathcal{L}(\mathcal{H})$, let $M_m(\mathcal{R})$ denote the algebra of $m \times m$ operator matrices with entries in $\mathcal{R}$, acting as operators on $\mathcal{H}^{(m)}$, the direct sum of m copies of $\mathcal{H}$. $D_m \subseteq M_m(\mathcal{R})$ will denote the subalgebra of diagonal matrices from $M_m(\mathcal{R})$ and $\mathcal{B}_m$ will denote the open unit ball of D_m. Suppose $A \in M_m(\mathcal{R})$. Clearly if $\|A\| \leq 1$ then for all $T \in \mathcal{B}_m$, $I - TA$ is invertible. Suppose there exists $X \in (D_m)'$, the commutant of D_m, such that X is invertible and $\|XAX^{-1}\| < 1$. Then, even if $\|A\| \geq 1$, it follows from the observation that $I - TA$ is invertible if and only if $X(I - TA)X^{-1} = I - TXAX^{-1}$ is, that $I - TA$ is invertible for all $T \in \mathcal{B}_m$. This motivates the introduction of the following number:

$$\hat{\mu}_{D_m}(A) = \inf\{\|XAX^{-1}\| : X \in (D_m)', \ X \text{ invertible}\}$$

and we will write $\mu_{D_1}(A)$ and $\hat{\mu}_{D_1}(A)$ simply as $\mu_{\mathcal{R}}(A)$, $\hat{\mu}_{\mathcal{R}}(A)$. It is always the case that $\mu_{D_m}(A) \leq \hat{\mu}_{D_m}(A)$. For $\mathcal{R} = \mathcal{L}(\mathcal{H})$ and $m = 1$, both these numbers are just $\|A\|$ and equality is immediate. However even for this algebra, when $m > 1$ the question: does $\mu_{D_m}(A) = \hat{\mu}_{D_m}(A)$? is non-trivial.

This was solved for finite dimensional $\mathcal{H}$, first by Doyle [5] (see also [12], [13]) and then in [6] as well as in [2]. The paper [6] defines Hermitian forms similar to those used in this paper. As far as we can tell these were introduced first in [15] which studied the invariance of the unit ball of operators under a fractional linear transformation whose coefficients are compact operators. This result was proved in [19] without the compactness assumption using the same Hermitian forms. The approach used in [2], operator theoretic as well, is quite different and uses the idea of a lifting. Common to all these papers is the use of some generalization (or consequence) of the Hausdorff-Toeplitz theorem on the convexity of the numerical range of an operator, and convexity will play a major role in our paper as well.

In the finite dimensional case equality holds for $m = 2, \ 3$, and Doyle [5] gives an example in the case $m = 4$ for which $\mu_{D_4}(A) < \hat{\mu}_{D_4}(A)$.

In section 3 we give a new proof for these results which holds for the case that $\mathcal{H}$ is infinite dimensional. As far as we have been able to ascertain, this is the first proof which holds in this case. We will also present a more complete analysis of the case $m = 4$. We mention that the case $m = 2$ is closely related to a result of Yu. Shmul'yan [19] on fractional linear transformations acting on the unit ball of $\mathcal{L}(\mathcal{H})$ and will discuss this fact as well.

The second type of algebra that we will discuss is a nest algebra. To simplify matters we will considered the simplest nest algebra on an infinite dimensional Hilbert space. $\mathcal{H}$ will be the Hilbert space of one-sided infinite square-summable sequences with entries in $\mathbf{C}^k$, $k < \infty$. P_n will denote the orthogonal projection onto the subspace $\{\langle a_0, \ldots, a_n, \ 0, \ldots\rangle\}$ and $\mathcal{C}$ will denote the nest algebra,

$$\mathcal{C} = \{T \in \mathcal{L}(\mathcal{H}) : \ P_n T P_n = P_n T, \ \ n \geq 0\}$$

of "lower triangular matrices". As is well known [8], $\mathcal{C}$ is identifiable with the stable linear systems whose inputs and outputs are from $\mathcal{H}$. The choice of this nest is not critical. It simply allows us to concentrate on the major issues while avoiding (sometimes non-trivial) technicalities.

For this algebra even the case $m = 1$ is non-trivial. This case was considered in [17] and [7]. We present the proof from [7] since it contains a fundamental idea used for $m > 1$. In this case the norm of A is replaced by $\|A\|_{\text{ess}}$, the essential norm of A. For $A \in \mathcal{C}$, $\mu_{\mathcal{C}}(A) = \|A\|_{\text{ess}}$.

Interestingly, the main result holds for all $m > 1$. The result given here has already been stated in [18]. Our proof seems to be both simpler and to clarify the difference between the algebras $\mathcal{L}(\mathcal{H})$ and $\mathcal{C}$. In $\mathcal{L}(\mathcal{H})$ the joint numerical range of two or three selfadjoint operators is a convex set ([1], [4]), while this breaks down for four or more operators. For $\mathcal{C}$ the appropriate notion is the joint essential numerical range (see [9]) and this is convex for any finite number of operators (see [3]).

The case $m = 2$ is of particular interest. A consequence of this case is the formulation of the result of Shmul'yan [19] in $\mathcal{L}(\mathcal{H})$ appropriate to the algebra $\mathcal{C}$.

2 Preliminaries

Let $A \in \mathcal{L}(\mathcal{H})$ and $\mathcal{R} \in \mathcal{L}(\mathcal{H})$ be a Banach subalgebra with identity, with $A^* \in \mathcal{R}$. Suppose that $I - TA$ is invertible for all $T \in \mathcal{R}$ with $\|T\| < 1$. If $\|A\| = \rho > 1$, let $T = \frac{1}{\rho^2}A^* \in \mathcal{R}$. Then $\|T\| < 1$ and $I - TA = \frac{1}{\rho^2}(\rho^2 I - A^*A)$. Since $\rho^2 \in \sigma(A^*A)$, we have shown that if $A^* \in \mathcal{R}$, then $\mu_{\mathcal{R}}(A) = \|A\|$. This holds trivially for $\mathcal{R} = \mathcal{L}(\mathcal{H})$.

It easily checked that if $\mathcal{R}_1 \subseteq \mathcal{R}_2$ then $\mu_{\mathcal{R}_1}(A) \leq \mu_{\mathcal{R}_2}(A)$. Take $\mathcal{R}$ to be the smallest possible Banach subalgebra with identity of $\mathcal{L}(\mathcal{H})$. Then $\mathcal{R} = \{\lambda I : \lambda \in \mathbf{C}\}$. It is a simple exercise to show that for $A \in \mathcal{L}(\mathcal{H})$, $\mu_{\mathcal{R}}(A) = \rho(A)$. Here $\rho(A)$ denotes the spectral radius of A. Thus for any subalgebra $\mathcal{R} \subset \mathcal{L}(\mathcal{H})$ and any $A \in \mathcal{L}(\mathcal{H})$ we have

$$\rho(A) \leq \mu_{\mathcal{R}}(A) \leq \|A\|.$$

In particular if A satisfies $\|A\| = \rho(A)$, then $\mu_{\mathcal{R}}(A) = \|A\|$, independently of $\mathcal{R}$. This holds, for example, if A is an (scalar valued) analytic Toeplitz operator, a class which is of interest in linear systems theory. It easily follows that for this case, if $\mathcal{R}$ is the algebra of analytic Toeplitz operators, then $\mu_{\mathcal{R}}(A) = \hat{\mu}_{\mathcal{R}}(A)$.

3 Finding $\mu_{D_m}(A)$ for $\mathcal{R} = \mathcal{L}(\mathcal{H})$

We have seen that $\mu_{\mathcal{L}(\mathcal{H})}(A) = \hat{\mu}_{\mathcal{L}(\mathcal{H})}(A)$. This is the case $m = 1$. It is instructive to discuss separately the cases $m = 2,\ 3$. We begin with $m = 2$.

Consider $A = \begin{bmatrix} A_{11} & A_{12} \\ A_{21} & A_{22} \end{bmatrix}$, $A_{ij} \in \mathcal{L}(\mathcal{H})$. If $\Delta \in \mathcal{L}(\mathcal{H})$ and $I - A_{11}\Delta$ is invertible, we can define the following *linear fractional transformation*

$$\mathcal{F}_A(\Delta) = A_{22} + A_{21}\Delta(I - A_{11}\Delta)^{-1}A_{12}.$$

Note that $\mathcal{F}$ is invariant under the following action: if $\lambda \in \mathbf{C}$, $\lambda \neq 0$, and

$$A_\lambda = \begin{bmatrix} I & 0 \\ 0 & \lambda^{-1}I \end{bmatrix} \begin{bmatrix} A_{11} & A_{12} \\ A_{21} & A_{22} \end{bmatrix} \begin{bmatrix} I & 0 \\ 0 & \lambda I \end{bmatrix} = \begin{bmatrix} A_{11} & \lambda A_{12} \\ \lambda^{-1}A_{21} & A_{22} \end{bmatrix},$$

then $\mathcal{F}_{A_\lambda} = \mathcal{F}_A$.

Also, suppose $\|A\| \leq 1$ and $\Delta \in \overline{\mathcal{B}}_1$, the closed unit ball of $\mathcal{L}(\mathcal{H})$, with $I - A_{11}\Delta$ invertible. We show that $\|\mathcal{F}_A(\Delta)\| \leq 1$. The argument is a special case of [10, p. 433, Lemma 1.2].

For arbitrary vector $x \in \mathcal{H}$ define $y = (I - A_{11}\Delta)^{-1}A_{12}x$, $z = \Delta y$, $u = A_{22}x + A_{21}z$. Then we have $u = \mathcal{F}_\Delta(A)x$ and

$$\begin{bmatrix} A_{11} & A_{12} \\ A_{21} & A_{22} \end{bmatrix} \begin{bmatrix} z \\ x \end{bmatrix} = \begin{bmatrix} y \\ u \end{bmatrix}.$$

Hence

$$\|x\|^2 - \|\mathcal{F}_\Delta(A)x\|^2 = \|x\|^2 - \|u\|^2 = \left\| \begin{bmatrix} z \\ x \end{bmatrix} \right\|^2 - \left\| \begin{bmatrix} y \\ u \end{bmatrix} \right\|^2 + \|y\|^2 - \|z\|^2$$

$$= \left\| \begin{bmatrix} z \\ x \end{bmatrix} \right\| - \left\| A \begin{bmatrix} z \\ x \end{bmatrix} \right\|^2 + \|y\|^2 - \|\Delta y\|^2 = \left\| D_A \begin{bmatrix} z \\ x \end{bmatrix} \right\|^2 + \|D_\Delta y\|^2 \geq 0,$$

where $D_S = (I - S^*S)^{\frac{1}{2}}$ for a contraction S. This implies the required inequality $\|\mathcal{F}_A(\Delta)\| \leq 1$.

We have seen that if $\|A_\lambda\| \leq 1$ for some $\lambda \neq 0$, then $\|\mathcal{F}_A(\Delta)\| \leq 1$ for all $\Delta \in \mathcal{B}_1$, the open unit ball. Shmul'yan [19] proved the converse.

Theorem 3.1 *Suppose that $\mathcal{F}_A$ is defined on whole $\mathcal{B}_1$ and transforms it into $\overline{\mathcal{B}_1}$. If $\mathcal{F}_A$ is non-trivial ($A_{12} \neq 0$, $A_{21} \neq 0$), then there exists $\lambda \neq 0$ (in fact $\lambda > 0$) such that $\|A_\lambda\| \leq 1$.*

As a consequence we obtain:

Theorem 3.2 *For $\mathcal{R} = \mathcal{L}(\mathcal{H})$ and $A \in M_2(\mathcal{R})$, $\mu_{D_2}(A) = \hat{\mu}_{D_2}(A)$.*

Proof: Since $(D_2)' = \{\text{diag}\,[\lambda_1 I, \lambda_2 I] : \lambda_i \in \mathbf{C}\}$ and $\hat{\mu}_{D_2}(A)$ depends on the invertible elements of $(D_2)'$ alone, we obtain, by normalizing, that $\hat{\mu}_{D_2}(A) = \inf\{\|A_\lambda\| : \lambda \neq 0\}$. We always have that $\hat{\mu} \geq \mu$. So we must show that $\hat{\mu} \leq \mu$. Since $I - \Delta A$ is invertible if and only if $I - A\Delta$ is, it suffices to show that if $I - A\Delta$ is invertible for all $\Delta \in \mathcal{B}_2$, then $\inf\{\|A_\lambda\|, \lambda \neq 0\} \leq 1$. We consider two cases.

Case 1 $\mathcal{F}_A$ is non-trivial; $A_{12} \neq 0$, $A_{21} \neq 0$. By assumption $I - A_{11}\Delta_1$ is invertible for all $\Delta_1 \in \mathcal{B}_1$ (just take $\Delta = \text{diag}\,[\Delta_1, 0]$) and then $\|A_{11}\| \leq 1$. Now

$$I - A\Delta = \begin{bmatrix} I - A_{11}\Delta_1 & -A_{12}\Delta_2 \\ -A_{21}\Delta_1 & I - A_{22}\Delta_2 \end{bmatrix}.$$

We write this as a product $A_1A_2A_3$ as follows:

$$A_1 = \begin{bmatrix} I & 0 \\ -A_{21}\Delta_1(I - A_{11}\Delta_1)^{-1} & I \end{bmatrix},$$

$$A_2 = \begin{bmatrix} I - A_{11}\Delta_1 & 0 \\ 0 & (I - A_{22}\Delta_2) - A_{21}\Delta_1(I - A_{11}\Delta_1)^{-1}A_{12}\Delta_2 \end{bmatrix},$$

$$A_3 = \begin{bmatrix} I & -(I - A_{11}\Delta_1)^{-1}A_{12}\Delta_2 \\ 0 & I \end{bmatrix}.$$

Since this product is invertible it follows that

$$(I - A_{22}\Delta_2) - A_{21}\Delta_1(I - A_{11}\Delta_1)^{-1}A_{12}\Delta_2$$

is. Rewrite this as

$$I - [A_{22} + A_{21}\Delta_1(I - A_{11}\Delta_1)^{-1}A_{12}]\Delta_2 = I - \mathcal{F}_A(\Delta_1)\Delta_2.$$

Since this is invertible for all $\Delta_2 \in \mathcal{B}_1$, it follows that $\|\mathcal{F}_A(\Delta_1)\| \leq 1$ for all $\Delta_1 \in \mathcal{B}_1$ and therefore, by Theorem 3.1, that there exists $\lambda \neq 0$ with $\|A_\lambda\| \leq 1$.

Case 2 At least one of A_{12}, A_{21} is zero. Assume A_{12} is zero. Then

$$I - A\Delta = \begin{bmatrix} I - A_{11}\Delta_1 & 0 \\ -A_{21}\Delta_1 & I - A_{22}\Delta_2 \end{bmatrix}$$

and as above, $\|A_{ii}\| \leq 1$, $i = 1, 2$. Taking $\{\lambda_n\}$ such that $\lambda_n \to \infty$ gives $\inf \|A_{\lambda_n}\| \leq 1$. □

Remarks 3.3 The difficult part of these results is on showing that invertibility for all $\Delta \in \mathcal{B}_2$ or that $\mathcal{F}_A$ maps $\mathcal{B}_1$ into $\overline{\mathcal{B}}_1$, implies that there exists $\lambda > 0$ such that $\|A_\lambda\| < 1$. Shmul'yan proves this as follows. From A define two Hermitian forms on $\mathcal{H}^{(2)}$ as follows: for $f = \begin{bmatrix} f_1, \\ f_2 \end{bmatrix}$,

$$\Omega_1(f) = \left\| [A_{11}\ A_{22}] \begin{bmatrix} f_1 \\ f_2 \end{bmatrix} \right\|^2 - \|f_1\|^2,$$

$$\Omega_2(f) = \|f_2\| - \left\| [A_{21}\ A_{22}] \begin{bmatrix} f_1 \\ f_2 \end{bmatrix} \right\|^2.$$

The assumption on $\mathcal{F}_A$ gives that $\Omega_1(f) \geq 0$ implies $\Omega_2(f) \geq 0$. Shmul'yan then uses the following property of such forms [11]. (The statement here is slightly stronger and more appropriate for our purposes.)

Lemma 3.4 *Suppose S and T are selfadjoint operators on $\mathcal{H}$ such that S is not negative semi-definite, and $(Sf, f) \geq 0$ implies $(Tf, f) \geq 0$. Then there exists $\rho \geq 0$ such that $(Tf, f) \geq \rho(Sf, f)$ for all $f \in \mathcal{H}$.*

This gives the existence of $\lambda > 0$ such that $\|A_\lambda\| \le 1$. In fact Lemma 3.4 is a special case of the Hausdorff-Toeplitz theorem that the numerical range of an operator on a Hilbert space is convex. For, consider $L = S + iT$. By hypothesis, the numerical range of L, $W(L)$, satisfies

$$W(L) \cap \{z = \langle x, y\rangle : x \ge 0,\ y < 0\} = \emptyset.$$

By convexity, there exists a line through the origin which separates $W(L)$ from $\{z : x \ge 0,\ y < 0\}$. The hypothesis with regard to S gives that this line can't be $x = 0$. Thus it has the form $y = \rho x$ for some $\rho \ge 0$. Thus the points in $W(L)$ satisfy $y \ge \rho x$. This proves the lemma.

We present this fact because it is important to understand the role of the convexity property of the numerical range of an operator in the proofs of Theorems 3.1 and 3.2. This becomes more transparent in the proof of the case $m = 3$. We begin the proof in this case by recalling the notion of the *joint numerical range* for a set $\{A_1, \dots, A_m\}$ of bounded selfadjoint operators on $\mathcal{H}$:

$$W(A_1, \dots, A_m) = \{\langle (Af, f), \dots, (A_m f, f)\rangle : \|f\| = 1\} \subseteq \mathbf{R}^m.$$

It has been shown ([1], [4]) that for $m = 3$, $W(A_1, A_2, A_3)$ is a convex set in $\mathbf{R}^3$. We need the following consequence of this fact.

Lemma 3.5 *If $(A_k f, f) \ge 0$ for $k = 1, 2, 3$ implies that $f = 0$, then there exist non-negative numbers $\alpha_1, \alpha_2, \alpha_3$ such that $\alpha_1 + \alpha_2 + \alpha_3 > 0$ and $\alpha_1 A_1 + \alpha_2 A_2 + \alpha_3 A_3 \le 0$.*

Proof: Let

$$S_+ = \{\langle x_1, x_2, x_3\rangle \in \mathbf{R}^3 : x_k \ge 0,\ k = 1, 2, 3\}.$$

By hypothesis $S_+ \cap W(A_1, A_2, A_3) = \emptyset$. By the convexity of $W(A_1, A_2, A_3)$ there exists a plane $\alpha_1 x_1 + \alpha_2 x_2 + \alpha_3 x_3 = 0$ in $\mathbf{R}^3$ such that

$$W(A_1, A_2, A_3) \subset \{\langle x_1, x_2, x_3\rangle : \alpha_1 x_1 + \alpha_2 x_2 + \alpha_3 x_3 \le 0\}$$

and

$$S_+ \subset \{\langle x_1, x_2, x_3\rangle : \alpha_1 x_1 + \alpha_2 x_2 + \alpha_3 x_3 \ge 0\}.$$

This gives that $\alpha_1 A_1 + \alpha_2 A_2 + \alpha_3 A_3 \le 0$. By the previous condition we can assume that $\alpha_k \ge 0$, $k = 1, 2, 3$. □

Theorem 3.6 *For $A \in M_3(\mathcal{L}(\mathcal{H}))$, $\mu_{D_3}(A) = \hat{\mu}_{D_3}(A)$.*

Proof: For $\lambda_1, \lambda_2 \ne 0$ define

$$A_{\lambda_1\lambda_2} = \begin{bmatrix} I & 0 & 0 \\ 0 & \lambda_1^{-1} I & 0 \\ 0 & 0 & \lambda_2^{-1} I \end{bmatrix} \begin{bmatrix} A_{11} & A_{12} & A_{13} \\ A_{21} & A_{22} & A_{23} \\ A_{31} & A_{32} & A_{33} \end{bmatrix} \begin{bmatrix} I & 0 & 0 \\ 0 & \lambda_1 I & 0 \\ 0 & 0 & \lambda_2 I \end{bmatrix}.$$

Since $\mu_{D_3}(A) \leq \hat{\mu}_{D_3}(A)$ always holds, we show the inverse inequality. It is sufficient to prove that if $I - \Delta A$ is invertible for all $\Delta \in \overline{\mathcal{B}}_3$, then $\inf\{\|A_{\lambda_1\lambda_2}\| : \lambda_1, \lambda_2 \neq 0\} \leq 1$.

For

$$f = \begin{bmatrix} f_1 \\ f_2 \\ f_3 \end{bmatrix} \in \mathcal{H}^{(3)}$$

define the Hermitian forms

$$\Omega_k(f) = \left\| \sum_{j=1}^{3} A_{kj} f_j \right\|^2 - \|f_k\|^2, \quad k = 1,\ 2,\ 3.$$

We first show that if $I - \Delta A$ is invertible for all $\Delta \in \overline{\mathcal{B}}_3$, then $\Omega_k(f) \geq 0,\ k = 1, 2, 3$ implies that $f = 0$.

Suppose there exists a non-zero vector $f \in \mathcal{H}^{(3)}$ such that $\Omega_k(f) \geq 0$ for $k = 1, 2, 3$. Rewrite this as

$$\|A_{k1}f_1 + A_{k2}f_2 + A_{k3}f_3\| \geq \|f_k\|, \quad k = 1,\ 2,\ 3.$$

Then for each such k there exists $\Delta_k \in \overline{\mathcal{B}}_1$ such that

$$\Delta_k[A_{k1}f_1 + A_{k2}f_2 + A_{k3}f_3] = f_k$$

and this implies that $(I - \Delta A)f = 0$ for $\Delta = \operatorname{diag}[\Delta_1,\ \Delta_2,\ \Delta_3] \in \overline{\mathcal{B}}_3$, which contradicts our assumption.

Now apply Lemma 3.5. There exists numbers $\alpha_1, \alpha_2, \alpha_3 \leq 0$, with $\alpha_1 + \alpha_2 + \alpha_3 > 0$, such that $\alpha_1\Omega_1(f) + \alpha_2\Omega_2(f) + \alpha_3\Omega_3(f) \leq 0$. Equivalently,

$$\left\| \left[\sum_{k=1}^{3} A_{jk} \alpha_j^{\frac{1}{2}} f_k \right]_{j=1}^{3} \right\| \leq \left\| \left[\alpha_j^{\frac{1}{2}} f_j \right]_{j=1}^{3} \right\|.$$

Now we consider different cases.

Case 1 $\alpha_1 > 0,\ \alpha_2 > 0,\ \alpha_3 > 0$.

Let $g_k = \alpha_k^{-\frac{1}{2}} f_k$ and rewrite the inequality as

$$\left\| \left[\sum_{k=1}^{3} \frac{\alpha_j^{\frac{1}{2}}}{\alpha_k^{\frac{1}{2}}} A_{jk} g_k \right]_{j=1}^{3} \right\| \leq \left\| [g_j]_{j=1}^{3} \right\|.$$

Taking $\lambda_1 = \alpha_1^{\frac{1}{2}} \alpha_2^{-\frac{1}{2}}$, $\lambda_2 = \alpha_2^{\frac{1}{2}} \alpha_3^{-\frac{1}{2}}$ gives the required result.

Case 2 Exactly one of $\alpha_1,\ \alpha_2,\ \alpha_3$ is zero. Suppose $\alpha_3 = 0,\ \alpha_1 > 0,\ \alpha_2 > 0$. Then, in the above inequality take $f_1 = f_2 = 0$. This gives

$$\alpha_1 \|A_{13} f_3\|^2 + \alpha_2 \|A_{23} f_3\|^2 \leq 0.$$

It follows that $A_{13} = A_{23} = 0$. In this case we take $\lambda_1 = \alpha_1^{\frac{1}{2}} \alpha_2^{-\frac{1}{2}}$ and consider the sequence X_n of diagonal operators of the form

$$X_n = \begin{bmatrix} I & & \\ & \lambda_1 I & \\ & & \lambda_{2,n} I \end{bmatrix}$$

where $\lambda_{2,n} \to \infty$ as $n \to \infty$. Since

$$\left\| \begin{bmatrix} A_{11} & \lambda_1 A_{12} \\ \lambda_1^{-1} A_{21} & A_{22} \end{bmatrix} \right\| \leq 1 \text{ and } \|A_{33}\| < 1$$

we have

$$\inf \|X_n^{-1} A X_n\| \leq 1.$$

Case 3 Two of $\alpha_1,\ \alpha_2,\ \alpha_3$ are zero. Suppose $\alpha_2 = \alpha_3 = 0,\ \alpha_1 > 0$ and let $f_1 = 0$ in the above inequality. Then $\alpha_1 \|A_{12} f_2 + A_{13} f_3\|^2 \leq 0$ for arbitrary $f_2,\ f_3$ which gives $A_{12} = A_{13} = 0$. Using Theorem 3.2 for the operator $[A_{j,k}]_{j,k=2}^3$ we find a sequence λ_n such that

$$\inf \left\| \begin{bmatrix} I & \\ & \lambda_n^{-1} I \end{bmatrix} \begin{bmatrix} A_{22} & A_{23} \\ A_{32} & A_{33} \end{bmatrix} \begin{bmatrix} I & \\ & \lambda_n I \end{bmatrix} \right\| \leq 1.$$

Now choose a sequence μ_n such that $\mu_n \to \infty$ and $\mu_n \lambda_n \to \infty$, and set

$$X_n = \operatorname{diag}\,[I,\ \mu_n I,\ \lambda_n \mu_n I].$$

It is easy to check that again

$$\inf \|X_n^{-1} A X_n\| \leq 1.$$

Remarks 3.7 (1) Of course the case $m = 2$ can be proved as above without the use of Theorem 3.1. However, the given proof of Theorem 3.2 is instructive and we return to it at the end of the paper.

(2) In the pioneering work of Doyle [5] another approach, which holds apparently only for $\dim \mathcal{H} < \infty$, is used. It reduces the question about validity of the equality

$$\mu_{D_m}(A) = \hat{\mu}_{D_m}(A)$$

to the problem of convexity of the joint numerical range of $m - 1$ instead of m selfadjoint operators in a subspace $\mathcal{H}_0$. The dimension of $\mathcal{H}_0$ is equal to the multiplicity of the maximal singular value of the operator $X_0 A X_0^{-1}$ where

$$\|X_0 A X_0^{-1}\| = \inf \{\|X A X^{-1}\| : X \in (D_m)'\} = \hat{\mu}_{D_m}(A).$$

(For simplicity we assume here that the infimum is attained). For the case $m = 3$ (Theorem 3.6 for $\dim \mathcal{H} < \infty$) Doyle [5] used the convexity of the joint numerical range of two selfadjoint operators, i.e. the Hausdorff-Toeplitz theorem.

We now turn to the case $m = 4$. In this case, for finite dimensional $\mathcal{H}$ (in fact for $\dim \mathcal{H} = 1$) an example was given in [5], [12] for which $\mu_{D_4}(A) < \hat{\mu}_{D_4}(A)$. We will not discuss the example here in detail and restrict ourselves to a number of remarks. For this example it was shown that the infimum in the definition of $\hat{\mu}(A)$ is in fact a minimum and is attained for some $X_0 \in (D_4)'$, $\hat{\mu}_{D_4}(A) = \|X_0 A X_0^{-1}\|$, and the multiplicity of the maximal singular value of $X_0 A X_0^{-1}$ is 2. But this is exactly the dimension for which the convexity of the joint numerical range of 3 selfadjoint operators fails (see e.g. [4]). It can be shown that for an operator A and $X_0 \in (D_4)'$ for which $\|XAX^{-1}\|$ attains its minimum, if the maximal singular value of $X_0 A X_0^{-1}$ has multiplicity 1 or ≥ 3, then in fact $\mu_{D_4}(A) = \hat{\mu}_{D_4}(A)$. Thus the failure of the theorem in all its generality is a restricted phenomenon and more research is required to understand fully even the finite dimensional case.

We complete this discussion by using the example of [12] to produce an example for the four-block case when $\mathcal{H}$ is infinite dimensional.

Example 3.8 Here $\mathcal{H}$ is infinite dimensional and we will represent operators by matrix representations with respect to a given orthonormal basis.

Suppose A is the 4×4 matrix with scalar entries from [12] for which $1 = \mu_{D_4}(A) < \hat{\mu}_{D_4}(A) = \sigma$. If $A = [a_{jk}]_{j,k=1}^4$, let M_{jk} be the infinite matrix with a_{jk} in its upper left corner and all other entries zero, and let $M = [M_{jk}]_{j,k=1}^4$. Clearly $\hat{\mu}_{D_4}(M) = \sigma > 1$. We show that for all $\Delta = \operatorname{diag}[\Delta_1, \Delta_2, \Delta_3, \Delta_4] \in \mathcal{B}_4$, $I - \Delta M$ is invertible. If Δ_i has matrix representation $[\Delta^i_{j,k}]_{j,k=1}^\infty$, $i = 1, \ldots, 4$, then $I - \Delta M = [S_{jk}]_{j,k=1}^4$ where

$$S_{jk} = \delta_{jk} I - a_{jk} \begin{bmatrix} \Delta^j_{11} & 0 & 0 & \cdots \\ \Delta^j_{21} & 0 & 0 & \cdots \\ \Delta^j_{31} & 0 & 0 & \cdots \\ \cdots & \cdots & \cdots & \cdots \end{bmatrix}$$

Note that all the entries of $I - \Delta M$, as well as $I - \Delta M$ itself, are compact perturbations of I. Thus $I - \Delta M$ is invertible if $\operatorname{Ker}(I - \Delta M) = \{0\}$. We show that this is the case. Suppose

$$f = \begin{bmatrix} f^1 \\ f^2 \\ f^3 \\ f^4 \end{bmatrix} \text{ where } f^i = \begin{bmatrix} f^i_1 \\ f^i_2 \\ \vdots \\ f^i_n \\ \vdots \end{bmatrix} \in \mathcal{H}.$$

If $(I - \Delta M)f = 0$ then the vector

$$\begin{bmatrix} f_1^1 \\ f_1^2 \\ f_1^3 \\ f_1^4 \end{bmatrix} \quad (\in \mathbf{C}^4)$$

satisfies the corresponding equation for A. But $|\Delta_{11}^i| < 1$ and $\mu_{D_4}(A) = 1$ give that $f_1^k = 0$ for $k = 1, 2, 3, 4$. But then $Mf = 0$ and this implies that $f = 0$.

4 Finding $\mu_{D_m}(A)$ for $\mathcal{R} = \mathcal{C}$

It is easy to see that there exists an operator $T \in \mathcal{C}$ such that $\|T\| > 1$ and $I - \Delta T$ is invertible for all $\Delta \in \mathcal{C}$ with $\|\Delta\| \leq 1$. Let T be any compact strictly lower triangular operator. Then, for any $\Delta \in \mathcal{C}$, ΔT is strictly lower triangular and compact. Thus ([14], Theorem 5.12) the spectrum of ΔT is just $\{0\}$ and $I - \Delta T$ is invertible.

We give a condition which is both necessary and sufficient and is appropriate to $\mathcal{C}$. In fact the example in the previous paragraph is at the heart of the issue. We give the following lemma just for the record. $\|T\|_{\text{ess}}$ denotes the *essential norm* of T, i.e.

$$\|T\|_{\text{ess}} = \inf\{\|T - K\|, \quad K \text{ compact}\}.$$

Lemma 4.1 *For $T \in \mathcal{L}(\mathcal{H})$,*

$$\|T\|_{\text{ess}} = \inf_{n \geq 0} \|T(I - P_n)\| = \lim_{n \longrightarrow \infty} \|T(I - P_n)\|.$$

Proof: The second equality follows from the fact that the sequence $\{\|T(I - P_n)\|\}$ is decreasing in n,

$$\begin{aligned} \|T(I - P_{n+1})\| &= \|T(I - P_n)(I - P_{n+1})\| \leq \|T(I - P_n)\| \|I - P_{n+1}\| \\ &= \|T(I - P_n)\|. \end{aligned}$$

For the first equality, note that for each n, $T = T(I - P_n) + F_n$, where F_n is finite rank. Thus

$$\inf_{n \geq 0} \|T(I - P_n)\| = \inf_{n \geq 0} \|T - F_n\| \geq \|T\|_{\text{ess}}.$$

On the other hand, since $(I - P_n)$ converges to zero strongly as $n \to \infty$, $\|K(I - P_n)\| \to 0$ for K compact. Therefore

$$\begin{aligned} \inf_{n \geq 0} \|T(I - P_n)\| &= \lim_{n \to \infty} \|T(I - P_n)\| = \lim_{n \to \infty} \|(T + K)(I - P_n)\| \\ &\leq \|T + K\| \end{aligned}$$

for any compact K. Thus

$$\inf_{n \geq 0} \|T(I - P_n)\| \leq \|T\|_{\text{ess}}.$$

□

The proof of sufficiency in the next theorem requires that $T \in \mathcal{C}$ be strictly lower triangular. This assumption plays no role in the proof of necessity, which is the main issue here.

Theorem 4.2 *Suppose that $T \in \mathcal{C}$ is strictly lower triangular. Then $I - \Delta T$ is invertible for all $\Delta \in \mathcal{C}$ with $\|\Delta\| \leq 1$ if and only if $\|T\|_{\text{ess}} < 1$.*

Proof: Suppose $\|T\|_{\text{ess}} < 1$. By the previous lemma there exists n such that $\|T(I - P_n)\| < 1$. Decompose $\mathcal{H}$ as $P_n\mathcal{H} \oplus (I - P_n)\mathcal{H}$, and $I - \Delta T$ with respect to this decomposition:

$$I - \Delta T = \begin{bmatrix} P_n - P_n \Delta T P_n & 0 \\ X & (I - P_n) - \Delta T(I - P_n) \end{bmatrix}.$$

The upper left entry is a finite dimensional lower triangular matrix with identity on the main diagonal and is therefore invertible. The lower right entry is of the form $I - S$ where $\|S\| < 1$ and is invertible. It follows that $I - \Delta T$ is invertible.

Now suppose $\|T\|_{\text{ess}} > q \geq 1$. Since $\|T\| \geq \|T\|_{\text{ess}}$ there exists $u_1 \in \mathcal{H}$ such that $\frac{\|T_{u_1}\|}{\|u_1\|} > q$. Now $\{P_n\}$ converges strongly to I. Thus we can choose ϵ_1, arbitrarily small, and $n_1 > 0$ so that

$$\frac{\|P_{n_1} T P_{n_1} u_1\|}{\|P_{n_1} u_1\|} > q, \quad \frac{\|(I - P_{n_1}) T P_{n_1} u_1\|}{\|P_{n_1} u_1\|} < \epsilon_1.$$

Letting $v = \frac{P_{n_1} u_1}{\|P_{n_1} u_1\|}$, we can rewrite this as

$$\|P_{n_1} T v_1\| > q, \quad \|(I - P_{n_1}) T v_1\| < \epsilon_1.$$

Now consider the restriction of T to the invariant subspace $(I - P_{n_1})\mathcal{H}$. Since

$$\inf_{n \geq n_1} \|T(I - P_n)\| \geq \inf_{n \geq 0} \|T(I - P_n)\| > q,$$

we can choose ϵ_2, arbitrarily small, $n_2 > n_1$, and $u_2 \in (I - P_{n_1})\mathcal{H} > 0$ so that for $v_2 = \frac{P_{n_2} u_2}{\|P_{n_2} u_2\|}$ we have

$$\|P_{n_2} T P_{n_2} v_2\| > q, \quad \|(I - P_{n_2}) T P_{n_2} v_2\| < \epsilon_2.$$

Continuing in this fashion, we construct a sequence of vectors $\{v_k\}$ and of arbitrarily small numbers $\{\epsilon_k\}$ such that, for each k,

(1) $v_k \in (P_{n_k} - P_{n_{k-1}})\mathcal{H}$;

(2) $\|v_k\| = 1$;

(3) $T v_k = y_k + z_k$, with

$$\begin{aligned} y_k &= P_{n_k} T v_k \in (P_{n_k} - P_{n_{k-1}})\mathcal{H}, \quad \|y_k\| > q, \\ z_k &= (I - P_{n_k}) T v_k, \quad \|z_k\| < \epsilon_k. \end{aligned}$$

Define Δ on $\mathcal{H}$ by

$$\Delta = \sum_{k=1}^{\infty} \frac{(\cdot, y_k)}{\|y_k\|^2} v_{k+1}.$$

Then $\|\Delta\| < 1$, and for all $n \geq 0$,

$$\Delta(I - P_n) = (I - P_n)\Delta(I - P_n),$$

that is, $\Delta \in \mathcal{C}$. Also

$$\Delta T v_k = \Delta y_k + \Delta z_k = v_{k+1} + \Delta z_k.$$

Therefore,

$$\Delta T \left(\sum_{k=1}^{n} v_k \right) = \sum_{k=1}^{n} v_{k+1} + \Delta \left(\sum_{k=1}^{n} z_k \right),$$

and,

$$(I - \Delta T) \left(\sum_{k=1}^{n} v_k \right) = v_1 - v_{n+1} - \Delta \left(\sum_{k=1}^{n} z_k \right).$$

It follows that

$$\left\| (I - \Delta T) \sum_{k=1}^{n} v_k \right\| \leq 2 + \sum_{k=1}^{n} \|z_k\| \leq 2 + \sum_{k=1}^{n} \epsilon_k.$$

However, $\{v_k\}$ is an orthogonal sequence and therefore $\| \sum_{k=1}^{n} v_k \| = \sqrt{n}$. Thus $I - \Delta T$ is not invertible and the proof is complete. □

Remarks 4.3 (1) The proof given here is essentially the same as that of Theorem 4.2 of [7]. A different proof is given in [17].
(2) If we take all $\|\Delta\| < 1$ the condition required is $\|T\|_{\text{ess}} \leq 1$.
(3) Theorem 4.2 remains true for an arbitrary operator T from $\mathcal{C}$, if we consider only strictly lower triangular Δ. This holds also for the previous remark.

We now turn to the case $m > 1$. We recall a special case of a notion introduced in [3] (see also [9]). For self-adjoint operators $A_1, \ldots, A_m$ on $\mathcal{H}$ their *joint essential numerical range* is

$$W_{\text{ess}}(A_1, \ldots, A_m) = \{\lambda \in \mathbf{R}^m, \ \lambda = \lim_{n \to \infty} \langle (A_k f_n, f_n) \rangle_{k=1}^m : \ \|f_n\| = 1,$$
$$f_n \longrightarrow 0 \text{ weakly}\}.$$

Several equivalent definitions of this notion can be found (cp. [9]).

It was shown in [3] (in a more general context) that $W_{\text{ess}}(A_1, \ldots, A_m)$ is a convex compact subset of $\mathbf{R}^m$. In our case it is clearly nonempty.

For $1 \leq k \leq m$ we will denote by $A_k^{(n)}$ the operator $(I - P_n)A_k(I - P_n)|(I - P_n)\mathcal{H}$. We need the following property.

Lemma 4.4 *The following are equivalent:*
(a) $\lambda \in W_{\text{ess}}(A_1, \ldots, A_m)$;
(b) $\lambda = \lim_{n\to\infty} \lambda_n$ *where* $\lambda_n \in W(A_1^{(n)}, \ldots, A_m^{(n)})$.

Proof: $(b) \Rightarrow (a)$. Let $\lambda_n = \langle (A_k^{(n)} f_n, f_n)\rangle_{k=1}^m$ with $\|f_n\| = 1$, $f_n \in (I - P_n)\mathcal{H}$. Choose a subsequence $\{f_{n_s}\}$ such that $\|(I - P_{n_{s+1}}) f_{n_s}\| \to 0$ as $n_s \to \infty$. Then $f_{n_s} \to 0$ weakly and $\lambda \in W_{\text{ess}}(A_1, \ldots, A_m)$.

$(a) \Rightarrow (b)$. Let $\lambda = \lim_{n\to\infty} \langle (A_k^{(n)} f_n, f_n)\rangle_{k=1}^m$ with $\|f_n\| = 1$ and $f_n \to 0$ weakly. Choose a subsequence $\{f_{n_s}\}_{s=1}^\infty$ and a subsequence $\{P_{n'_s}\}_{s=1}^\infty$ such that

$$\|(I - P_{n'_s}) f_{n_s}\| < \frac{1}{s} \quad \text{and} \quad \|P_{n'_s} f_{n_{s+1}}\| < \frac{1}{s}.$$

(This is done as follows: set $n_1 = 1$ and choose n'_1 such that $\|(I - P_{n'_1}) f_1\| < 1$. Then choose n_2 such that $\|P_{n'_1} f_{n_2}\| < 1$, etc.) If $h_s = \frac{P_{n'_s} f_{n_s}}{\|P_{n'_s} f_{n_s}\|}$, then

$$\lim_{s\to\infty} \langle (A_k^{(n'_s)} h_s, h_s)\rangle_{k=1}^m = \lambda.$$

This gives (b). □

Remark 4.5 It follows from Lemma 4.4 that

$$W_{\text{ess}}(A_1, \ldots, A_m) = \bigcap_n \overline{W(A_1^{(n)}, \ldots, A_m^{(n)})}.$$

The next lemma will be crucial. It supplies the right convex set for $m \geq 4$.

Lemma 4.6 *Let* $S_+^m = \{x \in \mathbf{R}^m : x_k \geq 0,\ k = 1, \ldots, m\}$. *If for some n,*

$$W(A_1^{(n)}, \ldots, A_m^{(n)}) \bigcap S_+^m = \emptyset$$

then there exist non-negative numbers $\alpha_1, \ldots, \alpha_m$, $\sum_{k=1}^m \alpha_k > 0$, *and a sequence* $\{\epsilon_n\}$ *of non-negative numbers with* $\lim_{n\to\infty} \epsilon_n = 0$, *such that for all n,*

$$W(A_1^{(n)}, \ldots, A_m^{(n)}) \subset \left\{ x \in \mathbf{R}^m : \sum_{k=1}^m \alpha_k x_k \leq \epsilon_n \right\}.$$

Proof: We first observe that

$$W_{\text{ess}}(A_1, \ldots, A_m) \bigcap \text{Int } S_+^m = \emptyset.$$

If not, take $\lambda \in W_{\text{ess}}(A_1, \ldots, A_m) \cap \text{Int } S_+^m$. By Lemma 4.4 there is a sequence $\lambda_n \in W(A_1^{(n)}, \ldots, A_m^{(n)})$ with $\lim_{n\to\infty} \lambda_n = \lambda$. Thus there exists N such that for $n > N$, $\lambda_n \in \text{Int } S_+^m$.

Since W_{ess} and S_+^m are convex, there exists a plane, $\sum_{k=1}^m \alpha_k x_k = 0$, which separates W_{ess} and S_+^m. More precisely,

$$S_+^m \subset \left\{x : \sum_{k=1}^m \alpha_k x_k \geq 0\right\}, \quad W_{\text{ess}}(A_1, \ldots, A_m) \subset \left\{x : \sum_1^m \alpha_k x_k \leq 0\right\}.$$

By the definition of S_+^m it follows that $\alpha_1, \ldots, \alpha_m$ are non-negative. The existence of the sequence $\{\epsilon_n\}$ with the required property follows from Lemma 4.4. □

The next lemma extends the main argument in the proof of Theorem 4.2 to the m-block case. We introduce some necessary notation. For $n \geq 0$, $\mathcal{P}_n^{(m)}$ will be the orthogonal projection diag $[P_n, \ldots, P_n]$ (m terms) acting on $\mathcal{H}^{(m)}$ and $\Pi_k^{(m)}$ will be the orthogonal projection from $\mathcal{H}^{(m)}$ onto the k-th co-ordinate space, $k = 1, \ldots, m$. Since m is fixed, we will not write it explicitly and just write $\mathcal{P}_n$ and Π_k. From this point onward $\mathcal{B}_m$, D_m will be from $\mathcal{C}$ (and not from $\mathcal{L}(\mathcal{H})$ as in Section 3).

Lemma 4.7 *If, for some $q > 1$ and for any $n \in \mathbf{N}$, there exists $f_n \in$ Im $(I - \mathcal{P}_n)$ such that*

$$\|\Pi_k A f_n\| > q\|\Pi_k f_n\|, \quad k = 1, \ldots, m,$$

then there exists $\Delta \in \mathcal{B}_m$ such that $I - \Delta A$ is not invertible.

Proof: We use induction on m. The case $m = 1$ is in the proof of Theorem 4.2. So assume the lemma holds for $1, 2, \ldots, m-1$, and consider the case of m co-ordinates.

Exactly as in the proof of Theorem 4.2 we construct a sequence $\{r_n\}$ of integers, $\{g_n\}$ of vectors, with $g_n \in$ Im $(P_{r_n} - P_{r_{n-1}})$ and $\{\epsilon_n\}$ of arbitrarily small positive numbers such that $\|g_n\| = 1$, $\|P_{r_n}\Pi_k A g_n\| > q\|\Pi_k g_n\|$, and $\|(I - P_{r_n})\Pi_k A g_n\| \leq \epsilon_n\|\Pi_k g_n\|$, for $k = 1, \ldots, m$ and $n = 1, 2, \ldots$.

Now consider two cases:

Case 1 $\lim_{n\to\infty} \inf \|\Pi_k g_n\| > 0$ for $k = 1, \ldots, m$. Then there exists $\delta > 0$ with $\|\Pi_k g_n\| \geq \delta > 0$ for all $1 \leq k \leq m$, $n \geq n_0$. Choose a number $q' \in (1, q)$ and divide the segment $[\delta, 1]$ into a finite number of segments $E_1, \ldots, E_l$ such that if $x_1, x_2 \in E_s$, $x_1 < q' x_2$. This defines a partition of the m-dimensional cube

$$\{x \in \mathbf{R}^m : \delta \leq x_k \leq 1; \ k = 1, \ldots, m\}$$

into l^m subcubes and there exists at least one such subcube which contains the entire subsequence $\langle\|\Pi_1 g_{n_j}\|, \|\Pi_2 g_{n_j}\|, \ldots, \|\Pi_m g_{n_j}\|\rangle$, $j = 1, 2, \ldots$. Thus

$$\|\Pi_k g_{n_{j+1}}\| < q'\|\Pi_k g_{n_j}\|, \quad 1 \leq k \leq m, \ j = 1, 2, \ldots.$$

Let $h_j = g_{n_j}$ and $y_{jk} = P_{r_{n_j}} \Pi_k A g_{n_j}$, and define the operators

$$\Delta_k = \sum_{j=1}^{\infty} \frac{(\cdot, y_{jk})}{\|y_{jk}\|^2} \Pi_k h_{j+1}, \quad 1 \le k \le m.$$

These are (in fact strictly) lower triangular and

$$\|\Delta_k\| = \sup_j \frac{\|\Pi_k h_{j+1}\|}{\|y_{jk}\|} \le \sup_j \frac{q'\|\Pi_k h_j\|}{q\|\Pi_k h_j\|} < 1.$$

Thus $\Delta = \operatorname{diag}[\Delta_1, \dots, \Delta_m] \in \mathcal{B}_m$. The argument that $I - \Delta A$ is not invertible is identical to that given in the proof of Theorem 4.2.

Case 2 $\lim_{n\to\infty} \inf \|\Pi_k g_n\| = 0$ for some k. We can assume that this is the case for $k = m$ and that $\lim_{n\to\infty} \|\Pi_m g_n\| = 0$. Then the condition of the lemma holds for the operator $[A_{jk}]_{j,k=1}^{m-1}$. By the induction hypothesis the proof is complete.

□

Remark 4.8 It is enough to assume that for any $q' > 1$ and for some $n \in \mathbf{N}$, there exists a non-zero $f_n \in \operatorname{Im}(I - \mathcal{P}_n)$ such that $\|\Pi_k A f_n\| \ge q'\|\Pi_k f_n\|$, $k = 1, \dots, m$. Indeed, if these inequalities hold, then the required inequalities of Lemma 4.7 follow for any $q \in (1, q')$.

The next lemma is just a normalization procedure.

Lemma 4.9 *Let $A = [A_{jk}]_{j,k=1}^m$ and consider the Hermitian forms*

$$\Omega_k(x) = \left\| \sum_{j=1}^m A_{kj} x_j \right\|^2 - \|x_k\|^2, \quad k = 1, \dots, m.$$

If there exist positive numbers δ, $\alpha_1, \dots, \alpha_m$ such that

$$\sum_{k=1}^m \alpha_k \Omega_k(x) \le \delta \|x\|^2 \ for\ all\ x \in \mathcal{H}^{(m)},$$

then there exist positive numbers $\zeta_1, \dots, \zeta_m$ such that

$$\|[\zeta_j \zeta_k^{-1} A_{jk}]_{j,k=1}^m\| \le \max_k \left(1 + \frac{\delta}{\alpha_k}\right)^{\frac{1}{2}}.$$

Proof: We have

$$\sum_{k=1}^m \left\| \sum_{j=1}^m \alpha_k^{\frac{1}{2}} A_{kj} x_j \right\|^2 \le \sum_1^m \|\alpha_k^{\frac{1}{2}} x_k\|^2 + \delta \sum_1^m \|x_k\|^2.$$

Let $y_j = \alpha_k^{\frac{1}{2}} x_k$ and rewrite this as

$$\sum_{k=1}^{m} \left\| \sum_{j=1}^{m} \frac{\alpha_k^{\frac{1}{2}}}{\alpha_j^{\frac{1}{2}}} A_{kj} y_j \right\|^2 \leq \sum_{k=1}^{m} \left(1 + \frac{\delta}{\alpha_k}\right) \|y_k\|^2.$$

Take $\zeta_j = \alpha_j^{\frac{1}{2}}$ and obtain the required inequality. □

We can now state the main result, which extends Theorem 4.2 to the m-block case.

Theorem 4.10 *If $I - \Delta A$ is invertible for all $\Delta \in \mathcal{B}_m$, then*

$$\inf_{\zeta_k > 0} \|[\zeta_i \zeta_k^{-1} A_{jk}]_{j,k=1}^{m}\|_{\text{ess}} \leq 1.$$

Proof: We will use induction on m. The case $m = 1$ is Theorem 4.2, where we have a slightly stronger statement. Assume we have proved the theorem for $l < m$ and consider the m co-ordinate case. From Remark 4.8 it follows that for any $q > 1$ there exists $n = n(q)$ such that if $f \in \text{Im}\,(I - \mathcal{P}_n)$ with $\|\Pi_k A f\| \geq q \|\Pi_k f\|$ for $k = 1, \dots, m$, then $f = 0$.

Fix $q > 1$ and define the Hermitian forms

$$\Omega_k(f) = \|\Pi_k q^{-1} A f\|^2 - \|\Pi_k f\|^2, \quad k = 1, \dots, m$$

for $f \in \mathcal{H}^{(m)}$. These can be rewritten as $\Omega_k(f) = (L_k f, f)$ where $\{L_k\}_{k=1}^{m}$ are the appropriate selfadjoint operators on $\mathcal{H}^{(m)}$. Our conclusion from Remark 4.8 can be reformulated as

$$W(L_1^{(n)}, \dots, L_m^{(n)}) \bigcap S_+^m = \emptyset,$$

and by Lemma 4.6 there exist non-negative numbers $\alpha_1, \dots, \alpha_m$ with $\sum_{k=1}^{m} \alpha_k > 0$ and a non-negative sequence $\{\epsilon_n\}$ with $\lim_{n\to\infty} \epsilon_n = 0$ such that

$$W(L_1^{(n)}, \dots, L_m^{(n)}) \subset \left\{ x \in \mathbf{R}^m : \sum_{1}^{m} \alpha_k x_k \leq \epsilon_n \right\}.$$

Equivalently,

$$\sum_{k=1}^{m} \alpha_k \Omega_k(f) \leq \epsilon_n \|f\|^2$$

for any $f \in \text{Im}\,(I - \mathcal{P}_n)$.

Again, we consider two cases.

Case 1 Suppose all $\alpha_k > 0$. By Lemma 4.9 there exist positive numbers $\zeta_1, \ldots, \zeta_m$ (depending on q and n) such that

$$\|[\zeta_j \zeta_k^{-1}(I - P_n)q^{-1}A_{jk}(I - P_n)]_{j,k=1}^m\| \leq \max_k \left(1 + \frac{\epsilon_n}{\alpha_k}\right).$$

(Here we have used the fact that since $A_{jk} \in \mathcal{C}$,

$$((I - \mathcal{P}_n)L_k f, f) = \|(I - P_n)\Pi_k q^{-1}Af\|^2 - \|(I - P_n)\Pi_k f\|^2.)$$

Thus,

$$\|(I - \mathcal{P}_n)[\zeta_j \zeta_k^{-1} A_{jk}]_{j,k=1}^m (I - \mathcal{P}_n)\| \leq q \max_k \left(1 + \frac{\epsilon_n}{\alpha_k}\right).$$

Thus for any $\epsilon > 0$ there exist $\zeta_1, \ldots, \zeta_m > 0$ such that

$$\|[\zeta_j \zeta_k^{-1} A_{jk}]_{j,k}^m\|_{\text{ess}} < 1 + \epsilon.$$

This gives (b).

Case 2 Some of the numbers $\alpha_1, \ldots, \alpha_m$ equal to zero. Suppose that $\alpha_1 > 0, \ldots, \alpha_k > 0$, $\alpha_j = 0$ $(k < j \leq m)$. Then we have

$$\sum_{i=1}^k \alpha_i \left\|\sum_{j=1}^m q^{-1}A_{ij} f_j\right\|^2 \leq \sum_{i=1}^k \|f_i\|^2 + \epsilon_n \sum_{j=1}^m \|f_j\|^2$$

for any $f_i \in \text{Im}\,(I - P_n)$. Set $f_1 = \cdots = f_k = 0$.

$$\sum_{i=1}^k \alpha_i \left\|\sum_{j=k+1}^m A_{ij} f_j\right\|^2 \leq \epsilon_n q^2 \sum_{j=k+1}^n \|f_j\|^2 \quad (f_j \in \text{Im}\,(I - P_n)).$$

From this inequality and the condition $\epsilon_n \to 0$ it follows that the operators A_{ij} are compact if $1 \leq i \leq k$ and $k < j \leq m$. By the induction hypothesis for the operators

$$[A_{ij}]_{i,j=1}^k \quad \text{and} \quad [A_{ij}]_{i,j=k+1}^m$$

we can for any $\epsilon > 0$ find numbers $\zeta_1, \ldots, \zeta_k$ and $\mu_{k+1}, \ldots, \mu_m$ (all non-zero) such that

$$\|[\zeta_i \zeta_j^{-1} A_{ij}]_{i,j=1}^k\|_{\text{ess}} < 1 + \epsilon, \quad \|[\mu_i \mu_j^{-1} A_{ij}]_{i,j=k+1}^m\|_{\text{ess}} < 1 + \epsilon.$$

Put $\lambda_i = \zeta_i$ $(i = 1, \ldots, k)$, $\lambda_i = n\mu_i$ $(i = k+1, \ldots, m)$. Now it is easy to check that for n sufficiently large

$$\|[\lambda_i \lambda_j^{-1} A_{ij}]_{i,j=1}^m\|_{\text{ess}} < 1 + \epsilon.$$

□

We state one immediate consequence of this theorem. This is just a reformulation in terms of the sructured norm. It follows from the equality $X(I - A\Delta)X^{-1} = I - XAX^{-1}\Delta$ for $\Delta \in \mathcal{B}_m$ and $X \in (D_m)'$ and the sufficiency argument of Theorem 4.2 that $\mu_{D_m}(A) \leq \inf\{\|XAX^{-1}\|_{\text{ess}} : X \in (D_m)', X \text{ invertible}\}$. Since for the algebra $\mathcal{C}$, $(D_m)'$ is the algebra of diagonal $m \times m$ operator matrices with scalars on the diagonal, Theorem 4.10 gives the opposite inequality.

Corollary 4.11 *For $\mathcal{R} = \mathcal{C}$, and $A \in M_m(\mathcal{C})$*

$$\mu_{D_m}(A) = \inf\{\|XAX^{-1}\|_{\text{ess}} : X \in (D_m)', \ X \text{ invertible}\}.$$

We recall that Shmul'yan's theorem (Theorem 3.1) was used to prove the equality of μ and $\hat{\mu}$ for the algebra $\mathcal{L}(\mathcal{H})$ when $m = 2$. We give the analogue of Theorem 3.1 for the algebra $\mathcal{C}$.

Theorem 4.12 *Suppose $A \in M_2(\mathcal{C})$, $\|A_{11}\| \leq 1$ and both A_{12} and A_{21} are non-compact. If for each $\Delta \in \mathcal{B}_1$, the open unit ball in $\mathcal{C}$, $\|\mathcal{F}_A(\Delta)\|_{\text{ess}} \leq 1$, then there exists $\lambda \neq 0$ such that $\|A_\lambda\|_{\text{ess}} \leq 1$.*

Proof: We begin by showing that the condition $\|\mathcal{F}_A(\Delta)\|_{\text{ess}} \leq 1$ is equivalent to the condition that $I - AT$ is invertible for all $T = \text{diag}\,[\Delta_1, \Delta_2]$, $\Delta_i \in \mathcal{B}_1$. As in the proof of Theorem 3.2, we write $I - AT$ as the product of three matrices $A_1A_2A_3$ where A_1 and A_3 are obviously invertible, and

$$A_2 = \begin{bmatrix} I - A_{11}\Delta_1 & 0 \\ 0 & I - \mathcal{F}_A(\Delta_1)\Delta_2 \end{bmatrix}$$

is invertible for all $T \in \mathcal{B}_2$ if and only if $I - \mathcal{F}_A(\Delta_1)\Delta_2$ is invertible for all $\Delta_2 \in \mathcal{B}_1$. By Remark 4.3 (2) this is equivalent to requiring $\|\mathcal{F}_A(\Delta_1)\|_{\text{ess}} \leq 1$ for all $\Delta_1 \in \mathcal{B}_1$.

We can therefore assume that $I - AT$ is invertible for all $T \in \mathcal{B}_2$ and apply Remark 4.8. Thus, for any $q > 1$ there exists $n_0 \in \mathbf{N}$ such that for any $f \in \text{Im}\,(I - \mathcal{P}_{n_0})$, $\|\Pi_1 Af\| \geq q\|\Pi_1 f\|$ implies $\|\Pi_2 Af\| \leq q\|\Pi_2 f\|$. Let $\hat{A}_{ij} = q^{-1}A_{ij}$ and define two Hermitian forms Ω_1, Ω_2 on $\text{Im}\,(I - \mathcal{P}_{n_0})$ as follows: for $f = \begin{bmatrix} f_1 \\ f_2 \end{bmatrix}$,

$$\Omega_1(f) = \|\hat{A}_{11}f_1 + \hat{A}_{12}f_2\|^2 - \|f_1\|^2,$$

$$\Omega_2(f) = \|f_2\|^2 - \|\hat{A}_{21}f_1 + \hat{A}_{22}f_2\|^2.$$

Then $\Omega_1(f) \geq 0$ implies $\Omega_2(f) \geq 0$ for $\in \text{Im}\,(I - \mathcal{P}_{n_0})$. Since A_{12} and A_{21} are non-compact there exist g_1, $g_2 \in \text{Im}\,(I - P_{n_0})$ such that $A_{12}g_2 \neq 0$, $A_{21}g_1 \neq 0$. Thus for $f_0 = \begin{bmatrix} 0 \\ g_2 \end{bmatrix}$, $\Omega_1(f) = \|\hat{A}_{12}g_2\|^2 > 0$ and for $g_0 = \begin{bmatrix} g_1 \\ 0 \end{bmatrix}$, $\Omega_2(g) = -\|\hat{A}_{21}g_1\|^2 < 0$.

By the fact that $\Omega_1(f_0) > 0$ and Lemma 3.4, there exists a non-negative real number ρ such that $\Omega_2(f) \geq \rho\Omega_1(f)$ for all $f \in \text{Im}\,(I - \mathcal{P}_{n_0})$ and, since $\Omega_2(g_0) < 0$, we can conclude that in fact $\rho > 0$. We thus have obtained that for all $f = \begin{bmatrix} f_1 \\ f_2 \end{bmatrix} \in \text{Im}\,(I - \mathcal{P}_{n_0})$,

$$\|f_2\|^2 - \|\hat{A}_{21}f_1 + \hat{A}_{22}f_2\|^2 \geq \rho\|\hat{A}_{11}f_1 + \hat{A}_{12}f_2\|^2 - \rho\|f_1\|^2.$$

We rewrite this as

$$\left\|\begin{bmatrix} \lambda f_1 \\ f_2 \end{bmatrix}\right\| \geq \left\|\begin{bmatrix} \lambda(\hat{A}_{11}f_1 + \hat{A}_{12}f_2) \\ \hat{A}_{21}f_1 + \hat{A}_{22}f_2 \end{bmatrix}\right\|^2,$$

with $\lambda = \sqrt{\rho} > 0$, for $f \in \text{Im}\,(I - \mathcal{P}_{n_0})$. Equivalently,

$$\left\|\begin{bmatrix} \hat{A}_{11} & \lambda\hat{A}_{12} \\ \lambda^{-1}\hat{A}_{21} & \hat{A}_{22} \end{bmatrix}(I - \mathcal{P}_{n_0})\right\| \leq 1$$

and therefore

$$\left\|\begin{bmatrix} A_{11} & \lambda A_{12} \\ \lambda^{-1}A_{21} & A_{22} \end{bmatrix}\right\|_{\text{ess}} \leq q,$$

which holds for any $q > 1$ and for some $\lambda > 0$ depending on q.

Since both A_{12} and A_{21} are non-compact, the set of such λ is bounded above and also bounded away from zero. Thus there exists a sequence $q_m \to 1$ such that the corresponding $\lambda_m \to \lambda_0$, $0 < \lambda_0 < \infty$, and we obtain

$$\left\|\begin{bmatrix} A_{11} & \lambda_0 A_{12} \\ \lambda_0^{-1}A_{21} & A_{22} \end{bmatrix}\right\|_{\text{ess}} \leq 1.$$

□

From the proof of Theorem 4.12 and from Remark 4.3 (3) it is easy to see that the following necessary and sufficient variant of Theorem 4.12 holds.

Theorem 4.13 *Suppose $A \in M_2(\mathcal{C})$ and both A_{12} and A_{21} are non-compact. The following statement are equivalent:*
(1) The transformation $\mathcal{F}_A(\Delta)$ is defined for all strictly lower triangular Δ with $\|\Delta\| < 1$, and $\|\mathcal{F}_A(\Delta)\|_{\text{ess}} \leq 1$;
(2) There exists $\lambda \neq 0$ such that $\|A_\lambda\|_{\text{ess}} \leq 1$.

Remarks 4.14 (1) We have considered a particular nest algebra. This made the proof of sufficiency straightforward because the orthogonal projections P_n have finite-dimensional range. The main contribution of this paper is to show (in Theorems 3.2, 3.6, 4.2, 4.10) that the invertibility of $I - \Delta A$, for Δ from an appropriate set, implies a size restriction on the norm or essential norm of A.

This holds for all nest algebras with the essential norm replaced by the quantity $\lim_{n\to\infty} \|A(I - \mathcal{P}_n)\|$ (or equivalently $\inf_{n\geq 0} \|A(I - \mathcal{P}_n)\|$). To obtain the converse, a condition that will guarantee the invertibility of $\mathcal{P}_n(I - \Delta A)\mathcal{P}_n$ for all n is required. This is a standard requirement and in linear systems theory is referred to as a "well-posedness" condition.

(2) A special case of interest, especially in linear systems theory, is the case where for $A \in M_m(\mathcal{C})$ the entries of A are scalar valued analytic Toeplitz operators (see [3], [18]). In this case the essential norm and operator norm are identical. Thus Corollary 4.11 can be restated in terms of the operator norm:

$$\mu_{D_m}(A) = \hat{\mu}_{D_m}(A).$$

A more interesting question is when $\mathcal{R} = \mathcal{H}^\infty$, the algebra of scalar valued analytic Toeplitz operators. We have shown in Section 2 that the case $m = 1$ is trivial. However, already for $m = 2$ this becomes a difficult problem. The reason for this is that $(D_m)'$ is not the diagonal scalar matrices but $(D_m)' = D_m$. Thus here $\hat{\mu}_{D_m}(A) = \{\inf \|XAX^{-1}\| : X \in D_m,\ X \text{ invertible}\}$. We don't know if for this case the equality $\mu_{D_m}(A) = \hat{\mu}_{D_m}(A)$ holds even for $m = 2$. We note that in this case $W(A_1, \ldots, A_m) = W_{\text{ess}}(A_1, \ldots, A_m)$ for A_k selfadjoint Toeplitz.

(3) From our proof, it is clear why the main result fails for $\mathcal{R} = \mathcal{L}(\mathcal{H})$ and holds for $\mathcal{R} = \mathcal{C}$ in the cases $m \geq 4$. The joint numerical range for four or more selfadjoint operators is in general not convex. The joint *essential* numerical range is.

References

[1] Y. Au-Yeung and N. Tsing, *An extension of the Hausdorff-Toeplitz Theorem on the numerical range*, Proc. A. M. S. **89** (1983), 315–318.

[2] H. Bercovici, C. Foias, P. Khargonekar and A. Tannenbaum, *On a lifting theorem for the structural singular value*, J. Math. Anal. Appl. **187** (1994), 617–627.

[3] H. Bercovici, C. Foias and A. Tannenbaum, *The structured singular value for linear input—output operators*, SIAM J. Cont. Optim. **34** (1996), 1392–1404.

[4] P. Binding, *Hermitian forms and the fibration of spheres*, Proc. A. M. S. **94** (1985), 581–584.

[5] J.C. Doyle, *Analysis of feedback systems with structured uncertainties*, IEE Proc., Pt. D **129** (1982), 242–250.

[6] M. Fan and A. Tits, *m-Form numerical range and the computation of structured singular value*, IEEE Trans. Aut. Contr. **33** (1988), 284–289.

[7] A. Feintuch, *Robustness of time-varying systems*, MCSS **6** (1993), 247–263.

[8] A. Feintuch and K. Saeks, *System Theory: A Hilbert Space Approach*, Academic Press, 1982.

[9] P. Fillmore, J. Stampfli and J. Williams, *On the essential numerical range, the essential spectrum and a problem of Halmos*, Acta Sci. Math. (Szeged) **33** (1972), 179–192.

[10] C. Foias and A. Frazho, *The Commutant Lifting Approach to Interpolation Problems*, OT44, Birkhausen, 1990.

[11] M. Krein and Yu. Shmul'yan, *On linear fractional transformations with operator coefficients*, A. M. S. Translations **103** (1974), 125–152.

[12] A. Packard, *What's new with μ: structured uncertainty in multivariable control*, Ph. D. Dissertation, Univ. of California, Berkley, 1988.

[13] A. Packard and J. C. Doyle, *The complex structured singular value*, Automatica **29** (1993), 71–110.

[14] H. Radjavi and P. Rosenthal, *Invariant Subspaces*, Springer-Verlag, 1972.

[15] R. Redheffer, *On a certain linear fractional transformation*, J. Math. Physics **39** (1960), 269–286.

[16] M.G. Safonov, *Stability Robustness of Multivariable Feedback Systems*, M. I. T. Press, 1980.

[17] J. Shamma, *The necessity of the small gain theorem for time-varying and non-linear systems*, IEEE Trans. Autom. Contr. **36** (1991), 1138–1147.

[18] J. Shamma, *Robust stability with time-varying structured uncertainty*, IEEE Trans. Autom. Contr. **39** (1994), 714–724.

[19] Yu. Shmul'yan, *General linear fractional transformations of operator balls*, Siberian Math. Jour. **19** (1978), 293–298.

Department of Math and Comp. Sciences
Ben-Gurion University of the Negev
Beer-Sheva 84105
Israel

MSC 1991: 93B35, 93C05, 93C25, 93C50, 47A10.

Operator Theory:
Advances and Applications, Vol. 115

Order Preserving Operator Function via Furuta Inequality "$A \geq B \geq 0$ Ensures $(A^{\frac{r}{2}} A^p A^{\frac{r}{2}})^{\frac{1+r}{p+r}} \geq (A^{\frac{r}{2}} B^p A^{\frac{r}{2}})^{\frac{1+r}{p+r}}$ for $p \geq 1$ and $r \geq 0$"

Takayuki Furuta, Takeaki Yamazaki and Masahiro Yanagida

Dedicated to Professor Sterling K. Berberian with respect and affection

By using the inequality made in its title, we shall prove the following results. Let S and T be positive invertible operators on a Hilbert space such that $S^{\beta_0} \geq (S^{\frac{\beta_0}{2}} T^{\alpha_0} S^{\frac{\beta_0}{2}})^{\frac{\beta_0}{\alpha_0+\beta_0}}$ holds for fixed $\alpha_0 > 0$ and $\beta_0 > 0$. Then for fixed $\delta \geq -\beta_0$,

$$f(\alpha, \beta) = S^{\frac{-\beta}{2}} (S^{\frac{\beta}{2}} T^{\alpha} S^{\frac{\beta}{2}})^{\frac{\delta+\beta}{\alpha+\beta}} S^{\frac{-\beta}{2}}$$

is a decreasing function of both α and β for $\alpha \geq max\{\delta, \alpha_0\}$ and $\beta \geq \beta_0$. As a simple corollary of this result, here we give a simplified proof of the following result; let $A \geq B \geq 0$ with $A > 0$ and let $t \in [0, 1]$ and $q \geq 0$. Then

$$G_{p,q,t}(A, B, r, s) = A^{\frac{-r}{2}} \{A^{\frac{r}{2}} (A^{\frac{-t}{2}} B^p A^{\frac{-t}{2}})^s A^{\frac{r}{2}}\}^{\frac{q-t+r}{(p-t)s+r}} A^{\frac{-r}{2}}$$

is decreasing for $r \geq t$ and $s \geq 1$ if $p \geq max\{q, t\}$. Also when the hypothesis $A \geq B \geq 0$ in this result is replaced by $logA > logB$ or $logA \geq logB$, the corresponding results are given. Moreover we give a direct proof of the following inequality; if $A \geq B \geq 0$ with $A > 0$, then for each $t \in [0, 1]$, $q \in [t, 1]$ and $p \geq q$,

$$A^{q-t+r} \geq \{A^{\frac{r}{2}} (A^{\frac{-t}{2}} B^p A^{\frac{-t}{2}})^s A^{\frac{r}{2}}\}^{\frac{q-t+r}{(p-t)s+r}}$$

holds for any $s \geq 1$ and $r \geq t$. Parallel results are given when $A \geq B \geq 0$ is replaced by $logA > logB$ or $logA \geq logB$. Finally we give a simplified proof of the following result; $logA \geq logB$ *holds if and only if for fixed* $q \geq 0$,

$$F_q(p, r) = A^{\frac{-r}{2}} (A^{\frac{r}{2}} B^p A^{\frac{r}{2}})^{\frac{q+r}{p+r}} A^{\frac{-r}{2}}$$

is decreasing for $p \geq q$ and $r \geq 0$.

Research supported in part by Grand-in-Aid for Scientific research.

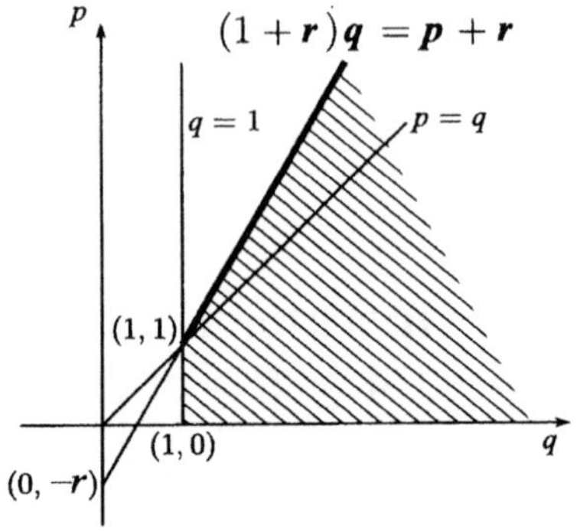

Figure 1

1 Introduction

A capital letter means a bounded linear operator on a complex Hilbert space H. An operator T is said to be positive (in symbol : $T \geq 0$) if $(Tx, x) \geq 0$ for all $x \in H$ and also an operator T is strictly positive (in symbol : $T > 0$) if T is positive and invertible. The following Theorem F is an extension of the celebrated Löwner-Heinz theorem [15][17].

Theorem F (Furuta inequality) [7]. *If $A \geq B \geq 0$, then for each $r \geq 0$*

(i) $$(B^{\frac{r}{2}} A^p B^{\frac{r}{2}})^{\frac{1}{q}} \geq (B^{\frac{r}{2}} B^p B^{\frac{r}{2}})^{\frac{1}{q}}$$

and

(ii) $$(A^{\frac{r}{2}} A^p A^{\frac{r}{2}})^{\frac{1}{q}} \geq (A^{\frac{r}{2}} B^p A^{\frac{r}{2}})^{\frac{1}{q}}$$

hold for $p \geq 0$ and $q \geq 1$ with $(1+r)q \geq p+r$.

We remark that Theorem F is essentially the same as the inequality made in its title and Theorem F yields the Löwner-Heinz theorem when we put $r = 0$ in (i) or (ii) stated above: $A \geq B \geq 0$ ensures $A^\alpha \geq B^\alpha$ for any $\alpha \in [0, 1]$. Alternative proofs of Theorem F are given in [3] [8] and [16] and also an elementary one page proof in [9]. It is shown in [18] that the domain drawn for p, q and r in the Figure 1 is the best possible one for Theorem F. In [12, Theorem 1.1] we established the following Theorem A as extensions of Theorem F.

Theorem A [12]. *If $A \geq B \geq 0$ with $A > 0$, then for each $t \in [0, 1]$ and $p \geq 1$,*

$$F_{p,t}(A, B, r, s) = A^{\frac{-r}{2}} \{A^{\frac{r}{2}} (A^{\frac{-t}{2}} B^p A^{\frac{-t}{2}})^s A^{\frac{r}{2}}\}^{\frac{1-t+r}{(p-t)s+r}} A^{\frac{-r}{2}}$$

is decreasing for $r \geq t$ *and* $s \geq 1$ *and* $F_{p,t}(A, A, r, s) \geq F_{p,t}(A, B, r, s)$, *that is, for each* $t \in [0, 1]$ *and* $p \geq 1$,

$$A^{1-t+r} \geq \{A^{\frac{r}{2}}(A^{\frac{-t}{2}} B^p A^{\frac{-t}{2}})^s A^{\frac{r}{2}}\}^{\frac{1-t+r}{(p-t)s+r}} \tag{1.1}$$

holds for any $s \geq 1$ *and* $r \geq t$.

Recently a nice mean theoretic proof of Theorem A is shown in [6]. Ando-Hiai [2] established excellent log majorization results and proved the useful inequality equivalent to the main log majorization theorem as follows; If $A \geq B \geq 0$ with $A > 0$, then

$$A^r \geq \{A^{\frac{r}{2}}(A^{\frac{-1}{2}} B^p A^{\frac{-1}{2}})^r A^{\frac{r}{2}}\}^{\frac{1}{p}}$$

holds for any $p \geq 1$ and $r \geq 1$. Theorem A interpolates the inequality stated above by Ando-Hiai and Theorem F itself and also extends results of [4] [10] and [11].

We write $A \gg B$ if $\log A \geq \log B$ for invertible positive operator A and B which is called the chaotic order [4] and related results on chaotic order are discussed in [1] and [4].

Very recently the following results are obtained as extensions of Theorem A.

Theorem B [14]. *Let* $A \geq B \geq 0$ *with* $A > 0$. *For each* $t \in [0, 1]$, $q \geq 0$ *and* $p \geq \max\{q, t\}$,

$$G_{p,q,t}(A, B, r, s) = A^{\frac{-r}{2}}\{A^{\frac{r}{2}}(A^{\frac{-t}{2}} B^p A^{\frac{-t}{2}})^s A^{\frac{r}{2}}\}^{\frac{q-t+r}{(p-t)s+r}} A^{\frac{-r}{2}}$$

is decreasing for $r \geq t$ *and* $s \geq 1$.

Theorem C [14]. *If* $A \geq B \geq 0$ *with* $A > 0$, *then for each* $t \in [0, 1]$, $q \in [t, 1]$ *and* $p \geq q$,

$$F_{p,q,t}(A, B, r, s) = A^{\frac{-r}{2}}\{A^{\frac{r}{2}}(A^{\frac{-t}{2}} B^p A^{\frac{-t}{2}})^s A^{\frac{r}{2}}\}^{\frac{q-t+r}{(p-t)s+r}} A^{\frac{-r}{2}}$$

is decreasing for $r \geq t$ *and* $s \geq 1$. *Moreover for each* $t \in [0, 1]$, $q \in [t, 1]$ *and* $p \geq q$, $F_{p,q,t}(A, A, r, s) \geq F_{p,q,t}(A, B, r, s)$, *that is,*

$$A^{q-t+r} \geq \{A^{\frac{r}{2}}(A^{\frac{-t}{2}} B^p A^{\frac{-t}{2}})^s A^{\frac{r}{2}}\}^{\frac{q-t+r}{(p-t)s+r}} \tag{1.2}$$

holds for any $s \geq 1$ *and* $r \geq t$.

In this paper, we prove Theorem 1 under below by an application of Theorem F and we show Theorem B as an immediate consequence of Theorem 1. Also when the hypothesis $A \geq B \geq 0$ in Theorem B is replaced by $\log A > \log B$ or $A \gg B$, the corresponding results are given. Moreover we give a simple proof of Theorem C and also we give a parallelism of Corollary 4 under below as an extension of [13, Remark 3.4]. Finally we give a simplified proof of the following characterization of chaotic order; $A \gg B$ *holds if and only if for fixed* $q \geq 0$,

$$F_q(p, r) = A^{\frac{-r}{2}}(A^{\frac{r}{2}} B^p A^{\frac{r}{2}})^{\frac{q+r}{p+r}} A^{\frac{-r}{2}}$$

is decreasing for $p \geq q$ *and* $r \geq 0$.

2 Results

Theorem 1 *Let S and T be positive invertible operators on a Hilbert space such that $S^{\beta_0} \geq (S^{\frac{\beta_0}{2}} T^{\alpha_0} S^{\frac{\beta_0}{2}})^{\frac{\beta_0}{\alpha_0+\beta_0}}$ holds for fixed $\alpha_0 > 0$ and $\beta_0 > 0$. Then for fixed $\delta \geq -\beta_0$,*

$$f(\alpha, \beta) = S^{\frac{-\beta}{2}} (S^{\frac{\beta}{2}} T^{\alpha} S^{\frac{\beta}{2}})^{\frac{\delta+\beta}{\alpha+\beta}} S^{\frac{-\beta}{2}} \tag{2.1}$$

is a decreasing function of both α and β for $\alpha \geq max\{\delta, \alpha_0\}$ and $\beta \geq \beta_0$.

The following Theorem 2 is an extension of [13, Theorem 3.2] and also Theorem 3 is an extension of [13, Theorem 3.1] respectively.

Theorem 2 *If $\log A > \log B$, then there exists an $\alpha \in (0, 1]$ and for each $t \in (0, \alpha]$, $q \geq 0$ and $p \geq max\{q, t\}$,*

$$G_{p,q,t}(A, B, r, s) = A^{\frac{-r}{2}} \{A^{\frac{r}{2}} (A^{\frac{-t}{2}} B^p A^{\frac{-t}{2}})^s A^{\frac{r}{2}}\}^{\frac{q-t+r}{(p-t)s+r}} A^{\frac{-r}{2}}$$

is decreasing for $r \geq t$ and $s \geq 1$.

Theorem 3 *If $A \gg B$, then for any $\delta > 0$, there exists an $\alpha = \alpha_\delta \in (0, 1]$, for each $t \in (0, \alpha]$, $q \geq 0$ and $p \geq max\{q, t\}$,*

$$H_{p,q,t}(A, B, r, s) = e^{\frac{-(q-t+r)ps\delta}{(p-t)s+r}} A^{\frac{-r}{2}} \{A^{\frac{r}{2}} (A^{\frac{-t}{2}} B^p A^{\frac{-t}{2}})^s A^{\frac{r}{2}}\}^{\frac{q-t+r}{(p-t)s+r}} A^{\frac{-r}{2}}$$

is decreasing for $r \geq t$ and $s \geq 1$.

The following parallelism is an extension of [13, Remark 3.4].

Corollary 4 (i) *If $A \geq B \geq 0$ with $A > 0$, then for each $t \in [0, 1]$, $q \in [t, 1]$ and $p \geq q$,*

$$A^{q-t+r} \geq \{A^{\frac{r}{2}} (A^{\frac{-t}{2}} B^p A^{\frac{-t}{2}})^s A^{\frac{r}{2}}\}^{\frac{q-t+r}{(p-t)s+r}} \tag{2.2}$$

holds for any $s \geq 1$ and $r \geq t$.
(ii) *If $\log A > \log B$, then there exists an $\alpha \in (0, 1]$ and for each $t \in [0, \alpha]$, $q \in [t, \alpha]$ and $p \geq q$,*

$$A^{q-t+r} \geq \{A^{\frac{r}{2}} (A^{\frac{-t}{2}} B^p A^{\frac{-t}{2}})^s A^{\frac{r}{2}}\}^{\frac{q-t+r}{(p-t)s+r}} \tag{2.3}$$

holds for any $s \geq 1$ and $r \geq t$.
(iii) *If $A \gg B$, then for any $\delta > 0$, there exists an $\alpha = \alpha_\delta \in (0, 1]$, for each $t \in [0, \alpha]$, $q \in [t, \alpha]$ and $p \geq q$,*

$$e^{\frac{(q-t+r)ps\delta}{(p-t)s+r}} A^{q-t+r} \geq \{A^{\frac{r}{2}} (A^{\frac{-t}{2}} B^p A^{\frac{-t}{2}})^s A^{\frac{r}{2}}\}^{\frac{q-t+r}{(p-t)s+r}} \tag{2.4}$$

holds for any $s \geq 1$ and $r \geq t$.

Also Theorem 1 implies the following result which is an extension of [10].

Corollary 5 [4][11]. *$A \gg B$ holds if and only if for fixed $q \geq 0$,*

$$F_q(p,r) = A^{\frac{-r}{2}}(A^{\frac{r}{2}}B^pA^{\frac{r}{2}})^{\frac{q+r}{p+r}}A^{\frac{-r}{2}}$$

is decreasing for $p \geq q$ and $r \geq 0$.

3 Proofs of results

We need the following lemmas to give proofs of the results in §2.

Lemma 1 [12]. *Let $A > 0$ and B be an invertible operator. For any real number λ*

$$(BAB^*)^{\lambda} = BA^{\frac{1}{2}}(A^{\frac{1}{2}}B^*BA^{\frac{1}{2}})^{\lambda-1}A^{\frac{1}{2}}B^*.$$

Lemma 2 *Let S and T be positive invertible operators on a Hilbert space such that $S^{\beta_0} \geq (S^{\frac{\beta_0}{2}}T^{\alpha_0}S^{\frac{\beta_0}{2}})^{\frac{\beta_0}{\alpha_0+\beta_0}}$ holds for fixed $\alpha_0 > 0$ and $\beta_0 > 0$. Then the following inequality holds*

$$S^{\beta} \geq (S^{\frac{\beta}{2}}T^{\alpha}S^{\frac{\beta}{2}})^{\frac{\beta}{\alpha+\beta}} \quad \textit{for all } \alpha \geq \alpha_0 \textit{ and } \beta \geq \beta_0. \tag{3.1}$$

Proof: Applying Theorem F to $S^{\beta_0} \geq (S^{\frac{\beta_0}{2}}T^{\alpha_0}S^{\frac{\beta_0}{2}})^{\frac{\beta_0}{\alpha_0+\beta_0}}$, we have

$$S^{\beta_0(1+t)} \geq \{S^{\frac{\beta_0 t}{2}}(S^{\frac{\beta_0}{2}}T^{\alpha_0}S^{\frac{\beta_0}{2}})^{\frac{\beta_0 p}{\alpha_0+\beta_0}}S^{\frac{\beta_0 t}{2}}\}^{\frac{1+t}{p+t}} \quad \text{for any } p \geq 1 \text{ and } t \geq 0. \tag{3.2}$$

Putting $p = \frac{\alpha_0+\beta_0}{\beta_0} \geq 1$ in (3.2), we have

$$S^{\beta_0(1+t)} \geq (S^{\frac{\beta_0}{2}(1+t)}T^{\alpha_0}S^{\frac{\beta_0}{2}(1+t)})^{\frac{(1+t)\beta_0}{\alpha_0+\beta_0+\beta_0 t}} \quad \text{for any } t \geq 0. \tag{3.3}$$

Put $\beta = (1+t)\beta_0 \geq \beta_0$ in (3.3). Then we have

$$S^{\beta} \geq (S^{\frac{\beta}{2}}T^{\alpha_0}S^{\frac{\beta}{2}})^{\frac{\beta}{\alpha_0+\beta}} \quad \text{for } \beta \geq \beta_0. \tag{3.4}$$

(3.4) is equivalent to the following (3.5) by Lemma 1

$$(T^{\frac{\alpha_0}{2}}S^{\beta}T^{\frac{\alpha_0}{2}})^{\frac{\alpha_0}{\alpha_0+\beta}} \geq T^{\alpha_0} \quad \text{for } \beta \geq \beta_0. \tag{3.5}$$

Again applying Theorem F to (3.5), we have

$$\{T^{\frac{\alpha_0 t}{2}}(T^{\frac{\alpha_0}{2}}S^{\beta}T^{\frac{\alpha_0}{2}})^{\frac{\alpha_0 p}{\alpha_0+\beta}}T^{\frac{\alpha_0 t}{2}}\}^{\frac{1+t}{p+t}} \geq T^{\alpha_0(1+t)} \quad \text{for any } p \geq 1 \text{ and } t \geq 0. \tag{3.6}$$

Putting $p = \frac{\alpha_0+\beta}{\alpha_0} \geq 1$ in (3.6), we have

$$(T^{\frac{\alpha_0}{2}(1+t)} S^{\beta} T^{\frac{\alpha_0}{2}(1+t)})^{\frac{(1+t)\alpha_0}{\alpha_0+\beta+\alpha_0 t}} \geq T^{\alpha_0(1+t)} \quad \text{for any } t \geq 0. \tag{3.7}$$

Put $\alpha = (1+t)\alpha_0 \geq \alpha_0$ in (3.7). Then we have

$$(T^{\frac{\alpha}{2}} S^{\beta} T^{\frac{\alpha}{2}})^{\frac{\alpha}{\alpha+\beta}} \geq T^{\alpha} \quad \text{for } \alpha \geq \alpha_0 \text{ and } \beta \geq \beta_0, \tag{3.8}$$

hence proof of Lemma 2 is complete since (3.8) is equivalent to (3.1) by Lemma 1. □

Proof of Theorem 1: (a) *Proof of the result that $f(\alpha, \beta)$ is a decreasing function of α for $\alpha \geq max\{\delta, \alpha_0\}$.*

The hypothesis in Theorem 1 ensures (3.1) by Lemma 2

$$S^{\beta} \geq (S^{\frac{\beta}{2}} T^{\alpha} S^{\frac{\beta}{2}})^{\frac{\beta}{\alpha+\beta}} \quad \text{for all } \alpha \geq \alpha_0 \text{ and } \beta \geq \beta_0 \tag{3.1}$$

and (3.1) is equivalent to the following (3.8) by Lemma 1

$$(T^{\frac{\alpha}{2}} S^{\beta} T^{\frac{\alpha}{2}})^{\frac{\alpha}{\alpha+\beta}} \geq T^{\alpha} \text{ for all } \alpha \geq \alpha_0 \text{ and } \beta \geq \beta_0. \tag{3.8}$$

(3.8) yields the following (3.9) by Löwner-Heinz theorem

$$(T^{\frac{\alpha}{2}} S^{\beta} T^{\frac{\alpha}{2}})^{\frac{u}{\alpha+\beta}} \geq T^{u} \text{ for all } \alpha \geq \alpha_0,\ \beta \geq \beta_0 \text{ and any } u \text{ such that } \alpha \geq u \geq 0. \tag{3.9}$$

Then we have

$$\begin{aligned} g(\alpha) &= (S^{\frac{\beta}{2}} T^{\alpha} S^{\frac{\beta}{2}})^{\frac{\delta+\beta}{\alpha+\beta}} \\ &= \{(S^{\frac{\beta}{2}} T^{\alpha} S^{\frac{\beta}{2}})^{\frac{\alpha+u+\beta}{\alpha+\beta}}\}^{\frac{\delta+\beta}{\alpha+u+\beta}} \\ &= \{S^{\frac{\beta}{2}} T^{\frac{\alpha}{2}} (T^{\frac{\alpha}{2}} S^{\beta} T^{\frac{\alpha}{2}})^{\frac{u}{\alpha+\beta}} T^{\frac{\alpha}{2}} S^{\frac{\beta}{2}}\}^{\frac{\delta+\beta}{\alpha+u+\beta}} \quad \text{by Lemma 1} \\ &\geq (S^{\frac{\beta}{2}} T^{\frac{\alpha}{2}} T^{u} T^{\frac{\alpha}{2}} S^{\frac{\beta}{2}})^{\frac{\delta+\beta}{\alpha+u+\beta}} \\ &= (S^{\frac{\beta}{2}} T^{\alpha+u} S^{\frac{\beta}{2}})^{\frac{\delta+\beta}{\alpha+u+\beta}} = g(\alpha+u) \end{aligned}$$

and the last inequality holds by (3.9) and Löwner-Heinz theorem since $\frac{\delta+\beta}{\alpha+u+\beta} \in [0, 1]$ holds by the hypothesis on α, β and δ. Hence $f(\alpha, \beta) = S^{\frac{-\beta}{2}} g(\alpha) S^{\frac{-\beta}{2}}$ is a decreasing function of α for $\alpha \geq max\{\delta, \alpha_0\}$.

(b) *Proof of the result that $f(\alpha, \beta)$ is a decreasing function of β for $\beta \geq \beta_0$.*

(3.1) yields the following (3.10) by Löwner-Heinz theorem

$$S^{v} \geq (S^{\frac{\beta}{2}} T^{\alpha} S^{\frac{\beta}{2}})^{\frac{v}{\alpha+\beta}} \text{ for all } \alpha \geq \alpha_0,\ \beta \geq \beta_0 \text{ and any } v \text{ such that } \beta \geq v \geq 0. \tag{3.10}$$

Then we have

$$\begin{aligned}
f(\alpha,\beta) &= S^{\frac{-\beta}{2}}(S^{\frac{\beta}{2}}T^{\alpha}S^{\frac{\beta}{2}})^{\frac{\delta+\beta}{\alpha+\beta}}S^{\frac{-\beta}{2}} \\
&= T^{\frac{\alpha}{2}}(T^{\frac{\alpha}{2}}S^{\beta}T^{\frac{\alpha}{2}})^{\frac{\delta-\alpha}{\beta+\alpha}}T^{\frac{\alpha}{2}} \quad \text{by Lemma 1} \\
&= T^{\frac{\alpha}{2}}\{(T^{\frac{\alpha}{2}}S^{\beta}T^{\frac{\alpha}{2}})^{\frac{\beta+v+\alpha}{\beta+\alpha}}\}^{\frac{\delta-\alpha}{\beta+v+\alpha}}T^{\frac{\alpha}{2}} \\
&= T^{\frac{\alpha}{2}}\{T^{\frac{\alpha}{2}}S^{\frac{\beta}{2}}(S^{\frac{\beta}{2}}T^{\alpha}S^{\frac{\beta}{2}})^{\frac{v}{\beta+v}}S^{\frac{\beta}{2}}T^{\frac{\alpha}{2}}\}^{\frac{\delta-\alpha}{\beta+v+\alpha}}T^{\frac{\alpha}{2}} \quad \text{by Lemma 1} \\
&\geq T^{\frac{\alpha}{2}}(T^{\frac{\alpha}{2}}S^{\frac{\beta}{2}}S^{v}S^{\frac{\beta}{2}}T^{\frac{\alpha}{2}})^{\frac{\delta-\alpha}{\beta+v+\alpha}}T^{\frac{\alpha}{2}} \\
&= T^{\frac{\alpha}{2}}(T^{\frac{\alpha}{2}}S^{\beta+v}T^{\frac{\alpha}{2}})^{\frac{\delta-\alpha}{\beta+v+\alpha}}T^{\frac{\alpha}{2}} = f(\alpha,\beta+v)
\end{aligned}$$

and the last inequality holds by (3.10) and Löwner-Heinz theorem since $\frac{\delta-\alpha}{\beta+v+\alpha} \in [-1,0]$, and taking inverses. Hence $f(\alpha,\beta)$ is a decreasing function of β for $\beta \geq \beta_0$.

Consequently we have finished a proof of Theorem 1 by (a) and (b).

Proof of Theorem B: We consider the case $p > t > 0$. Put $X = (A^{\frac{-t}{2}}B^{p}A^{\frac{-t}{2}})^{\frac{1}{p-t}}$. Then we have $A^{\frac{t}{2}}X^{p-t}A^{\frac{t}{2}} = B^{p}$ and $A \geq (A^{\frac{t}{2}}X^{p-t}A^{\frac{t}{2}})^{\frac{1}{p}}$ by the hypothesis $A \geq B \geq 0$. Put $\beta_0 = t \in (0,1]$ and $\alpha_0 = p - t > 0$. Then $A \geq (A^{\frac{\beta_0}{2}}X^{\alpha_0}A^{\frac{\beta_0}{2}})^{\frac{1}{\alpha_0+\beta_0}}$, so that

$$A^{\beta_0} \geq (A^{\frac{\beta_0}{2}}X^{\alpha_0}A^{\frac{\beta_0}{2}})^{\frac{\beta_0}{\alpha_0+\beta_0}}$$

holds by Löwner-Heinz theorem. Put $\alpha = (p-t)s$, $\beta = r$ and $\delta = q - t$. The hypotheses $t \in (0,1]$, $q \geq 0$ and $p \geq max\{q,t\}$ in Theorem B satisfy the conditions required on α, β and δ in Theorem 1, that is, $\delta \geq -\beta_0, \alpha \geq max\{\alpha_0,\delta\}$ and $\beta \geq \beta_0$. Applying Theorem 1,

$$\begin{aligned}
f(\alpha,\beta) &= A^{\frac{-\beta}{2}}(A^{\frac{\beta}{2}}X^{\alpha}A^{\frac{\beta}{2}})^{\frac{\delta+\beta}{\alpha+\beta}}A^{\frac{-\beta}{2}} \\
&= G_{p,q,t}(A,B,r,s) \\
&= A^{\frac{-r}{2}}\{A^{\frac{r}{2}}(A^{\frac{-t}{2}}B^{p}A^{\frac{-t}{2}})^{s}A^{\frac{r}{2}}\}^{\frac{q-t+r}{(p-t)s+r}}A^{\frac{-r}{2}}
\end{aligned}$$

is decreasing for $r \geq t$ and $s \geq 1$, so the proof in the case $p > t > 0$ is complete.

In the case $t = 0$, Theorem B easily follows by [10, Theorem 3].

In the case $p = t \geq q \geq 0$. Let $C = A^{\frac{-t}{2}}B^{t}A^{\frac{-t}{2}}$. Then $I \geq C \geq 0$ for $t \in [0,1]$ by Löwner-Heinz theorem, so that $A^{r} \geq A^{\frac{r}{2}}C^{s}A^{\frac{r}{2}}$ holds since $I \geq C \geq 0$ and $s \geq 1$, and by again Löwner-Heinz theorem

$$A^{u} \geq (A^{\frac{r}{2}}C^{s}A^{\frac{r}{2}})^{\frac{u}{r}} \quad \text{for } r \geq u \geq 0. \tag{3.11}$$

Then we obtain

$$
\begin{aligned}
G_{t,q,t}(A,B,r,s) &= A^{\frac{-r}{2}}(A^{\frac{r}{2}}C^{s}A^{\frac{r}{2}})^{\frac{q-t+r}{r}}A^{\frac{-r}{2}}\\
&= C^{\frac{s}{2}}(C^{\frac{s}{2}}A^{r}C^{\frac{s}{2}})^{\frac{q-t}{r}}C^{\frac{s}{2}} \quad \text{by Lemma 1}\\
&= C^{\frac{s}{2}}\{(C^{\frac{s}{2}}A^{r}C^{\frac{s}{2}})^{\frac{r+u}{r}}\}^{\frac{q-t}{r+u}}C^{\frac{s}{2}} \qquad (3.12)\\
&= C^{\frac{s}{2}}\{C^{\frac{s}{2}}A^{\frac{r}{2}}(A^{\frac{r}{2}}C^{s}A^{\frac{r}{2}})^{\frac{u}{r}}A^{\frac{r}{2}}C^{\frac{s}{2}}\}^{\frac{q-t}{r+u}}C^{\frac{s}{2}} \quad \text{by Lemma 1}\\
&\geq C^{\frac{s}{2}}(C^{\frac{s}{2}}A^{\frac{r}{2}}A^{u}A^{\frac{r}{2}}C^{\frac{s}{2}})^{\frac{q-t}{r+u}}C^{\frac{s}{2}}\\
&= C^{\frac{s}{2}}(C^{\frac{s}{2}}A^{r+u}C^{\frac{s}{2}})^{\frac{q-t}{r+u}}C^{\frac{s}{2}} = G_{t,q,t}(A,B,r+u,s)
\end{aligned}
$$

and the last inequality holds by (3.11) and Löwner-Heinz theorem since $\frac{q-t}{r+u} \in [-1,0]$ and taking inverses. Consequently $G_{t,q,t}(A,B,r,s)$ is a decreasing function of both $r \geq t$ and $s \geq 1$ because $G_{t,q,t}(A,B,r,s)$ is decreasing of $s \geq 1$ by (3.12) since $1 \geq C \geq 0$.

Whence the proof of Theorem B is complete.

Proof of Theorem C: We have only to show the latter part since the former part follows by Theorem B. For each $t \in [0,1]$, $q \in [t,1]$ and $p \geq q$, $G_{p,q,t}(A,B,t,1) \geq G_{p,q,t}(A,B,r,s)$ holds by Theorem B, that is,

$$
\begin{aligned}
A^{q-t} &\geq A^{\frac{-t}{2}}B^{q}A^{\frac{-t}{2}} \quad \text{by Löwner-Heinz theorem}\\
&\geq G_{p,q,t}(A,B,r,s)
\end{aligned}
$$

holds for any $s \geq 1$ and $r \geq t$, and we obtain (1.2), so the proof is complete.

We cite the following two nice results to give proofs of Theorem 2 and Theorem 3.

Theorem D [5]. *$\log A > \log B$ holds if and only if there exists an $\alpha \in (0,1]$ such that $A^{\alpha} > B^{\alpha}$.*

Theorem E [5]. *$A \gg B$ holds if and only if for any $\delta > 0$ there exists an $\alpha = \alpha_{\delta} \in (0,1]$ such that $(e^{\delta}A)^{\alpha} > B^{\alpha}$.*

Proof of Theorem 2: The hypothesis $\log A > \log B$ holds iff there exists an $\alpha \in (0,1]$ such that $A^{\alpha} > B^{\alpha}$ by Theorem D. Put $A_1 = A^{\alpha}$ and $B_1 = B^{\alpha}$. As $A_1 > B_1$ holds, Theorem B yields that for each $t_1 \in [0,1]$, $q_1 \geq 0$ and $p_1 \geq max\{q_1, t_1\}$,

(3.13) $\quad G_{p_1,q_1,t_1}(A_1,B_1,r_1,s)$ is decreasing for $r_1 \geq t_1$ and $s \geq 1$.

Put $r_1 = \frac{r}{\alpha}$, $t_1 = \frac{t}{\alpha}$, $q_1 = \frac{q}{\alpha}$ and $p_1 = \frac{p}{\alpha}$. Then $t_1 \in [0,1]$, $r_1 \geq t_1$ and $p_1 \geq max\{q_1, t_1\}$ by the hypothesis on p, q, t, α and r, so that (3.13) ensures that

$$G_{p,q,t}(A,B,r,s) = A^{\frac{-r}{2}}\{A^{\frac{r}{2}}(A^{\frac{-t}{2}}B^{p}A^{\frac{-t}{2}})^{s}A^{\frac{r}{2}}\}^{\frac{q-t+r}{(p-t)s+r}}A^{\frac{-r}{2}}$$

is decreasing for $r \geq t$ and $s \geq 1$.

Proof of Theorem 3: The hypothesis $A \gg B$ holds iff for any $\delta > 0$, there exists an $\alpha = \alpha_\delta \in (0, 1]$ such that $A^\alpha > (e^{-\delta}B)^\alpha$ by Theorem E. Put $A_1 = A^\alpha$ and $B_1 = (e^{-\delta}B)^\alpha$. As $A_1 > B_1$ holds, Theorem B yields that for each $t_1 \in [0, 1]$, $q_1 \geq 0$ and $p_1 \geq max\{q_1, t_1\}$,

(3.14) $\quad H_{p_1,q_1,t_1}(A_1, B_1, r_1, s)$ is decreasing for $r_1 \geq t_1$ and $s \geq 1$.

Put $r_1 = \frac{r}{\alpha}$, $t_1 = \frac{t}{\alpha}$, $q_1 = \frac{q}{\alpha}$ and $p_1 = \frac{p}{\alpha}$. Then $t_1 \in [0, 1]$, $r_1 \geq t_1$ and $p_1 \geq max\{q_1, t_1\}$ by the hypothesis on p, q, t, α and r, so that (3.14) ensures that

$$H_{p,q,t}(A, B, r, s) = e^{\frac{-(q-t+r)ps\delta}{(p-t)s+r}} A^{\frac{-r}{2}}\{A^{\frac{r}{2}}(A^{\frac{-t}{2}}B^pA^{\frac{-t}{2}})^sA^{\frac{r}{2}}\}^{\frac{q-t+r}{(p-t)s+r}}A^{\frac{-r}{2}}$$

is decreasing for $r \geq t$ and $s \geq 1$.

Proof of Corollary 4: (i) (2.2) is (1.2) itself in Theorem C.

(ii) Let $q \in [t, \alpha]$ in Theorem 2. As seen in the proof of Theorem 2, there exists an $\alpha \in (0, 1]$ such that $A^\alpha > B^\alpha$, so that $A^q > B^q$ by Löwner-Heinz theorem since $\frac{q}{\alpha} \in [0, 1]$. Then for $p \geq q$, $G_{p,q,t}(A, B, r, s)$ is decreasing for $r \geq t$ and $s \geq 1$ by Theorem 2, so that $G_{p,q,t}(A, B, t, 1) \geq G_{p,q,t}(A, B, r, s)$ holds, that is,

$$\begin{aligned} A^{q-t} &\geq A^{\frac{-t}{2}}B^qA^{\frac{-t}{2}} \quad \text{by } A^q > B^q \\ &\geq G_{p,q,t}(A, B, r, s) \end{aligned}$$

holds for $s \geq 1$ and $r \geq t$, so the proof of (2.3) is complete.

(iii) Let $q \in [t, \alpha]$ in Theorem 3. As seen in the proof of Theorem 3, for any $\delta > 0$ there exists $\alpha = \alpha_\delta \in (0, 1]$ such that $A^\alpha > (e^{-\delta}B)^\alpha$, so that $A^q > (e^{-\delta}B)^q$ by Löwner-Heinz theorem since $\frac{q}{\alpha} \in [0, 1]$. Then for $p \geq q$, $H_{p,q,t}(A, B, r, s)$ is decreasing for $r \geq t$ and $s \geq 1$ by Theorem 3, so that $H_{p,q,t}(A, B, t, 1) \geq H_{p,q,t}(A, B, r, s)$ holds, that is,

$$\begin{aligned} A^{q-t} &\geq A^{\frac{-t}{2}}(e^{-\delta}B)^qA^{\frac{-t}{2}} \quad \text{by } A^q > (e^{-\delta}B)^q \\ &\geq H_{p,q,t}(A, B, r, s) \end{aligned}$$

holds for $s \geq 1$ and $r \geq t$, so the proof of (2.4) is complete.

Proof of Corollary 5: We recall the following result in [4][11], which is an extension of [1] :

$$A \gg B \; holds \; if \; and \; only \; if \; A^r \geq (A^{\frac{r}{2}}B^pA^{\frac{r}{2}})^{\frac{r}{p+r}} \; for \; all \; p \geq 0 \; and \; r \geq 0. \tag{3.15}$$

"only if" part of Corollary 5 follows by Theorem 1 and (3.15). Conversely we shall show "if" part. Assume that $F_0(p, r)$ is decreasing for $r \geq 0$. Then $F_0(p, 0) \geq F_0(p, r)$ holds, that is, $I \geq A^{\frac{-r}{2}}(A^{\frac{r}{2}}B^pA^{\frac{r}{2}})^{\frac{r}{p+r}}A^{\frac{-r}{2}}$, so that $A^r \geq (A^{\frac{r}{2}}B^pA^{\frac{r}{2}})^{\frac{r}{p+r}}$ for all $p \geq 0$ and $r \geq 0$, which is equivalent to $A \gg B$ by (3.15). Hence the proof of Corollary 5 is complete.

Further extensions of this paper and related applications will appear in the forthcoming paper: Operator functions implying generalized Furuta inequality.

References

[1] T. Ando, *On some operator inequalities*, Math. Ann. **279** (1987), 157–159.

[2] T. Ando and F.Hiai, *Log-majorization and complementary Golden-Thompson type inequalities*, Linear Alg. and Its Appl. **197,198** (1994), 113–131.

[3] M. Fujii, *Furuta's inequality and its mean theoretic approach*, J. Operator Theory **23** (1990), 67–72.

[4] M. Fujii, T.Furuta and E.Kamei, *Furuta's inequality and its application to Ando's theorem*, Linear Alg. and Its Appl. **179** (1993), 161–169.

[5] M. Fujii, J-F.Jiang and E.Kamei, *Characterization of chaotic order and its application to Furuta inequality*, Proc. Amer. Math. Soc., to appear.

[6] M. Fujii and E.Kamei, *Mean theoretic approach to the grand Furuta inequality*, Proc. Amer. Math. Soc. **124** (1996), 2751–2756.

[7] T. Furuta, $A \geq B \geq 0$ *assures* $(B^r A^p B^r)^{1/q} \geq B^{(p+2r)/q}$ *for* $r \geq 0$, $p \geq 0$, $q \geq 1$ *with* $(1+2r)q \geq p+2r$, Proc. Amer. Math. Soc. **101** (1987), 85–88.

[8] T. Furuta, *A proof via operator means of an order preserving inequality*, Linear Alg. and Its Appl. **113** (1989), 129–130.

[9] T. Furuta, *An elementary proof of an order preserving inequality*, Proc. Japan Acad. **65** (1989), 126.

[10] T. Furuta, *Two operator functions with monotone property*, Proc. Amer. Math. Soc. **111** (1991), 511–516.

[11] T. Furuta, *Applications of order preserving operator inequalities*, Operator Theory: Advances and Applications **59** (1992), 180–190.

[12] T. Furuta, *Extension of the Furuta inequality and Ando-Hiai log-majorization*, Linear Alg. and Its Appl. **219** (1995), 139–155.

[13] T. Furuta, *Characterizations of chaotic order via generalized Furuta inequality*, J. of Inequal. and Appl. **1** (1997), 11–24.

[14] T. Furuta and D.Wang, *A decreasing operator function associated with the Furuta inequality*, preprint.

[15] E. Heinz, *Beiträge zur Störungstheorie der Spektralzerlegung*, Math. Ann. **123** (1951), 415–438.

[16] E. Kamei, *A satellite to Furuta's inequality*, Math. Japon. **33** (1988), 883–886.

[17] K. Löwner, *Über monotone Matrixfunktionen*, Math. Z. **38** (1934), 177–216.

[18] K. Tanahashi, *Best possibility of the Furuta inequality*, Proc. Amer. Math. Soc. **124** (1996), 141–146.

Department of Applied Mathematics
Faculty of Science
Science University of Tokyo
Kagurazaka, Shinjuku 162, Tokyo
Japan

1991 *Mathematics Subject Classification.* Primary 47A63.

Operator Theory:
Advances and Applications, Vol. 115

On a Nonlinear Causal Commutant Lifting Theorem*

Caixing Gu

In this note, we extend the local nonlinear commutant lifting theorem of J. Ball, C. Foias, J. W. Helton, and A. Tannenbaum [2, 3] to causal input/output operators. This is accomplished by utilizing the causal dilation theory developed by C. Foias and A. Tannenbaum [14] and C. Foias and A. Tannenbaum and the author [10, 11], in order to guarantee the causality of the optimal interpolant derived in the nonlinear framework. This theory is moreover applied to a nonlinear H^{∞} synthesis problem in robust control.

1991 *Mathematics Subject Classifications.* Primary 47A20, Secondary 47A99, 93B35, 93C05.

Keywords and phrases. Causality, nonlinear input/output operators, H^{∞} optimization, commutant lifting theorem, interpolation theory, Hankel operators.

1 Introduction

In this paper we give an extension of a local nonlinear commutant lifting theorem of J. Ball, C. Foias, J. W. Helton, and A. Tannenbaum [2, 3] to causal operators. Indeed, in those papers [2, 3] a generalization of the classical commutant lifting theorem [21, 9] to nonlinear input/output operators was formulated. In addition to the mathematical interest, we were motivated in this work to try to extend the linear H^{∞} theory from robust control (whose mathematics can be based on the linear commutant lifting theorem [9]) to nonlinear systems.

Now a fundamental problem with the results of [2, 3] is that they were not able to guarantee the *causality* of the resulting nonlinear dilation, even if they initially started with causal data. (They were able to guarantee only "weak causality.") From a controls point of view, this is a fatal flaw since one can not physically implement noncausal systems. Thus these results remained just of theoretical interest.

On the other hand, in recent years an encouraging development occurred in that for the first time, we were able to explicitly include a causality constraint in the commutant lifting framework, see C. Foias and A. Tannenbaum [14] and C. Foias and A. Tannenbaum and the author [10, 11]. In this paper therefore, we will use causal dilation theory to formulate and prove a causal nonlinear commutant lifting result. (See Theorem 2 in Section 3 below.) This theorem should be much more applicable now to real control systems as we will indicate in our discussion in Section 5.

---

*This work was supported in part by a grant from the National Science Foundation.

In the theory presented below, we will consider analytic input/output operators of fading memory defined on a ball around the origin in a complex Hilbert space. (See Sections 2 and 3 for the precise definitions). Our class of operators also includes Volterra series of fading memory. A key part of our strategy will be to express each n-linear term of the Taylor expansion of such an operator as a linear operator on a certain tensor space to which we can apply the causal linear commutant lifting theorem from [14, 10, 11]. It is precisely this result which makes it possible to extend the nonlinear commutant lifting theorem from [2, 3] to the causal framework.

We now outline the contents of this paper. In Section 2, we give some of the relevant theory about causal analytic operators. In Section 3, we state and prove the Nonlinear Causal Commutant Lifting Theorem. Next in Section 4, we give a notion of nonlinear causal Hankel operators, and finally in Section 5 we apply our theory to give a nonlinear version of interpolation theory together with the application to nonlinear robust control.

2 Causal Analytic Mappings

In this section, we will define the class of nonlinear input/output operators which we will study in this paper. In order to do this, we will first need to discuss a few standard results about analytic mappings on Hilbert spaces. See [2], [3], [12], [13] and the references therein for complete details.

Let $\mathcal{G}$ and $\mathcal{H}$ denote complex separable Hilbert spaces. Set

$$B_{r_o}(\mathcal{G}) := \{g \in \mathcal{G} : \|g\| < r_o\}$$

(the open ball of radius r_o in $\mathcal{G}$ about the origin). Then we say that a mapping $\phi : B_{r_o}(\mathcal{G}) \mapsto \mathcal{H}$ is *analytic* if the complex function $(z_1, \ldots, z_n) \mapsto \langle \phi(z_1 g_1 + \cdots + z_n g_n), h\rangle$ is analytic in a neighborhood of $(1, 1, \ldots, 1) \in \mathbf{C}^n$ as a function of the complex variables $z_1, \ldots, z_n$ for all $g_1, \ldots, g_n \in \mathcal{G}$ such that $\|g_1 + \cdots + g_n\| < r_o$, for all $h \in \mathcal{H}$, and for all $n > 0$.

We will now assume that $\phi(0) = 0$. It is easy to see that if $\phi : B_{r_o}(\mathcal{G}) \mapsto \mathcal{H}$ is analytic, then ϕ admits a convergent Taylor series expansion, i.e.,

$$\phi(g) = \phi_1(g) + \phi_2(g, g) + \cdots + \phi_n(g, \ldots, g) + \cdots$$

where $\phi_n : \mathcal{G} \times \cdots \times \mathcal{G} \mapsto \mathcal{H}$ is an n-linear map. We will call ϕ_n the n-th power term of ϕ. Clearly, without loss of generality we may assume that the n-linear map $(g_1, \ldots, g_n) \mapsto \phi(g_1, \ldots, g_n)$ is symmetric in the arguments $g_1, \ldots, g_n$. This assumption will be made throughout this paper for the various analytic maps which we consider. For ϕ a Volterra series, ϕ_n is basically the n-th Volterra kernel.

Now set

$$\hat{\phi}_n(g_1 \otimes \cdots \otimes g_n) := \phi_n(g_1, \ldots, g_n).$$

Then $\hat{\phi}_n$ extends in a unique manner to a dense set of $\mathcal{G}^{\otimes n} := \mathcal{G} \otimes \cdots \otimes \mathcal{G}$ (tensor product taken n times). Notice by $\mathcal{G}^{\otimes n}$ we mean the Hilbert space completion of the algebraic tensor product of the $\mathcal{G}$'s. Clearly if $\hat{\phi}_n$ has finite norm on this dense set, then $\hat{\phi}_n$ extends by continuity to a bounded linear operator $\hat{\phi}_n : \mathcal{G}^{\otimes n} \mapsto \mathcal{H}$. By abuse of notation, we will set $\phi_n := \hat{\phi}_n$. We can recover ϕ_n from ϕ via the Cauchy formula. Indeed, by expanding $\phi(z_1 g_1 + \cdots + z_n g_n)$ in powers of $z_1, \ldots, z_n$, it is easy to see that the coefficient of $z_1, \ldots, z_n$ is precisely $n!\phi_n(g_1 \otimes \cdots \otimes g_n)$. Thus by the Cauchy formula we have that

$$n!\phi_n(g_1 \otimes \cdots \otimes g_n) \tag{1}$$
$$= \left(\frac{1}{2\pi}\right)^n \int_0^{2\pi} \cdots \int_0^{2\pi} \phi(e^{i\theta_1} g_1 + \cdots + e^{i\theta_n} g_n) e^{-i(\theta_1 + \cdots + \theta_n)} d\theta_1 \cdots d\theta_n$$

for all $g_1, \cdots g_n \in \mathcal{G}$ with $\|g_1 + \cdots + g_n\| < r_0$.

We now recall the following standard definitions:

Definitions 1

(i) Notation as above. By a *majorizing sequence* for the analytic map ϕ, we mean a positive sequence of numbers α_n $n = 1, 2, \ldots$ such that $\|\phi_n\| \leq \alpha_n$ for $n \geq 1$. Suppose that $\rho := \limsup {\alpha_n}^{1/n} < \infty$. Then it is completely standard that the Taylor series expansion of ϕ converges at least on the ball $B_r(\mathcal{G})$ of radius $r = 1/\rho$.

(ii) If ϕ admits a majorizing sequence as in (i), then we will say that ϕ is *majorizable*.

Let S be an isometry on $\mathcal{G}$. Let T be a contraction on $\mathcal{H}$ with a minimal isometric dilation $U : \mathcal{K} \to \mathcal{K}$. In our commutant lifting framework to study analytic mappings from $\mathcal{G}$ into $\mathcal{H}$, we need to consider analytic maps from $\mathcal{G}$ into $\mathcal{K}$. We will need the following in the sequel.

Definition 2

(i) Let $\psi : \mathcal{G} \to \mathcal{K}$ be an analytic map. ψ is said to be *time-invariant* (with respect to S and U) if

$$\psi \circ S = U\psi. \tag{2}$$

(ii) ψ is said to be *causal* (with respect to S and U) if

$$(I - U^j U^{*j})\psi = (I - U^j U^{*j})\psi \circ (I - S^j S^{*j}) \text{ for all } j \geq 0. \tag{3}$$

Roughly, causality means that for a given input/output map ψ the past output is independent of the future inputs.

As noted above if $\psi(0) = 0$ and $\psi : \mathcal{G} \to \mathcal{K}$ is analytic, then

$$\psi(g) = \psi_1(g) + \psi_2(g, g) + \cdots + \psi_n(g, \ldots, g) + \cdots \tag{4}$$

where $\psi_n : \mathcal{G}^{\otimes n} \to \mathcal{K}$ is a bounded linear operator. Let $S^{\otimes n} = S \otimes \cdots \otimes S$ where S appears n times. By the Cauchy formula (1), ψ is time-invariant in terms of its Taylor series expansion if and only if

$$\psi_n S^{\otimes n} = U\psi_n, \ \forall n \geq 1.$$

(We will also say each ψ_n is time-invariant.) Let $P_{nj} = (I - S^j S^{*j})^{\otimes n}$. ψ is causal if and only if

(5) $$(I - U^j U^{*j})\psi_n = (I - U^j U^{*j})\psi_n P_{nj} \;\; \forall j \geq 0, n \geq 1.$$

(We will also say each ψ_n is causal.)

Next we give a concrete example of causal analytic mappings which is also the natural class of input/output mappings that appear in engineering applications. Let $H^2_K(D^n)$ denote the standard Hardy space of $\mathbf{C}^K$–valued analytic functions on the n–disc D^n (D denotes the unit disc) with square integrable boundary values. We set $H^2_K := H^2_K(D)$ and $H^2 := H^2_1$. We denote the shift on $H^2_K(D^n)$ by $S_{(n)}$. Note that $S_{(n)}$ is defined by multiplication by the function $(z_1, \ldots, z_n)$.

We now consider an analytic map ϕ with $\mathcal{G} = \mathcal{H} = H^2_k$. S on H^2_k is the multiplication by the function z and $U = S$. Note that

(6) $$H^2_k \otimes \cdots \otimes H^2_k = (H^2_k)^{\otimes n} \cong H^2_K(D^n) \;\text{ with } K = k^n$$

where we map $1 \otimes \cdots \otimes z \otimes \cdots \otimes 1$ (z in the i-th place) to z_i, $i = 1, \ldots, n$. Clearly, $S^{\otimes n}$ corresponds to $S_{(n)}$ under this identification.

Note that $\phi_n : H^2_K(D^n) \to H^2_k$ is causal if for $F(z_1, \ldots, z_n) \in H^2_K(D^n)$,

$$F(z_1, \ldots, z_n) = \sum_{i_1, \ldots, i_n \geq 0} F_{i_1, \ldots, i_n} z_1^{i_1}, \ldots, z_n^{i_n}, \quad \phi_n(F)(z) := \sum_{m \geq 0} f_m z^m,$$

each f_m only depends on

$$\{F_{i_1, \ldots, i_n} : 0 \leq i_1, \ldots, i_n \leq m\}.$$

We would now like to discuss the relationship between time-invariance and causality. For simplicity, we assume $k = 1$, i.e., we work with SISO systems. Let $\phi : H^2 \to H^2$ be linear and time-invariant (i.e., intertwines with the shift). Then it is easy to see that ϕ is causal. Indeed, $\phi U = U \phi$ implies

$$\begin{aligned} U^m U^{*m} \phi U^m U^{*m} &= U^m U^{*m} U^m \phi U^{*m} \\ &= U^m \phi U^{*m} \\ &= \phi U^m U^{*m} \end{aligned}$$

which immediately implies

$$(I - U^m U^{*m})\phi = (I - U^m U^{*m})\phi(I - U^m U^{*m}), \quad \forall m \geq 0,$$

that is, ϕ is causal.

In the nonlinear setting however, time-invariance may not imply causality. See [14] for a concrete example. Thus we must explicitly put in a causality constraint when dealing with nonlinear input/output operators.

We conclude this section by giving a version of causal commutant lifting theorem appropriate for the purpose here; see [14], [10] and [18] for more general versions.

Theorem 1 (Causal Commutant Lifting Theorem [14]) *Let $\mathcal{G}_1$ and $\mathcal{H}_1$ be two complex separable Hilbert spaces. Let S_1 on $\mathcal{G}_1$ be an isometry, T_1 on $\mathcal{H}_1$ be an invertible contraction, and U_1 on $\mathcal{K}_1$ be the minimal isometric dilation of T_1. Let $P_j,\ j \geq 0$ be a sequence of projections on $\mathcal{G}_1$ satisfying*

$$0 = P_0 \leq P_1 \leq P_2 \leq \cdots \leq P_j \cdots$$

$$P_j \leq I - S_1^j S_1^{*j}, \text{ or equivalently } S_1^j S_1^{*j} P_j = 0, \quad j \geq 0$$

$$P_{j+1} S_1 (I - P_j) = 0, \quad j \geq 0.$$

(We call $P_j,\ j \geq 0$ a causal structure.) Let A be a linear bounded operator from $\mathcal{G}_1$ into $\mathcal{H}_1$ such that $AS_1 = T_1 A$. Then there exists a linear bounded operator B from $\mathcal{G}_1$ into $\mathcal{K}_1$ satisfying

1. $P_{\mathcal{H}_1} B = A,$
2. $BS_1 = U_1 B,$
3. $(I - S_1^j S_1^{*j})B = (I - S_1^j S_1^{*j}) B P_j,\ j \geq 1$

if and only if

$$\mu(A) := \sup_{j \geq 0} \|T_1^{-j} A (I - P_j)\| < \infty.$$

If this is the case, then

$$\nu(A) := \inf\{\|B\| : \ B \text{ satisfies (1)–(3) as above}\} = \mu(A).$$

3 Nonlinear Causal Commutant Lifting Theorem

In this section, we use the same notation as in Section 2. In particular, S on $\mathcal{G}$ is an isometry, T on $\mathcal{H}$ is a contraction and U on $\mathcal{K}$ is the minimal isometric dilation of T. For simplicity we will assume throughout the paper that T is invertible. We start with the following result.

Proposition 1 *Suppose $\phi : B_{r_0}(\mathcal{G}) \to \mathcal{H}$ is an analytic map satisfying $\phi(Sg) = T\phi(g)$ for all $g \in B_{r_0}(\mathcal{G})$ (in short $\phi \circ S = T\phi$) and $\phi(0) = 0$. Suppose moreover that $\{\alpha_n\}_{n=1}^{\infty}$ is a majorizing sequence for all $T^{-j}(\phi - \phi \circ (I - S^j S^{*j}))$, $j \geq 1$. (Here for $j = 0$, we set $T^{-j}(\phi - \phi \circ (I - S^j S^{*j})) = \phi$,) and that $\beta :=$ limsup$_{n \to \infty} (\alpha_n)^{1/n} < \infty$. Set $r = 1/\beta$. (Without loss of generality, we may assume that $r < r_0$.) Then there exists a causal analytic map $\psi : B_r(\mathcal{G}) \to \mathcal{K}$ satisfying*

1. $\psi(0) = 0$;
2. $\psi(Sg) = U\psi(g)$ *for all* $g \in B_r(\mathcal{G})$;

3. $\phi(g) = P_{\mathcal{H}}\psi(g)$ *for all* $g \in B_r(\mathcal{G})$;
4. $\{\alpha_n\}_{n=1}^{\infty}$ *is a majorizing sequence for* ψ.

Proof: We prove the result by considering the Taylor series expansion of ϕ and applying Causal Commutant Lifting Theorem to each n-th power term of ϕ. Namely, let

$$\phi(g) = \phi_1(g) + \phi_2(g, g) + \cdots + \phi_n(g, \ldots, g) + \cdots, \quad g \in B_{r_0}(\mathcal{G})$$

where ϕ_n is a bounded linear operator from $\mathcal{G}^{\otimes n}$ into $\mathcal{H}$. By the assumption that $\phi \circ S = T\phi$ and the Cauchy formula (1), we have that ϕ_n is time-invariant, i.e.

$$\phi_n S^{\otimes n} = T\phi_n.$$

Next we apply Causal Commutant Lifting Theorem to the case where $S_1 = S^{\otimes n}$, $T_1 = T$, $U = U_1$, $P_j = P_{nj} := (I - S^j S^{*j})^{\otimes n}$ and $A = \phi_n$. It is immediate that

$$0 = P_{n0} \leq P_{n1} \leq P_{n2} \leq \cdots$$

We then check that

$$S_1^j S_1^{*j} P_{nj} = S_1^j (S^{*j}(I - S^j S^{*j}))^{\otimes n} = 0, \quad j \geq 0,$$

and

$$\begin{aligned} P_{nj+1} S^{\otimes n}(I - P_{nj}) &= ((I - S^{j+1} S^{*j+1})S)^{\otimes n}(I - P_{nj}) \\ &= S^{\otimes n} P_{nj}(I - P_{nj}) = 0, \quad j \geq 0. \end{aligned}$$

Therefore $P_{nj} = (I - S^j S^{*j})^{\otimes n}$, $j \geq 0$ is a *causal structure*. Note also by our assumption that $T^{-j}(\phi - \phi \circ (I - S^j S^{*j})) : B_{r_0}(\mathcal{G}) \to \mathcal{H}$ is analytic and its Taylor series expansion is

$$\begin{aligned} T^{-j}(\phi(g) - \phi((I - S^j S^{*j})g)) &= T^{-j}\phi_1((I - P_{1j})g) + T^{-j}\phi_2((I - P_{2j})g) \\ &+ \cdots + T^{-j}\phi_n((I - P_{nj})g) + \cdots \end{aligned}$$

Thus $\|T^{-j}\phi_n(I - P_{nj})\| \leq \alpha_n$ for all $j \geq 0$. By Causal Commutant Lifting Theorem, there exists a causal bounded linear operator ψ_n from $\mathcal{G}^{\otimes n}$ into $\mathcal{K}$ satisfying $\psi_n S^{\otimes n} = U\psi_n$, $\|\psi_n\| \leq \alpha_n$, and $P_{\mathcal{H}}\psi_n = \phi_n$ for all $n \geq 1$. Let $\psi : B_r(\mathcal{G}) \to \mathcal{K}$ be defined

$$\psi(g) = \phi_1(g) + \psi_2(g, g) + \cdots + \psi_n(g, \ldots, g) + \cdots, \quad g \in B_r(\mathcal{G})$$

It follows from above analysis that ψ is time-invariant, causal, $P_{\mathcal{H}}\psi = \phi$ and $\{\alpha_n\}_{n=1}^{\infty}$ is a majorizing sequence for ψ. The proof is complete. □

Remark: We call such a causal analytic mapping ψ satisfying (1)–(3) as in the above proposition a *causal dilation* of ϕ. We remark that the assumption that $\{\alpha_n\}_{n=1}^{\infty}$ is a majorizing sequence for all $T^{-j}(\phi - \phi \circ (I - S^j S^{*j}))$, $j \geq 0$ is also necessary for ϕ to have a causal dilation ψ as in the above proposition. Indeed, let $\psi : B_r(\mathcal{G}) \to \mathcal{K}$ be a time-invariant, causal analytic map, and let $\{\alpha_n\}_{n=1}^{\infty}$ be a majorizing sequence for ψ. Let $\phi = P_{\mathcal{H}}\psi$. Since $\psi \circ S = U\psi$, we have $\phi \circ S = T\phi$. Furthermore, by virtue of

$$(I - U^j U^{*j})\psi = (I - U^j U^{*j})\psi \circ (I - S^j S^{*j}),\ \forall j \geq 0,$$

we get that

$$\psi - \psi \circ (I - S^j S^{*j}) = U^j U^{*j}(\psi - \psi \circ (I - S^j S^{*j})).$$

By taking the projection on $\mathcal{H}$, we have that

$$\phi - \phi \circ (I - S^j S^{*j}) = T^j P_{\mathcal{H}} U^{*j}(\psi - \psi \circ (I - S^j S^{*j})).$$

Hence for all $j \geq 0, n \geq 1$

$$T^{-j}\phi_n(I - P_{nj}) = P_{\mathcal{H}} U^{*j}\psi_n(I - P_{nj}).$$

Therefore $\{\alpha_n\}_{n=1}^{\infty}$ is a majorizing sequence for all $T^{-j}(\phi - \phi \circ (I - S^j S^{*j}))$, $j \geq 0$.

A natural question is to compute explicitly a majorizing sequence for all mappings $T^{-j}(\phi - \phi \circ (I - S^j S^{*j}))$, $j \geq 0$. Indeed next we give explicit formulas of a majorizing sequence for an important class of input/output maps.

In order to do this, we first recall that W on $\mathcal{G}$ is a linear *Hilbert-Schmidt operator* if

$$\sum_{i=1}^{\infty} \|We_i\|^2 < \infty \tag{7}$$

for an orthonormal basis $\{e_i\}_{i=1}^{\infty}$ of $\mathcal{G}$. Moreover in this case the quantity in (7) is independent of the choice of the orthonormal basis of $\mathcal{G}$ and is defined to the $\|W\|_2^2$.

Definition 3 Let $\phi : B_r(\mathcal{G}) \to \mathcal{H}$ be an analytic map satisfying $\phi(0) = 0$. We say ϕ has *fading memory* if its nonlinear part $\phi - \phi'(0)$ admits a factorization

$$\phi - \phi'(0) = \chi \circ W$$

where χ is an analytic mapping defined in some neighborhood of $0 \in \mathcal{G}$ into $\mathcal{H}$ and W is a linear Hilbert-Schmidt operator on $\mathcal{G}$. See [7] for more details about this important class of input/output mappings.

Lemma 1 *Let $\phi : B_{r_0}(\mathcal{G}) \to \mathcal{H}$ be an analytic mapping with $\phi(0) = 0$ and fading memory. That is, as in Definition 3, the nonlinear part $\phi - \phi'(0)$ of ϕ admits a* *factorization*

$$\phi - \phi'(0) = \chi \circ W, \tag{8}$$

where $\chi : B_{r_1}(\mathcal{G}) \to B_{r_2}(\mathcal{H})$ *and* W *is a linear Hilbert-Schmidt operator on* $\mathcal{G}$. *Suppose moreover that*

$$\sup_{j\geq 0} \|T^{-j}\| \, \|WS^jS^{*j}\|_2 = M < \infty, \tag{9}$$

where $\|\cdot\|_2$ *denote the Hilbert-Schmidt norm of an operator. Then the sequence*

$$\alpha_1 := \|\phi'(0)\|, \tag{10}$$

$$\alpha_n := \sqrt{n}M\frac{r_2}{r_1^n}\frac{n^n}{n!}\|W\|_2^{n-1}, \quad n \geq 2, \tag{11}$$

is a majorizing sequence for all $T^{-j}(\phi - \phi \circ (I - S^jS^{*j}))$, $j \geq 0$.

Proof: Let $j \geq 1$ be a fixed positive integer. Let e_i, $i \in I$ be an orthonormal basis of $(I - S^jS^{*j})\mathcal{G}$, where the index set I may be finite or infinite. Let $\{e_k\}$, $k \in J$ be an orthonormal basis of $S^jS^{*j}\mathcal{G}$, where J is some index set. Let $n \geq 2$ be a fixed positive integer. For $k = 1, \ldots, n$, let

$$I_k = \{(i_1, \ldots, i_n) : i_k \in J \text{ and } i_1, \ldots, i_{k-1}, i_{k+1}, \ldots, i_n \in I \cup J\}.$$

It is easy to see that with this notation

$$e_{i_1} \otimes \cdots \otimes e_{i_n} : (i_1, \ldots, i_n) \in I_1 \cup \cdots \cup I_n,$$

is an orthonormal basis for $(I - P_{nj})\mathcal{G}^{\otimes n}$ where we recall that $P_{nj} = (I - S^jS^{*j})^{\otimes n}$. We also note that the n-th power term of $\phi - \phi \circ (I - S^jS^{*j})$ is $\phi_n(I - P_{nj})$. By the factorization (8) we have that

$$\phi_n(I - P_{nj}) = \chi_n W^{\otimes n}(I - P_{nj}).$$

We next write $g \in (I - P_{nj})\mathcal{G}^{\otimes n}$ as

$$g = \sum_{k=1}^{n} \sum_{(i_1,\ldots,i_n)\in I_k} g_{i_1,\ldots,i_n} e_{i_1} \otimes \cdots \otimes e_{i_n}$$

with

$$\|g\|^2 = \sum_{k=1}^{n} \sum_{(i_1,\ldots,i_n)\in I_k} |g_{i_1,\ldots,i_n}|^2 < \infty.$$

From the above, we have the following estimates:

$$\left\|\phi_n\left(\sum_k \sum_{(i_1,\ldots,i_n)\in I_k} g_{i_1,\ldots,i_n} e_{i_1} \otimes \cdots \otimes e_{i_n}\right)\right\|$$

$$
\begin{aligned}
&= \left\| \chi_n W^{\otimes n} \sum_k \sum_{(i_1,\dots,i_n)\in I_k} g_{i_1,\dots,i_n} e_{i_1} \otimes \cdots \otimes e_{i_n} \right\| \\
&= \left\| \chi_n \sum_k \sum_{(i_1,\dots,i_n)\in I_k} g_{i_1,\dots,i_n} W e_{i_1} \otimes \cdots \otimes W e_{i_n} \right\| \\
&\leq \sum_k \sum_{(i_1,\dots,i_n)\in I_k} |g_{i_1,\dots,i_n}| \, \|W e_{i_1}\| \cdots \|W e_{i_n}\| \\
&\quad \left\| \chi_n \left(\frac{W e_{i_1}}{\|W e_{i_1}\|} \otimes \cdots \otimes \frac{W e_{i_n}}{\|W e_{i_n}\|} \right) \right\| \\
&\leq \frac{n^n}{r_1^n} \sum_k \sum_{(i_1,\dots,i_n)\in I_k} |g_{i_1,\dots,i_n}| \, \|W e_{i_1}\| \cdots \|W e_{i_n}\| \\
&\quad \left\| \chi_n \left(\frac{r_1 W e_{i_1}}{n\|W e_{i_1}\|} \otimes \cdots \otimes \frac{r_1 W e_{i_n}}{n\|W e_{i_n}\|} \right) \right\| \\
&\leq \frac{n^n}{r_1^n} \frac{r_2}{n!} \sum_k \sum_{(i_1,\dots,i_n)\in I_k} |g_{i_1,\dots,i_n}| \, \|W e_{i_1}\| \cdots \|W e_{i_n}\| \\
&\leq \frac{n^n}{r_1^n} \frac{r_2}{n!} \left(\sum \sum |g_{i_1,\dots,i_n}|^2 \right)^{1/2} \left(\sum \sum \|W e_{i_1}\|^2 \cdots \|W e_{i_n}\|^2 \right)^{1/2} \\
&= \sqrt{n} \frac{n^n}{r_1^n} \frac{r_2}{n!} \|W S^j S^{*j}\|_2 \|W\|_2^{n-1} \left\| \sum_k \sum_{(i_1,\dots,i_n)\in I_k} g_{i_1,\dots,i_n} e_{i_1} \otimes \cdots \otimes e_{i_n} \right\| .
\end{aligned}
$$

The second inequality follows from the fact that $\chi : B_{r_1}(\mathcal{G}) \to B_{r_2}(\mathcal{H})$ implies

$$
\left\| \chi_n \left(\frac{r_1 W e_{i_1}}{n\|W e_{i_1}\|} \otimes \cdots \otimes \frac{r_1 W e_{i_n}}{n\|W e_{i_n}\|} \right) \right\| \leq \frac{r_2}{n!}.
$$

In the last equality we have used the fact that

$$
\|W\|_2^2 = \sum_{i\in I} \|W e_i\|^2 + \sum_{k\in J} \|W e_k\|^2,
$$

$$
\|W S^j S^{*j}\|_2^2 = \sum_{k\in J} \|W e_k\|^2.
$$

Next note that the n-th power term of $T^{-j}(\phi - \phi \circ (I - S^j S^{*j}))$ is $T^{-j}\phi_n(I - P_{nj})$. Now the result follows from the fact that

$$
\begin{aligned}
&\|T^{-j}\phi_n(I - P_{nj})\| \leq \|T^{-j}\| \|\phi_n(I - P_{nj})\| \\
&\leq \|T^{-j}\| \|W S^j S^{*j}\|_2 \sqrt{n} \frac{n^n}{r_1^n} \frac{r_2}{n!} \|W\|_2^{n-1} \leq \sqrt{n} M \frac{n^n}{r_1^n} \frac{r_2}{n!} \|W\|_2^{n-1}.
\end{aligned}
$$

This completes the proof. □

Remark: The following example shows that without the assumption (9), Lemma 1 may not hold. Let $\mathcal{G} = \mathcal{H} = H^2$. Let W a linear Hilbert-Schmidt operator on H^2 be defined by

$$Wz^n = \beta^n z^n, \quad n \geq 0,$$

for some β such that $0 < \beta < 1$. Let T be a contraction on H^2 which is defined by $Tf = \alpha f$ for some positive number α such that

$$0 < \alpha < \beta^2. \tag{12}$$

Let χ on H^2 be the square map, i.e.

$$\chi(f \otimes f) = \sum_{n=0}^{\infty} f_n^2 z^n, \quad f = \sum_{n=0}^{\infty} f_n z^n \in H^2.$$

Now set $\psi = \chi \circ W$. First note that $\psi : B_1(H^2) \to B_1(H^2)$. Note also that $T^{-j}(\phi - \phi \circ (I - S^j S^{*j}))$ is the second power term of itself. Moreover

$$\begin{aligned} &\|T^{-j}(\phi - \phi \circ (I - S^j S^{*j}))\| \\ &\geq \|T^{-j}(\phi(z^j) - \phi((I - S^j S^{*j})z^j)\| = \alpha^{-j}\beta^{2j} = \left(\frac{\beta^2}{\alpha}\right)^j \to \infty \end{aligned}$$

as $j \to \infty$. This shows that there is no uniform majorizing sequence for all $T^{-j}(\phi - \phi \circ (I - S^j S^{*j}))$, $j \geq 0$. It is easy to check that

$$\sup_{j \geq 0} \|T^{-j}\| \, \|WS^j S^{*j}\|_2 = \sup_{j \geq 0} \left(\frac{\beta}{\alpha}\right)^j \frac{1}{\sqrt{1 - \beta^2}} = \infty,$$

i.e. (9) does not hold.

Now by combining Proposition 1 and Lemma 1 we can state the following nonlinear causal commutant lifting theorem.

Theorem 2 *Let $\phi : B_{r_0}(\mathcal{G}) \to \mathcal{H}$ be an analytic map satisfying $\phi(Sg) = T\phi(g)$ for all $g \in B_{r_0}(\mathcal{G})$ and $\phi(0) = 0$. Suppose that ϕ has fading memory, i.e., $\phi - \phi'(0)$ admits a factorization*

$$\phi - \phi'(0) = \chi \circ W$$

where $\chi : B_{r_0}(\mathcal{G}) \to \mathcal{H}$ and W is a linear Hilbert-Schmidt operator on $\mathcal{G}$. Suppose moreover that

$$\sup_{j \geq 0} \|T^{-j}\| \, \|WS^j S^{*j}\|_2 = M < \infty. \tag{13}$$

Then ϕ has a causal dilation $\psi : B_r(\mathcal{G}) \to \mathcal{K}$, where r can be chosen as

$$r = \frac{r_1}{e\|W\|_2} \quad (\leq r_0).$$

Furthermore for any $r_3 < r$, ψ maps $B_{r_3}(\mathcal{G})$ into $B_{r_4}(\mathcal{K})$, where r_4 can be chosen as

$$r_4 = \|\psi'(0)\| r_3 + \frac{M r_2 \|W\|_2 e^2 r_3^2 (2r_1 - e\|W\|_2 r_3)}{r_1 (r_1 - e\|W\|_2 r_3)^2}.$$

Proof: By Lemma 1, α_n defined by (10) and (11) is a majorizing sequence for all $T^{-j}(\phi - \phi \circ (I - S^j S^{*j}))$, $j \geq 0$. So $\rho = \limsup(\alpha_n)^{1/n} = e\|W\|_2 / r_1$, i.e., $r = 1/\rho = r_1 / e\|W\|_2$. Now by Proposition 1, there exists a causal dilation ψ of ϕ from $B_r(\mathcal{G})$ into $\mathcal{K}$ and α_n is a majorizing sequence for ψ. Next for $r_3 < r$, we note that if $\|g\| \leq r_3 < r$, then

$$\begin{aligned}
\|\psi(g)\| &\leq \|\psi'(0)g\| + \sum_{n=2}^{\infty} \|\psi_n(g \otimes \cdots \otimes g)\| \\
&\leq \|\psi'(0)g\| + \sum_{n=2}^{\infty} \sqrt{n} M \frac{r_2}{r_1^n} \frac{n^n}{n!} \|W\|_2^{n-1} r_3^n \\
&\leq \|\psi'(0)\| r_3 + \frac{M r_2 \|W\|_2 e^2 r_3^2 (2r_1 - e\|W\|_2 r_3)}{r_1 (r_1 - e\|W\|_2 r_3)^2}.
\end{aligned}$$

Here we have used the inequalities

$$\frac{n^n}{n!} \leq e^n, \ \sum_{n=2}^{\infty} \sqrt{n} x^n \leq \sum_{n=2}^{\infty} n x^n = \frac{2x^2 - x^3}{(1-x)^2}, \quad \text{for } 0 < x = \frac{e\|W\|_2 r_3}{r_1} < 1.$$

The proof is complete. □

4 Nonlinear (Causal) Hankel Mappings

The power of the linear commutant lifting theorem lies in the fact the linear time-invariant operators are easy to characterize and well understood. Fortunately, we have a similar characterization of time-invariant and causal analytic mappings from $\mathcal{G}$ into $\mathcal{K}$. In the linear case this characterization essentially reduces to the fact that linear bounded operators intertwining shifts are multiplication operators.

Proposition 2 *Notations as above. Assume S on $\mathcal{G}$ is a unilateral shift (i.e., $S^{*n} g \to 0$ as $n \to \infty$ for all $g \in \mathcal{G}$). Let $\psi : B_r(\mathcal{G}) \to \mathcal{K}$ denote a time-invariant and causal analytic mapping with $\psi(0) = 0$. Then there exists an analytic mapping $\Delta : B_r((I - SS^*)\mathcal{G}) \to \mathcal{L}((I - SS^*)\mathcal{G}, \mathcal{K})$ satisfying*

$$\begin{aligned}
&(I - U^j U^{*j}) \Delta(g)(I - S^j S^{*j}) g \\
&= (I - U^j U^{*j}) \Delta((I - S^j S^{*j}) g)(I - S^j S^{*j}) g, \ j \geq 0
\end{aligned} \tag{14}$$

(here $\mathcal{L}((I-SS^*)\mathcal{G},\mathcal{K})$ denotes the space of all bounded linear operators from $(I-SS^*)\mathcal{G}$ into $\mathcal{K}$) such that

$$(15)\qquad \psi(g)=\sum_{n=0}^{\infty}U^n\Delta(S^{*n}g)(I-SS^*)S^{*n}g$$

for all $g\in B_r(\mathcal{G})$. The series given in (15) *is boundedly weakly convergent in $\mathcal{K}$ for all $g\in B_r(\mathcal{G})$. Conversely, if there is Δ satisfying* (14) *and the series as in* (15) *such that it is bounded weakly convergent in $\mathcal{K}$ for all $g\in B_r(\mathcal{G})$, then the map $\psi: B_r(\mathcal{G})\to\mathcal{K}$ defined as in* (15) *is a time-invariant and causal analytic mapping with $\psi(0)=0$.*

Proof: We follow closely the proof of Proposition 4.1 in [2]. Let $\psi: B_r(\mathcal{G})\to\mathcal{K}$ be a time-invariant and causal analytic mapping with $\psi(0)=0$. Define an analytic map $\Gamma: B_r(\mathcal{G})\to\mathcal{K}$ by

$$(16)\qquad \Gamma(g)=\psi(g)-\psi(SS^*g).$$

Γ is also causal since for any $g\in B_r(\mathcal{G})$,

$$\begin{aligned}
&(I-U^jU^{*j})\Gamma(g)\\
=\ &(I-U^jU^{*j})\psi(g)-(I-U^jU^{*j})\psi(SS^*g)\\
=\ &(I-U^jU^{*j})\psi((I-S^jS^{*j})g)-(I-U^jU^{*j})\psi((I-S^jS^{*j})SS^*g)\\
=\ &(I-U^jU^{*j})\psi((I-S^jS^{*j})g)-(I-U^jU^{*j})\psi(SS^*(I-S^jS^{*j})g)\\
=\ &(I-U^jU^{*j})\Gamma((I-S^jS^{*j})g),\quad j\geq 0.
\end{aligned}$$

The second equality above follows from the fact that ψ is causal. The other equalities follow from the definition of Γ and the fact that $(I-S^jS^{*j})SS^*=SS^*(I-S^jS^{*j})$. Let

$$\begin{aligned}
\Delta(g):\ =\ &\int_0^1\Gamma'(tg+(1-t)SS^*g)dt\\
=\ &\int_0^1\Gamma'(SS^*g+t(I-S^jS^{*j})g)dt,\quad g\in B_r(\mathcal{G}).
\end{aligned}$$

It is easy to see that $\Delta: B_r(\mathcal{G})\to B((I-S^jS^{*j})\mathcal{G},\mathcal{K})$ is analytic. Furthermore by the operator-valued Newton-Leibnitz formula we have

$$(17)\qquad \Gamma(g)=\Delta(g)(I-S^jS^{*j})g.$$

Therefore by the fact that Γ is causal, Δ satisfies (14). Now by combining (16) and (17) we have the relation

$$\psi(g)=\Gamma(g)+\psi(SS^*g)=\Delta(g)(I-SS^*)g+U\psi(S^*g).$$

By iterating the above relation we get

$$\psi(g) = \sum_{n=0}^{K} U^n \Delta(S^{*n}g)(I - SS^*)S^{*n}g + U^{K+1}\psi(S^{*K+1}g).$$

By the assumption that U is an isometry and S is a shift, we have

$$\|U^{K+1}\psi(S^{*K+1}g)\| = \|\psi(S^{*K+1}g)\| \to \|\psi(0)\| = 0$$

as $K \to \infty$. Therefore ψ admit the series representation (15).

Conversely, if ψ admits the series representation (15) with Δ satisfying (14), then it is immediate that $\psi(0) = 0$. It is also straightforward to check that ψ is time-invariant by noting that

$$\begin{aligned}\psi(Sg) &= \sum_{n=0}^{\infty} U^n \Delta(S^{*n}Sg)(I - SS^*)S^{*n}Sg \\ &= \sum_{n=1}^{\infty} U^n \Delta(S^{*n-1}g)(I - SS^*)S^{*n-1}g = U\psi(g).\end{aligned}$$

To see that ψ is causal we note that

$$\begin{aligned}&(I - U^jU^{*j})\psi(g) \\ &= (I - U^jU^{*j})\sum_{n=0}^{j-1} U^n \Delta(S^{*n}g)(I - SS^*)S^{*n}g \\ &= \sum_{n=0}^{j-1} U^n(I - U^{j-n}U^{*j-n})\Delta(S^{*n}g)(I - SS^*)S^{*n}g \\ &= \sum_{n=0}^{j-1} U^n(I - U^{j-n}U^{*j-n})\Delta((I - S^{j-n}S^{*j-n})S^{*n}g)(I - SS^*) \\ &\qquad (I - S^{j-n}S^{*j-n})S^{*n}g \\ &= \sum_{n=0}^{j-1} U^n(I - U^{j-n}U^{*j-n})\Delta(S^{*n}(I - S^jS^{*j})g)(I - SS^*)(I - S^jS^{*j})S^{*n}g \\ &= (I - U^jU^{*j})\psi((I - S^jS^{*j})g), \quad j \geq 0\end{aligned}$$

where the second equality follows from (14), and the other equalities follow from (15) and the fact that

$$\begin{aligned}(I - U^jU^{*j})U^n &= U^n(I - U^{j-n}U^{*j-n}), \\ (I - SS^*)(I - S^{j-n}S^{*j-n}) &= (I - SS^*).\end{aligned}$$

This completes the proof. □

Remark: We remark that if the shift S has multiplicity one, then Δ is uniquely determined by ψ; see Remark 4.2 in [2].

Now we specialize Proposition 2 to the case where $\mathcal{G} = \mathcal{K} = H^2$ and $S = U = M_z$ on H^2. In this case we call the mapping ψ as in Proposition 2 a *nonlinear (causal) analytic Toeplitz operator*, since in the linear case such a ψ is precisely a classical analytic Toeplitz operator. Note that in this case $(I - SS^*)H^2 = \mathbf{C}$, where $\mathbf{C}$ denotes the complex numbers. So $\Delta(g) : B_r(H^2) \to \mathcal{L}(\mathbf{C}, H^2)$ is given by a mapping $\Delta(g, z) : B_r(H^2) \to H^2$ in the following way,

$$\Delta(g)a = a\Delta(g, z), \quad a \in \mathbf{C} \tag{18}$$

for

$$g = g_0 + g_1 z + g_2 z^2 + \cdots \in B_r(H^2).$$

Moreover, if

$$\Delta(g, z) = \delta_0(g) + \delta_1(g)z + \delta_2(g)z^2 + \cdots$$

then Δ satisfies (14) if and only if

$$\delta_n(g) = \delta_n(g_0 + g_1 z + \cdots + g_n z^n), \quad n = 0, 1, 2, \cdots.$$

With this notation, ψ admits the following series representation

$$\psi(g) = \sum_{n=0}^{\infty} \sum_{m=n}^{\infty} z^m g_n \delta_{m-n}(g_n + g_{n+1}z + \cdots + g_{m-n}z^{m-n}).$$

To define our nonlinear analogue of classical Hankel operators, we first recall some basic facts about classical Hankel operators. Let m be a nonconstant inner function. Let $H(m) := H^2 \ominus mH^2$, $U = M_z$ and $T = S(m) := PU \mid H(m)$ the compressed shift, and $P : H^2 \to H(m)$ the projection. Then a (linear) Hankel operator $A : H^2 \to H(m)$ is a linear bounded operator satisfying $AS = S(m)A$. Our definition of a Hankel operator is equivalent to the usual definition as given, e.g., in [19] via the unitary equivalence between $H(m)$ and the closed subspace $\bar{m}H^2 \ominus H^2$ of the Lebesque space L^2 on the unit circle.

A fundamental result is that the Hankel operator admits a symbol, that is, there exists $a \in H^\infty$ such that

$$A = PM_a,$$

where $M_a : H^2 \to H^2$ denotes the multiplication by a. Furthermore an important theorem of Sarason [20] states that a can be chosen such that $\|a\|_\infty = \|A\|$.

Now the nonlinear (causal) analogue of classical Hankel operators is defined to be the analytic mapping ϕ as in Proposition 1. By Proposition 1, ϕ has a causal dilation ψ; and by Proposition 2, ψ admits a series representation as in (15). Similar to the linear case we call $\Delta(g, z)$ a symbol of the nonlinear (causal) analytic Toeplitz mapping ψ. To summarize, we have the following basic result for nonlinear (causal) Hankel mappings, which is a natural extension of Sarason's theorem [20] mentioned above to the nonlinear case. (Note that $T = S(m)$.)

Theorem 3 *Let $\phi : B_{r_0}(H^2) \to H(m)$ be such that $\phi(Sg) = T\phi(g)$ for all $g \in B_{r_0}(H^2)$ and $\phi(0) = 0$. Suppose that $\{\alpha_n\}_{n=1}^{\infty}$ is a majorizing sequence for all $T^{-j}(\phi - \phi \circ (I - S^j S^{*j}))$, $j \geq 0$, and that $\beta := limsup_{n\to\infty}(\alpha_n)^{1/n} < \infty$. Set $r = 1/\beta$. Then ϕ has a causal dilation $\psi : B_r(\mathcal{G}) \to \mathcal{K}$ which admits the series representation*

$$\psi(g) = \sum_{n=0}^{\infty} z^n g_n \Delta(S^{*n} g, z)$$

for

$$g = g_0 + g_1 z + g_2 z^2 + \cdots \in B_r(H^2),$$

where $\Delta(g, z) : B_r(H^2) \to H^2$ (as in (18)) satisfies

$$\Delta(g, z) = \delta_0(g_0) + \delta_1(g_0 + g_1 z)z + \delta_2(g_0 + g_1 z + g_2 z^2)z^2 + \cdots.$$

Moreover,

$$\Delta(0, z) \in H^{\infty}, \quad \|\Delta(0, z)\|_{\infty} = \|\psi'(0)\|. \tag{19}$$

Proof: It follows from above analysis that we need only to show (19). We note that

$$\Delta(0, z) = \frac{\partial \psi}{\partial g}\,|_{g=0} = \psi'(0)1.$$

But $\psi \circ S = S\psi$ implies that $\psi'(0)S = S\psi'(0)$. Thus $\psi'(0)$ is the multiplication in H^2 by the function $\Delta(0, z)$. Therefore $\Delta(0, z) \in H^{\infty}$ and $\|\Delta(0, z)\|_{\infty} = \|\psi'(0)\|$. The proof is complete. □

5 Nonlinear Interpolation and Robust Control

We conclude this paper with a possible application of the nonlinear causal commutant lifting theorem to a nonlinear version of weighted sensitivity H^{∞}-minimization in robust control; see [22, 16, 15, 8] and references therein for the linear case of sensitivity minimization. We will first describe the physical robust control problem in which we are interested. In our treatment which follows, we will add the causality constraint to the results of [2], and thereby propose a physically realizable nonlinear optimization method.

Let $m(z)$ be the transfer function of a given discrete-time, linear time-invariant SISO system. For simplicity we will assume that $m(z)$ is stable and all-pass. This means that $m(z) \in H^{\infty}$ and is an inner function. In control, m is called the *plant.* We now make the following definition. We say an analytic input/output operator ϕ defined in a ball around the orgin in H^2 is *admissible* if it is causal, time-invariant, majorizable, and $\phi(0) = 0$. We denote

$$\mathcal{C}_w := \{\text{space of admissible operators}\}.$$

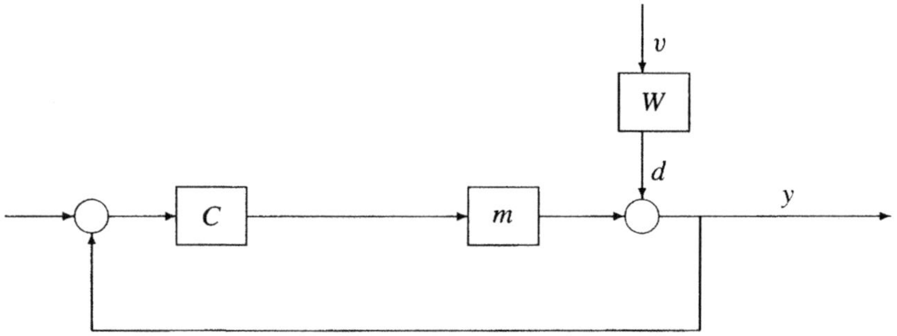

Figure 1: Standard Feedback Configuration

Since the theory we are considering is local, the notion of admissibility is sufficient for all of the applications we have in mind.

Now, referring to Figure 1, in our problem, we are required to construct an admissible operator C (called a *feedback compensator*) in such a way as to stabilize the close loop system and attenuate the effect of the filtered disturbances (filtered by the "weight" W) d. The unfiltered disturbances v are assumed to have energy (i.e., 2-norm) bounded by some fixed constant. This leads to following kind of mathematical problem. See [12] and [13] for more details.

Let $W : B_{r_0}(H^2) \to H^2$ be an invertible admissible operator. Then we say that the feedback compensator C *stabilizes* the closed loop if the operators $(I+mC)^{-1}$ and $C\circ(I+mC)^{-1}$ are well-defined and admissible. One can show that C stabilizes the closed loop if and only if

$$C = \hat{q} \circ (I - m\hat{q})^{-1} \tag{20}$$

for some $\hat{q} \in \mathcal{C}_w$. (See [12], [13] and the references therein.) Note that the *weighted sensitivity* $(I + mC)^{-1} \circ W$ can be written as $W - mq$, where $q := \hat{q} \circ W$. This is precisely the operator relating the disturbance v to the output y. (Since W is invertible, the data q and $\hat{q}$ are equivalent.) In this context, we will call such a q, a *compensating parameter.* Note that from the compensating parameter q, we get a stabilizing compensator C via the formula (20).

Now for the weighted sensitivity operator

$$S_q := W - mq, \quad q \in \mathcal{C}_w$$

and for $u < r_0$, define $v_q > 0$ to be the smallest number such that S_q maps $B_u(H^2)$ into $B_{v_q}(H^2)$. Then a nonlinear version of the weighted sensitivity H^∞ minimization considered in [22] is to compute or more realistically find a non-trivial upper bound of

$$\mu := \inf_{q \in \mathcal{C}_w} v_q; \tag{21}$$

since for input signals whose energy is bounded by u, μ is the minimal weighted sensitivity over the admissible set of feedback compensators.

We remark that in the linear case, W and q are multiplication by some functions in H^∞. By abuse of the notation, we still denote these functions by W and q, respectively. Then (21) becomes

$$\mu := \inf_{q \in H^\infty} \|W - mq\|_\infty u. \tag{22}$$

To compute μ as above is precisely the generalized interpolation problem considered in [20]. For example, if $m(z) = z^n$ for some positive integer, then (22) corresponds to the Caratheodory interpolation problem; for $m(z)$ a finite Blaschke product such that all its zeros with multiplicity one, (22) corresponds to the Nevanlinna-Pick interpolation problem. See [9] and [20] references therein for more details. Hence our problem (21) can be viewed as a type of generalized nonlinear interpolation problem.

But mathematically this is precisely the kind of problem we have been considering. To see this, let $\phi : B_{r_0}(H^2) \to H(m) := H^2$ be the analytic mapping defined by

$$\phi(g) := PW(g), g \in B_{r_0}(H^2),$$

where P is the projection from H^2 onto $H(m)$. It is easy to check that ψ is a causal dilation of ϕ if and only if ψ is of the form $W - mq$ for some $q \in C_w$. Now by Proposition 1 we can give an upper bound of μ, and moreover an explicit formula for the upper bound of μ is given in Theorem 1, if we make the natural assumption that ϕ is of fading memory.

Acknowledegments

I would like to thank C. Foias and A. Tannenbaum for several very helpful suggestions.

References

[1] W. Arveson, "Interpolation problems in nest algebras," *J. Functional Analysis* **20** (1975), 208–233.

[2] J. Ball, C. Foias, J.W. Helton and A. Tannenbaum, "On a local nonlinear commutant lifting theorem," *Indiana J. Mathematics* **36** (1987), 693–709.

[3] J. Ball, C. Foias, J.W. Helton and A. Tannenbaum, "Nonlinear interpolation theory in H^∞," in *Modelling, Robustness, and Sensitivity in Control Systems* (edited by Ruth Curtain), NATO-ASI Series, Springer-Verlag, New York, 1987.

[4] J. Ball, C. Foias, J. W. Helton and A. Tannenbaum, "A Poincaré-Dulac approach to a nonlinear Beurling-Lax-Halmos theorem," *Journal of Math. Anal. and Applications* **139** (1989), 496–514.

[5] J. Ball and J. W. Helton, "Shift invariant manifold and nonlinear analytic function theory," *Integral Equations Operator Theory* **11** (1988), 615–725.

[6] J. Ball and J. W. Helton, "Inner-outer factorization of nonlinear operators," *J. Functional Analysis* **104** (1992), 363–413.

[7] S. Boyd and L. Chua, "Fading memory and the problem of approximating nonlinear operators with Volterra series," *IEEE Trans. Circuits and Systems* **CAS-32** (1985), 1150–1161.

[8] J. Doyle, B. Francis and A. Tannenbaum, *Feedback Control Theory*, MacMillan, New York, 1991.

[9] C. Foias and A. Frazho, *The Commutant Lifting Approach to Interpolation Problems*, Birhauser-Verlag, Boston, 1990.

[10] C. Foias, C. Gu and A. Tannenbaum, "Intertwining dilations, intertwining extensions and causality," *Acta Sci. Math (Szeged)* **57** (1993), 101–123.

[11] C. Foias, C. Gu and A. Tannenbaum, "Nonlinear H^∞ optimization: A causal power series approach," *SIAM J. Control and Optimization* **33** (1995), 185–207.

[12] C. Foias and A. Tannenbaum, "Iterated cummutant lifting for systems with rational symbol," *Operator Theory: Advances and Applications* **41** (1989), 255–277.

[13] C. Foias and A. Tannenbaum, "Weighted optimization theory for nonlinear systems," *SIAM J. on Control and Optimization* **27** (1989), 842–860.

[14] C. Foias and A. Tannenbaum, "Causality in commutant lifting theory," *J. Functional Analysis* **118** (1993), 407–441.

[15] C. Foias, A. Tannenbaum and G. Zames, "Some explicit formulae for the singular values of certain Hankel operators with factorizable symbol," *SIAM J. Math. Analysis* **19** (1988), 1081–1089.

[16] B. Francis, *A Course in H^∞ Control Theory*, McGraw-Hill, New York, 1981.

[17] B. Francis and A. Tannenbaum, "Generalized interpolation theory in control," *Mathematical Intelligencer* **10** (1988), 48–43.

[18] C. Gu, "On causality in commutant lifting theorem. I," *Integral Equations and Operator Theory* **16** (1993), 82–97.

[19] S.C. Power, *Hankel Operators on Hilbert Spaces*, Pitman Advanced Publishing Program, Boston, 1982.

[20] D. Sarason, "Generalized interpolation in H^∞," *Trans. AMS* **127** (1967), 179–203.

[21] B. Sz.-Nagy and C. Foias, *Harmonic Analysis of Operators on Hilbert Space*, North-Holland Publishing Company, Amsterdam, 1970.

[22] G. Zames, "Feedback and optimal sensitivity: model reference transformations, multiplicative seminorms, and approximate inverses," *IEEE Trans. Automat. Control* **AC-26** (1981), 301–320.

Department of Mathematics
California Polytechnic
State University San Luis Obispo,
CA 93407

Operator Theory:
Advances and Applications, Vol. 115

Parametrization of Solutions of the Nehari Problem and Nonorthogonal Dynamics

A. Kheifets

This paper deals with operator Nehari problem. The main objective is to adopt the feedback coupling method developed earlier in analytic (orthogonal scattering) case to nonanalytic (nonorthogonal scattering) case. The exposition goes in terms slightly different from the ones of Adamjan and Arov and leads to a parametrization of the solution set in general case. The parametrization formula is analogue of Arov–Grossman formula for nonanalytic case. A generalization of Adamjan–Arov–Krein theorem is formulated in the last section for general case (the classical theorem concerns the completely indeterminate matrix case only). The proof will be published elsewhere. It depends upon studying the residual part of the feedback coupling of scattering systems.

Introduction

Adamjan and Arov extended the Lax-Philips scattering scheme to nonorthogonal situation (nonanalytic scattering function) in paper [AA], see also [Ad]. The machinery was applied to solving the Nehari problem in famous series of papers due to Adamjan, Arov and Krein (see [AAK1], [AAK2], [Ad]) in late 60's–early 70's. The Abstract Interpolation Problem method was developed in [KKY] and [Kh1] (see also [KY]) for analytic interpolation problems (i.e. the ones for the Schur class functions). The parametrization formula which was obtained there was exactly the Arov–Grossman formula [AG] for scattering functions of unitary extensions of an isometric colligation. In [Kh2] a generalization of the Adamjan–Arov–Krein theorem was obtained for this class of problems. Our goal here is to adopt the method to nonanalytic case in the context of the Nehari problem. The author was also influenced by the paper [Aroc] doing this work. Although the list of references does not claim to be complete the papers with a little different approaches should also be mentioned as the ones of close relevance: M. Moran (1989), C. Foias and A. Frazo (1986, 1988, 1995), S. Kupin (1996), as well as some recent papers by R. Arocena.

1 Setting of the Nehari Problem

Let $\gamma_{-1}, \gamma_{-2}, \ldots$ be a specified sequence of linear operators of the space E_1 to the space E_2 (E_1 and E_2 are separable Hilbert spaces). One wants to find all functions $w(t)$, $t \in \mathbb{T}$ (unit circle) such that $w(t)\colon\ E_1 \to E_2$ is a contraction for a.a. $t \in \mathbb{T}$

and Fourier series of $w(t)$ is

$$w(t) \sim \sum_{k=-\infty}^{+\infty} \gamma_k t^k, \qquad t \in \mathbb{T}$$

where γ_k are specified for $k \leq -1$ and are unknown for $k \geq 0$.

We will also consider harmonic continuation of such a function $w(t)$ inside the unit disk $\mathbb{D}$:

$$w(\zeta) = \sum_{k=-\infty}^{-1} \gamma_k \overline{\zeta}^{|k|} + \sum_{k=0}^{\infty} \gamma_k \zeta^k, \qquad |\zeta| < 1.$$

This continuation is contractive valued as well.

2 Associated Isometry

Associate first a Hankel operator $\Gamma\colon H^2_+(E_1) \to H^2_-(E_2)$ to the data of the Nehari problem:

$$\Gamma\colon t^k e_1 \to \sum_{j=-\infty}^{-k-1} \gamma_j t^{j+k} e_1, \qquad k \geq 0.$$

If $w(t)$ is a solution of the Nehari problem then

$$\Gamma(t^k e_1) = P_- w(t) t^k e_1, \qquad e_1 \in E_1, \quad k \geq 0.$$

Hence Γ is extended on $H^2_+(E_1)$ by the same formula

$$\Gamma h_+ = P_- w h_+$$

provided a solution w exists. Thus, contractivity of Γ is necessary for the Nehari problem to be solvable. We are assuming Γ is a contraction in what follows and will see that this condition is also sufficient for solvability of the Nehari problem.

Remark: Γ is a Hankel operator means that $\Gamma(th_+) = P_- t \Gamma h_+, h_+ \in H^2_+(E_1)$.

Consider linear space

$$X = \begin{bmatrix} H^2_-(E_2) \\ H^2_+(E_1) \end{bmatrix} = H^2_-(E_2) \times H^2_+(E_1)$$

to be cartesian product of $H^2_-(E_2)$ and $H^2_+(E_1)$, i.e.

$$X = \left\{ x = \begin{bmatrix} x_2 \\ x_1 \end{bmatrix} : x_2 \in H^2_-(E_2),\ x_1 \in H^2_+(E_1) \right\}.$$

There is the natural inner product in X. We also will need another quadratic form in X:

$$D(x', x'') \stackrel{\text{def}}{=} \left\langle \begin{bmatrix} \mathbf{1} & \Gamma \\ \Gamma^* & \mathbf{1} \end{bmatrix} x', x'' \right\rangle_X,$$

where Γ is the Hankel operator associated with data of the Nehari problem. $D(x, x) \geq 0$ if and only if Γ is a contraction.

We denote by $\vec{x}$ the equivalence class of x with respect to the form D:

$$x \sim 0 \text{ if } D(x, x) = 0 \Longleftrightarrow D(x, y) = 0, \qquad \forall y \in X.$$

Let $\vec{X}$ stands for the space of these equivalence classes

$$\vec{X} = \{\vec{x},\ x \in X\}.$$

The inner product

$$\langle \vec{x}', \vec{x}'' \rangle \stackrel{\text{def}}{=} D(x', x'')$$

is well defined in $\vec{X}$. H_0 will stand for the Hilbert space which is the completion of $\vec{X}$ with respect to this inner product. Let

$$x \in d_V \stackrel{\text{def}}{=} \begin{bmatrix} H^2_-(E_2) \\ tH^2_+(E_1) \end{bmatrix} \subset X, \text{ then } \bar{t}x \in \Delta_V \stackrel{\text{def}}{=} \begin{bmatrix} \bar{t}H^2_-(E_2) \\ H^2_+(E_1) \end{bmatrix} \subset X.$$

Because Γ is a Hankel operator

$$D(\bar{t}x, \bar{t}x) = D(x, x), \qquad \forall x \in d_V.$$

This insures that the operator

$$V \colon \vec{x} \to \overrightarrow{(\bar{t}x)}$$

is well defined isometry on H_0 with domain clos $\overrightarrow{d_V} \subseteq H_0$ and range clos $\overrightarrow{\Delta_V} \subseteq H_0$, where $\overrightarrow{d_V}$ and $\overrightarrow{\Delta_V}$ stand for equivalence classes of vectors of d_V and Δ_V respectively.

3 Unitary Extensions of V and All

Let H be a Hilbert space that contains H_0 (as a closed subspace), $H \supseteq H_0$ and let $\mathcal{U}$ be a unitary operator on H that extends V, i.e.

$$\mathcal{U}\big|_{\overrightarrow{d_V}} = V.$$

There exists a natural isometric mapping of the space E_1 into H_0:

$$\rho_1\colon e_1 \longmapsto \overrightarrow{\begin{bmatrix} 0 \\ e_1 \end{bmatrix}} \in \vec{X} \subseteq H_0, \qquad e_1 \in E_1,$$

and the one of the space E_2 into H_0:

$$\rho_2\colon e_2 \longmapsto \overrightarrow{\begin{bmatrix} \overline{t}e_2 \\ 0 \end{bmatrix}} \in \vec{X} \subseteq H_0, \qquad e_2 \in E_2.$$

Definition: Harmonic function $w(\zeta)\colon E_1 \to E_2$, $|\zeta| < 1$

$$w_{\mathcal{U}}(\zeta) \stackrel{\text{def}}{=} \rho_2^* \mathcal{U}[(\mathbf{1}_H - \zeta\mathcal{U})^{-1} + (\mathbf{1}_H - \overline{\zeta}\mathcal{U}^*)^{-1} - \mathbf{1}_H]\rho_1$$

is called a scattering function of the unitary extension $\mathcal{U}$.

Lemma 1 *$w(\zeta)$ is a contractive valued function for any unitary extension of V.*

Theorem 1 *$w(\zeta)\colon E_1 \to E_2$ is a solution of the Nehari problem if and only if $w(\zeta)$ is the scattering function of a unitary extension $\mathcal{U}$ of V.*

Proof of Lemma 1: Let $e_1 \in E_1$, $e_2 \in E_2$. Consider the following expression

$$\langle e_2, e_2\rangle_{E_2} + \langle w(\zeta)e_1, e_2\rangle_{E_2} + \langle w(\zeta)^* e_2, e_1\rangle_{E_1} + \langle e_1, e_1\rangle_{E_1}. \tag{3.1}$$

Observe that

$$\langle e_1, e_1\rangle = \langle [(\mathbf{1}_H - \zeta\mathcal{U})^{-1} + (\mathbf{1}_H - \overline{\zeta}\mathcal{U}^*)^{-1} - \mathbf{1}_H]\rho_1 e_1, \rho_1 e_1\rangle_H. \tag{3.2}$$

In fact, by the definition

$$\rho_1 e_1 = \overrightarrow{\begin{bmatrix} 0 \\ e_1 \end{bmatrix}} = V\overrightarrow{\begin{bmatrix} 0 \\ te_1 \end{bmatrix}} = \mathcal{U}\overrightarrow{\begin{bmatrix} 0 \\ te_1 \end{bmatrix}}.$$

Hence

$$\overrightarrow{\begin{bmatrix} 0 \\ t^k e_1 \end{bmatrix}} = \mathcal{U}^{*k}\rho_1 e_1, \qquad k \geq 0.$$

Since

$$\left\langle \overrightarrow{\begin{bmatrix} 0 \\ t^k e_1 \end{bmatrix}}, \overrightarrow{\begin{bmatrix} 0 \\ e_1 \end{bmatrix}} \right\rangle_{H_0} = \left\langle \begin{bmatrix} \mathbf{1} & \Gamma \\ \Gamma^* & \mathbf{1} \end{bmatrix}\begin{bmatrix} 0 \\ t^k e_1 \end{bmatrix}, \begin{bmatrix} 0 \\ e_1 \end{bmatrix} \right\rangle_X$$

$$= \langle t^k e_1, e_1\rangle_{H^2_+(E_1)} = 0, \quad k > 0$$

we obtain

$$\langle \mathcal{U}^{*k}\rho_1 e_1, \rho_1 e_1\rangle_H = 0, \qquad k > 0.$$

I.e. $\rho_1(E_1)$ is a wandering subspace for $\mathcal{U}$. Therefore

$$\langle(\mathbf{1}_H - \bar{\zeta}\mathcal{U}^*)^{-1}\rho_1 e_1, \rho_1 e_1\rangle_H = \langle\rho_1 e_1, \rho_1 e_1\rangle_H = \langle e_1, e_1\rangle_{E_1}.$$

Further,

$$\langle(\mathbf{1}_H - \zeta\mathcal{U})^{-1}\rho_1 e_1, \rho_1 e_1\rangle_H = \langle\rho_1 e_1, (\mathbf{1}_H - \bar{\zeta}\mathcal{U}^*)^{-1}\rho_1 e_1\rangle_H = \langle e_1, e_1\rangle_{E_1}.$$

Hence, the right hand side of (3.2) is equal to

$$\langle e_1, e_1\rangle + \langle e_1, e_1\rangle - \langle e_1, e_1\rangle.$$

Relation (3.2) follows. Similarly to (3.2) one can verify that

$$\text{(3.3)}\quad \langle e_2, e_2\rangle = \left\langle[(\mathbf{1}_H - \zeta\mathcal{U})^{-1} + (\mathbf{1}_H - \bar{\zeta}\mathcal{U}^*)^{-1} - \mathbf{1}_H]\mathcal{U}^*\rho_2 e_2, \mathcal{U}^*\rho_2 e_2\right\rangle_H,$$

and $\rho_2(E_2)$ is also a wandering subspace for $\mathcal{U}$. We can re-express (3.1) now by means of (3.2) and (3.3)

$$\begin{aligned}&\langle e_2, e_2\rangle_{E_2} + \langle w(\zeta)e_1, e_2\rangle_{E_2} + \langle w(\zeta)^* e_2, e_1\rangle_{E_1} + \langle e_1, e_1\rangle_{E_1}\\ &= \big\langle[(\mathbf{1}_H - \zeta\mathcal{U})^{-1} + (\mathbf{1}_H - \bar{\zeta}\mathcal{U}^*)^{-1} - \mathbf{1}_H](\rho_1 e_1 + \mathcal{U}^*\rho_2 e_2),\\ &\quad(\rho_1 e_1 + \mathcal{U}^*\rho_2 e_2)\big\rangle = (1 - |\zeta|^2)\langle(\mathbf{1}_H - \zeta\mathcal{U})^{-1}(\rho_1 e_1 + \mathcal{U}^*\rho_2 e_2),\\ &\quad(\mathbf{1}_H - \zeta\mathcal{U})^{-1}(\rho_1 e_1 + \mathcal{U}^*\rho_2 e_2)\rangle_H \ge 0.\end{aligned}$$

We used here the relation

$$(\mathbf{1}_H - \zeta\mathcal{U})^{-1} + (\mathbf{1}_H - \bar{\zeta}\mathcal{U}^*)^{-1} - \mathbf{1}_H = (1 - |\zeta|^2)(\mathbf{1}_H - \bar{\zeta}\mathcal{U}^*)^{-1}(\mathbf{1}_H - \zeta\mathcal{U})^{-1}.$$

Since (3.1) can be re-expressed as

$$\left\langle\begin{bmatrix}\mathbf{1} & w(\zeta)\\ w(\zeta)^* & \mathbf{1}\end{bmatrix}\begin{bmatrix}e_2\\ e_1\end{bmatrix}, \begin{bmatrix}e_2\\ e_1\end{bmatrix}\right\rangle_{E_2\oplus E_1} \ge 0$$

Lemma 1 follows.

Proof of Theorem 1:
(1) Let $w(\zeta)$ be a solution of the Nehari problem. We are going to show it is the scattering function of a unitary extension of V. Consider space $Y = \begin{bmatrix}L^2(E_2)\\ L^2(E_1)\end{bmatrix}$. Define quadratic form $\mathcal{K}$ on Y:

$$\mathcal{K}(y', y'') = \int_{\mathbb{T}}\left\langle\begin{bmatrix}\mathbf{1} & w(t)\\ w(t)^* & \mathbf{1}\end{bmatrix}\begin{bmatrix}y_2'(t)\\ y_1'(t)\end{bmatrix}, \begin{bmatrix}y_2''(t)\\ y_1''(t)\end{bmatrix}\right\rangle_{E_2\oplus E_1} m(dt),$$

where $m(dt)$ is the normalized Lebesgue measure. Considering the space of equivalence classes $\vec{Y}$ and doing completion we will arrive at a Hilbert space H. Multiplication by $\bar{t}$ on Y defines a unitary operator $\mathcal{U}$ on H

$$\mathcal{U}\colon \vec{y} \longmapsto \overrightarrow{(\bar{t}y)}\,.$$

(a) We are going to show that $H_0 \subseteq H$ and $\mathcal{U}$ extends V.

$$X = \begin{bmatrix} H_-^2(E_2) \\ H_+^2(E_1) \end{bmatrix} \subset \begin{bmatrix} L^2(E_2) \\ L^2(E_1) \end{bmatrix} = Y.$$

Let $x = \begin{bmatrix} x_2 \\ x_1 \end{bmatrix} \in X$, then

$$\begin{aligned}
\mathcal{K}(x,x) &= \langle x_2, x_2\rangle_{L^2(E_2)} + \langle w x_1, x_2\rangle_{L^2(E_2)} + \langle w^* x_2, x_1\rangle_{L^2(E_1)} + \langle x_1, x_1\rangle_{L^2(E_1)} \\
&= \langle x_2, x_2\rangle_{H_-^2(E_2)} + \langle P_- w x_1, x_2\rangle_{H_-^2(E_2)} + \langle P_+ w^* x_2, x_1\rangle_{H_+^2(E_1)} \\
&\quad + \langle x_1, x_1\rangle_{H_+^2(E_1)}
\end{aligned}$$

because w is a solution

$$= \langle x_2, x_2\rangle + \langle \Gamma x_1, x_2\rangle + \langle \Gamma^* x_2, x_1\rangle + \langle x_1, x_1\rangle = D(x,x).$$

Hence $H_0 \subseteq H$. Extension property of $\mathcal{U}$ follows from definitions of V and $\mathcal{U}$.
(b) We are going to show that the scattering function of $\mathcal{U}$ equals $w(\zeta)$. By the definition

$$\rho_1 e_1 = \begin{bmatrix} 0 \\ e_1 \end{bmatrix}$$

$$\mathcal{U}^* \rho_2 e_2 = \mathcal{U}^* \overrightarrow{\begin{bmatrix} \bar{t} e_2 \\ 0 \end{bmatrix}} = \overrightarrow{\begin{bmatrix} e_2 \\ 0 \end{bmatrix}}$$

and

$$\mathcal{U}\vec{y} = \overrightarrow{(\bar{t}y)}\,, \qquad y \in Y.$$

Hence,

$$\begin{aligned}
(\mathbf{1}_H - \zeta\mathcal{U})^{-1}\vec{y} &= \overrightarrow{\left(\frac{y}{1-\zeta\bar{t}}\right)}, \\
(\mathbf{1}_H - \bar{\zeta}\mathcal{U}^*)^{-1}\vec{y} &= \overrightarrow{\left(\frac{y}{1-\bar{\zeta}t}\right)}
\end{aligned}$$

and

$$[(\mathbf{1}_H - \zeta\mathcal{U})^{-1} + (\mathbf{1}_H - \bar{\zeta}\mathcal{U}^*)^{-1} - \mathbf{1}_H]\vec{y} = \overrightarrow{\left(\frac{1-|\zeta|^2}{|1-\zeta\bar{t}|^2}\,y\right)}.$$

Therefore

$$\langle w_{\mathcal{U}}(\zeta)e_1, e_2\rangle = \langle [(\mathbf{1}_H - \zeta\mathcal{U})^{-1} + (\mathbf{1}_H - \bar{\zeta}\mathcal{U}^*)^{-1} - \mathbf{1}_H]\rho_1 e_1, \mathcal{U}^*\rho_2 e_2\rangle$$

$$= \left\langle \overrightarrow{\left(\frac{1-|\zeta|^2}{|1-\zeta\bar{t}|^2}\begin{bmatrix}0\\ e_1\end{bmatrix}\right)}, \overrightarrow{\begin{bmatrix}e_2\\ 0\end{bmatrix}} \right\rangle_H = \mathcal{K}\left(\frac{1-|\zeta|^2}{|1-\bar{\zeta}t|^2}\begin{bmatrix}0\\ e_1\end{bmatrix}, \begin{bmatrix}e_2\\ 0\end{bmatrix}\right)$$

$$= \int_{\mathbb{T}} \frac{1-|\zeta|^2}{|1-\zeta\bar{t}|^2}\langle w(t)e_1, e_2\rangle m(dt).$$

But this is the Poison integral, i.e., $w_{\mathcal{U}}(\zeta)$ is the harmonic continuation of $w(t)$ on the unit disk $\mathbb{D}$. Claim (b) follows.

(2) Let $H \supseteq H_0$ and let $\mathcal{U}$ be a unitary operator on H extending V, $\mathcal{U}\big|_{\overrightarrow{d_V}} = V$. Let $w_{\mathcal{U}}(\zeta)$ be the scattering function of $\mathcal{U}$. We are going to show that $w_{\mathcal{U}}$ solves the Nehari problem. To this end we have to compute "negative" Fourier coefficients of $w_{\mathcal{U}}$. By the definition of $w_{\mathcal{U}}$, coefficient with $\bar{\zeta}^k$ $(k \geq +1)$ equals

$$\langle w_{-k}e_1, e_2\rangle = \langle (\mathcal{U}^*)^{k-1}\rho_1 e_1, \rho_2 e_2\rangle_H = \langle \rho_1 e_1, \mathcal{U}^{k-1}\rho_2 e_2\rangle$$

$$= \left\langle \overrightarrow{\begin{bmatrix}0\\ e_1\end{bmatrix}}, \mathcal{U}^{k-1}\overrightarrow{\begin{bmatrix}\bar{t}e_2\\ 0\end{bmatrix}} \right\rangle_H.$$

But $\begin{bmatrix}\bar{t}e_2\\ 0\end{bmatrix} \in d_V$, hence,

$$\mathcal{U}\overrightarrow{\begin{bmatrix}\bar{t}e_2\\ 0\end{bmatrix}} = V\overrightarrow{\begin{bmatrix}\bar{t}e_2\\ 0\end{bmatrix}} = \overrightarrow{\begin{bmatrix}\bar{t}^2e_2\\ 0\end{bmatrix}}.$$

Since $\begin{bmatrix}\bar{t}^2e_2\\ 0\end{bmatrix}$ again sits in d_V we can iterate action of $\mathcal{U}$. Continuing this process we will obtain

$$\mathcal{U}^{k-1}\overrightarrow{\begin{bmatrix}\bar{t}e_2\\ 0\end{bmatrix}} = \overrightarrow{\begin{bmatrix}\bar{t}^k e_2\\ 0\end{bmatrix}}, \qquad k \geq 1.$$

Finally,

$$\langle w_{-k}e_1, e_2\rangle_{E_2} = \left\langle \overrightarrow{\begin{bmatrix}0\\ e_1\end{bmatrix}}, \overrightarrow{\begin{bmatrix}\bar{t}^k e_2\\ 0\end{bmatrix}} \right\rangle_H = \left\langle \overrightarrow{\begin{bmatrix}0\\ e_1\end{bmatrix}}, \overrightarrow{\begin{bmatrix}\bar{t}^k e_2\\ 0\end{bmatrix}} \right\rangle_{H_0} = D\left(\begin{bmatrix}0\\ e_1\end{bmatrix}, \begin{bmatrix}\bar{t}^k e_2\\ 0\end{bmatrix}\right)$$

$$= \left\langle \begin{bmatrix}\mathbf{1} & \Gamma\\ \Gamma^* & \mathbf{1}\end{bmatrix}\begin{bmatrix}0\\ e_1\end{bmatrix}, \begin{bmatrix}\bar{t}^k e_2\\ 0\end{bmatrix} \right\rangle_X = \langle \Gamma e_1, \bar{t}^k e_2\rangle_{H^2_-(E_2)}$$

$$= \left\langle \sum_{j=-\infty}^{-1} \gamma_j \bar{t}^{|j|}e_1, \bar{t}^k e_2 \right\rangle = \langle \gamma_{-k}e_1, e_2\rangle_{E_2}.$$

Thus, $w_{-k} = \gamma_{-k}$, $k \geq 1$. The theorem follows.

4 Motivation of the Definition of the Scattering Function

Let $\mathcal{U}$ be a unitary mapping on a Hilbert space H. Consider associated dynamics

$$\mathcal{U}h(k) = h(k+1), \qquad k \in \mathbb{Z}$$

$h(k) \in H$, with arbitrary initial condition $h(0)$. We can also consider "frequency" representation of the dynamics instead of the "time" one:

$$\mathcal{U}\widetilde{h}(\zeta) = \frac{\widetilde{h}^+(\zeta) - h(0)}{\zeta} + \overline{\zeta}\big(\widetilde{h}^-(\zeta) + h(0)\big), \qquad |\zeta| < 1,$$

where

$$\widetilde{h}(\zeta) = \widetilde{h}^-(\zeta) + \widetilde{h}^+(\zeta) = \sum_{k=-\infty}^{-1} \overline{\zeta}^{|k|}\mathcal{U}^k h(0) + \sum_{k=0}^{\infty} \zeta^k \mathcal{U}^k h(0).$$

One can easily see that

$$\widetilde{h}(\zeta) = \big[(\mathbf{1} - \zeta\mathcal{U})^{-1} + (\mathbf{1} - \overline{\zeta}\mathcal{U}^*)^{-1} - \mathbf{1}\big]h(0).$$

Fix now two subspaces E_1 and E_2 in H, call them "input" and "output" respectively and try to observe the system through the output subspace E_2:

$$e_2(k) \stackrel{\text{def}}{=} P_{E_2} h(k+1) = P_{E_2}\mathcal{U}h(k).$$

Passing to the spectral representation

$$\widetilde{e_2}(\zeta) = \widetilde{e_2}^-(\zeta) + \widetilde{e_2}^+(\zeta) = \sum_{k=-\infty}^{-1} \overline{\zeta}^{|k|} e_2(k) + \sum_{k=0}^{\infty} \zeta^k e_2(k),$$

we obtain

$$\begin{aligned}
\widetilde{e_2}(\zeta) &= P_{E_2}\mathcal{U}\widetilde{h}(\zeta) \\
&= P_{E_2}\mathcal{U}\big[(\mathbf{1} - \zeta\mathcal{U})^{-1} + (\mathbf{1} - \overline{\zeta}\mathcal{U}^*)^{-1} - \mathbf{1}\big]h(0) \\
&= P_{E_2}\big[\mathcal{U}(\mathbf{1} - \zeta\mathcal{U})^{-1} + \overline{\zeta}(\mathbf{1} - \overline{\zeta}\mathcal{U}^*)^{-1}\big]h(0).
\end{aligned}$$

Pick up $h(0) = e_1 \in E_1$ then $\widetilde{e_2}(\zeta) = w(\zeta)e_1$, where

$$w(\zeta) \stackrel{\text{def}}{=} P_{E_2}\mathcal{U}\big[(\mathbf{1} - \zeta\mathcal{U})^{-1} + (\mathbf{1} - \overline{\zeta}\mathcal{U}^*)^{-1} - \mathbf{1}\big]\Big|_{E_1}$$

$w(\zeta)\colon E_1 \to E_2$. It is called a scattering function of the system. If the subspaces E_1 and E_2 are wandering for $\mathcal{U}$ (i.e. $\mathcal{U}^n E_1 \perp \mathcal{U}^m E_1$ and $\mathcal{U}^n E_2 \perp \mathcal{U}^m E_2$ if

$m \neq n$) then $w(\zeta)$ is a contraction for all ζ, $|\zeta| < 1$ (see Lemma 1, Section 3). Actually we had this definition in section 3 modulo natural inclusions of E_1 and E_2 in H.

Remark: Any minimal extension (i.e., without reducing subspace in $H \ominus H_0$) of the isometry V considered above is equivalent to the one of the type of Theorem 1 (first part) with the same scattering function, i.e., there exists unitary mapping $G: H \to H'$ such that

$$G\big|_{H_0} = id,$$

$$G\mathcal{U} = \mathcal{U}'G,$$

$$\rho_1' = \rho_1,$$

$$\rho_2' = \rho_2.$$

Two last properties are consistent with the first one because ρ_1 and ρ_2 map in H_0.

5 Structure of Minimal Unitary Extensions of the Isometry V

Let H_0 be a Hilbert space, let V be an isometry on H_0 with domain $\vec{d_V}$ and range $\vec{\Delta_V}$. Let $\vec{d_V}^{\perp}$ and $\vec{\Delta_V}^{\perp}$ be orthogonal complements of $\vec{d_V}$ and $\vec{\Delta_V}$ respectively. Let N_1 be a copy of $\vec{d_V}^{\perp}$ and let N_2 be a copy of $\vec{\Delta_V}^{\perp}$. Consider the following unitary colligation:

$$A_0\colon\ H_0 \oplus N_2 \to H_0 \oplus N_1$$

which acts by the rule

$$A_0\big|_{\vec{d_V}} = V$$

$A_0\big|_{\vec{d_V}^{\perp}}$ is a unitary mapping of $\vec{d_V}^{\perp}$ onto N_1

$A_0\big|_{N_2}$ is a unitary mapping of N_2 onto $\vec{\Delta_V}^{\perp}$.

Let A_1 be an arbitrary unitary colligation

$$A_1\colon\ H_1 \oplus N_1 \to H_1 \oplus N_2.$$

A_1 is an arbitrary unitary mapping, N_1 and N_1 are fixed and they are the same as in the definition of A_0.

Define unitary operator $\mathcal{U}$: $H_0 \oplus H_1 \to H_0 \oplus H_1$ as follows

$$\mathcal{U}\begin{bmatrix} h_0 \\ h_1 \end{bmatrix} \stackrel{\text{def}}{=} \begin{bmatrix} h'_0 \\ h'_1 \end{bmatrix}$$

where $\begin{bmatrix} h'_0 \\ h'_1 \end{bmatrix}$ is determined from the following system

$$A_0\begin{bmatrix} h_0 \\ n_2 \end{bmatrix} = \begin{bmatrix} h'_0 \\ n_1 \end{bmatrix} \tag{5.1}$$

$$A_1\begin{bmatrix} h_1 \\ n_1 \end{bmatrix} = \begin{bmatrix} h'_1 \\ n_2 \end{bmatrix}. \tag{5.2}$$

This operator $\mathcal{U}$ is said to be a *coupling* of the untiary colligations A_0 and A_1. The coupling means that n_1 and n_2 that come into the first and into the second equations are the same. $\begin{vmatrix} h'_0 \\ h'_1 \end{vmatrix}$ is determined *uniqualy* from the system (5.1)–(5.2) for arbitrary $\begin{vmatrix} h_0 \\ h_1 \end{vmatrix}$ because $P_{N_1} A_0\big|_{N_2} = 0$ (see the definition of A_0).

Lemma 2 *The coupling $\mathcal{U}$ of the unitary colligation A_0: $H_0 \oplus N_2 \to H_0 \oplus N_1$ with an arbitrary unitary colligation A_1: $H_1 \oplus N_1 \to H_1 \oplus N_2$ is a unitary extension of V. Arbitrary unitary extension of V is of this form.*

Proof of Lemma 2:
(1) Let $\mathcal{U}$ be the coupling and let $h_0 \in \overrightarrow{d_V}$, $h_1 = 0$. Take $h'_0 = Vh_0$, $h'_1 = 0$, $n_1 = 0$, $n_2 = 0$ and see that (5.1)–(5.2) are met. Hence

$$\mathcal{U}\begin{bmatrix} h_0 \\ 0 \end{bmatrix} = \begin{bmatrix} Vh_0 \\ 0 \end{bmatrix}$$

when $h_0 \in \overrightarrow{d_V}$, i.e., $\mathcal{U}$ is an extension of V.
(2) Let $H \supseteq H_0$ and let $\mathcal{U}$ be a unitary extension of V on H.

Let $H_1 = H \ominus H_0$, then $H_1 \oplus \overrightarrow{d_V}^{\perp} = H \ominus \overrightarrow{d_V}$ and $H_1 \oplus \overrightarrow{\Delta_V}^{\perp} = H \ominus \overrightarrow{\Delta_V}$. Because $\mathcal{U}$ is an extension of V it maps $\overrightarrow{d_V}$ onto $\overrightarrow{\Delta_V}$. Because $\mathcal{U}$ is a unitary operator it has to map $H \ominus \overrightarrow{d_V}$ onto $H \ominus \overrightarrow{\Delta_V}$ unitarily. Thus, one can consider a unitary colligation

$$\widehat{A}_1 \stackrel{\text{def}}{=} \mathcal{U}\Big|_{H \ominus \overrightarrow{d_V}}, \qquad \widehat{A}_1\colon H_1 \oplus \overrightarrow{d_V}^{\perp} \longrightarrow H_1 \oplus \overrightarrow{\Delta_V}^{\perp}.$$

Define unitary colligation

$$A_1\colon H_1 \oplus N_1 \to H_1 \oplus N_2$$

as follows

$$P_{H_1} A_1 \mid H_1 = P_{H_1} \widehat{A} \mid H_1$$

$$P_{H_1} A_1 \mid N_1 = P_{H_1} \widehat{A}_1 A_0^* \mid N_1$$

$$P_{N_2} A_1 \mid H_1 = A_0^* P_{\overrightarrow{\Delta}_V^{\perp}} \widehat{A}_1 \mid H_1$$

$$P_{N_2} A_1 \mid N_1 = A_0^* P_{\overrightarrow{\Delta}_V^{\perp}} \widehat{A}_1 A_0^* \mid N_1.$$

These formulas mean that A_1 is obtained from $\widehat{A}_1$ by means of identifying $\overrightarrow{d}_V^{\perp}$ with N_1 and $\overrightarrow{\Delta}_V^{\perp}$ with N_2 in the same way that A_0 does it!
Let $\mathcal{U}\begin{bmatrix} h_0 \\ h_1 \end{bmatrix} = \begin{bmatrix} h_0' \\ h_1' \end{bmatrix}$. Pick up

$$n_1 = A_0 P_{\overrightarrow{\Delta}_V^{\perp}} h_0$$

and

$$n_2 = A_0^* P_{\overrightarrow{\Delta}_V^{\perp}} h_0'.$$

Verify that the system (5.1)–(5.2) holds true for these vectors. The following diagram allows one to see that without computation.

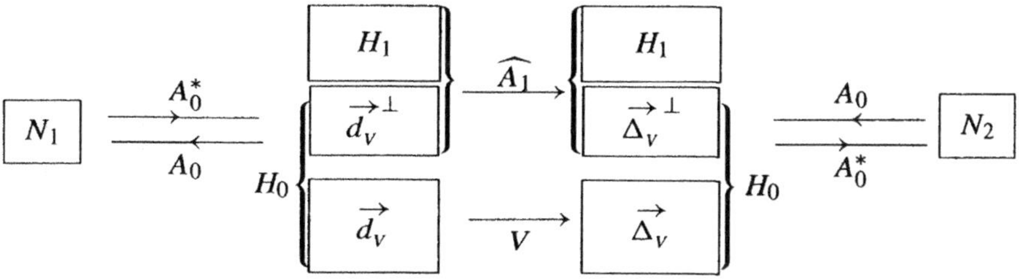

$$H_0 \oplus H_1 = H.$$

This means that $\mathcal{U}$ is the coupling of A_0 and A_1. Lemma 2 follows.

Remark: Unitary extension $\mathcal{U}$ is minimal if and only if A_1 is simple (i.e., has no nontrivial reducing subspaces in H_1).

6 Scattering Function of the Colligation A_0 (motivation)

Let A_0: $H_0 \oplus N_2 \to H_0 \oplus N_1$ be the unitary colligation defined above (A_0 comes from the isometry V). Consider the following dynamcis

(6.1) $$A_0 \begin{bmatrix} h_0(k) \\ n_2(k) \end{bmatrix} = \begin{bmatrix} h_0(k+1) \\ n_1(k) \end{bmatrix}, \qquad k \in \mathbb{Z}$$

where $h_0(k) \in H_0, n_2(k) \in N_2, n_1(k) \in N_1$; $h_0(0)$ is given initial state, $n_2(k)$ are given for $k \geq 0$, $n_1(k)$ are given for $k \leq -1$. We can re-express this dynamics in spectral terms:

$$A_0 \begin{bmatrix} \widetilde{h}_0(\zeta) \\ \widetilde{n}_2(\zeta) \end{bmatrix} = \begin{bmatrix} \frac{\widetilde{h}_0^+(\zeta)-h_0(0)}{\zeta} + \overline{\zeta}\left(\widetilde{h}_0^-(\zeta) + h_0(0)\right) \\ \widetilde{n}_1(\zeta) \end{bmatrix} \tag{6.2}$$

$|\zeta| < 1$, where

$$\widetilde{h}_0(\zeta) = \widetilde{h}_0^-(\zeta) + \widetilde{h}_0^+(\zeta) = \sum_{k=-\infty}^{-1} h_0(k)\overline{\zeta}^{|k|} + \sum_{k=0}^{\infty} h_0(k)\zeta^k$$

$$\widetilde{n}_1(\zeta) = \widetilde{n}_1^-(\zeta) + \widetilde{n}_1^+(\zeta) = \sum_{k=-\infty}^{-1} n_1(k)\overline{\zeta}^{|k|} + \sum_{k=0}^{\infty} n_1(k)\zeta^k$$

$$\widetilde{n}_2(\zeta) = \widetilde{n}_2^-(\zeta) + \widetilde{n}_2^+(\zeta) = \sum_{k=-\infty}^{-1} n_2(k)\overline{\zeta}^{|k|} + \sum_{k=0}^{\infty} n_2(k)\zeta^k$$

$h_0(0)$, $\widetilde{n}_2^+(\zeta)$ and $\widetilde{n}_1^-(\zeta)$ are given.

We pick up again $h_0(0) = \rho_1 e_1 = \overrightarrow{\begin{bmatrix} 0 \\ e_1 \end{bmatrix}}$ like in section 4. Since $\widetilde{n}_1^-(\zeta)$ is arbitrary we can pick it up to be zero, $\widetilde{n}_1^-(\zeta) = 0$, i.e., $n_1(k) = 0$, $k \leq -1$. Since $h_0(0) = \rho_1 e_1 = \overrightarrow{\begin{bmatrix} 0 \\ e_1 \end{bmatrix}} \in \overrightarrow{\Delta_V}$,

$$A_0^* \begin{bmatrix} \overrightarrow{\begin{bmatrix} 0 \\ e_1 \end{bmatrix}} \\ 0_{N_1} \end{bmatrix} = \begin{bmatrix} V^* \overrightarrow{\begin{bmatrix} 0 \\ e_1 \end{bmatrix}} \\ 0_{N_2} \end{bmatrix} = \begin{bmatrix} \overrightarrow{\begin{bmatrix} 0 \\ te_1 \end{bmatrix}} \\ 0_{N_2} \end{bmatrix} \in \overrightarrow{\Delta_V}$$

again, i.e., $h_0(-1) \in \overrightarrow{\Delta_V}$ and $n_2(-1) = 0$. Iterating this we will see that

$$h_0(-k) = \overrightarrow{\begin{bmatrix} 0 \\ t^k e_1 \end{bmatrix}} \in \overrightarrow{\Delta_V}$$

and $n_2(-k) = 0$, for $k \geq +1$. In particular, this means that $\widetilde{n}_2^-(\zeta) = 0$. Thus, we took $h_0(0) = \rho_1 e_1$, $\widetilde{n}_2^+(\zeta)$ arbitrary, and $\widetilde{n}_1^-(\zeta) = 0$. Then the dynamics forces $\widetilde{n}_2^-(\zeta) = 0$. So, $\widetilde{n}_2(\zeta) = \widetilde{n}_2^+(\zeta)$, $\widetilde{n}_1(\zeta) = \widetilde{n}_1^+(\zeta)$ in this case.

We define the output of the system

$$\widetilde{e}_2(\zeta) = \rho_2^* A_0 \begin{bmatrix} \widetilde{h}_0(\zeta) \\ \widetilde{n}_2^+(\zeta) \end{bmatrix} \tag{6.4}$$

and want to express

$$\begin{bmatrix} \widetilde{n}_1^+(\zeta) \\ \widetilde{e}_2(\zeta) \end{bmatrix} \quad \text{through the} \quad \begin{bmatrix} \widetilde{n}_2^+(\zeta) \\ e_1 \end{bmatrix}.$$

We introduce notations

$$\begin{bmatrix} \widetilde{n}_1^+(\zeta) \\ \widetilde{e}_2(\zeta) \end{bmatrix} \stackrel{\text{def}}{=} S(\zeta) \begin{bmatrix} \widetilde{n}_2^+(\zeta) \\ e_1 \end{bmatrix} \stackrel{\text{def}}{=} \begin{bmatrix} s(\zeta) & s_1(\zeta) \\ s_2(\zeta) & s_0(\zeta) \end{bmatrix} \begin{bmatrix} \widetilde{n}_2^+(\zeta) \\ e_1 \end{bmatrix} \tag{6.5}$$

and call $S(\zeta)$ a scattering function of the colligation A_0.

7 Scattering Function of the Colligation A_0 (formulas)

To obtain formulas for entries of the scattering matrix $S(\zeta)$ we will break operator A_0 into four blocks according to the decomposition of spaces:

$$A_0 = \begin{bmatrix} A^{11} & A^{12} \\ A^{21} & A^{22} \end{bmatrix} : H_0 \oplus N_2 \to H_0 \oplus N_1,$$

i.e.,

$$A^{11}\colon H_0 \to H_0, \qquad A^{12}\colon N_2 \to H_0,$$
$$A^{21}\colon H_0 \to N_1, \qquad A^{22}\colon N_2 \to N_1.$$

Then we can rewrite (6.2) as follows

$$\begin{aligned} & A^{11}\widetilde{h}_0(\zeta) + A^{12}\widetilde{n}_2^+(\zeta) = \frac{\widetilde{h}_0^+(\zeta) - h_0(0)}{\zeta} + \overline{\zeta}\big(\widetilde{h}_0^- + h_0(0)\big) \\ & A^{21}\widetilde{h}_0(\zeta) + A^{22}\widetilde{n}_2^+(\zeta) = \widetilde{n}_1^+(\zeta) \end{aligned} \tag{7.1}$$

or (separating "+" and "−" parts)

$$\begin{aligned} &\text{(i)} && A^{11}\widetilde{h}_0^-(\zeta) = \overline{\zeta}\big(\widetilde{h}_0^-(\zeta) + h_0(0)\big) \\ &\text{(ii)} && A^{11}\widetilde{h}_0^+(\zeta) + A^{12}\widetilde{n}_2^+(\zeta) = \frac{\widetilde{h}_0^+(\zeta) - h_0(0)}{\zeta} \\ &\text{(iii)} && A^{21}\widetilde{h}_0^+(\zeta) + A^{22}\widetilde{n}_2^+(\zeta) = \widetilde{n}_1^+(\zeta) \\ &\text{(iv)} && A^{21}\widetilde{h}_0^-(\zeta) = 0. \end{aligned} \tag{7.2}$$

Relation (iv) of (7.2) was actually verified earlier: (6.3) implies $\widetilde{h}_0^-(\zeta) \in \overrightarrow{d_V}$, hence $A_0\widetilde{h}_0^-(\zeta) \in \overrightarrow{\Delta_V} \subseteq H_0$, therefore $A^{21}\widetilde{h}_0^-(\zeta) = P_{N_1} A_0\widetilde{h}_0^-(\zeta) = 0$.

Relation (i) of (7.2) can be re-expressed (if we multiply (7.1) by A_0^*) as

$$\widetilde{h}_0^-(\zeta) = A^{11*}\overline{\zeta}\big(\widetilde{h}_0^-(\zeta) + h_0(0)\big).$$

Hence

$$(7.3)\qquad \begin{aligned}\widetilde{h}_0^-(\zeta) &= \overline{\zeta}(\mathbf{1}_{H_0} - A^{11*}\overline{\zeta})^{-1}A^{11*}h_0(0) = \overline{\zeta}(\mathbf{1}_{H_0} - A^{11*}\overline{\zeta})^{-1}A^{11*}\rho_1 e_1\\ &= (\mathbf{1}_{H_0} - A^{11*}\overline{\zeta})^{-1}\rho_1 e_1 - \rho_1 e_1.\end{aligned}$$

We will express now $\widetilde{h}_0^+(\zeta)$ and $\widetilde{n}_1^+(\zeta)$ from (ii) and (iii) of (7.2). From (ii) one has

$$(\mathbf{1}_{H_0} - \zeta A^{11})\widetilde{h}_0^+(\zeta) = \zeta A^{12}\widetilde{n}_2^+(\zeta) + h_0(0),$$

or

$$(7.4)\qquad \begin{aligned}\widetilde{h}_0^+(\zeta) &= (\mathbf{1}_{H_0} - \zeta A^{11})^{-1}\zeta A^{12}\widetilde{n}_2^+(\zeta) + (\mathbf{1}_{H_0} - \zeta A^{11})^{-1}h_0(0)\\ &= \zeta(\mathbf{1}_{H_0} - \zeta A^{11})^{-1}A^{12}\widetilde{n}_2^+(\zeta) + (\mathbf{1}_{H_0} - \zeta A^{11})^{-1}\rho_1 e_1,\end{aligned}$$

and from (iii) one has

$$(7.5)\qquad \begin{aligned}\widetilde{n}_1^+(\zeta) &= A^{22}\widetilde{n}_2^+(\zeta) + A^{21}\widetilde{h}_0^+(\zeta)\\ &\qquad \text{(after substituting (7.4))}\\ &= A^{22}\widetilde{n}_2^+(\zeta) + \zeta A^{21}(\mathbf{1}_{H_0} - \zeta A^{11})^{-1}A^{12}\widetilde{n}_2^+(\zeta)\\ &\quad + A^{21}(\mathbf{1}_{H_0} - \zeta A^{11})^{-1}\rho_1 e_1\\ &= \big(A^{22} + \zeta A^{21}(\mathbf{1}_{H_0} - \zeta A^{11})^{-1}A^{12}\big)\widetilde{n}_2^+(\zeta)\\ &\quad + A^{21}(\mathbf{1}_{H_0} - \zeta A^{11})^{-1}\rho_1 e_1.\end{aligned}$$

Combining (7.3) and (7.4) we obtain

$$(7.6)\qquad \begin{aligned}\widetilde{h}_0(\zeta) &= \zeta(\mathbf{1}_{H_0} - \zeta A^{11})^{-1}A^{12}\widetilde{n}_2^+(\zeta)\\ &\quad + \big[(\mathbf{1}_{H_0} - \zeta A^{11})^{-1} + (\mathbf{1}_{H_0} - \overline{\zeta}A^{11*})^{-1} - \mathbf{1}_{H_0}\big]\rho_1 e_1.\end{aligned}$$

Thus we can read the entries of $S(\zeta)$ from (7.5) and (7.6) (see definitions (6.4) and (6.5))

$$S(\zeta) = \begin{bmatrix} s(\zeta) & s_1(\zeta)\\ s_2(\zeta) & s_0(\zeta)\end{bmatrix} : \begin{bmatrix} N_2\\ E_1\end{bmatrix} \to \begin{bmatrix} N_1\\ E_2\end{bmatrix}$$

where

$$(7.7)\qquad \begin{aligned} s(\zeta) &= A^{22} + \zeta A^{21}(\mathbf{1}_{H_0} - \zeta A^{11})^{-1}A^{12}\\ s_1(\zeta) &= A^{21}(\mathbf{1}_{H_0} - \zeta A^{11})^{-1}\rho_1\\ s_2(\zeta) &= \rho_2^* A_0 \begin{bmatrix} \zeta(\mathbf{1}_{H_0} - \zeta A^{11})^{-1}A^{12}\\ \mathbf{1}_{N_2}\end{bmatrix}\\ s_0(\zeta) &= \rho_2^* A_0 \begin{bmatrix} ((\mathbf{1}_{H_0} - \zeta A^{11})^{-1} + (\mathbf{1}_{H_0} - \overline{\zeta}A^{11*})^{-1} - \mathbf{1}_{H_0})\rho_1\\ 0_{E_1\to N_2}\end{bmatrix}\end{aligned}$$

and

$$\rho_1 \colon E_1 \to H_0, \qquad \rho_1 e_1 = \overrightarrow{\begin{bmatrix} 0 \\ e_1 \end{bmatrix}}$$

$$\rho_2 \colon E_2 \to H_0, \qquad \rho_2 e_2 = \overrightarrow{\begin{bmatrix} \bar{t} e_2 \\ 0 \end{bmatrix}}.$$

Entries of the matrix S can also be re-expressed as follows

$$\begin{aligned} s(\zeta) &= P_{N_1} A_0 (\mathbf{1}_{H_0 \oplus N_2} - \zeta P_{H_0} A_0)^{-1} \mid N_2 \\ s_1(\zeta) &= P_{N_1} A_0 (\mathbf{1}_{H_0 \oplus N_2} - \zeta P_{H_0} A_0)^{-1} \rho_1 \\ s_2(\zeta) &= \rho_2^* A_0 (\mathbf{1}_{H_0 \oplus N_2} - \zeta P_{H_0} A_0)^{-1} \mid N_2 \\ s_0(\zeta) &= \rho_2^* A_0 \big((\mathbf{1}_{H_0 \oplus N_2} - \zeta P_{H_0} A_0)^{-1} \\ &\quad + (\mathbf{1}_{H_0 \oplus N_2} - \bar{\zeta} A_0^* P_{H_0})^{-1} - \mathbf{1}_{H_0 \oplus N_2} \big) \rho_1 \end{aligned} \tag{7.8}$$

Remark: We would like to emphasize some points of these formulas:

(1) Function $S(\zeta)$ depends on the colligation A_0 only and, hence, on the isometry V only.

(2) Entries s, s_1 and s_2 are analytic inside the unit disk $\mathbb{D}$.

(3) $s(0) = 0$, (because $A_0 N_2 = \Delta_V^\perp \subseteq H_0$, hence $P_{N_1} A_0 \mid N_2 = 0$).

Putting together all the entries of S we can rewrite (7.8) as

$$\begin{aligned} S(\zeta) &= \begin{bmatrix} s(\zeta) & s_1(\zeta) \\ s_2(\zeta) & s_0(\zeta) \end{bmatrix} \\ &= \begin{bmatrix} P_{N_1} \\ \rho_2^* \end{bmatrix} \big(A_0 (\mathbf{1}_{H_0 \oplus N_2} - \zeta P_{H_0} A_0)^{-1} \\ &\quad + \bar{\zeta} P_{H_0} (\mathbf{1}_{H_0 \oplus N_2} - \bar{\zeta} A_0^* P_{H_0})^{-1} \big) [\ \mid N_2, \quad \rho_1\ ], \end{aligned}$$

where ρ_1 and ρ_2 are the same as above, P_{N_1} is the orthogonal projection on N_1 in $H_0 \oplus N_1$, $\mid N_2$ is restriction on N_2 in $H_0 \oplus N_2$. The operator in parentheses is the same one as in the formula (7.8) for $s_0(\zeta)$. Obviously the second (antianalytic) term cancels for entries $s(\zeta)$, $s_1(\zeta)$ and $s_2(\zeta)$ because of orthogonality of N_1 to H_0 and of N_2 to H_0.

Lemma 3 *$S(\zeta)$ is a contractive matrix-function of $\begin{bmatrix} N_2 \\ E_1 \end{bmatrix}$ to $\begin{bmatrix} N_1 \\ E_2 \end{bmatrix}$ if inner products on these cartesian products are defined component-wise.*

Proof: Similar to the proof of Lemma 1. □

8 Parametrization of the Scattering Functions of the Unitary Extensions of the Isometry V (i.e., solutions of the Nehari problem)

Theorem 4 *The set of scattering functions (see Secs. 3, 4) of unitary extensions of isometry V (see Sec. 2 for definition) is described as follows:*

$$w(\zeta) = s_0(\zeta) + s_2(\zeta)\omega(\zeta)\big(\mathbf{1}_{N_1} - s(\zeta)\omega(\zeta)\big)^{-1} s_1(\zeta) \qquad (|\zeta| < 1) \tag{8.1}$$

where s, s_1, s_2, s_0 are entries of the scattering function S of the colligation A_0 associated to V (see Secs. 5, 6, 7) and $\omega(\zeta)$ runs through the class of all contractive-valued analytic in the unit disc $\mathbb{D}$ functions, $\omega(\zeta)$: $N_1 \to N_2$.

Proof: To "compute" scattering function $w(\zeta)$ of a unitary extension $\mathcal{U}$ on $H_0 \oplus H_1$ we will use representation of $\mathcal{U}$ as the coupling of colligations A_0 and A_1 (see Sec. 5). By the definition (see Sec. 4)

$$w(\zeta) = \rho_2^* \mathcal{U}\left(\sum_{k=-\infty}^{-1} \bar{\zeta}^{|k|}\mathcal{U}^k + \sum_{k=0}^{\infty} \zeta^k \mathcal{U}^k\right)\begin{bmatrix} h_0(0) \\ h_1(0) \end{bmatrix}, \tag{8.2}$$

where $h_0(0) = \rho_1 e_1$, $h_1(0) = 0$.

Let

$$\begin{bmatrix} h_0(k) \\ h_1(k) \end{bmatrix} = \mathcal{U}^k \begin{bmatrix} h_0(0) \\ h_1(0) \end{bmatrix}.$$

Then $\begin{bmatrix} h_0(k) \\ h_1(k) \end{bmatrix}$ is determined from the system (coupling representation of $\mathcal{U}$):

$$A_0 \begin{bmatrix} h_0(k) \\ n_2(k) \end{bmatrix} = \begin{bmatrix} h_0(k+1) \\ n_1(k) \end{bmatrix} \tag{8.3}$$

$$A_1 \begin{bmatrix} h_1(k) \\ n_1(k) \end{bmatrix} = \begin{bmatrix} h_1(k+1) \\ n_2(k) \end{bmatrix}. \tag{8.4}$$

Now all the objects in this system are determined uniquely by $\begin{bmatrix} h_0(0) \\ h_1(0) \end{bmatrix}$. If $h_0(0) = \rho_1 e_1$ and $h_1(0) = 0$, then (as we have seen in Sec. 6) $n_1(k) = 0$ $(k \le -1)$ and $n_2(k) = 0$ $(k \le -1)$ fit the dynamics of A_0 (8.3). Obviously they fit the dynamics of A_1 (8.4) together with $h_1(k) = 0$ $(k \le 0)$. Since everything is defined uniquely by $h_0(0)$ and $h_1(0)$ we can conclude that if $h_0(0) = \rho_1 e_1$ and $h_1(0) = 0$, then coupling conditions (8.3), (8.4) force $n_1(k) = 0$, $n_2(k) = 0$, $h_1(k) = 0$ for $k \le -1$.

Emphasize here, that it is not our choice now like it was in Sec. 6 when we considered dynamics of A_0 separately and chose $n_1(k) = 0$ for $k \le -1$ but we were able to choose $n_1(k)$, $k \le -1$ in arbitrary way. Now we can not!

We can rewrite coupling dynamics in spectral representation ($|\zeta| < 1$) using these observations

$$(8.5)\qquad A_0\begin{bmatrix} \widetilde{h}_0(\zeta) \\ \widetilde{n}_2^+(\zeta) \end{bmatrix} = \begin{bmatrix} \frac{\widetilde{h}_0^+(\zeta)-h_0(0)}{\zeta} + \bar{\zeta}\big(\widetilde{h}_0^-(\zeta) + h_0(0)\big) \\ \widetilde{n}_1^+(\zeta) \end{bmatrix}$$

$$(8.6)\qquad A_1\begin{bmatrix} \widetilde{h}_1^+(\zeta) \\ \widetilde{n}_1^+(\zeta) \end{bmatrix} = \begin{bmatrix} \frac{\widetilde{h}_1^+(\zeta)-h_1(0)}{\zeta} \\ \widetilde{n}_2^+(\zeta) \end{bmatrix}.$$

Actually, $h_1(0) = 0$ by our asssumption. Then the coupling $\mathcal{U}$ acts as follows

$$\mathcal{U}\begin{bmatrix} \widetilde{h}_0(\zeta) \\ \widetilde{h}_1^+(\zeta) \end{bmatrix} = \begin{bmatrix} \frac{\widetilde{h}_0^+(\zeta)-h_0(0)}{\zeta} + \bar{\zeta}\big(\widetilde{h}_0^-(\zeta) + h_0(0)\big) \\ \frac{\widetilde{h}_1^+(\zeta)-h_1(0)}{\zeta} \end{bmatrix}.$$

Define $\widetilde{e}_2(\zeta) = \rho_2^* A_0\begin{bmatrix} \widetilde{h}_0(\zeta) \\ \widetilde{n}_2^+(\zeta) \end{bmatrix}$ like in (6.4), then (see Sec. 6, formula (6.5))

$$(8.7)\qquad \begin{bmatrix} \widetilde{n}_1^+(\zeta) \\ \widetilde{e}_2(\zeta) \end{bmatrix} = \begin{bmatrix} s(\zeta) & s_1(\zeta) \\ s_2(\zeta) & s_0(\zeta) \end{bmatrix}\begin{bmatrix} \widetilde{n}_2^+(\zeta) \\ e_1 \end{bmatrix}.$$

On the other hand, from the dynamics A_1 we obtain

$$(8.8)\qquad \widetilde{n}_2^+(\zeta) = \omega(\zeta)\widetilde{n}_1^+(\zeta),$$

where

$$\omega(\zeta) = P_{N_2} A_1(\mathbf{1}_{H_1\oplus N_1} - \zeta P_{H_1} A_1)^{-1} \mid N_1$$

is the characteristic function of the unitary colligation A_1.

Since ρ_2 maps into H_0 we have

$$\begin{aligned}\widetilde{e}_2(\zeta) &= \rho_2^* A_0\begin{bmatrix} \widetilde{h}_0(\zeta) \\ \widetilde{n}_2^+(\zeta) \end{bmatrix} = \rho_2^*\Big(\frac{\widetilde{h}_0^+(\zeta) - h_0(0)}{\zeta} + \bar{\zeta}(\widetilde{h}_0^-(\zeta) + h_0(0))\Big) \\ &= \rho_2^*\mathcal{U}\begin{bmatrix} \widetilde{h}_0(\zeta) \\ \widetilde{h}_1^+(\zeta) \end{bmatrix}.\end{aligned}$$

Since $h_0(0) = \rho_1 e_1$ and $h_1(0) = 0$, the latter expression for $\widetilde{e}_2(\zeta)$ means that

$$(8.9)\qquad \widetilde{e}_2(\zeta) = w(\zeta)e_1$$

according to (8.2). Then (8.7), (8.8) and (8.9) imply

$$(8.10)\qquad \begin{bmatrix} \widetilde{n}_1^+(\zeta) \\ w(\zeta)e_1 \end{bmatrix} = \begin{bmatrix} s(\zeta) & s_1(\zeta) \\ s_2(\zeta) & s_0(\zeta) \end{bmatrix}\begin{bmatrix} \omega(\zeta)\widetilde{n}_1^+(\zeta) \\ e_1 \end{bmatrix}$$

from where we obtain

$$\widetilde{n}_1^+(\zeta) = \left(\mathbf{1}_{N_1} - s(\zeta)\omega(\zeta)\right)^{-1} s_1(\zeta)e_1 \tag{8.11}$$

and

$$w(\zeta)e_1 = \left[s_0(\zeta) + s_2(\zeta)\omega(\zeta)\left(\mathbf{1}_{N_1} - s(\zeta)\omega(\zeta)\right)^{-1} s_1(\zeta)\right]e_1. \tag{8.12}$$

Since A_1 runs through the class of all unitary colligations with input and output spaces N_1 and N_2 respectively when $\mathcal{U}$ runs through the class of all unitary extensions of V, then $\omega(\zeta)$ runs through the class of all contractive-valued, analytic in the unit disc $\mathbb{D}$ functions, $\omega(\zeta)$: $N_1 \to N_2$.

Hence, the formula (8.12) gives parametrization of the scattering functions of all unitary extensions of V. By Theorem 1 this gives a parametrization of the solution set of the Nehari problem.

As we have seen in Sec. 7, colligation A_0 depends on the data of the problem only (via isometry V), hence $S(\zeta)$ depends on the data only. Subspaces N_1 and N_2 also are defined uniquely by the data of the problem, and $\omega(\zeta)$ is an arbitrary parameter.

9 Some Properties of the Matrix-function $S(\zeta)$ (without proofs)

Theorem 5 $s(0) = 0$; *$s_1(\zeta)$ is an outer function $s_1(\zeta)$: $E_1 \to N_1$; $s_2(\zeta)$ is $*$-outer function, $s_2(\zeta)$: $N_2 \to E_2$.*

Corollary 6 *Formula* (8.1) *gives a one to one correspondence between parameters $\omega(\zeta)$ and solutions $w(\zeta)$ of the Nehari problem.*

Theorem 7

(1) *$s_1(\zeta)$ is the maximal solution of the following factorization problem*

$$\mathbf{1} - s_0^* s_0 \geq r_1^* r_1, \text{ a.e. on } \mathbb{T}, \; r_1 \in H_+^\infty.$$

I.e.,

$$s_1(\zeta)^* s_1(\zeta) \geq r_1(\zeta)^* r_1(\zeta), \; |\zeta| \leq 1$$

for any solution r_1 of the factorization problem.

(2) *$s_2(\zeta)$ is the maximal solution of the following factorization problem*

$$\mathbf{1} - s_0 s_0^* \geq r_2 r_2^*, \text{ a.e. on } \mathbb{T}, \; r_2 \in H_+^\infty.$$

I.e.,

$$s_2(\zeta) s_2(\zeta)^* \geq r_2(\zeta) r_2(\zeta)^*, \; |\zeta| \leq 1$$

for any solution r_2 of the factorization problem.

(3) *The following factorization problems*

$$\mathbf{1} - S^*S \geq \Phi^*\Phi, \ a.e. \ on \ \mathbb{T}, \ \Phi \in H_+^\infty$$

and

$$\mathbf{1} - SS^* \geq \Psi\Psi^*, \ a.e. \ on \ \mathbb{T}, \ \Psi \in H_+^\infty$$

have no nontrivial solutions.

Theorem 8 *For arbitrary parameter* $\omega(\zeta)\colon N_1 \to N_2$

$$\mathring{a}_\omega(\zeta) - \begin{bmatrix} 0 & \omega(0) \\ -\omega(0)^* & 0 \end{bmatrix} - \int_{\mathbb{T}} \frac{t+\zeta}{t-\zeta} \begin{bmatrix} \psi^* & \omega\varphi \\ \omega^*\psi^* & \varphi \end{bmatrix} \begin{bmatrix} \mathbf{1}_{E_2} & w \\ w^* & \mathbf{1}_{E_1} \end{bmatrix}^{[-1]} \begin{bmatrix} \psi & \psi\omega \\ \varphi^*\omega^* & \varphi^* \end{bmatrix} m(dt)$$
$$\equiv 0, \quad |\zeta| < 1$$

where

$$\mathring{a}_\omega(\zeta) \stackrel{\text{def}}{=} \frac{\mathbf{1}_{N_2\oplus N_1} + \begin{bmatrix} 0 & \omega \\ s & 0 \end{bmatrix}}{\mathbf{1}_{N_2\oplus N_1} - \begin{bmatrix} 0 & \omega \\ s & 0 \end{bmatrix}} = \begin{bmatrix} \frac{\mathbf{1}_{N_2}+\omega s}{\mathbf{1}_{N_2}-\omega s} & 2(\mathbf{1}_{N_2} - \omega s)^{-1}\omega \\ 2(\mathbf{1}_{N_1} - s\omega)^{-1}s & \frac{\mathbf{1}_{N_1}+s\omega}{\mathbf{1}_{N_1}-s\omega} \end{bmatrix},$$

$$\varphi = (\mathbf{1}_{N_1} - s\omega)^{-1}s_1, \qquad \psi = s_2(\mathbf{1}_{N_2} - \omega s)^{-1}$$

w is the solution correspondent to the parameter ω

$$w = s_0 + s_2\omega(\mathbf{1}_{N_1} - s\omega)^{-1}s_1.$$

Corollary 9 *The Herglotz measure correspondent to the function* $\mathring{a}_\omega(\zeta)$ *is absolutely continuous.*

Corollary 10

$$\begin{bmatrix} \mathbf{1}_{N_2} & s^* \\ s & \mathbf{1}_{N_1} \end{bmatrix} = \begin{bmatrix} s_2^* & 0 \\ 0 & s_1 \end{bmatrix} \begin{bmatrix} \mathbf{1}_{E_2} & s_0 \\ s_0^* & \mathbf{1}_{E_1} \end{bmatrix}^{[-1]} \begin{bmatrix} s_2 & 0 \\ 0 & s_1^* \end{bmatrix}$$

a.e. on the unit circle $\mathbb{T}$.

In the case of the maximal rank of indeterminacy, i.e., if $s_1(t)$ and $s_2(t)$ are invertible, a.e. on $\mathbb{T}$, this property is equivalent to the property of S to be inner

$$\mathbf{1} - SS^* = 0 \quad \text{a.e. on } \mathbb{T}$$
$$\mathbf{1} - S^*S = 0 \quad \text{a.e. on } \mathbb{T}$$

In the general case S need not be inner, but the property of Corollary 10 holds true.

Corollary 11 *If dim $E_1 < \infty$ and dim $E_1 < \infty$ then the property of Corollary 9 is equivalent to each one of the following three*

$$\operatorname{rank}\begin{bmatrix} \mathbf{1}_{N_1 \oplus E_2} & S \\ S^* & \mathbf{1}_{N_2 \oplus E_1} \end{bmatrix} = \operatorname{rank}\begin{bmatrix} \mathbf{1}_{E_2} & s_0 \\ s_0^* & \mathbf{1}_{E_1} \end{bmatrix}$$

$$\operatorname{rank}(\mathbf{1} - SS^*) = \operatorname{rank}(\mathbf{1} - s_0 s_0^*) - \dim N_2$$

$$\operatorname{rank}(\mathbf{1} - S^* S) = \operatorname{rank}(\mathbf{1} - s_0^* s_0) - \dim N_1$$

a.e. on $\mathbb{T}$.

Acknowledgements

Author would like to thank Professors C. Foias and H. Bercovici, Mathematics Department, University of Indiana, Bloomington, Indiana for support in participating the IWOTA-96 and Sara Fleigelmann, Secretary, Mathematics Department, Weizmann Institute of Science, Rehovot, Israel for the perfect typing of the manuscript.

References

[AA] V.M. Adamjan and Arov D.Z., On unitary couplings of semiunitary operators, *Matem. Issled.* **1:2** (1966) 3–64 [in Russian]. English Translation: Amer. Math. Soc. Transl., Ser. 2 **95** (1970), 75–129.

[AAK1] V.M. Adamjan, D.Z. Arov and M.G. Krein, Infinite Hankel matrices and generalized problems of Caratheodory–Fejer and I. Schur, *Funktsional. Anal. i Prilozhen* **2:4** (1968) 1–17 [in Russian]. English transl.: Functional Anal. Appl. 2 (1968), 269–281.

[AAK2] V.M. Adamjan, D.Z. Arov and Krein M.G., Infinite block Hankel matrices and related continuation problems, *Izv. Akad. Nauk Armyan. SSR, Ser. Matem.* **2–3** (1971) 87–112 [in Russian]. English transl.: Amer. Math. Soc. Transl., Ser. 2 **111** (1978), 133–156.

[Ad] V.M. Adamjan, Non-degenerate unitary couplings of semiunitary operators, *Funkt. Anal. Prilozhen.* **7:4** (1973), 1–16 [in Russian]. English translation: Functional Analysis and Applications **7** (1973), 255–267.

[AG] D.Z. Arov and L.Z. Grossman, *Scattering Matrices in extension theory of isometric operators*, Dokl. Akad. Nauk SSSR **270** (1983), 17–20 [in Russian]. English Translation: Soviet Math. Dokladi, **27** (1983), 518–522.

[KKY] V.E. Katsnelson, Kheifets, A.Ya., Yuditskii, P. M., Abstract Interpolation Problem and Isometric Operators Extension Theory, *Operators in Functional Spaces and Questions of Function Theory*, Kiev (1987), 83–96, [in Russian]. English transl.: *Operator Theory: Advances and Applications*, Birkhäuser Verlag, Basel, (1997).

[Kh1] A.Ya. Kheifets, Parseval equality in abstract interpolation problem and coupling of open systems, *Teor. Funk., Funk. Anal. i ikh Prilozhen* **49** (1988) 112–120, **50** (1988) 98–103 [in Russian]. English transl.: *J. Sov. Math.* **49**, 4 (1990), 1114–1120, **49**, 6 (1990), 1307–1310.

[Kh2] A.Ya. Kheifets, Nevanlinna–Adamjan–Arov–Krein theorem in semi–determinate case, *Teor. Funkt., Funkt. Anal. i ikh Prilozhen.* **56** (1991), 128–137 [in Russian]. English transl.: *Journal of Mathematical Sciences* **76**, 4 (1995), 2542–2549.

[KY] A. Ya. Kheifets and P.M. Yuditskii, An analysis and extension of V.P. Potapov's approach to interpolation problems with applications to the generalized bi-tangential Schur– Nevanlinna–Pick problem and j-inner-outer factorization, *in Operator Theory: Advances and Applications*, **72** (1994), 133–161, Birkhäuser Verlag, Basel.

[Aroc] R. Arocena, Unitary colligations and parametrization formulas, *Ukrainian Math. Journal*, Volume dedicated to the memory of M.G. Krein, **46** 3 (1994).

[Mor] M.D. Moran, On Intertwining Dilations, *Journal of Mathematical Analysis and Applications* **141** 1 (1989), 219–234.

[FF1] C. Foias and A. Frazo, On the Schur Representation in the Commutant Lifting Theorem I, *Operator Theory: Advances and Applications* **18** (1986), 207–217, Birkhäuser Verlag, Basel.

[FF2] C. Foias and A. Frazo, On the Schur Representation in the Commutant Lifting Theorem II, *Operator Theory: Advances and Applications* **29** (1988), 171–179, Birkhäuser Verlag, Basel.

[FF3] C. Foias and A. Frazo, Constructing the Schur Contraction in the Commutant Lifting Theorem, *Acta Sci. Math. (Szeged)* **61** (1995), 425–442.

[Kup] S. Kupin, Lifting Theorem as a Special Case of Abstract Interpolation Problem, *Zeitshrift für Analysis und ihre Anwendungen (Journal for Analysis and its Applications)* **15** 4 (1996), 789–798.

Mathematics Department
Weizmann Institute
of Science
Rehovot, Israel

Mathematics Department
Institute for Low Temperatures
47 Lenin Ave., Kharkov 310164
Ukraine
kheifets@ilt.kharkov.ua

Operator Theory:
Advances and Applications, Vol. 115

On Zeros of Defining Functions for some Hilbert Spaces of Polynomials

Xian-Jin Li

Finite-dimensional examples are given about the location and the simplicity of zeros of the defining polynomials for certain reproducing kernel Hilbert spaces of polynomials. These examples are related to de Branges' theory of the Riemann zeta function and Dirichlet L-functions [2] [3] [4].

1991 *Mathematics Subject Classification.* Primary 30C15, 33C45, 46E22

Keywords and phrases. A positivity condition and zeros of some polynomials, reproducing kernel Hilbert spaces, Pollaczek's polynomials

1 Introduction

The hypergeometric series $F(a, b; c, z)$ is given by

$$F(a, b; c, z) = \frac{\Gamma(c)}{\Gamma(a)\Gamma(b)} \sum_{n=0}^{\infty} \frac{\Gamma(a+n)\Gamma(b+n)}{n!\Gamma(c+n)} z^n$$

for z inside the unit disk, where c is not zero or a negative integer. Let h be a real number greater than $\frac{1}{2}$. The polynomials

$$S_n(z) = i^{-n} F(-n, h + iz; 2h; 2)$$

for $n = 0, 1, \ldots$ are called the Pollaczek polynomials. Finite dimensional Hilbert spaces of entire function related to Pollaczek's polynomials are studied in de Branges [1]. The following theorem is due to Louis de Branges in an unpublished work, and the author wishes to thank Louis de Branges for suggesting him to write out a proof of this theorem. This note is mainly devoted to that proof.

Theorem 1 *Let $E_n(z) = (2h + n)S_{n+1}(z) - i(n + 1)S_n(z)$ for all nonnegative integers n. Then $|E_n(\bar{z})| < |E_n(z)|$ for z in the upper half-plane, the functional identity*

$$\bar{E}_n(\bar{z}) = E_n(z - i)$$

holds for all complex numbers z, and the zeros of $E_n(z)$ lie on the line $Im\, z = -\frac{1}{2}$ and are simple for every nonnegative integer n.

Remark: Let

$$E(z) = iz(1-iz)\pi^{-\frac{1-iz}{2}}\Gamma\left(\frac{1-iz}{2}\right)\zeta(1-iz),$$

where $\zeta(s)$ is the Riemann zeta function. Then $|E(\bar{z})| < |E(z)|$ for z in the upper half-plane, and the functional identity

$$\bar{E}(\bar{z}) = E(z-i)$$

holds for all complex numbers z. The Riemann hypothesis says that the zeros of $E(z)$ lie on the line $Im\, z = -\frac{1}{2}$. It is also believed that the zeros of $E(z)$ are all simple.

2 Properties of Hypergeometric Series

Consider the inner product on polynomials which is defined by

$$\langle F, G\rangle = \frac{2^{2h-1}}{\pi\Gamma(2h)}\int_{-\infty}^{\infty} F(t)\bar{G}(t)|\Gamma(h+it)|^2 dt \tag{2.1}$$

for all polynomials $F(z)$ and $G(z)$. Define operators L_+ and L_- on polynomials by

$$(L_+F)(z) = -\frac{i}{2}(h-iz)F(z+i) - zF(z) + \frac{i}{2}(h+iz)F(z-i)$$

and

$$(L_-F)(z) = \frac{i}{2}(h-iz)F(z+i) - zF(z) - \frac{i}{2}(h+iz)F(z-i)$$

for all polynomials $F(z)$.

Lemma 2 *The identity*

$$\langle L_-F, G\rangle = \langle F, L_+G\rangle$$

holds for all polynomials $F(z)$ and $G(z)$.

Proof: Let $F(z)$ and $G(z)$ be polynomials. By moving the lines of integration we obtain that

$$\begin{aligned}
&\pi\Gamma(2h)2^{2h-1}\langle L_-F, G\rangle \\
&= \int_{-\infty}^{\infty}\left\{\frac{i}{2}(h-it)F(t+i) - tF(t) - \frac{i}{2}(h+it)F(t-i)\right\}\bar{G}(t)|\Gamma(h+it)|^2 dt \\
&= \int_{-\infty}^{\infty} F(t)\left\{\frac{i}{2}(h+it)\bar{G}(t+i) - t\bar{G}(t) - \frac{i}{2}(h-it)\bar{G}(t-i)\right\}|\Gamma(h+it)|^2 dt \\
&= \pi\Gamma(2h)2^{2h-1}\langle F, L_+G\rangle.
\end{aligned}$$

The stated identity then follows. □

Lemma 3 *We have* $L_- S_n = n S_{n-1}$ *and* $L_+ S_n = (2h+n) S_{n+1}$ *for all nonnegative integers* n, *where* $S_{-1}(z) = 0$.

Proof: By the Gauss contiguous relation of hypergeometric series

$$(c-b)F(a,b-1;c;z) + [2b-c-(b-a)z]F(a,b;c;z) \\ + b(z-1)F(a,b+1;c;z) = 0,$$

we obtain that

$$\begin{aligned} (h-iz)F(-n,h-1+iz;2h;2) - 2(h+n)F(-n,h+iz;2h;2) \\ + (h+iz)F(-n,h+iz;2h;2) = 0. \end{aligned} \tag{2.2}$$

By definition we have

$$\begin{aligned} (L_- S_n)(z) = \frac{i}{2}(h-iz)i^{-n}F(-n,h-1+iz;2h;2) \\ - i^{-n}zF(-n,h+iz;2h;2) \\ - \frac{i}{2}(h+iz)i^{-n}F(-n,h+1+iz;2h;2). \end{aligned}$$

It follows from (2.2) that

$$\begin{aligned} (L_- S_n)(z) = -(h+iz)i^{1-n}F(-n,h+1+iz;2h;2) \\ + (n+h+iz)i^{1-n}F(-n,h+iz;2h;2). \end{aligned} \tag{2.3}$$

By the Gauss contiguous relation of hypergeometric series

$$(b-a)F(a,b;c;z) + aF(a+1,b;c;z) - bF(a,b+1;c;z) = 0,$$

we obtain that

$$\begin{aligned} (n+h+iz)F(-n,h+iz;2h;2) - nF(1-n,h+iz;2h;2) \\ - (h+iz)F(-n,h+1+iz;2h;2) = 0. \end{aligned} \tag{2.4}$$

It follows from (2.3) and (2.4) that $L_- S_n = n S_{n-1}$ for every positive integer n.

By definition we have $(L_+ F)(z) = -(L_- F)(z) - 2zF(z)$ for all polynomials $F(z)$, and hence

$$(L_+ S_n)(z) = -nS_{n-1}(z) - 2zi^{-n}F(-n,h+iz;2h;2). \tag{2.5}$$

By the Gauss contiguous relation of hypergeometric series

$$(c-a)F(a-1,b;c;z) + [2a-c-(a-b)z]F(a,b;c;z) \\ + a(z-1)F(a+1,b;c;z) = 0,$$

we obtain that

$$\begin{aligned} (n+2h)F(-n-1,h+iz;2h;2) + 2izF(-n,h+iz;2h;2) \\ - nF(1-n,h+iz;2h;2) = 0. \end{aligned} \tag{2.6}$$

It follows from (2.5) and (2.6) that $L_+ S_n = (2h+n)S_{n+1}$ for all nonnegative integers n. This completes the proof of the lemma. □

Theorem 4 *The polynomials $S_n(z)$, $n = 0, 1, \dots$, are orthogonal with respect to the inner product* (2.1), *and the self-product of $S_n(z)$ with respect to* (2.1) *is equal to $n!\Gamma(2h)/\Gamma(2h+n)$ for every nonnegative integer n.*

Proof: We know that $S_0(z) = 1$ and $S_1(z) = -z/h$. It is easy to see that $\langle S_0, S_0\rangle = 1$ and $\langle S_0, S_1\rangle = 0$. Assume that

$$\langle S_k, S_l\rangle = \begin{cases} k!\Gamma(2h)/\Gamma(2h+k), & \text{if } k = l; \\ 0, & \text{if } k \neq l \end{cases}$$

for $k, l = 0, 1, \dots, n$. Then it follows from Lemma 2 and Lemma 3 that

$$\begin{aligned} \langle S_k, S_{n+1}\rangle &= \frac{1}{2h+n}\langle S_k, L_+ S_n\rangle \\ &= \frac{1}{2h+n}\langle L_- S_k, S_n\rangle \\ &= \frac{k}{2h+n}\langle S_{k-1}, S_n\rangle \\ &= \begin{cases} (n+1)!\Gamma(2h)/\Gamma(2h+n+1), & \text{if } k = n+1; \\ 0, & \text{if } k \neq n+1 \end{cases} \end{aligned}$$

for $k = 0, 1, \dots, n+1$. The proof of the theorem is then complete by induction. □

3 Proof of Theorem 1

Let $E(z)$ be an entire function satisfying $|E(\bar{z})| < |E(z)|$ for z in the upper half-plane. A Hilbert space of entire functions $\mathcal{H}(E)$ (cf. de Branges [1]) is the set of all entire functions $F(z)$ such that

$$\|F\|^2 \overset{\text{def.}}{=} \int_{-\infty}^{\infty} |F(t)/E(t)|^2 dt < \infty$$

and such that

$$|F(z)|^2 \le \|F\|^2 K(z, z)$$

for all complex numbers z, where

$$K(w, z) = \frac{E(z)\bar{E}(w) - \bar{E}(\bar{z})E(\bar{w})}{2\pi i(\bar{w} - z)}.$$

The following theorem is due to de Branges (cf.[2] [3] [4]). Since it is also important to know the simplicity of zeros, we state the theorem slightly different from his.

Theorem 5 *Let $\mathcal{H}(E)$ be a given space such that $E(z)$ has no real zeros and such that $\bar{E}(\bar{z}) = \epsilon E(z-i)$ for a constant ϵ of absolute value one. If the real part of $\langle F(z), F(z+i)\rangle_{\mathcal{H}(E)}$ is positive for every nonzero element $F(z)$ in $\mathcal{H}(E)$ such that $F(z+i)$ belongs to the space, then the zeros of $E(z)$ lie on the line $Im\, z = -\frac{1}{2}$ and are simple.*

Proof: Let w be a zero of $E(z)$. Since $\bar{E}(\bar{z}) = \epsilon E(z-i)$ with $|\epsilon| = 1$, the identity

$$\bar{E}(w+i)K(w, z+i) = -\bar{E}(w-i)K(w+i, z)$$

holds for all complex numbers z. Since $E(z)$ has no real zeros and since $|E(\bar{z})| < |E(z)|$ for z in the upper half-plane, $E(w+i)$ and $E(w-i)$ are nonzero. It follows that $K(w, z)$ is a nonzero element of $\mathcal{H}(E)$ such that $K(w, z+i)$ belongs to the space. Assume that $F(z)$ is an element in $\mathcal{H}(E)$ such that $F(z+i)$ belongs to the space. Then

$$\langle F(z+i), K(w, z)\rangle_{\mathcal{H}(E)} + \langle F(z), K(w, z+i)\rangle_{\mathcal{H}(E)}$$
$$= \frac{E(w+i) - E(w-i)}{E(w+i)} F(w+i).$$

Since the real part of $\langle F(z+i), F(z)\rangle_{\mathcal{H}(E)}$ is positive for every nonzero element $F(z)$ in $\mathcal{H}(E)$ such that $F(z+i)$ belongs to the space, we have

$$\left| Re \left\{ \frac{E(w+i) - E(w-i)}{E(w+i)} F(w+i) \right\} \right|^2 \tag{3.1}$$
$$\leq 4 Re\, K(w, w+i) Re\, \langle F(z+i), F(z)\rangle_{\mathcal{H}(E)}$$

by the Schwartz inequality, where the equality holds in the inequality if and only if $F(z)$ and $K(w, z)$ are linearly dependent. Since $F(z)/(z - z_0)$ belongs to $\mathcal{H}(E)$ whenever $F(z)$ belongs to the space and vanishes at z_0, there exists a nonzero element $F(z)$ in $\mathcal{H}(E)$ such that $F(w+i) \neq 0$ and such that $F(z+i)$ belongs to the space. If $K(w, w+i) = 0$, then the strict inequality holds in (3.1), and hence $Re\, K(w, w+i) > 0$. This is a contradiction. Therefore, $K(w, w+i)$ is not equal to zero. Since $\bar{E}(\bar{z}) = \epsilon E(z-i)$, if $Im\, w \neq -\frac{1}{2}$ then $K(w, w+i) = 0$. This contradicts to the fact that $K(w, w+i)$ is not equal to zero, and therefore $Im\, w = -\frac{1}{2}$. Since $K(w, w+i) = E'(w)\bar{E}(w-i)/2\pi i$ and since $K(w, w+i)$ is not equal to zero, we have $E'(w) \neq 0$. Thus, we have proved that the zeros of $E(z)$ lie on the line $Im\, z = -\frac{1}{2}$ and are simple.

This completes the proof of the theorem. □

Proof of Theorem 1: Let $\mathcal{H}_n$ be the Hilbert space of polynomials of degree at most n, considered with the inner product (2.1). By Theorem 4 and by Theorem 3.2.2

of Szegö [6], the reproducing kernel function for the space $\mathcal{H}_n$ is given by the expression

$$
\begin{aligned}
&\sum_{k=0}^{n} \frac{\Gamma(2h+k)}{k!\Gamma(2h)} S_k(z)\bar{S}_k(w) \\
(3.2)\qquad &= \frac{\Gamma(2h+n+1)\sqrt{2h+n}}{2\Gamma(2h)n!\sqrt{n+1}} \frac{S_n(z)\bar{S}_{n+1}(w) - S_{n+1}(z)\bar{S}_n(w)}{z-\bar{w}}.
\end{aligned}
$$

Define

$$K_n(w,z) = \frac{E_n(z)\bar{E}_n(w) - \bar{E}_n(\bar{z})E_n(\bar{w})}{2\pi i(\bar{w}-z)}.$$

We have

$$K_n(w,z) = \frac{(n+1)(2h+n)}{2\pi} \frac{S_n(z)\bar{S}_{n+1}(w) - S_{n+1}(z)\bar{S}_n(w)}{z-\bar{w}}.$$

By (3.2) we have

$$(3.3)\qquad \sum_{k=0}^{n} \frac{\Gamma(2h+k)}{k!\Gamma(2h)} S_k(z)\bar{S}_k(w) = \frac{\pi\Gamma(2h+n)\sqrt{2h+n}}{(n+1)!\Gamma(2h)} K_n(w,z).$$

It follows from (3.3) that $|E_n(\bar{z})| < |E_n(z)|$ for z in the upper half-plane, and that the spaces $\mathcal{H}_n$ and $\mathcal{H}(E_n)$ contain the same elements. Moreover the identity

$$(3.4)\qquad \langle F, G\rangle_{\mathcal{H}_n} = \frac{(n+1)!\Gamma(2h)}{\pi\Gamma(2h+n)\sqrt{2h+n}} \langle F, G\rangle_{\mathcal{H}(E_n)}$$

holds for all polynomials $F(z)$ and $G(z)$ of degree at most n. Since $\mathcal{H}_n$ contains constant functions, $E_n(z)$ has no real zeros. By moving the lines of integration we find that

$$
\begin{aligned}
&\langle F(z+i), G(z)\rangle_{\mathcal{H}_n} + \langle F(z), G(z+i)\rangle_{\mathcal{H}_n} \\
(3.5)\qquad &= \frac{(2h-1)2^{2h-1}}{\pi\Gamma(2h)} \int_{-\infty}^{\infty} F\left(t+\frac{i}{2}\right) \bar{G}\left(t+\frac{i}{2}\right) \left|\Gamma\left(h-\frac{1}{2}+it\right)\right|^2 dt
\end{aligned}
$$

for all polynomials $F(z)$ and $G(z)$. By (3.5), the real part of

$$\langle F(z+i), F(z)\rangle_{\mathcal{H}_n}$$

is positive for every nonzero element $F(z)$ in $\mathcal{H}_n$. It follows from (3.4) that the real part of

$$\langle F(z+i), F(z)\rangle_{\mathcal{H}(E_n)}$$

is positive for every nonzero element $F(z)$ in $\mathcal{H}(E_n)$.

If $F(z)$ is real for real z, then $(L_+F)(z)$ is also real for real z by definition. It follows from Lemma 3 that $S_n(z)$ has real coefficients for every nonnegative integer n. The identity $\bar{E}_n(\bar{z}) = E_n(z-i)$ can then be written as

$$\begin{aligned}(2h+n)F(-n-1,h+iz;2h;2) - (n+1)F(-n,h+iz;2h;2)\\ = (2h+n)F(-n-1,h+1+iz;2h;2)\\ + (n+1)F(-n,h+1+iz;2h;2).\end{aligned} \tag{3.6}$$

By the Gauss contiguous relation of hypergeometric series

$$\begin{aligned}(a-c)F(a-1,b;c;z) + (c-a-b)F(a,b;c;z)\\ +b(1-z)F(a,b+1;c;z) = 0,\end{aligned}$$

we obtain that

$$\begin{aligned}-(2h+n)F(-n-1,h+iz;2h;2) + (n+h-iz)\\ F(-n,h+iz;2h;2) - (h+iz)F(-n,h+1+iz;2h;2) = 0.\end{aligned} \tag{3.7}$$

By the Gauss contiguous relation of hypergeometric series

$$\begin{aligned}(a-c)F(a-1,b;c;z) + (c-b)F(a,b-1;c;z)\\ +(b-a)(1-z)F(a,b;c;z) = 0,\end{aligned}$$

we obtain that

$$\begin{aligned}-(2h+n)F(-n-1,h+1+iz;2h;2)\\ + (h-1-iz)F(-n,h+iz;2h;2)\\ - (n+h+1+iz)F(-n,h+1+iz;2h;2) = 0.\end{aligned} \tag{3.8}$$

Hence, the left side of (3.7) is equal to the left side of (3.8), and the resulting identity is (3.6). Thus, we have proved that $E_n(z)$ satisfies the functional identity $\bar{E}_n(\bar{z}) = E_n(z-i)$. Theorem 1 then follows from Theorem 5.

This completes the proof of Theorem 1.

4 A Second Proof of Theorem 4

A nice proof of Theorem 4 using the Mellin transform is implicit in Bump, Choi, Kurlberg and Vaaler [5]. For the convenience of readers, we give it here. The author wishes to thank Jeffrey D. Vaaler for giving him a copy of their preprint [5].

Proof of Theorem 4: Let $\alpha = 2h-1$. The Laguerre polynomials $L_n^{(\alpha)}(x)$, $n = 0, 1, \ldots$, are given by (cf. (5.1.6) of Szegö [6])

$$L_n^{(\alpha)}(x) = \sum_{k=0}^{n} \binom{n+\alpha}{n-k} \frac{(-x)^k}{k!}, \tag{4.1}$$

and they satisfy (cf. (5.1.1) of Szegö [6])

$$\int_0^\infty e^{-x}x^\alpha L_m^{(\alpha)}(x)L_n^{(\alpha)}(x)dx = \begin{cases} \frac{\Gamma(n+2h)}{n!}, & \text{if } m = n; \\ 0, & \text{if } m \neq n \end{cases} \tag{4.2}$$

for all nonnegative integers m and n. By (4.1) we have

$$\int_0^\infty e^{-\frac{1}{2}x}x^{h+iz-1}L_n^{(\alpha)}(x)dx = 2^{h+iz}\sum_{k=0}^{n}(-2)^k\binom{n+\alpha}{n-k}\frac{\Gamma(h+iz+k)}{k!}. \tag{4.3}$$

Since

$$\begin{aligned} S_n(z) &= \frac{i^{-n}\Gamma(2h)}{\Gamma(h+iz)}\sum_{k=0}^{n}\frac{(-n+k-1)\cdots(-n)\Gamma(h+iz+k)}{k!\Gamma(2h+k)}2^k \\ &= \frac{i^{-n}\Gamma(2h)n!}{\Gamma(h+iz)\Gamma(n+2h)}\sum_{k=0}^{n}\frac{(n+2h-1)\cdots(k+2h)\Gamma(h+iz+k)}{(n-k)!k!}(-2)^k \\ &= \frac{i^{-n}\Gamma(2h)n!}{\Gamma(h+iz)\Gamma(n+2h)}\sum_{k=0}^{n}\binom{n+\alpha}{n-k}\frac{\Gamma(h+iz+k)}{k!}(-2)^k, \end{aligned}$$

by (4.3) we have

$$\int_0^\infty e^{-\frac{1}{2}x}x^{h+iz-1}L_n^{(\alpha)}(x)dx = i^n 2^{h+iz}\Gamma(h+iz)\frac{\Gamma(n+2h)}{n!\Gamma(2h)}S_n(z) \tag{4.4}$$

for all nonnegative integers n. It follows from (4.4) and the Parseval-Plancherel identity that

$$\begin{aligned} &\int_0^\infty e^{-x}x^\alpha L_m^{(\alpha)}(x)L_n^{(\alpha)}(x)dx \\ &\quad = \frac{2^\alpha}{\pi}\frac{\Gamma(m+2h)\Gamma(n+2h)}{m!n!\Gamma(2h)^2}\int_{-\infty}^\infty S_m(x)\bar{S}_n(x)|\Gamma(h+ix)|^2dx \end{aligned} \tag{4.5}$$

for all nonnegative integers m and n. By (4.2) and (4.5), we have that $S_n(z)$, $n = 0, 1, \ldots$, are orthogonal with respect to the inner product (2.1), and that

$$\frac{2^{2h-1}}{\pi\Gamma(2h)}\int_{-\infty}^\infty S_n(x)\bar{S}_n(x)|\Gamma(h+ix)|^2dx = \frac{n!\Gamma(2h)}{\Gamma(n+2h)}$$

for every nonnegative integer n.

This completes the second proof of Theorem 4.

References

1. Louis de Branges, *Hilbert Spaces of Entire Functions*, Prentice-Hall, Englewood Cliffs, 1968.
2. Louis de Branges, *A conjecture which implies the Riemann hypothesis*, J. Functional Analysis **121** (1994), 117–184.
3. Louis de Branges, *The Riemann hypothesis for Hilbert spaces of entire functions*, Bull. Amer. Math. Soc. **15** (1986), 1–17.
4. Louis de Branges, *The convergence of Euler products*, J. Functional Analysis **107** (1992), 122–210.
5. Daniel Bump, K.K.S. Choi, Pär Kurlberg and Jeffrey Vaaler, *A local Riemann hypothesis, I*, Preprint, 1996.
6. Gábor Szegö, *Orthogonal Polynomials*, Colloquium Publications, volume 23, American Mathematical Society, Providence, 1939.

Department of Mathematics
The University of Texas at Austin
Austin, TX 78712 USA

Current address:
Department of Mathematics
Brigham Young University
Provo, Utah 84602

Operator Theory:
Advances and Applications, Vol. 115

One or Two Parameter Algebraic Scattering Systems and the Commutant Lifting Theorem

María Dolores Morán

M. Cotlar and C. Sadosky proved (cfr.[C-S1], [C-S2] and [C-S3]) that all bounded Hankel forms defined in a Algebraic Scattering Systems have a bounded Toeplitz extension, and also do some bounded Hankel forms defined in a two-parameter Algebraic Scattering Systems. In this paper we labell the set of such extensions by a set of some analytic operator-valued functions. Finally, we deal with a version of the Sz.-Nagy-Foias Commutant Lifting Theorem (cfr.[F-Fr1] and [F-Fr2]).

1 Introduction

R. Arocena proved the Commutant Lifting Theorem (cfr.[Ar]), using the tool of Generalized Toeplitz Kernels due to himself, M. Cotlar and C. Sadosky (cfr.[Ar-C-S]), the method was useful for finding liftings with the same norm equal as the intertwining operator (the one two be lifted).

Since 1987, M. Cotlar and C. Sadosky (see for example [C-S1], [C-S2] and [C-S3]) have introduced the notion of Algebraic Scattering Systems, idea which generalizes Lax and Phillips Scattering Systems (cfr.[L-P]) and Adamjan and Arov Scattering Systems (cfr.[A-Aro]). In this paper we find out that the Algebraic Scattering Systems approach leads to a version of the Sz.-Nagy-Foias Commutant Lifting Theorem (cfr.[F-Fr1] and [F-Fr2]), which permits find liftings with norm bigger than the norm intertwing operator.

We use the next definitions:

Definition 1 $[V; W_1, W_2; \Upsilon]$ is a discrete algebraic scattering system if and only if V is a vector space, W_1, W_2 are subspaces of V, $\Upsilon : V \to V$ is an algebraic isomorphism, such that $\Upsilon (W_1) \subset W_1$ and $\Upsilon^{-1} (W_2) \subset W_2$.

Example 1 Let $\mathbb{T}$ be the unit circle, V the set of polynomials, W_1 the set of analytical polynomials, W_2 the set of antianalytical polynomials (i.e. $f \in W_2 \Leftrightarrow \bar{f} \in W_1$) and $\Upsilon : V \to V$, $\Upsilon (P)(z) = zP(z)$.

If $[V; W_1, W_2; \Upsilon]$ is a discrete scattering algebraic system such that V is a Hilbert space W_1, W_2 are two closed subspaces of V, and Υ verifies

$$V = \left[\bigvee_{n\in\mathbb{Z}} \Upsilon^n W_1\right] \vee \left[\bigvee_{n\in\mathbb{Z}} \Upsilon^n W_2\right] \text{ and } \bigcap_{n=1}^{\infty} \Upsilon^n (W_1) = \bigcap_{n=1}^{\infty} \Upsilon^{-n} (W_2) = \{0\},$$

then $[V; W_1, W_2; \Upsilon]$ is a Adamjan-Arov scattering system. If, in addition, $W_1 \perp W_2$ the classical Lax-Phillips scattering system is obtained.

Definition 2 B^0 is a Hankel form defined in the discrete algebraic scattering system $[V; W_1, W_2; \Upsilon]$ if and only if $B^0 : W_1 \times W_2 \to \mathbb{C}$ is a sesquilinear form and $B^0(\Upsilon w_1, w_2) = B^0(w_1, \Upsilon^{-1} w_2)$ for all $(w_1, w_2) \in W_1 \times W_2$.

We set for $i = 1, 2$ $B_i : V \times V \to \mathbb{C}$ a positive sesquilinear form such that

$$B_i (\Upsilon f, \Upsilon g) = B_i (f, g) \text{ for all } (f, g) \in V \times V.$$

Definition 3 B^0 is a (B_1, B_2)-bounded form defined in $[V; W_1, W_2; \Upsilon]$ if and only if $B^0 : W_1 \times W_2 \to \mathbb{C}$ is a sesquilinear form and for all $(w_1, w_2) \in W_1 \times W_2$, $\left|B^0 (w_1, w_2)\right| \leq B_1^{1/2} (w_1, w_1) B_2^{1/2}(w_2, w_2)$. In this case we denote $B^0 \prec (B_1, B_2)$.

Setting for $i = 1, 2$ V_i the minimal vector space which contains $\Upsilon^n (W_i)$ for each $n \in \mathbb{Z}$, we have that $\Upsilon (V_i) = V_i$.

Definition 4 B is a Toeplitz form defined in $[V; W_1, W_2; \Upsilon]$ if and only if $B : V_1 \times V_2 \to \mathbb{C}$ is a sesquilinear form and

$$B (\Upsilon f_1, \Upsilon f_2) = B (f_1, f_2) \text{ for all } (f_1, f_2) \in V_1 \times V_2.$$

Definition 5 B is a (B_1, B_2)-bounded form defined in $[V; W_1, W_2; \Upsilon]$ if and only if $B : V_1 \times V_2 \to \mathbb{C}$ is a sesquilinear form and for all $(f_1, f_2) \in V_1 \times V_2$, $|B (f_1, f_2)| \leq B_1^{1/2} (f_1, f_1) B_2^{1/2} (f_2, f_2)$. In this case we denote $B \leq (B_1, B_2)$.

Finally we want to describe:

$$\beta^0_{ext} = \{B : V_1 \times V_2 \to \mathbb{C} : \text{Toeplitz form, } B \leq (B_1, B_2) \; B\,|_{W_1 \times W_2} = B^0\}$$

Remark 1 Having a discrete algebraic scattering system $[V; W_1, W_2; \Upsilon]$ and B^0 a Hankel (B_1, B_2)-bounded form defined in $[V; W_1, W_2; \Upsilon]$ since $B^0 \prec (B_1, B_2)$ the form C defined in $W_1 \times W_2$ by

$$C((w_1, w_2), (w_1', w_2')) = B_1(w_1, w_1') + B^0(w_1, w_2') + \overline{B^0(w_1', w_2)} + B_2(w_2, w_2')$$

is non negative and the operator $\tilde{\Upsilon} = \begin{pmatrix} \Upsilon & 0 \\ 0 & \Upsilon \end{pmatrix}$ verifies: for all

$$(w_1, w_2), \left(w_1', w_2'\right) \in W_1 \times W_2,$$

such that

$$\tilde{\Upsilon}(w_1, w_2), \tilde{\Upsilon} \left(w_1', w_2'\right) \in W_1 \times W_2,$$

we have

$$C(\tilde{\Upsilon}(w_1, w_2), \tilde{\Upsilon}(w_1', w_2')) = C((w_1, w_2), (w_1', w_2')).$$

So defining $E_0 = \{w \in W_1 \times W_2 : C(w, w) = 0\}$, $E = W_1 \times W_2/E_0$, $\pi : W_1 \times W_2 \to E$ the canonical projection and the scalar product in E as usual, we know that there exists a Hilbert space H and a isometry $i : E \to H$ with dense range. Setting $D_T = \overline{i\pi(W_1 \times \Upsilon^{-1} W_2)}^H$, $R_T = \overline{i\pi(\Upsilon W_1 \times W_2)}^H$, $X = \overline{i\pi(W_1 \times \{0\})}^H$, $Y = \overline{i\pi(\{0\} \times W_2)}^H$ and $T : D_T \to R_T$ as the unique isometry which satisfies $Ti\pi(w_1, \Upsilon^{-1}w_2) = i\pi(\Upsilon w_1, w_2)$ we have that D_T, R_T are closed subspaces of H, and X, Y are closed subspaces of D_T, R_T (resp.) such that $T(X) \subset X$, $T^{-1}(Y) \subset Y$. Also there are two natural inmersions $J_1 : W_1 \to H$, $J_1(w_1) = i\pi(w_1, 0)$ and $J_2 : W_2 \to H$, $J_2(w_2) = i\pi(0, w_2)$.

Having a $B : V_1 \times V_2 \to \mathbf{C}$ sesquilinear Toeplitz (B_1, B_2)-bounded form, and reproducing the precedent construction we obtain a Hilbert space F and a unitary operator U defined in F. Results $B \in \beta_{ext}^0 \Leftrightarrow H$ is a closed subspace of F and U is a minimal unitary extension of T.

2 Description of β_{ext}^0

As a consequence of Arov-Grossman result (cfr. [Aro-G]) which gives a description of all minimal unitary extensions of a isometry by $\beta(H \ominus D_T, H \ominus R_T)$, the unitary ball of analytic functions operator value $L(H \ominus D_T, H \ominus R_T)$, we obtain:

Theorem 1 *The function $\Psi : \beta(H \ominus D_T, H \ominus R_T) \to \beta_{ext}^0$ defined by $\Psi(\theta) = B$ iff $B : V_1 \times V_2 \to \mathbb{C}$ is a Toeplitz form such that: $B(w_1, w_2) = B^0(w_1, w_2)$ if $(w_1, w_2) \in W_1 \times W_2$ and $B(w_1, (I - z\Upsilon)^{-1}\Upsilon w_2) = \langle J_1 w_1, A(z) J_2 w_2 \rangle_H$ where $A(z) = (TP_{D_T} + \theta(z)(I - P_{D_T}))[I - z(TP_{D_T} + \theta(z)(I - P_{D_T}))]^{-1}$ is a bijection.*

As an immediate consequence, since 0 is always a contractive analytic operator-valued function, we have that β_{ext}^0 isn't empty.

3 Commutant Lifting Theorem

We deal with a special version of Sz.-Nagy-Foias Commutant Lifting Theorem (cfr.[F-Fr1], [F-Fr2]) which permits find a lifting of 0 with norm different of 0.

For $i = 1, 2$ E_i denotes a Hilbert space, $L(E_1, E_2)$ will denote the set of all linear bounded operators from E_1 into E_2, $L(E_i) = L(E_i, E_i)$.

Theorem 2 (Sz.-Nagy and Foias) *Let E_1, E_2 be Hilbert spaces, $T_1 \in L(E_1)$ a isometric operator, $T_2 \in L(E_2)$ a contractive operator and $V_2 \in L(K_2)$ be its minimal isometric dilation. For all $A \in L(E_1, E_2)$ such that $AT_1 = T_2 A$ and σ a real number such that $\sigma \geq \|A\|$ there exists $Y \in L(E_1, K_2)$ such that $YT_1 = V_2 Y$, $\|Y\| \leq \sigma$ and $P_{E_2}^{K_2} Y = A$.*

Proof: For $i = 1, 2$ $U_i \in L(G_i)$ will denote the minimal unitary dilation of T_i, to the relation $AT_1 = T_2A$ we asociete the next Algebraic Scattering System

$$V = G_1 \times G_2, W_1 = E_1 x \{0\}, W_2 = \{0\} \times \bigvee_{n=0}^{\infty} U_2^{-n} E_2, \Upsilon = \begin{pmatrix} U_1 & 0 \\ 0 & U_2 \end{pmatrix}$$

And we define the next forms

$$\begin{aligned} \beta_1 &: V \times V \longrightarrow \mathbb{C}, \ \beta_1((g_1, g_2), (g_1', g_2')) \\ &= \sigma^2 (g_1, g_1')_{G_1}, \forall (g_1, g_2), (g_1', g_2') \in V \\ \beta_2 &: V \times V \longrightarrow \mathbb{C}, \ \beta_2((g_1, g_2), (g_1', g_2')) \\ &= (g_2, g_2')_{G_2}, \forall (g_1, g_2), (g_1', g_2') \in V \\ \beta^0 &: W_1 \times W_2 \longrightarrow \mathbb{C}, \ \beta^0((e_1, 0), (0, g_2)) \\ &= (Ae_1, g_2)_{G_2}, \forall e_1 \in E_1, g_2 \in \bigvee_{n=0}^{\infty} U_2^{-n} E_2. \end{aligned}$$

It's easy to check that β_1, β_2 are positive sesquilinear forms and for $i = 1, 2$ and for all $(g_1, g_2), (g_1', g_2') \in V$ $\beta_i(\Upsilon(g_1, g_2), \Upsilon(g_1', g_2')) = \beta_i((g_1, g_2), (g_1', g_2'))$.

β^0 is Hankel and $\beta^0 \prec (\beta_1, \beta_2)$ is a consequence of the fact that $\sigma \geq \|A\|$. Given $\beta \in \beta^0_{ext}$, the operator $Y \in L(E_1, K_2)$ defined by $\langle Ye_1, k_2 \rangle = B((e_1, 0), (0, k_2))$ satisfies $YT_1 = V_2 Y$, $\|Y\| \leq \sigma$ and $P_{E_2}^{K_2} Y = A$ and that relation induces a bijection from β^0_{ext} onto the set

$$\{Y \in L(E_1, K_2) : YT_1 = V_2 Y, \ \|Y\| \leq \sigma \text{ and } P_{E_2}^{K_2} Y = A\}.$$

Finally using theorem 1, the description will be more elegant if we realize that $H \ominus D_T$ and $H \ominus R_T$ unitary equivalent to

$$\mathcal{D}_A^\sigma \oplus \mathcal{D}_{T_2} \ominus \{D_A^\sigma h \oplus D_{T_2} Ah : h \in E_1\} \text{ and } \mathcal{D}_A^\sigma \ominus \overline{D_A^\sigma T_1 E_1}$$

respectively, where

$$\begin{aligned} \mathcal{D}_A^\sigma &= \overline{(\sigma^2 - A^* A)^{\frac{1}{2}} (E_1)}, \ D_A^\sigma = (\sigma^2 - A^* A)^{\frac{1}{2}} \\ \mathcal{D}_{T_2} &= \overline{(I - T_2^* T_2)^{\frac{1}{2}} (E_2)}, \ D_{T_2} = (I - T_2^* T_2)^{\frac{1}{2}}. \end{aligned}$$

The last statement is a consequence the fact that $\Lambda : W_1 \times W_2 \to \mathcal{D}_A^\sigma \oplus \bigvee_{n=0}^{\infty} U_2^{-n} E_2$, defined by $\Lambda((e_1, 0), (0, g_2)) = D_A^\sigma e_1 \oplus Ae_1 + g_2$ for all $((e_1, 0), (0, g_2)) \in W_1 \times W_2$ is a isometry with dense range such that

$$\Lambda(R_T) = \overline{\{\Lambda((U_1 e_1, 0), (0, g_2)) : ((e_1, 0), (0, g_2)) \in W_1 \times W_2\}}$$

$$= \overline{\left\{ D_A^\sigma T_1 e_1 \oplus A T_1 e_1 + g_2 : (e_1, g_2)\, E_1 \times \bigvee_{n=0}^{\infty} U_2^{-n} E_2 \right\}}$$

$$= D_A^\sigma T_1 E_1 \oplus \bigvee_{n=0}^{\infty} U_2^{-n} E_2,$$

and so $H \ominus R_T$ is unitary equivalent to $\mathcal{D}_A^\sigma \ominus \overline{D_A^\sigma T_1 E_1}$. In the other hand $\Omega : \mathcal{D}_A^\sigma \oplus E_2 \oplus \mathcal{D}_{T_2^*} \to \mathcal{D}_A^\sigma \oplus E_2 \oplus \mathcal{D}_{T_2}$ defined by $\Omega\,(a+b+c) = a + U_2\,(b+c)$ for all $a \in \mathcal{D}_A^\sigma, b \in E_2, c \in \mathcal{D}_{T_2^*}$ is unitary, $\Omega\left(\left\{D_A^\sigma e_1 + Ae_1 : e_1 \in E_1\right\} \vee U_2^* E_2\right) = \left\{D_A^\sigma e_1 + U_2 A e_1 : e_1 \in E_1\right\} \vee E_2 = \left\{D_A^\sigma e_1 + D_{T_2} Ae : e_1 \in E_1\right\} \vee E_2$ and $\Omega\Lambda\,(H \ominus D_T) = \mathcal{D}_A^\sigma \oplus \mathcal{D}_{T_2} \ominus \left\{D_A^\sigma h \oplus D_{T_2} Ah : h \in E_1\right\}$.

If $\theta \in B\,(H \ominus D_T, H \ominus R_T)$, the corresponding lifting can be writting as

$$Ye_1 = Ae_1 + a\,(z)\,e_1 + b\,(z)\,\theta^*\,(\overline{z})\left[I - c\,(z)\,\theta^*\,(\overline{z})\right]^{-1} d\,(z)\,e_1,$$

where:

$$\begin{aligned}
a\,(z) &: E_1 \longrightarrow L_2, \\
a\,(z) &= P_{L_2} U_2 \Lambda P_L^H T^* P_{R_T}^H \left[I - zT^* P_{R_T}^H\right]^{-1}; \\
b\,(z) &: H \ominus D_T \longrightarrow L_2, \\
b(z) &= P_{L_2} U_2 \Lambda P_L^H \left\{ I + zT^* P_{R_T}^H \left[I - z\left(I - P_L^H\right) T^* P_{R_T}^H\right]^{-1} \left(I - P_L^H\right)\right\}, \\
c\,(z) &: H \ominus D_T \longrightarrow H \ominus R_T, \\
c\,(z) &= zP_{H\ominus R_T}^H \left[I - z\left(I - P_L^H\right) T^* P_{R_T}^H\right]^{-1} \left(I - P_L^H\right); \\
d\,(z) &: E_1 \longrightarrow H \ominus R_T, \\
d\,(z) &= P_{H\ominus R_T}^H \left[I - z\left(I - P_L^H\right) T^* P_{R_T}^H\right]^{-1},
\end{aligned}$$

$L_2 = \overline{(U_2 - T_2)E_2}$ and P_L^H denotes the orthogonal proyection from H to L, the closure of $\{((0,0), (0, (I - U_2^* T_2)e_2) : e_2 \in E_2)\}$. The technique to proof the last statement can be seen in [M1] □

4 Unitary *c*-commuting Extension of a Pair of Isometries

We know that a giving a pair of isometries isn't always possible to find a pair of commuting unitary extension, but the next theorem explains the situation. We recall the Arov and Grossman result to obtain:

Theorem 3 *Let H be a Hilbert space, D_1, R_1, D_2, and R_2 closed subspace of H, $T_1 : D_1 \to R_1$, $T_2 : D_2 \to R_2$ a pair of isometries and c a complex number such that $|c| = 1$. The next conditions are equivalent:*

a) *There exists a contractive analytic function operator valued function* $\theta_1 \in B(H \ominus D_1, H \ominus R_1)$ *such that*

$$\begin{aligned} & T_2 P^H_{D_2} c\left(T_1 P^H_{D_1} + \theta_1(cz)\left(I - P^H_{D_1}\right)\right) \\ & \quad \left[I - cz\left(T_1 P^H_{D_1} + \theta_1(cz)\left(I - P^H_{D_1}\right)\right)\right]^{-1} |D_2 \\ = \; & P^H_{R_2}\left(T_1 P^H_{D_1} + \theta_1(z)\left(I - P^H_{D_1}\right)\right) \\ & \quad \left[I - z\left(T_1 P^H_{D_1} + \theta_1(z)\left(I - P^H_{D_1}\right)\right)\right]^{-1} T_2 \end{aligned}$$

b) *There exists a Hilbert space* F *which contains* H *as a closed subspace,* U_1, U_2 *unitary extensions of* T_1, T_2 *defined in* F *such that* $U_1 U_2 = c U_2 U_1$.

c) *There exists a contractive analytic function operator valued function* $\theta_2 \in B(H \ominus D_2, H \ominus R_2)$ *such that*

$$\begin{aligned} & T_1 P^H_{D_1} \bar{c}\left(T_2 P^H_{D_2} + \theta_2(\bar{c}z)\left(I - P^H_{D_2}\right)\right) \\ & \quad \left[I - \bar{c}z\left(T_2 P^H_{D_2} + \theta_2(\bar{c}z)\left(I - P^H_{D_2}\right)\right)\right]^{-1} |D_1 \\ = \; & P^H_{R_1}\left(T_2 P^H_{D_2} + \theta_2(z)\left(I - P^H_{D_2}\right)\right) \\ & \quad \left[I - z\left(T_2 P^H_{D_2} + \theta_2(z)\left(I - P^H_{D_2}\right)\right)\right]^{-1} T_1 \end{aligned}$$

Proof: Let us assume that condition a) is true and set H' the Hilbert space such that $U \in L(H')$ the minimal unitary extension of T_1 corresponding to θ_1 by Arov and Grossman correspondence. Condition a) insures that $W : \bigvee_{n\in\mathbb{Z}} U^n(D_2) \to \bigvee_{n\in\mathbb{Z}} U^n(R_2)$ defined by $W \sum_{n\in\mathbb{Z}} U^n d_2(n) = \sum_{n\in\mathbb{Z}} \bar{c}^n U^n T_2 d_2(n)$ is a isometric operator. It is easy to prove that for all $d \in \bigvee_{n\in\mathbb{Z}} U^n(D_2)$, $WUd = \bar{c}UWd$. Let U_2 the minimal unitary dilation of $WP^{H'}_{\bigvee_{n\in\mathbb{Z}} U^n(D_2)}$, acting in a Hilbert space F, which contains H' then U_2 is a extension of T_2 and setting $U_1 : F \to F$ as $U_1 \sum_{n\in\mathbb{Z}} U_2^n h'(n) = \sum_{n\in\mathbb{Z}} c^n U_2^n U h'(n)$ it results that U_1 is a unitary extension of T_1 and $U_1 U_2 = c U_2 U_1$. Finally if condition b) holds set $U = U_{1|\bigvee_{n\in\mathbb{Z}} U_1^n(H)}$ and for

$$\theta_1(z) = P^{\bigvee_{n\in\mathbb{Z}} U_1^n(H)}_{H\ominus R_1} U(I - z(I - P^{\bigvee_{n\in\mathbb{Z}} U_1^n(H)}_{H})U)^{-1} |H \ominus D_1$$

condition a) holds. □

5 Two-parameter Algebraic Scattering System

Definition 6 $[V; W_1, W_2; \Upsilon, \Psi]$ is a two parameter discrete algebraic scattering system if and only if $[V; W_1, W_2; \Upsilon]$, $[V; W_1, W_2; \Psi]$ are two discrete algebraic scattering system and exists a complex number c, $|c| = 1$ such that $\Upsilon\Psi = c\Psi\Upsilon$.

Example 2 Let $\mathbb{T}$ be the unit circle, V the set of polynomials in $\mathbb{T}^2$, W_1 the set of analytical polynomials, W_2 the set of antianalytical polynomials, and $\Upsilon : V \to V$, $\Upsilon(P)(z, w) = zP(z, w)$ and $\Psi : V \to V$, $\Psi(P)(z, w) = wP(z, w)$.

Definition 7 B^0 is a Hankel form defined in the two-parameter discrete algebraic system $[V; W_1, W_2; \Upsilon, \Psi]$ if and only if B^0: is a Hankel form defined in both discrete algebraic scattering systems $[V; W_1, W_2; \Upsilon]$, $[V; W_1, W_2; \Psi]$.

We set for $i = 1, 2$ $B_i : V \times V \to \mathbb{C}$ a positive sesquilinear form such that $B_i(\Upsilon f, \Upsilon g) = B_i(f, g)$ and $B_i(\Psi f, \Psi g) = B_i(f, g)$ for all $(f, g) \in V \times V$.

Setting for $i = 1, 2$ V_i the minimal vector space which contains $\Psi^m \Upsilon^n(W_i)$ for each $(n, m) \in \mathbb{Z}^2$ we have that $\Upsilon(V_i) = V_i$ and $\Psi(V_i) = V_i$.

Definition 8 B is a Toeplitz form defined in $[V; W_1, W_2; \Upsilon, \Psi]$ if and only if $B : V_1 \times V_2 \to \mathbb{C}$ is a sesquilinear form and $B(\Upsilon f_1, \Upsilon f_2) = B(f_1, f_2)$ and $B(\Psi f_1, \Psi f_2) = B(f_1, f_2)$ for all $(f_1, f_2) \in V_1 \times V_2$.

Finally we want to describe

$$\beta^0_{ext} = \{B : V_1 \times V_2 \longrightarrow \mathbb{C} : \text{Toeplitz form } B \leq (B_1, B_2)\ B\,|_{W_1 \times W_2} = B^0\}$$

Remark 2 Having a two-parameter discrete (A.S.S) $[V; W_1, W_2; \Upsilon, \Psi]$ and B^0 a Hankel (B_1, B_2)-bounded form defined in $[V; W_1, W_2; \Upsilon, \Psi]$ we obtain a Hilbert space H, D_T, R_T, D_S, R_S closed subspace of H, and two isometries $T : D_T \to R_T$ (corresponding to Υ) and $S : D_S \to R_S$ (corresponding to Ψ). Having a $B : V_1 \times V_2 \to \mathbb{C}$ sesquilinear Toeplitz (B_1, B_2)-bounded form, reproducing the precedent construction we obtain a Hilbert space F which contains H and two unitary operators U_T and U_S defined in F, and results $B \in \beta^0_{ext} \Leftrightarrow H$ is a closed subspace of F, U_T is a minimal unitary extension of T, U_S is a minimal unitary extension of S and $U_T U_S = c U_S U_T$.

6 Description of β^0_{ext} for a Two-parameter Algebraic Scattering System

As consequence of the last remark and the result of the section Unitary c-commuting extension we can conclude that β^0_{ext} can be empty, but when T or S are unitary β^0_{ext} is not empty. If T is unitary then:

$$SP^H_{D_S^c}(I - czT)^{-1}\,T\,|_{D_S} = P^H_{R_S}(I - zT)^{-1}\,TS$$

and there is a bijection between β^0_{ext} and the set

$$\left\{\theta \in B(H \ominus D_S, H \ominus R_S) : T\overline{c}\theta(\overline{c}z)\left(I - P^H_{D_S}\right)_H = \theta(z)\left(I - P^H_{D_S}\right)T\right\}.$$

Finally β^0_{ext} is not empty if and only if one of the conditions of the theorem of the section Unitary c-commuting extension of a pair of isometries holds when $T_1 = T$ and $T_2 = S$ (none of then unitary), let us assume that condition a) holds, then there exists a contractive analytic function operator valued function $\theta \in B(H \ominus D_T, H \ominus R_T)$ such that

$$\begin{aligned} & SP^H_{D_S}c\left(TP^H_{D_T} + \theta(cz)\left(I - P^H_{D_T}\right)\right) \\ & \quad \left[I - cz\left(TP^H_{D_S} + \theta(cz)\left(I - P^H_{D_T}\right)\right)\right]^{-1} |D_S \\ = \; & P^H_{R_2}\left(TP^H_{D_S} + \theta(z)\left(I - P^H_{D_S}\right)\right)\left[I - z\left(TP^H_{D_T} + \theta(z)\left(I - P^H_{D_T}\right)\right)\right]^{-1} S. \end{aligned}$$

Let $U \in L(G)$, be the minimal unitary extension of T corresponding to $\theta \in B(H \ominus D_T, H \ominus R_T)$,

$$D = \bigvee_{n \in \mathbb{Z}} U^n D_S, \quad R = \bigvee_{n \in \mathbb{Z}} U^n R_S$$

and $W : D \to R$ defined by $W\left(\sum_{n \in \mathbb{Z}} U^n d_n\right) = \sum_{n \in \mathbb{Z}} \overline{c}^n U^n T d_n$ where $d_n \in D_S$, for all $n \in \mathbb{Z}$. For $(w_1, w_2) \in W_1 \times W_2, n, m \in \mathbb{Z}$ we define

$$B\left(w_1, \Psi^n \Upsilon^m w_2\right) = \begin{cases} \left\langle w_1, U^n \left(VP^G_D\right)^m w_2\right\rangle & \text{if } m \geq 0 \\ \overline{c}^{nm}\left\langle \Psi^{-m} w_1, U^n w_2\right\rangle & \text{if } m < 0 \end{cases}$$

and

$$\begin{aligned} & B\left(\sum_{(n,m) \in \mathbb{Z}^2} \Psi^n \Upsilon^m w_1(n,m), \sum_{(r,s) \in \mathbb{Z}^2} \Psi^r \Upsilon^s w_2(n,m)\right) \\ = \; & \sum_{(n,m) \in \mathbb{Z}^2} \sum_{(r,s) \in \mathbb{Z}^2} B(w_1(n,m), c^{-m(r-n)} \Psi^{r-n} \Upsilon^{s-m} w_2(n,m)), \end{aligned}$$

it results that $B \in \beta^0_{ext}$.

Finally we set some notation if $\theta \in B(H \ominus D_T, H \ominus R_T)$ such that

$$\begin{aligned} & SP^H_{D_S}c\left(TP^H_{D_T} + \theta(cz)\left(I - P^H_{D_T}\right)\right) \\ & \quad \left[I - cz\left(TP^H_{D_S} + \theta(cz)\left(I - P^H_{D_T}\right)\right)\right]^{-1} |D_S \\ = \; & P^H_{R_2}\left(TP^H_{D_S} + \theta(z)\left(I - P^H_{D_S}\right)\right)\left[I - z\left(TP^H_{D_T} + \theta(z)\left(I - P^H_{D_T}\right)\right)\right]^{-1} S. \end{aligned}$$

Let $U_\theta \in G_\theta$, be the minimal unitary extension of T corresponding to $\theta \in B(H \ominus D_T, H \ominus R_T)$, $D_\theta = \bigvee_{n \in \mathbb{Z}} U^n_\theta D_S$, $R_\theta = \bigvee_{n \in \mathbb{Z}} U^n_\theta R_S$ and $W_\theta : D_\theta \to R_\theta$

defined by $W_\theta\left(\sum_{n\in\mathbb{Z}} U_\theta^n d_n\right) = \sum_{n\in\mathbb{Z}} \bar{c}^n U_\theta^n S d_n$ with this notation there exists a bijection between β_{ext}^0 and

$$\bigcup_{\theta \text{ satisfies condition a)}} \{\alpha \in B(G_\theta \ominus D_\theta, G_\theta \ominus R_\theta) : U_\theta c\alpha(cz) = \alpha(z) U_\theta | G_\theta \ominus D_\theta\}.$$

References

[A-Aro] V.M. Adamjan and D.Z. Arov, *On unitary coupling of semiunitary operators*, Matem. Issledovanya (1966), 3–4 (in Russian).

[Ar] R. Arocena, *Generalized Toeplitz Kernels and Dilations of intertwining Operators*. Integral Equations and Operator Theory **6** (1983), 759–778.

[Ar-C-S] R. Arocena, M. Cotlar and C. Sadosky, *Weighted Inequalities in L^2 and Lifting Properties*. Adv. in Math. Suppl. Stud. A **7** (1981), 95–128. **6** (1983), 759–778.

[Aro-G] D.Z. Arov and L.Z. Grossmann, *Scattering matrices in the theory of dilations of isometric operators*. Soviet Math. Dokl. **27** (1983), 518–522.

[F-Fr1] C. Foias and A. Frazho, *The Commutant Lifting Approach to Interpolation Problems*, Birkhauser Verlag, Basel-Boston-Berlin, 1990.

[F-Fr2] C. Foias and A. Frazho, *Constructing the Schur Contration in the Commutant Lifting Theorem*. Acta Sci. Math. (Szeged) **61** (1995), 425–442.

[C-S1] M. Cotlar and C. Sadosky, *Integral Representations of Bounded Hankel Forms Defined in Scattering Systems with a Multiparametric Evolution Group Operator Theory*: Advances and Applications, **35** (1988), 357–375.

[C-S2] M. Cotlar and C. Sadosky, *The Generalized Bochner Theorem in Algebraic Scattering Systems*. Mathematical Sciences Research Institute, Berkeley California, 1987.

[C-S3] M. Cotlar and C. Sadosky, *Transference of Metrics Induced by Unitary Couplings*, a Sarason Theorem for the Bidimensional Torus, and Sz.-Nagy-Foias Theorem for Pairs of Dilations, Journal of Functional Analysis **11**, no. 2 (1993), 473–488.

[L-P] P. Lax and Phillips, *Scattering Theory*, Academic Press, New York 1967.

[M] M.D. Morán, *On intertwining Dilations*. Journal of Math. Anal. and App., **141**, no. 1 (1989), 219–234.

[N-F] B. Sz.-Nagy and C. Foias, *Analyse harmonique de operateurs de l'espace de Hilbert*, Masson-Akkad., Kiado, Paris-Budapest, 1970.

Dpto. de Matemática
Facultad de Ciencias
Universidad Central de Venezuela
Apartado 20513
Caracas 1020-A
Venezuela
mmoran@euler.ciens.ucv.ve

MSC numbers: Primary 42A82, Secondary 47D03

Operator Theory:
Advances and Applications, Vol. 115

On Interpolation Problems in the Complex Right Half Plane

M.J. Smith and A.E. Frazho

In this paper, an application of the commutant lifting theorem, is used to solve some mixed H^2 and H^∞ Sarason type problems for continuous time systems. A solution is presented that solves the H^∞ problem and a H^2 bound is obtained as a by product of our analysis. In contrast to the discrete time problem in [8], an additional weight is introduced on the H^2 norm that occurs naturally from the construction a certain shift operator in H^2. The theory is developed to include the irrational case, and state space formulas are provided for a certain rational Nehari problem.

1 Introduction

Recently, there has been an extensive amount of literature published on both the H^2 and H^∞ controls problem. Glover [11] was the first to initiate state space techniques to solve the rational Nehari problem. Since then, there has been an explosion of literature on both the H^2 and H^∞ problem; see [1, 2, 7, 9, 14, 18, 24] for some recent results and the history of this subject area. It is well known that H^2 optimal controllers yield nice closed loop performance properties, and H^∞ controllers are robust to uncertainties in the system. In practice, one has to compromise performance for a certain amount of robustness in the synthesis problem. To accomplish this, the synthesis problem consists of designing controllers for the closed loop system that yield an acceptable level of H^2 performance and inherits the H^∞ robustness. This leads to H^2-H^∞ type control problems, and the interested reader should consult references [4, 8, 10, 17, 19, 20] for a partial list of work accomplished in this area. The typical approach to solve these types of problems is based on Riccatti equations. In this paper, we will use commutant lifting techniques to solve a mixed H^2-H^∞ Sarason problem which naturally arises in control theory. One advantage to the commutant lifting approach is that it can be applied to both finite and infinite dimensional systems.

It is well known that one can use the Youla parameterization to convert many control problems to an equivalent Sarason or Nehari type optimization problem. For purely mathematical reasons, Kaftal, Larson and Weiss [15] obtained some nice H^2-H^∞ bounds for the Nehari problem in the unit disc. However, no computational algorithms were provided, and their techniques are not readily suitable for interpolation problems in the right half plane. By a direct application of the central solution to the commutant lifting theorem, it was shown in Foias and Frazho [6], and Foias, Frazho and Li [8] that the results in [15] could be extended to the general model matching problem arising in control theory. State space formulas for some rational H^2-H^∞ type interpolation problems are given in [7]. The commutant lifting theorem has been used extensively to solve interpolation problems in the unit disc; see [5, 7] for various applications. Here, we will show how one can use this theorem to directly solve some H^∞ problems in the right half plane. Although passage from the half plane to the unit disc and vise versa is standard, we prefer to use a functional passage from the time translation group to a shift operator in a certain H^2 space. This shift operator, which is multiplication by an explicit Mobius transformation, introduces a new complex parameter, which may allow more flexibility in design; see the example in the last section. Using this approach, we will obtain some mixed H^∞ and weighted H^2 bounds for a general model matching problem. Some explicit state space formulas in the rational case are also presented.

2 H^2 and H^∞ Optimization Problems

In this section, we will use the commutant lifting theorem to extend and tighten the Kaftal, Larson and Wise bounds [15] to the right half plane setting. Throughout, $L^2(\mathcal{E})$ is the space of all square integrable Lebesgue measurable functions $f(\imath w)$ for $-\infty < w < \infty$ with values in the Hilbert space $\mathcal{E}$, under the inner product $(f, g) = 1/(2\pi)\int(f, g)dw$. Let $\Re$ denote the real part of a complex number. The space $H^2(\mathcal{E})$ is the Hardy space of all analytic functions in the right half s-plane ($\Re(s) > 0$) with values in $\mathcal{E}$ whose norm

$$(1) \qquad ||f||_2^2 = \sup_{\sigma>0} \frac{1}{2\pi}\int_{-\infty}^{\infty} ||f(\sigma + \imath w)||^2 dw$$

is finite. By passing limits to the $\imath w$ axis a.e., $H^2(\mathcal{E})$ is viewed as a subspace of $L^2(\mathcal{E})$. Throughout, $K^2(\mathcal{E})$ is the orthogonal complement of $H^2(\mathcal{E})$ in $L^2(\mathcal{E})$. If $\mathcal{E}$ is the Hilbert space generated by the set of all Hilbert-Schmidt operators from $\mathcal{U}$ into $\mathcal{Y}$, then $L^2(\mathcal{E})$ is denoted by $L^2(\mathcal{U}, \mathcal{Y})$ and $H^2(\mathcal{E})$ is denoted by $H^2(\mathcal{U}, \mathcal{Y})$. As expected, $L^\infty(\mathcal{U}, \mathcal{Y})$ is the Banach space of all uniformly essentially bounded Lebesgue measurable functions on the interval $(-\infty, \infty)$, whose values are linear operators from $\mathcal{U}$ into $\mathcal{Y}$. The space $H^\infty(\mathcal{U}, \mathcal{Y})$ is the subspace of $L^\infty(\mathcal{U}, \mathcal{Y})$ corresponding to the set of all uniformly bounded analytic functions in $\Re(s) > 0$, whose values are linear operators from $\mathcal{U}$ into $\mathcal{Y}$.

To begin, let α be a scalar satisfying $\Re(\alpha) > 0$. Let ϕ be the function in H^2 defined by $\phi = \Re(\alpha)/(s + \bar{\alpha})$. Also let $||\cdot||_2$ denote the $L^2(\mathcal{U}, \mathcal{Y})$ norm and assume $\mathcal{U}$ is finite dimensional. Let F be a specified function in $L^\infty(\mathcal{U}, \mathcal{Y})$, and Θ an inner function in $H^\infty(\mathcal{E}, \mathcal{Y})$, that is, $\Theta(iw)$ is a.e. an isometry. Now consider the weighted H^2 optimization problem,

$$(2) \qquad d_{2,\alpha} = \inf\{||(F - \Theta H)\phi||_2 : H \in H^\infty(\mathcal{U}, \mathcal{E})\}.$$

For the solution to this problem, let P' denote the orthogonal projection onto the space $\mathcal{H}' := L^2(\mathcal{Y}) \ominus \Theta H^2(\mathcal{E})$. Then an application of the projection theorem yields $d_{2,\alpha} = \|P'F\phi\|_2$.

Now consider the H^∞ Sarason optimization problem in [22]:

$$d_\infty = \inf\{\|F - \Theta H\|_\infty : \ H \in H^\infty(\mathcal{U}, \mathcal{E})\}, \tag{3}$$

where F is a specified function in $L^\infty(\mathcal{U}, \mathcal{Y})$ and Θ is an inner function in $H^\infty(\mathcal{E}, \mathcal{Y})$. By using standard commutant lifting techniques [5, 7], it follows that the optimal error is given by $d_\infty = \|\Gamma\|$, where Γ is the operator from $H^2(\mathcal{U})$ into $\mathcal{H}'$ defined by $\Gamma = P'M_F|H^2(\mathcal{U})$. (Here, M_G denotes the multiplication operator from $L^2(\mathcal{U})$ into $L^2(\mathcal{Y})$ defined by $M_G u = Gu$ for u in $L^2(\mathcal{U})$ where G is in $L^\infty(\mathcal{U}, \mathcal{Y})$). In fact, this result will be a by product of our analysis. The optimization problems in (2) and (3) occur naturally in control theory; see [1, 2, 7, 9, 14]. The following is a right half plane version of some of the results in [7, 8, 15].

Theorem 1 *Let F be an element of $L^\infty(\mathcal{U}, \mathcal{Y})$, $\delta > 1$, and n the dimension of $\mathcal{U}$. Also let Θ be an inner function in $H^\infty(\mathcal{E}, \mathcal{Y})$. Then there exists an H_α in $H^\infty(\mathcal{U}, \mathcal{E})$ satisfying $\|F - \Theta H_\alpha\|_\infty \le \delta d_\infty$ and*

$$\|(F - \Theta H_\alpha)\phi\|_2 \le \delta d_{2,\alpha}(\delta^2 - 1 + 2d_{2,\alpha}^2/(n\Re(\alpha)d_\infty^2))^{-1/2}. \tag{4}$$

In particular, by choosing $\delta^2 = 2 - 2d_{2,\alpha}^2/(n\Re(\alpha)d_\infty^2)$, there exists an H_α in $H^\infty(\mathcal{U}, \mathcal{E})$ satisfying the following bounds

$$\|(F - \Theta H_\alpha)\phi\|_2 \le \sqrt{2}d_{2,\alpha}(1 - d_{2,\alpha}^2/(n\Re(\alpha)d_\infty^2))^{1/2} \tag{5}$$

$$\|F - \Theta H_\alpha\|_\infty \le \sqrt{2}d_\infty(1 - d_{2,\alpha}^2/(n\Re(\alpha)d_\infty^2))^{1/2}. \tag{6}$$

Notice that the results in Theorem 1 depend upon the choice of α. This α can be used as a weight in certain mixed H^2-H^∞ controls problems. Our proof of Theorem 1 will be based upon the central solution for the commutant lifting theorem; see [7, 8].

To complete this section, we will present some explicit formulas that enable us to construct an operator valued function H_α satisfying the properties of Theorem 1. To this end, let $\gamma = \delta\|\Gamma\|$ where $\delta > 1$, and Γ is the operator defined by $\Gamma = P'M_F|H^2(\mathcal{U})$. Let Φ be the operator from $\mathcal{U}$ into $H^2(\mathcal{U})$ defined by $\Phi u = \phi u$ for u in $\mathcal{U}$. Recall that ϕ has the following reproducing property, $(f, \phi) = \Re(\alpha)f(\alpha)$ for f in H^2. In particular, $\|\phi\|^2 = \Re(\alpha)/2$ and thus $\sqrt{2/\Re(\alpha)}\Phi$ is an isometry. Also let $D(s)$ be the analytic function in $\Re(s) > 0$, whose values are linear operators on $\mathcal{U}$ defined by

$$D(s)u = ((\gamma^2 I - \Gamma^*\Gamma)^{-1}\Phi u)(s) \quad (u \in \mathcal{U}). \tag{7}$$

Notice that for each $u \in \mathcal{U}$, the function $D(s)$ is in $H^2(\mathcal{U})$. In addition, it will be later shown that $((s + \bar{\alpha})D(s))^{-1}$ is an outer function in $H^\infty(\mathcal{U}, \mathcal{U})$. In particular,

$D(iw)$ is invertible a.e. Let $N(s)$ be the function, whose values are linear operators from $\mathcal{U}$ into $\mathcal{Y}$, defined by

$$N(s)u = (\Gamma Du)(s) \quad (u \in \mathcal{U}) \, . \tag{8}$$

Then clearly Nu is in $\mathcal{H}'$ for each u in $\mathcal{U}$. The following result uses the operator valued functions D and N, to construct a function H in $H^\infty(\mathcal{U}, \mathcal{Y})$, satisfying the bounds given in Theorem 1.

Theorem 2 *Assume that F is a function in $L^\infty(\mathcal{U}, \mathcal{Y})$. Let D and N be the operator valued functions defined in* (7) *and* (8), *respectively. Then $((s+\bar{\alpha})D)^{-1}$ is an outer function in $H^\infty(\mathcal{U}, \mathcal{U})$. In addition, $G = ND^{-1}$ admits a decomposition of the form $G = F - \Theta H$ where H is in $H^\infty(\mathcal{U}, \mathcal{E})$. Moreover, this G satisfies the bounds in Theorem* 1, *and $H = \Theta^*(F - ND^{-1})$.*

The proofs of these results are presented in the next section.

3 Proofs

To present the proofs of Theorems 1 and 2, we will first establish some nomenclature. Let T be an isometry on $\mathcal{H}$, T' a contraction on $\mathcal{H}'$, and Γ be an operator mapping $\mathcal{H}$ into $\mathcal{H}'$ satisfying $T'\Gamma = \Gamma T$. Recall that V on $\mathcal{K}(\supseteq \mathcal{H}')$ is an isometric lifting of T' if V is an isometry satisfying $V^*|\mathcal{H}' = T'^*$. An operator B mapping $\mathcal{H}$ into $\mathcal{K}$ is called an *intertwining lifting* of Γ if $VB = BT$ and $P'B = \Gamma$. The Sz.-Nagy-Foias commutant lifting theorem [5, 7, 23], states that there exists an intertwining lifting B of Γ satisfying $||B|| = ||\Gamma||$. By applying the commutant lifting theorem, it is well known that the optimal cost d_∞ defined in (3) is given by $d_\infty = ||\Gamma||$. Following some of the ideas in [7, 8], we will relax the constraint $||B|| = ||\Gamma||$, and construct an intertwining lifting B_γ that satisfies the bound $||B_\gamma|| \le \delta||\Gamma||$ for some scalar $\delta > 0$. This will allow us to construct intertwining lifting which also satisfy a H^2 constraint.

Recall that $l^2_+(\mathcal{U})$ is the Hilbert space formed by the set of all square summable unilateral sequences with components in $\mathcal{U}$. Now assume that $||\Gamma|| < \gamma$ and set $\Delta = (\gamma^2 I - \Gamma^*\Gamma)^{-1}$. Assume that Φ is an invertible operator from a Hilbert space $\mathcal{U}$ onto $\mathcal{L} = \ker T^*$. Let L from $l^2_+(\mathcal{U})$ into $\mathcal{H}$ and R from $l^2_+(\mathcal{U})$ into $\mathcal{K}$ be the linear maps defined by

$$L = \Big[\Delta\Phi, T\Delta\Phi, T^2\Delta\Phi, T^3\Delta\Phi, \cdots\Big] : l^2_+(\mathcal{U}) \to \mathcal{H} \tag{9}$$

$$R = \Big[\Gamma\Delta\Phi, V\Gamma\Delta\Phi, V^2\Gamma\Delta\Phi, \cdots\Big] : l^2_+(\mathcal{U}) \to \mathcal{K} \, . \tag{10}$$

Recall that an operator T on $\mathcal{H}$ is a unilateral shift, if T is an isometry and $\mathcal{H} = \oplus_0^\infty T^n\mathcal{L}$; see [5, 23]. Finally, let $||A||_{HS}$ represent the Hilbert-Schmidt norm of an operator A. This sets the stage for the following result which is taken from [8]; see also [7].

Theorem 3 ([8]) *Let T be a unilateral shift on $\mathcal{H}$ and T' a contraction on $\mathcal{H}'$. Assume $\mathcal{L} = \ker T^*$ is an n dimensional subspace. Let Γ be an operator from $\mathcal{H}$ into $\mathcal{H}'$ satisfying $\Gamma T = T'\Gamma$, and let V on $\mathcal{K}$ be an isometric lifting of T'. Also let $\delta > 1$ and γ the constant defined by $\gamma = \delta||\Gamma||$. Finally, let L and R be the linear maps defined in and* (9) *and* (10). *Then L^{-1} exists as a bounded linear operator from $\mathcal{H}$ into $l_+^2(\mathcal{U})$. Furthermore, $B_\gamma = RL^{-1}$ defines a bounded operator from $\mathcal{H}$ into $\mathcal{K}$, and this B_γ is an intertwining lifting of Γ satisfying $||B_\gamma|| \leq \delta||\Gamma||$ and*

$$||B_\gamma|\mathcal{L}||_{HS} \leq \delta||\Gamma|\mathcal{L}||_{HS}(\delta^2 - 1 + ||\Gamma|\mathcal{L}||^2_{HS}/(n||\Gamma||^2))^{-1/2}. \tag{11}$$

This $B_\gamma = RL^{-1}$ is referred to as the *central intertwining lifting* of Γ with *tolerance* γ. For a more detailed discussion of the central solution, see Chapter IV in [7].

Proof of Theorem 3: For completeness of presentation, let us present a proof of some of these results. By consulting [8], or Chapter IV in [7], we see that $\Delta\mathcal{L}$ is cyclic for T, that is, $\mathcal{H} = \bigvee_0^\infty T^n\Delta\mathcal{L}$. Hence, L is a linear map onto a dense set in $\mathcal{H}$. Let $f = \{f_n\}$ be any sequence in $l_+^2(\mathcal{U})$ with compact support. Then notice that

$$\begin{aligned}\gamma^2||Lf||^2 &= \gamma^2 \sum_{n,m=0}^{\infty} (T^n\Delta\Phi f_n, T^m\Delta\Phi f_m)\\ &= \gamma^2\sum_{n\geq m}(T^{n-m}\Delta\Phi f_n, \Delta\Phi f_m) + \gamma^2\sum_{n<m}(\Delta\Phi f_n, T^{m-n}\Delta\Phi f_m).\end{aligned} \tag{12}$$

Since V on $\mathcal{K}$ is an isometric lifting of T', we have $P'V = T'P'$ which shows that $P'V^k\Gamma = \Gamma T^k$ for all integers $k \geq 0$. Using this fact, a calculation similar to (12), yields

$$\begin{aligned}||Rf||^2 &= \sum_{n\geq m}(V^{n-m}\Gamma\Delta\Phi f_n, \Gamma\Delta\Phi f_m) + \sum_{n<m}(\Gamma\Delta\Phi f_n, V^{m-n}\Gamma\Delta\Phi f_m)\\ &= \sum_{n\geq m}(\Gamma^*\Gamma T^{n-m}\Delta\Phi f_n, \Delta\Phi f_m) + \sum_{n<m}(\Delta\Phi f_n, \Gamma^*\Gamma T^{m-n}\Delta\Phi f_m).\end{aligned} \tag{13}$$

By combining (12) and (13) along with $T^*\Phi = 0$, we obtain

$$\begin{aligned}&\gamma^2||Lf||^2 - ||Rf||^2 = \sum_{n\geq m}(\Delta^{-1}T^{n-m}\Delta\Phi f_n, \Delta\Phi f_m)\\ &+\sum_{n<m}(\Delta\Phi f_n, \Delta^{-1}T^{m-n}\Delta\Phi f_m) = \sum_{n=0}^{\infty}||\Delta^{1/2}\Phi f_n||^2 \geq \epsilon^2||f||^2.\end{aligned} \tag{14}$$

This holds for some $\epsilon > 0$, because Δ is positive and invertible. Thus, $\gamma||Lf|| \geq \epsilon||f||$ and hence L is one to one and onto a dense set in $\mathcal{H}$. So L^{-1} exists as a bounded operator. In fact, $||L^{-1}|| \leq \gamma/\epsilon$. Let B_γ be the linear map from a dense set in $\mathcal{H}$ to $\mathcal{K}$ defined by $B_\gamma Lf = Rf$ where f is any sequence in $l_+^2(\mathcal{U})$ with compact support. So if $g = Lf$, then (14) gives

$$\begin{aligned} \gamma^2\|g\|^2 - \|B_\gamma g\|^2 &= \gamma^2\|Lf\|^2 - \|Rf\|^2 \\ &= \|\text{diag}\,[\Delta^{1/2}\Phi, \Delta^{1/2}\Phi, \Delta^{1/2}\Phi, \cdots]L^{-1}g\|^2 \geq 0\,. \end{aligned} \tag{15}$$

Hence, B_γ is bounded by γ on a dense set in $\mathcal{H}$. So B_γ can be extended by continuity to a bounded operator on all of $\mathcal{H}$. This operator is also denoted by B_γ. Thus $B_\gamma = RL^{-1}$.

To verify that the lifting property $P'B_\gamma = \Gamma$, holds, notice that $\Gamma T = T'\Gamma$ and $T'P' = P'V$ gives

$$\begin{aligned} P'B_\gamma &= P'RL^{-1} = P'\Big[\Gamma\Delta\Phi, V\Gamma\Delta\Phi, V^2\Gamma\Delta\Phi, \cdots\Big]L^{-1} \\ &= \Big[\Gamma\Delta\Phi, \Gamma T\Delta\Phi, \Gamma T^2\Delta\Phi, \cdots\Big]L^{-1} = \Gamma LL^{-1} = \Gamma\,. \end{aligned} \tag{16}$$

Finally, to show that $VB_\gamma = B_\gamma T$, let S_+ be the standard unilateral shift on $l_+^2(\mathcal{U})$, that is, S_+ is the block matrix with the identity appearing directly below the main diagonal entries and zeros elsewhere. Then it is easy to verify that $TL = LS_+$ and $VR = RS_+$ on a dense set. Thus,

$$VB_\gamma L = VR = RS_+ = B_\gamma LS_+ = B_\gamma TL\,. \tag{17}$$

Hence $VB_\gamma = B_\gamma T$ on a dense set, and by continuity, are equal everywhere. Therefore, B_γ is an intertwining lifting of Γ satisfying $\|B_\gamma\| \leq \gamma$.

Since $\|B_\gamma\| \leq \gamma$ and $P'B_\gamma = \Gamma$, it follows that for all h in $\mathcal{H}$,

$$\begin{aligned} &\|(I - P')B_\gamma h\|^2 = \|B_\gamma h\|^2 - \|P'B_\gamma h\|^2 \\ &\leq \gamma^2\|h\|^2 - \|\Gamma h\|^2 = ((\gamma^2 I - \Gamma^*\Gamma)h, h) = \|\Delta^{-1/2}h\|^2\,. \end{aligned} \tag{18}$$

From (18) we conclude that there exists a contraction Y from $\mathcal{H}$ into $\mathcal{K}' \ominus \mathcal{H}'$ satisfying $(I - P')B_\gamma = Y\Delta^{-1/2}$. Since $P'B_\gamma = \Gamma$, we see that B_γ admits a decomposition of the form $B_\gamma = \Gamma + Y\Delta^{-1/2}$. Now let $\mathcal{F} = \Delta^{-1/2}T\mathcal{H}$ and $\mathcal{G} = \mathcal{H} \ominus \mathcal{F}$. Then we claim that $Y|\mathcal{G} = 0$. To verify this, notice that $\mathcal{G} = \Delta^{1/2}\mathcal{L} = \Delta^{1/2}\Phi\mathcal{U}$ where $\mathcal{L} = \ker T^*$. Therefore, any g in $\mathcal{G}$ is given by $g = \Delta^{1/2}\Phi u$ for some u in $\mathcal{U}$. For this g we have

$$\begin{aligned} Yg &= Y\Delta^{1/2}\Phi u = (B_\gamma - \Gamma)\Delta\Phi u = RL^{-1}\Delta\Phi u - \Gamma\Delta\Phi u \\ &= R(u \oplus 0 \oplus 0, \cdots) - \Gamma\Delta\Phi u = 0\,. \end{aligned} \tag{19}$$

Hence, $Y|\mathcal{G} = 0$ and thus $Y = YP_\mathcal{F}$. Because $\|Y\| \leq 1$ we have for any h in $\mathcal{H}$,

$$\begin{aligned} \|B_\gamma h\|^2 &= \|P'B_\gamma h\|^2 + \|(I - P')B_\gamma h\|^2 = \|\Gamma h\|^2 + \|YP_\mathcal{F}\Delta^{-1/2}h\|^2 \\ &\leq \|\Gamma h\|^2 + \|P_\mathcal{F}\Delta^{-1/2}h\|^2\,. \end{aligned} \tag{20}$$

Finally, using (20) and consulting [8] or Chapter IV in [7], we obtain (11). This completes the proof.

To prove Theorem 1, we need to introduce some isometries S and V on the appropriate H^2 and L^2 spaces. To this end, let α be any fixed scalar in the open right half complex plane, and b the Blaschke factor defined by

$$b(s) = \frac{s-\alpha}{s+\bar{\alpha}}. \tag{21}$$

Now, let S be the isometry on $H^2(\mathcal{U})$ defined by $Sf = bf$, and V the unitary operator on $L^2(\mathcal{Y})$ defined by $Vg = bg$. This sets the stage for the following fundamental result whose proof is omitted.

Lemma 1 *Let B be an operator form $H^2(\mathcal{U})$ into $L^2(\mathcal{Y})$. Then*

(i) $VB = BS$ *if and only if* $B = M_G|H^2(\mathcal{U})$ *where* G *is a function in* $L^\infty(\mathcal{U}, \mathcal{Y})$.

(ii) $VB = BS$ *and* $BH^2(\mathcal{U}) \subseteq H^2(\mathcal{Y})$ *if and only if* $B = M_G|H^2(\mathcal{U})$ *where* G *is a function in* $H^\infty(\mathcal{U}, \mathcal{Y})$.

(iii) *If* (i) *or* (ii) *holds, then* B *and* G *uniquely determine each other, and* $||B|| = ||G||_\infty$.

Proof of Theorem 1: As before, let S on $H^2(\mathcal{U})$ and V on $L^2(\mathcal{Y})$ be the isometry and unitary operator, respectively, defined by $S = M_b|H^2(\mathcal{U})$ and $V = M_b|L^2(\mathcal{Y})$. To convert Theorem 1 to the setting of the commutant lifting theorem, let $T = S$ and set $\mathcal{H}' := L^2(\mathcal{Y}) \ominus \Theta H^2(\mathcal{E})$. Let T' on $\mathcal{H}'$ be the operator defined by compressing V to $\mathcal{H}'$, that is, $T' = P'V|\mathcal{H}'$. Using the fact that $\Theta H^2(\mathcal{E})$ is an invariant subspace for V, it follows that V is an isometric lifting of T', that is, $P'V = T'P'$. Let Γ be the operator from $H^2(\mathcal{U})$ into $\mathcal{H}'$ defined by $\Gamma = P'M_F|H^2(\mathcal{U})$. By employing $P'V = T'P'$, we observe that Γ intertwines T with T', that is, $T'\Gamma = \Gamma T$. According to Theorem 3, there exists an intertwining lifting B_γ of Γ satisfying $||B_\gamma|| \leq \delta||\Gamma||$, and the bound given in (11). Since $VB_\gamma = B_\gamma S$, Lemma 1 shows that there exists a unique G in $L^\infty(\mathcal{U}, \mathcal{Y})$ such that $B_\gamma = M_G|H^2(\mathcal{U})$. Because $P'B_\gamma = \Gamma$, we have

$$P'Gh = P'B_\gamma h = \Gamma h = P'Fh \quad (h \in H^2(\mathcal{U})). \tag{22}$$

Hence, $P'(F - G)h = 0$ for h in $H^2(\mathcal{U})$, and it follows that $G = F - \Theta H_\alpha$ for some H_α in $H^\infty(\mathcal{U}, \mathcal{E})$. From this and the definition of d_∞ in (3), we obtain

$$d_\infty \leq ||F - \Theta H_\alpha||_\infty = ||G||_\infty = ||B_\gamma||_\infty \leq \delta||\Gamma||. \tag{23}$$

Clearly, for any H in $H^\infty(\mathcal{U}, \mathcal{E})$, we have

$$||\Gamma|| = ||P'(F - \Theta H)|H^2(\mathcal{U})|| \leq ||F - \Theta H||_\infty. \tag{24}$$

Hence, $||\Gamma|| \le d_\infty$. By combining this with (23), we obtain $||\Gamma|| \le d_\infty \le \delta||\Gamma||$ for any $\delta \ge 1$. Therefore, $||\Gamma|| = d_\infty$, as expected, and $||F - \Theta H_\alpha||_\infty \le \delta d_\infty$.

To verify (4), first note that a simple application of the Cauchy residue formula shows that $(f, \Phi u) = \Re(\alpha)(f, u)_{\mathcal{U}}$ where f is in $H^2(\mathcal{U})$ and u is in $\mathcal{U}$. In particular, this shows that $||\phi||^2 = (\phi, \phi) = \Re(\alpha)/2$. Hence, $\sqrt{2/\Re(\alpha)}\Phi$ is a unitary operator from $\mathcal{U}$ onto $\mathcal{L} = \ker T^* = \Phi\mathcal{U}$. Thus,

$$d_{2,\alpha} = ||P'F\phi||_2 = ||P'F\Phi||_{HS} = \sqrt{\Re(\alpha)/2}||\Gamma|\mathcal{L}||_{HS}\,. \tag{25}$$

Using the result in (25) and the fact that $B_\gamma = M_G|H^2(\mathcal{U})$, Theorem 3 yields

$$\begin{aligned} ||G\phi||_2 &= ||B_\gamma\Phi||_{HS} \le \delta||\Gamma\Phi||_{HS}(\delta^2 - 1 + ||\Gamma|\mathcal{L}||^2_{HS}/(nd^2_\infty))^{-1/2} \\ &= \delta d_{2,\alpha}(\delta^2 - 1 + 2d^2_{2,\alpha}/(n\Re(\alpha)d^2_\infty))^{-1/2}\,. \end{aligned} \tag{26}$$

Therefore (4) holds and the proof is complete.

Proof of Theorem 2: Recall that the central intertwining lifting $B_\gamma = M_G|H^2(\mathcal{U})$ where G is an element of $L^\infty(\mathcal{U}, \mathcal{Y})$. Moreover $G = F - \Theta H_\alpha$, satisfies the bounds in Theorem 1. It remains to show that $G = ND^{-1}$ where N and D are defined in (7) and (8). To this end, first note that $B_\gamma L = R$ implies that $B_\gamma\Delta\Phi = \Gamma\Delta\Phi$, thus

$$GD = G\Delta\Phi = B_\gamma\Delta\Phi = \Gamma\Delta\Phi = N\,. \tag{27}$$

Now let us demonstrate that D is an outer function and thus $G = ND^{-1}$, which proves Theorem 2. Let W be the operator from $l^2_+(\mathcal{U})$ onto $H^2(\mathcal{U})$ defined by

$$W = \left[\Phi, S\Phi, S^2\Phi, S^3\Phi, \cdots\right]. \tag{28}$$

Because $\Phi\mathcal{U}$ is a wandering subspace for S, it follows that $\sqrt{2/\Re(\alpha)}W$ is unitary. Now since $WS_+ = SW$ and $LS_+ = SL$ on a dense set, it is also true that $LW^{-1}S = SLW^{-1}$ on a dense set. Since L is invertible, WL^{-1} commutes with the unilateral shift S. According to Lemma 1, there exists a Q in $H^\infty(\mathcal{U}, \mathcal{U})$ satisfying $WL^{-1} = M_{\check{Q}}|H^2(\mathcal{U})$. Because L^{-1} is onto a dense set, Q is outer. To obtain a formula for Q, notice that for u in $\mathcal{U}$, we have

$$Q^{-1}\phi u = LW^{-1}\Phi u = L(u \oplus 0 \oplus 0 \oplus 0 \cdots) = \Delta\Phi u = Du\,. \tag{29}$$

Hence, $\Re(\alpha)Q^{-1} = (s + \bar{\alpha})D$. Thus $(s + \bar{\alpha})^{-1}D^{-1}$ is an outer function in $H^\infty(\mathcal{U}, \mathcal{U})$. Therefore, $D(\imath w)$ is a.e. invertible and $G = ND^{-1}$. This completes the proof.

4 Some State Space Formulas

For completeness, let us obtain some state space formulas for the rational Nehari problem associated with Theorem 1. For a state space solution to many interpolation problems; see [1, 7, 9, 14]. Throughout this section, F in Theorem 1 is rational and $\Theta = I$. In this case, Theorem 1 yields a continuous time version of the mixed H^2-H^∞ bounds given for the Nehari problem in [15]. An expression for D in (7) and N in (8) will be presented in terms of explicit state space realizations.

Let us recall some terminology from control theory. We say that F is a rational function in $K^\infty(\mathcal{U}, \mathcal{Y})$, if $F(s)$ is a proper rational function with all of its poles in the open right half complex plane. Let A be an operator on a finite dimensional vector space $\mathcal{X}$, and B an operator from $\mathcal{U}$ into $\mathcal{X}$, while C is an operator from $\mathcal{X}$ into $\mathcal{Y}$. Then $\{A, B, C\}$ is a *minimal anticausel realization* for F, if the pair $\{A, B\}$ is controllable, $\{C, A\}$ is observable, and

$$F(s) = C(-sI - A)^{-1}B\,. \tag{30}$$

For further details on realization theory, see [16, 21]. Throughout this section we assume F is in $K^\infty(\mathcal{U}, \mathcal{Y})$ and $\{A\,, B\,, C\}$ is a minimal anticausel realization for F. Since F is in $K^\infty(\mathcal{U}, \mathcal{Y})$, all of the eigenvalues of A are in the open left half plane. Because $\Theta = I$, the operator Γ defined in Section 2 is now the Hankel operator mapping $H^2(\mathcal{U})$ into $\mathcal{H}' = K^2(\mathcal{Y}) := L^2(\mathcal{Y}) \ominus H^2(\mathcal{Y})$ defined by $\Gamma = P'M_F|H^2(\mathcal{U})$. Here P' is the orthogonal projection onto the space $K^2(\mathcal{Y})$. Let W_o be the observability operator from $\mathcal{X}$ into $K^2(\mathcal{Y})$ and W_c be the controllability operator from $\mathcal{X}$ into $H^2(\mathcal{U})$ defined by

$$W_o x = C(-sI - A)^{-1}x \text{ and } W_c x = B^*(sI - A^*)^{-1}x \quad (x \in \mathcal{X})\,. \tag{31}$$

It is well known (see [1, 7, 9, 11, 14, 24]) that this Hankel operator admits a decomposition of the form $\Gamma = W_o W_c^*$. The controllability Gramian P and the observability Gramian Q are defined by $P = W_c^* W_c$ and $Q = W_o^* W_o$. Because $\{A, B, C\}$ is a controllable and observable realization, both P and Q are invertible positive definite operators. Futhermore, P and Q are the unique solutions to the following Lyapunov equations

$$AP + PA^* + BB^* = 0 \text{ and } A^*Q + QA + C^*C = 0\,. \tag{32}$$

The minimal realization $\{A, B, C\}$ along with its controllability and observability Gramians P and Q, will play a key role in our state space formulas for N and D. Finally, recall that $||\Gamma||^2 = d_\infty^2$ equals the largest eigenvalue of QP; see [1, 7, 9, 11, 14, 24].

Using the previous definitions, the H^2 optimization problem in (2) is solved by decomposing $F\Phi$ into orthogonal components. Notice that

$$F\Phi = \Re(\alpha)(C(\bar{\alpha}I - A)^{-1}(s + \bar{\alpha})^{-1}B + C(-sI - A)^{-1}(\bar{\alpha}I - A)^{-1}B)\,. \tag{33}$$

The first and second terms of (33) are contained in $H^\infty(\mathcal{U}, \mathcal{Y})$ and $K^\infty(\mathcal{U}, \mathcal{Y})$, respectively. So if H is defined by

$$H = C(\bar{\alpha}I - A)^{-1}B\,, \tag{34}$$

then $(F - H)\phi$ is in $K^\infty(\mathcal{U}, \mathcal{Y})$. Thus $d_{2,\alpha} = \|(F - H)\phi\|_2$. Moreover, since $Q = W_o^* W_o$, the optimal H^2 error is given by

$$
\begin{aligned}
d_{2,\alpha}^2 &= \|\Re(\alpha) W_o(\bar{\alpha} I - A)^{-1} B\|_2^2 \\
&= \Re(\alpha)^2 \operatorname{trace}\left(B^*(\alpha I - A^*)^{-1} Q(\bar{\alpha} I - A)^{-1} B\right).
\end{aligned} \tag{35}
$$

The following result provides us with state space realizations for the functions N and D in Theorem 2.

Theorem 4 *Let $\{A, B, C\}$ be a minimal anticausel realization for the function F in $K^\infty(\mathcal{U}, \mathcal{Y})$. Let P and Q be the solutions to the Lyapunov equations in* (32), *and let λ_{max} be the largest eigenvalue of QP. Then $d_\infty^2 = \lambda_{max}$. Set $\delta > 1$ and $\gamma = \delta d_\infty$. Let N_α in $K^\infty(\mathcal{U}, \mathcal{Y})$ and D_α in $H^\infty(\mathcal{U}, \mathcal{Y})$ be the rational functions given by*

$$N_\alpha = \gamma^2 W_o(\gamma^2 I - PQ)^{-1}(\bar{\alpha} I - A)^{-1} B, \tag{36}$$

$$D_\alpha = (s + \bar{\alpha})^{-1} I + W_c Q(\gamma^2 I - PQ)^{-1}(\bar{\alpha} I - A)^{-1} B. \tag{37}$$

Then $G_\alpha = N_\alpha D_\alpha^{-1}$ is a function in $L^\infty(\mathcal{U}, \mathcal{Y})$ satisfying the bounds in Theorem 1. *Finally, G_α admits a decomposition of the form $G_\alpha = F - H_\alpha$ for some H_α in $H^\infty(\mathcal{U}, \mathcal{Y})$.*

Proof: Following [12], a straight forward calculation involving $P = W_c^* W_c$ and $Q = W_o^* W_o$, along with $\Gamma = W_o W_c^*$ shows that

$$
\begin{aligned}
\Delta &= (\gamma^2 I - \Gamma^* \Gamma)^{-1} = (\gamma^2 I - W_c W_o^* W_o W_c^*)^{-1} \\
&= \gamma^{-2} I + \gamma^{-2} W_c(\gamma^2 Q^{-1} - P)^{-1} W_c^*.
\end{aligned} \tag{38}
$$

Thus, it follows that

$$
\begin{aligned}
D &= \Delta\Phi = \gamma^{-2}\Re(\alpha)((s + \bar{\alpha})^{-1} I + B^*(sI - A^*)^{-1} Q(\gamma^2 I - PQ)^{-1} \\
&\quad (\bar{\alpha} I - A)^{-1} B),
\end{aligned} \tag{39}
$$

$$
\begin{aligned}
N &= \Gamma\Delta\Phi = W_o W_c^*(\gamma^{-2}\Phi I + \gamma^{-2}\Re(\alpha) W_c(\gamma^2 Q^{-1} - P)^{-1} \\
&\quad (\bar{\alpha} I - A)^{-1} B) \\
&= \Re(\alpha) C(-sI - A)^{-1}(\gamma^2 I - PQ)^{-1}(\bar{\alpha} I - A)^{-1} B.
\end{aligned} \tag{40}
$$

By cancelling common terms, we obtain $ND^{-1} = N_\alpha D_\alpha^{-1}$ where N_α and D_α defined in (36) and (37), respectively. This completes the proof. □

Notice that if we let $\alpha \to \infty$ along the real axis, then (36) and (37) show that $G_\alpha = N_\alpha D_\alpha^{-1}$ approaches $G = \hat{N}\hat{D}^{-1}$, where $\hat{N}$ and $\hat{D}$ are the functions defined by

(41) $\hat{N} = \gamma^2 W_o(\gamma^2 I - PQ)^{-1}B$ and $\hat{D} = I + W_c Q(\gamma^2 I - PQ)^{-1}B$.

By passing limits in (35), we obtain $d_{2,\infty}^2 = d_2^2 = ||P'F||_2^2 = \text{trace } B^*QB$. Moreover, by (4) the function $G = \hat{N}\hat{D}^{-1} = F - H$ for H in $H^\infty(\mathcal{U}, \mathcal{Y})$ solves the classical Nehari problem with the bounds

$$(42) \qquad ||G||_\infty \le \delta d_\infty \text{ and } ||G||_2 \le \frac{\delta d_2}{\sqrt{\delta^2 - 1}},$$

where $\delta > 1$. Finally, let us note that this G agrees with the central or maximal entropy solution for the Nehari problem in [1, 9]. So we have shown that this central solution also satisfies the bounds given in [15].

5 Example

To see how the parameter α affects the solution of the Nehari interpolation problem, consider the function F in K^∞ given by

$$F(s) = \frac{0.114s + 3.7211}{s^2 - 0.01s + 100}.$$

By using Theorem 4 for values of $\delta = 1.001, \sqrt{2}, \infty$ and $\alpha = 1/10, 10, \infty$, we computed the function $G_\alpha = N_\alpha D_\alpha^{-1} = F - H_\alpha$. Then we plotted $|G_\alpha(\imath w)|$ on a log-log scale for $1/100 \le w \le 100$ in Figure 1. The top, middle, and lower graphs correspond to $\delta = 1.001, \sqrt{2}, \infty$, respectively. In all graphs the solid line corresponds to $|G_\alpha(\imath w)|$ for $\alpha = 1/10$, whereas the two dashed lines correspond to $|G_\alpha(\imath w)|$ for $\alpha = 10$ and $\alpha = \infty$. In all three cases, the magnitudes satisfy $|G_\infty(\imath w)| \le |G_{10}(\imath w)| \le |G_{1/10}(\imath w)|$ for large frequences w. Whereas, $|G_{1/10}(\imath w)| \le |G_{10}(\imath w)| \le |G_\infty(\imath w)|$ for small frequences w. So in this example, if one wants the error $|G_\alpha(\imath w)|$ to perform like a low pass filter (respectively, high pass filter), then one chooses α large (respectively, small for high pass). In this sense, the designer can vary α and δ to gain additional freedom in the design process. In many control problems, one would like to force the error to be small in some specified frequency range. By varying α and δ, the above method may be useful in signal processing and control design. Finally, in this example all of the bounds in Theorem 1 have been satisfied.

Acknowledgements

This research was supported in part by a grant from NASA Dryen Flight Research Center.

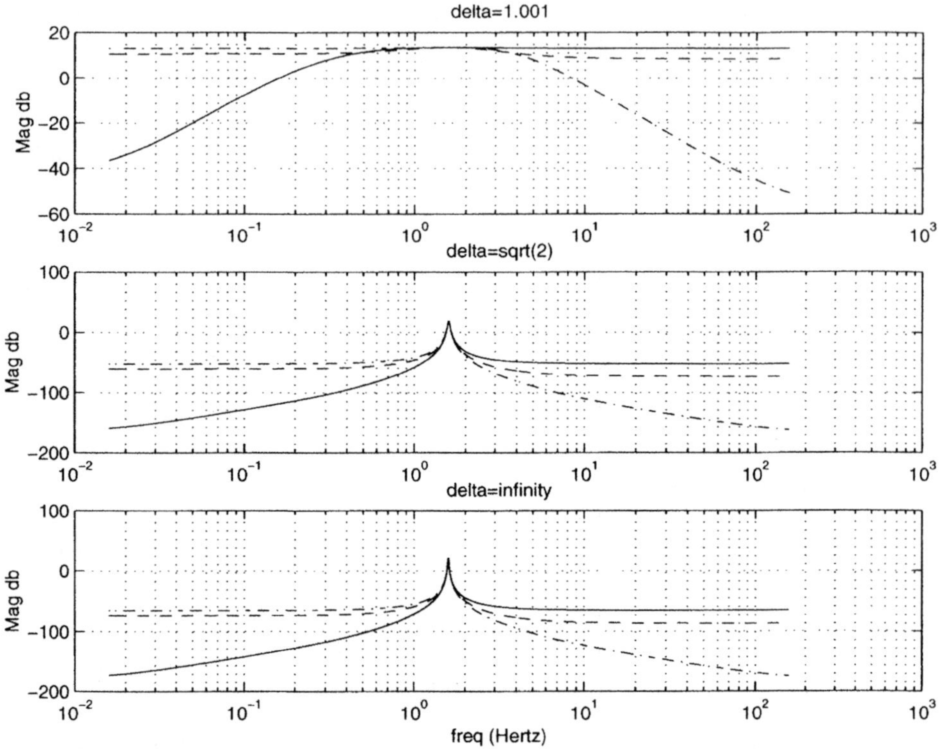

Figure 1

References

[1] J.A. Ball, I. Gohberg and L. Rodman, *Interpolation For Rational Matrix Functions*, Operator Theory: Advances and Applications, **45**, Birkhäuser Verlag, Basel, 1990.

[2] J.C. Doyle, B.A. Francis and A. Tannenbaum, *Feedback Control Theory*, MacMillan, New York, 1991.

[3] J.C. Doyle, K. Glover, P.P. Khargonekar and B.A. Francis, State-space solutions to standard H_2 and H_∞ control problems, *IEEE Trans. Autom. Contr.*, **34** (1988), pp. 831–847.

[4] J.C. Doyle, K. Zhou, K. Glover and B. Bodenheimer, Optimal control with mixed H_2 and H_∞ performance objectives, *IEEE Trans. Autom. Contr.*, **39** (1994), pp. 1575–1587.

[5] C. Foias and A.E. Frazho, *The Commutant Lifting Approach to Interpolation Problems*, Operator Theory Advances and Applications, **44**, Birkhäuser Verlag, Basel, 1990.

[6] C. Foias and A.E. Frazho, Commutant lifting and simultaneous H^∞ and L^2 suboptimization, *SIAM J. Math. Anal.*, **23**(4) (1992), pp. 984–994.

[7] C. Foias, A.E, Frazho, I. Gohberg and M.A. Kaashoek, *Metric Constrained Interpolation, Communtant Lifting and Systems*, Operator Theory: Advances and Applications, **100**, Birkhäuser Verlag, Basel, 1998.

[8] C. Foias, A.E. Frazho and W.S. Li, The exact H^2 estimate for the central H^∞ interpolant, in: *New aspects in interpolation and completion theories* (Ed. I Gohberg) Operator Theory Advances and Applications, **64**, Birkhäuser Verlag, Basel (1993), pp. 119–156.

[9] B.A. Francis, *A Course in Control Theory*, Lecture Notes in Control and Information Sciences, Springer-Verlag, New York, 1987.

[10] A.E. Frazho and M.A. Rotea, A remark on mixed L^2/L^∞ bounds, *Integral Equations and Operator Theory*, **15** (1992), pp. 343–348.

[11] K. Glover, All optimal Hankel-norm approximations of linear multivariable systems and their error bounds, *International J. Control*, **39** (1984), pp. 1115–1193.

[12] I. Gohberg, M.A. Kaashoek and F. Van Schagen, Rational contractive and unitary interpolants in realized form, *Integral Equations and Operator Theory*, **11** (1988), pp. 105–127.

[13] I. Gohberg, M.A. Kaashoek and H. J. Woerdeman, The band method for positive and strictly contractive extension problems: an alternative version and new applications, *Integral Equations and Operator Theory*, **12** (1989), pp. 343–382.

[14] M. Green and D. Limebeer, *Linear Robust Control*, Prentice Hall, Inc., 1995.

[15] V. Kaftal, D. Larson and G. Weiss, Quasitriangular subalgebras of semifinite Von Neumann algebras are closed, *J. Functional Analysis*, **107** (2) (1992), pp. 387–401.

[16] T. Kailath, *Linear Systems*, Englewood Cliffs: Prentice Hall, New Jersey, 1980.

[17] P.P. Khargonekar and M.A. Rotea, Mixed H^2/H^∞ control: a convex optimization approach, *IEEE Trans. Autom. Control*, **36** (7) (1991), pp. 824–837.

[18] D. Mustafa and K. Glover, Minimum Entropy H_∞ Control, Lecture notes in Control and Information Sciences, *Springer-Verlag*, New York, 1990.

[19] M. Peters and P. Iglesias, The minimum entropy controller as a mixed H_2/H_∞ controller: the time varying case, *Proc. of 34th Conference on Decision and Control*, December 1995.

[20] M. Rotea and A.E. Frazho, Bounds on solutions to H^∞ algebraic Riccati equations and properties of central solution, *Systems and Control Letters*, **19** (15) (1992), pp. 341–352.

[21] W.J. Rugh, *Linear System Theory*, Prentice Hall, Inc., New Jersey, 1993.

[22] D. Sarason, Generalized interpolating in H^∞, *Trans. American Math. Soc.*, **127** (1967), pp. 179–203.

[23] B. Sz.-Nagy and C. Foias, *Harmonic Analysis of Operators on Hilbert Space*, North-Holland Publishing Co., Amsterdam, 1970.

[24] K. Zhou, J.C. Doyle and K. Glover, *Robust and Optimal Control*, Prentice Hall, Inc., 1996.

School of Aeronautics and Astronautics
Purdue University
West Lafayette
IN 47907
U.S.A.

MSC 1992: Primary, 47A20, 47A57, 93B36.

Operator Theory:
Advances and Applications, Vol. 115

Maximum Entropy Truncable Positive Definite Sequences of Analytic Toeplitz Operators*

Ion Suciu

In this paper we present a geometric approach for studying positive definite sequence of analytic Toeplitz operators following the lines of the general framework of operator extrapolation theory based on choice sequences (cf.[1], [3], [6], [11]). The specific problem here appears is when the positive definite sequence $(T_n)_{n\geq 1}$ produced by a choice sequence $(R_n)_{n\geq 1}$ initiated on H^2 is a sequence of analytic Toeplitz operators. Using special matrix representation based on the choice sequence of the minimal isometric dilation $[K, V]$ of $(T_n)_{n\geq 1}$, we prove, in Section 2, that T_n is analytic Toeplitz operator for any n if and only if there exists an isometry U on K which commutes with V such that H^2, as the subspace of K, reduces U to the shift operator S on H^2 (Theorem 1). This connects our problem with the topics of Ando dilations (cf.[2], [7], [8]). In Section 3 we write some recurrent formulas appearing in this theory under the specific forms of our context. Section 4 deals with the maximum entropy truncable sequences i.e. the sequences $(T_n)_{n\geq 1}$ of analytic Toeplitz operators whose choice sequence $(R_n)_{n\geq 1}$ has the property that for each $p \geq 1$, the truncated choice sequence $\{R_1, R_2, \ldots, R_p, 0, 0, \ldots\}$ still produces positive definite sequence of analytic Toeplitz operators. We proved that this happens if and only if, in the matricial form based on the choice sequence, the isometry U is diagonal (Theorem 5). Then we are able to give an algorithm to produce, by a system of free parameters, the set of all maximum entropy truncable positive definite sequences of analytic Toeplitz operators . We end with some considerations concerning a special moment problem on $\mathbf{Z}^2$ connected with this results (Section 5).

In Section 1, following in principal the book "The Commutant Lifting Approach to Interpolation Problems" by C. Foias and A.E. Frazho, we present some preliminaries and notations.

1 Preliminaries

Let H be a Hilbert space and $(T_n)_{n\geq 1}$ be a sequence of operators on H. The sequence $(T_n)_{n\geq 1}$ is *positive definite* if for any $n \geq 1$ the attached Toeplitz matrix

*This work was supported in part by the Romanian Academy under contract GAR 5573/1996–1997.

$[T_{ij}]_{\substack{1\le i\le n\\ 1\le j\le n}} = [T_{i-j}]_{\substack{1\le i\le n\\ 1\le j\le n}}$, where $T_0 = I_H$ and $T_{-k} = T_k^*, k \ge 0$, defines a positive operator on the orthogonal sum of n copies of H. The Naymark dilation theorem (cf.[16], [11]) assures the existence and the uniqueness of the minimal isometric dilation $[K, V]$ of $(T_n)_{n\ge 1}$, i.e. K is a Hilbert space containing H as the closed subspace and V is an isometry on K such that

$$K = \bigvee_{n\ge 0} V^n H \tag{1.1}$$

$$T_n = P_H V^n | H, \quad n \ge 1. \tag{1.2}$$

It results that for any $n \ge 1$, T_n is a contraction on H.

A *choice sequence* on H (cf.[6], [11]) is a sequence $(R_n)_{n\ge 1}$, where R_1 is a contraction on H and for any $n \ge 2$, R_n is a contraction from $\mathcal{D}_{R_{n-1}}$ into $\mathcal{D}_{R^*_{n-1}}$. Here and all over the paper we denote by $D_A = [I - A^*A]^{1/2}$ the defect operator of the contraction A and by $\mathcal{D}_A = \overline{D_A H}$-the defect space.

In the recursive construction of a choice sequence $(R_n)_{n\ge 1}$ there are no restrictions in choosing R_n unless its domain $\mathcal{D}_{R_{n-1}}$ and its codomain $\mathcal{D}_{R^*_{n-1}}$. Hence any choice sequence $R = (R_n)_{n\ge 1}$ can be truncated, for any $p \ge 1$, to the choice sequence $R^{p|0} = (R_1, R_2, \dots, R_p, 0, 0, \dots)$.

To any choice sequence we can attach the pair $[K, V] = [K(R), V(R)]$ where $K = K(R)$ is the Hilbert space containing H as a closed subspace, defined by

$$K = K(R) = H \oplus \mathcal{D}_{R_1} \oplus \mathcal{D}_{R_2} \oplus \cdots \tag{1.3}$$

and $V = V(R)$ is the isometry on $K = K(R)$ defined by the matrix

$$V = \begin{bmatrix} R_1 & D_{R_1^*} R_2 & D_{R_1^*} D_{R_2^*} R_3 & \cdots \\ D_{R_1} & -R_1^* R_2 & -R_1^* D_{R_2^*} R_3 & \cdots \\ 0 & D_{R_2} & -R_2^* R_3 & \cdots \\ \cdot & \cdot & \cdot & \cdots \end{bmatrix}. \tag{1.4}$$

One of the most important results of the operator extrapolation theory (cf.[6], [11]) is the bijective correspondence between the set of all choice sequences $R = (R_n)_{n\ge 1}$ on H and the set of all positive definite sequences $T = (T_n)_{n\ge 1}$ on H given by $R \to T(R) = (T_n)_{n\ge 1}$, where

$$T_n = P_H V^n | H, \quad n \ge 1,$$

with $V = V(R)$ the isometry on $K = K(R)$ attached to R as above.

Clearly $[K, V] = [K(R), V(R)]$ is (coincides to) the minimal isometric dilation of $T(R) = (T_n)_{n\ge 1}$.

In this correspondence, for any $p \ge 1$ the string $\{T_1, T_2, \dots, T_p\}$ depends only on the string $\{R_1, R_2, \dots, R_p\}$ and conversely.

It results that for any $p \ge 1$ there exists a positive definite sequence $T(R^{p|0}) = (T_n^{p|0})_{n\ge 1}$, hawing as choice sequence $R^{p|0} = \{R_1, R_2, \dots, R_p, 0, 0, \dots\}$, which is an extension of the string $\{T_1, T_2, \ldots, T_p\}$. We call this extension the *maximum entropy extension* of $\{T_1, T_2, \ldots, T_p\}$ (cf.[5], [9], [11]). This denomination comes from prediction theory (cf.[4], [14]) and has a real support whenever the entropy of the attached spectral distribution is defined (in the scalar case, for example).

We shall use also the notation

$$K_n = \bigvee_{j=0}^{n-1} V^j H, \qquad V_n = P_{K_n} V | K_n, \qquad n \geq 1.$$

In the matricial form we have

$$K_n = H \oplus \mathcal{D}_{R_1} \oplus \ldots\ldots + \mathcal{D}_{R_{n-1}} \tag{1.5}$$

and

$$V_n = \begin{bmatrix} R_1 & D_{R_1^*} R_2 & \cdots & D_{R_1^*} D_{R_2^*} \cdots D_{R_{n-1}^*} R_n \\ D_{R_1} & -R_1^* R_2 & \cdots & -R_1^* D_{R_1^*} \cdots D_{R_{n-1}^*} R_n \\ 0 & D_{R_2} & \cdots & -R_2^* D_{R_3^*} \cdots D_{R_{n-1}^*} R_n \\ . & \cdot\ \cdots & & . \\ . & \cdot\ \cdots & & . \\ . & \cdot\ \cdots & & . \\ 0 & 0\ \cdots & & -R_{n-1}^* R_n \end{bmatrix}. \tag{1.6}$$

Note that, for $n \geq 1$, K_n depends only on $\{R_1, R_2 \ldots\ldots R_{n-1}\}$ and V_n depends only on $\{R_1, R_2, \ldots, R_n\}$. We can obtain $[K, V]$ from $[K_n, V_n]$ by inductive strong limit.

Note also that, for $n \geq 2$,

$$V | K_{n-1} = V_n | K_{n-1} \tag{1.7}$$

and that $T_n = T^n$ for same contraction T on H if and only if the corresponding choice sequence R has the form $R = \{T, 0, 0, \ldots\}$ (cf.[11]).

2 Sequences of Analytic Toeplitz Operators

Let now H be the scalar Hardy space H^2. Any function f in H^∞ defines, by pointwise multiplication, the bounded operator T_f on H^2-the analytic Toeplitz operator of symbol f. Multiplication by coordinate function on H^2 is the shift operator denoted by S. A bounded operator on H^2 is Toeplitz analytic if and only if it commutes with S (cf.[12],[13]).

We are interested in producing, by free parameters, positive definite sequences of analytic Toeplitz operators. Following the way of producing positive definite sequences by choice sequences, we try to give conditions under which a choice sequence on H^2 produces positive definite sequence of analytic Toeplitz operators.

We begin with the following geometric characterization of the choice sequences which produce positive definite sequences of analytic Toeplitz operators.

Theorem 1 *Let $R = (R_n)_{n\geq 1}$ be a choice sequence on H^2, $(T_n)_{n\geq 1}$ be its corresponding positive definite sequence and $[K, V] = [K(R), V(R)]$ be the minimal isometric dilation of $(T_n)_{n\geq 1}$. The following assertions are equivalent:*

(i) *For any $n \geq 1$, $T_n = T_{f_n}$ for some $f_n \in H^\infty$.*

(ii) *There exists an isometry $U = U(R)$ on K such that $UV = VU$ and H^2, as the subspace of K, reduces U to the shift operator S on H^2.*

Moreover, if (i) (or equivalently (ii)) holds then, for any $n \geq 2$, the subspace $K_n = \bigvee_{j=0}^{n-1} V^j H^2$ of K is invariant to U. Setting $U_n = U|K_n$ then, on $K_n = K_{n-1} \oplus \mathcal{D}_{R_{n-1}}$, the isometry U_n has the following recurrent matricial form:

$$U_n = \begin{bmatrix} U_{n-1} & \Gamma_n \\ 0 & Y_n \end{bmatrix}, \tag{2.1}$$

where Γ_n and Y_n are defined by:

$$\Gamma_2 = 0, \quad Y_2 D_{R_1} h = D_{R_1} S h, \quad h \in H^2 \tag{2.2}$$

and for $n \geq 3$

$$\Gamma_n D_{R_{n-1}} \dots D_{R_1} h = (V_{n-1}U_{n-1} - U_{n-1}V_{n-1}) \begin{bmatrix} 0 \\ \cdot \\ \cdot \\ \cdot \\ D_{R_{n-2}} \dots D_{R_1} h \end{bmatrix}, \quad h \in H^2, \tag{2.3}$$

$$Y_n D_{R_{n-1}} \dots D_{R_1} h = D_{R_{n-1}} Y_{n-1} D_{R_{n-2}} \dots D_{R_1} h, \quad h \in H^2. \tag{2.4}$$

We also have

$$U_n^* V_n U_n = V_n \tag{2.5}$$

i.e. V_n is a U_n-Toeplitz operator.

Proof: Suppose that the assertion (ii) holds. Then for any $n \geq 1$ and any $h \in H^2$ we have

$$T_n S h = P_{H^2} V^n S h = P_{H^2} V^n U h = P_{H^2} U V^n h = U P_{H^2} V^n h = S T_n h.$$

Hence T_n commutes with the shift S on H^2 which implies $T_n = T_{f_n}$ for some $f_n \in H^\infty$. The assertion (i) results.

Suppose now that (i) holds. Setting for any $n \geq 0$ and any $h_0, h_1, \dots, h_n$ in H^2

$$U \sum_{j=0}^{n} V^j h_j = \sum_{j=0}^{n} V^j S h_j, \tag{2.6}$$

we have

$$\begin{aligned}
\left\| U \sum_{j=0}^{n} V^j h_j \right\|_K^2 &= \left(\sum_{j=0}^{n} V^j S h_j,\ \sum_{k=0}^{n} V^k S h_k \right)_K \\
&= \sum_{j \geq k} (V^{j-k} S h_j,\ S h_k)_K + \sum_{j<k} (S h_j,\ V^{k-j} S h_k)_K \\
&= \sum_{j \geq k} (T_{f_{j-k}} S h_j,\ S h_k)_{H^2} + \sum_{j<k} (S h_j,\ T_{f_{k-j}} S h_k)_{H^2} \\
&= \sum_{j \geq k} (S T_{f_{j-k}} h_j,\ S h_k)_{H^2} + \sum_{j<k} (S h_j,\ S T_{f_{k-j}} h_k)_{H^2} \\
&= \sum_{j \geq k} (T_{f_{j-k}} h_j,\ h_k)_{H^2} + \sum_{j<k} (h_j,\ T_{f_{k-j}} h_k)_{H^2} = \cdots = \left\| \sum_{j=1}^{n} V^j h_j \right\|_K^2 .
\end{aligned}$$

It results that (2.6) defines an isometry U on K. Clearly U commutes with V and coincides with S on H^2.

For $h \in H^2$ with $h(0) = 0$ we have $h = Sg = Ug$ for some $g \in H^2$. Hence $U^* h = U^* U g = g \in H^2$. In case $h = c$ -a constant function in H^2 then we have

$$\begin{aligned}
(U^* h,\ V^n g)_K &= (h,\ U V^n g)_K = (h,\ V^n U g)_K \\
&= (h,\ P_{H^2} V^n S g)_{H^2} = (h,\ T_{f_n} S g)_{H^2} = 0.
\end{aligned}$$

We showed that H^2 is invariant to U^* hence it reduces U to S. The assertion (ii) is proved.

Clearly $K_n = \vee_{j=0}^{n} V^j H^2$ is invariant to any U as in (ii). It results that for any $n \geq 2$, on $K_n = K_{n-1} \oplus \mathcal{D}_{R_{n-1}}$, U_n can be written under the form (2.1).

Since H^2 reduces U, it results $\Gamma_2 = 0$ and from

$$0 = (UV - VU)h = U \begin{bmatrix} R_1 h \\ D_{R_1} h \end{bmatrix} - VSh = \begin{bmatrix} S R_1 h \\ Y_2 D_{R_1} h \end{bmatrix} - \begin{bmatrix} R_1 S h \\ D_{R_1} S h \end{bmatrix}$$

it results

$$Y_2 D_{R_1} h = D_{R_1} S h, \quad h \in H^2.$$

which is (2.2).

Using induction, for $n \geq 3$ and for any element k_{n-1} belonging to K_{n-1}, of the form $k_{n-1} = 0 \oplus \cdots \oplus 0 \oplus D_{R_{n-2}} \ldots D_{R_1}h$, $h \in H^2$, we obtain:

$$\begin{aligned} 0 &= (UV - VU)k_{n-1} = (U_n V_n - V_n U_n)k_{n-1} = U_n V_n k_{n-1} - V_n U_{n-1} k_{n-1} \\ &= \begin{bmatrix} U_{u-1} & \Gamma_n \\ 0 & Y_n \end{bmatrix} \begin{bmatrix} V_{n-1}k_{n-1} \\ D_{R_{n-1}} \cdots D_{R_1}h \end{bmatrix} - \begin{bmatrix} V_{n-1}U_{n-1}k_{n-1} \\ D_{R_{n-1}}Y_{n-1}D_{R_{n-1}} \cdots D_{R_1}h \end{bmatrix} \\ &= \begin{bmatrix} (U_{n-1}V_{n-1} - V_{n-1}U_{n-1})k_{n-1} + \Gamma_n D_{R_{n-1}} \cdots D_{R_1}h \\ Y_n D_{R_{n-1}} \cdots D_{R_1}h - D_{R_{n-1}}Y_{n-1}D_{R_{n-2}} \cdots D_{R_1}h \end{bmatrix}, \end{aligned}$$

which proves (2.3) and (2.4).

The Theorem is completely proved. □

Corollary 1 *For any $n \geq 2$ the isometry U_n has a diagonal form on the space $K_n = H^2 \oplus \mathcal{D}_{R_1} \oplus \cdots \oplus \mathcal{D}_{R_{n-1}}$ if and only if, on K_{n-1}, U_{n-1} commutes with V_{n-1}.*

We say that the choice sequence $(R_n)_{n\geq 1}$ on H^2 *produces* a positive definite sequence of analytic Toeplitz operator if one of the equivalent assertions (i) or (ii) of the Theorem 1 is fulfilled.

If $f \in H^\infty$, $\|f\| \leq 1$, then $(T_f^n)_{n\geq 1} = (T_{f^n})_{n\geq 1}$ is positive definite. As we know, the choice sequence corresponding to $(T_f^n)_{n\geq 1}$ is $\{T_f, 0, 0, \ldots\}$.

Hence a choice sequence on H^2 of the form $(R_1, 0, 0, \ldots)$ produces positive definite sequence of analytic Toeplitz operators if and only if $R_1 = T_f$ for some $f \in H^\infty$, $\|f\| \leq 1$.

3 Extensions to L^2

Consider now H^2 as the subspace of L^2 - the Lebesgue space on the unit circle $\mathbf{T}$ in the complex plane. Pointwise multiplication by a function r from L^∞ gives rise to a bounded operator M_r on L^2. Multiplication by the coordinate function on L^2 is the bilateral shift denoted here by $\hat{S}$ which is the unique minimal unitary extension of the unilateral shift S on H^2. An operator $\hat{R}$ on L^2 is of the form M_r with $r \in L^\infty$ if and only if $\hat{R}\hat{S} = \hat{S}\hat{R}$.

Theorem 2 *Let $(f_n)_{n\geq 1}$ be a sequence of functions in H^∞. The sequence $(T_{f_n})_{n\geq 1}$ of operators on H^2 is positive definite if and only if the corresponding sequence $(M_{f_n})_{n\geq 1}$ of operators on L^2 is positive definite.*

Proof: Positive definiteness of $(M_{f_n})_{n\geq 1}$ clearly implies positive definiteness of $(T_{f_n})_{n\geq 1}$.

Suppose that $(T_{f_n})_{n\geq 1}$ is positive definite and let $[K, V]$ be the minimal isometric dilation of $(T_{f_n})_{n\geq 1}$. According to Theorem 2, there exists the isometry U on K, which commutes with V, such that H^2-reduces U to the shift S. Let $[\hat{K}, \hat{U}]$ be the minimal unitary extension of $[K, U]$ (cf.[10]). This means that $\hat{K}$ is a Hilbert space, $K \subset \hat{K}$, K is invariant to $\hat{U}$, $\hat{U}|K = U$ and $\hat{K} = \bigvee_{n\geq 0} \hat{U}^{*n}K$. The isometry V which commutes with U can be uniquely extended to an isometry $\hat{V}$ on $\hat{K}$ such that $\hat{U}\hat{V} = \hat{V}\hat{U}$ and $V = \hat{V}|K$.

Let $\hat{K}_0 = \bigvee_{n\geq 0} \hat{U}^{*n}H^2$. Clearly $\hat{K}_0$ reduces $\hat{U}$. Let $\hat{U}_0 = \hat{U}|\hat{K}_0$. We have $H^2 \subset \hat{K}_0$, $\hat{K}_0 = \bigvee_{n\geq 0} \hat{U}_0^{*n}H^2$ and $\hat{U}_0|H^2 = S$. Hence $[\hat{K}_0, \hat{U}_0]$ is the minimal unitary extension of $[H^2, S]$ and can be identified to $[L^2, \hat{S}]$.

Define, for any $n \geq 1$, $\hat{T}_n = P_{L^2}\hat{V}^n|L^2$. We obtain a positive definite sequence $(\hat{T}_n)_{n\geq 1}$ of operators on L^2.

Since for any $\hat{h} \in L^2$ we have

$$\hat{T}_n\hat{S}\hat{h} = P_{L^2}^{\hat{K}}\hat{V}^n\hat{U}\hat{h} = P_{L^2}\hat{U}\hat{V}^n\hat{h} = \hat{U}P_{L^2}\hat{V}^n\hat{h} = \hat{S}\hat{T}_n\hat{h},$$

it results that $\hat{T}_n$ is multiplication operator. Since for any $h, g \in H^2$ and $p \geq 1$ we have

$$\begin{aligned}(\hat{T}_n h, \hat{S}^{*p}g) &= (\hat{S}^p\hat{T}_n h, g) = (\hat{S}^p P_{L^2}\hat{V}^n h, g) = (P_{L^2}\hat{U}^p\hat{V}^n h, g)\\ &= (\hat{U}^p\hat{V}^n h, g) = (\hat{V}^n\hat{U}^p h, g) = (V^n U^p h, g) = (T_{f_n}S^p h, g)\\ &= (S^p T_{f_n}h, g) = (\hat{S}^p T_{f_n}h, g) = (T_{f_n}h, \hat{S}^{*p}g),\end{aligned}$$

it results that $T_{f_n} = \hat{T}_n|H^2$. Clearly then $\hat{T}_n = M_{f_n}$. Hence $(M_{f_n})_{n\geq 1}$ is positive definite.

The Theorem is completely proved. □

We say that a choice sequence $(\hat{R}_n)_{n\geq 1}$ on L^2 *produces* a positive definite sequence of analytic Toeplitz operators if for any $n \geq 1$ there exists f_n in H^∞ such that the positive definite sequences corresponding to $(\hat{R}_n)_{n\geq 1}$ is $(M_{f_n})_{n\geq 1}$.

Theorem 2 clearly establishes a bijection between the set of all choice sequences $(\hat{R}_n)_{n\geq 1}$ on L^2 which produce positive definite sequences of analytic Toeplitz operators and the set of all choice sequences $(R_n)_{n\geq 1}$ on H^2 having the same property: $(\hat{R}_n)_{n\geq 1}$ and $(R_n)_{n\geq 1}$ are in this correspondence if and only if there exists the sequence $(f_n)_{n\geq 1}$, $f_n \in H^\infty$ such that the positive definite sequence corresponding to $(\hat{R}_n)_{n\geq 1}$, is $(M_{f_n})_{n\geq 1}$, and the positive definite sequence corresponding to $(R_n)_{n\geq 1}$ is $(T_{f_n})_{n\geq 1}$.

The next result will describe the connection between the geometric models attached to the corresponding choice sequences.

Theorem 3 *Let $(\hat{R}_n)_{n\geq 1}$ be a choice sequence on L^2, $(\hat{T}_n)_{n\geq 1}$ be its corresponding positive definite sequence and $[\hat{K}, \hat{V}]$ be the minimal isometric dilation of $(\hat{T}_n)_{n\geq 1}$. Let us denote*

$$K = \bigvee_{n\geq 0} \hat{V}^n H^2, V = \hat{V}|K, T_n = P_{H^2}V^n|H^2. \tag{3.1}$$

Suppose that, for any $n \geq 1$, $\hat{T}_n = M_{f_n}$ *for some* $f_n \in H^\infty$. *Then:*

a) We have

$$\hat{R}_1 \hat{S}_1 = \hat{S}_1 \hat{R}_1, \quad \hat{R}_n \hat{S} | \mathcal{D}_{\hat{R}_{n-1}} = \hat{S} \hat{R}_n, \quad n \geq 2. \tag{3.2}$$

b) We have

$$T_n = T_{f_n}, \quad n \geq 1. \tag{3.3}$$

c) If U *is the isometry on* K *corresponding to* $(T_n)_{n\geq 1}$ *as in the Theorem 1 and* $\hat{U}$ *is the unitary operator defined on*

$$\hat{K} = L^2 \oplus \mathcal{D}_{\hat{\mathcal{R}}_1} \oplus \mathcal{D}_{\hat{\mathcal{R}}_2} \oplus \cdots$$

by the matrix

$$\hat{U} = \begin{bmatrix} \hat{S} & 0 & 0 & \dots \\ 0 & \hat{Y}_2 & 0 & \dots \\ 0 & 0 & \hat{Y}_3 & \dots \\ \dots & \dots & \dots & \dots \end{bmatrix} \tag{3.4}$$

where, for any $n \geq 2$, $\hat{Y}_n$ *is the unitary operator on* $\mathcal{D}_{\hat{R}_{n-1}}$ *defined by*

$$\hat{Y}_n D_{\hat{R}_{n-1}} \dots D_{\hat{R}_1} \hat{h} = D_{\hat{R}_{n-1}} \dots D_{\hat{R}_1} \hat{S} \hat{h}, \quad \hat{h} \in L^2,$$

then $\hat{U}$ *on* $\hat{K}$ *is (can be identified to) the minimal unitary extension of the isometry* U *and* $\hat{U}\hat{V} = \hat{V}\hat{U}$, *i.e.* $\hat{V}$ *is the unique extension to a commutant of* $\hat{U}$ *of the commutant* V *of* U.

d) For $n \geq 2$, *the embedding* Z_n *of* $K_n = \bigvee_{j=1}^{n-1} V^j H^2$ *in* $\hat{K}_n = \bigvee_{j=0}^{n-1} \hat{V}^j L^2$, *as operator from* $K_n = K_{n-1} \oplus \mathcal{D}_{\mathcal{R}_{n-1}}$ *into* $\hat{K}_n = \hat{K}_{n-1} \oplus \mathcal{D}_{\hat{\mathcal{R}}_{n-1}}$ *has the following recursive matricial form*

$$Z_n = \begin{bmatrix} Z_{n-1} & a_{n-1} \\ 0 & b_{n-1} \end{bmatrix}, \tag{3.5}$$

where Z_1 *is the inclusion of* H^2 *into* L^2, $a_1 = 0$, $b_1 D_{R_1} h = D_{\hat{R}_1} h$, $h \in H^2$ *and, for* $n \geq 3$, *denoting* $k_{n-1} = 0 \oplus \cdots \oplus 0 \oplus D_{R_{n-2}} \cdots D_{R_1} h$, *we have*

$$a_{n-1} D_{R_{n-1}} \dots D_{R_1} h = (\hat{V}_{n-1} Z_{n-1} - Z_{n-1} V_{n-1}) k_{n-1}, \tag{3.6}$$

$$b_{n-1} D_{R_{n-1}} \dots D_{R_1} h = D_{\hat{R}_{n-1}} \dots D_{\hat{R}_1} h. \tag{3.7}$$

e) For any $n \geq 1$, $\hat{U}_n = \hat{U} | \hat{K}_n$ *is the minimal unitary extension of the isometry* U_n *on* K_n *and* $\hat{V}$ *is the unique contractive dilation to a commutant of* $\hat{U}_n$ *(the symbol) of the* U_n *- Toeplitz operator* V_n *on* K_n. *In the matricial form this means*

$$\hat{U}_n Z_n = Z_n U_n, \tag{3.8}$$

$$Z_n^* V_n Z_n = V_n. \tag{3.9}$$

Proof: Since $\hat{R}_1 = \hat{T}_1 = M_{f_1}$ is a normal operator commuting with $\hat{S}$, it results $\hat{S}D_{\hat{R}_1} = \mathcal{D}_{\hat{R}_1}\hat{S}$. Hence $\mathcal{D}_{\hat{R}_1}$ reduces $\hat{S}$ to the unitary operator $\hat{Y}_2$ defined by $\hat{Y}_2 D_{\hat{R}_1}\hat{h} = D_{\hat{R}_1}\hat{S}\hat{h}$, $\hat{h} \in L^2$.

Moreover from

$$M_{f_2}\hat{h} = P_{L^2}\hat{V}^2\hat{h} = \hat{R}_1^2\hat{h} + D_{\hat{R}_1^*}\hat{R}_2 D_{\hat{R}_1}\hat{h}, \quad \hat{h} \in L^2,$$

it results

$$D_{\hat{R}_1^*}\hat{R}_2 D_{\hat{R}_1}\hat{h} = M_{f_2}\hat{h} - \hat{R}_1^2\hat{h},$$

hence

$$\begin{aligned} D_{\hat{R}_1^*}\hat{R}_2\hat{Y}_2 D_{\hat{R}_1}\hat{h} &= D_{\hat{R}_1^*}\hat{R}_2 D_{\hat{R}_1}\hat{S}\hat{h} = M_{f_2}\hat{S}\hat{h} - \hat{R}_1^2\hat{S}\hat{h} = \hat{S}(M_{f_2}\hat{h} - \hat{R}_1^2\hat{h}) \\ &= \hat{S}D_{\hat{R}_1^*}\hat{R}_2 D_{\hat{R}_1}\hat{h} = D_{\hat{R}_1^*}\hat{S}\hat{R}_2 D_{\hat{R}_1}\hat{h} \end{aligned}$$

which implies $\hat{R}_2 Y_2 = \hat{S}\hat{R}_2$.

Suppose that for $n \geq 3$ and $2 \leq j \leq n-1$,

$$\hat{R}_j\hat{S}|\mathcal{D}_{\hat{R}_{j-1}} = \hat{S}\hat{R}_j.$$

It results that $\mathcal{D}_{\hat{R}_j}$ reduces $\hat{S}$ to a unitary operator $\hat{Y}_j = \hat{S}|\mathcal{D}_{\hat{R}_{j-1}}$ and $\hat{R}_j\hat{Y}_j = \hat{S}\hat{R}_j$. Clearly then for any $\hat{h} \in L^2$, $\hat{V}^{n-1}\hat{h}$ can be written under the form

$$\hat{V}^{n-1}\hat{h} = \begin{bmatrix} M_{f_{n-1}}\hat{h} \\ L_2\hat{h} \\ \cdot \\ \cdot \\ \cdot \\ L_{n-2}\hat{h} \\ D_{\hat{R}_{n-1}} \cdots\cdots D_{\hat{R}_1}\hat{h} \end{bmatrix},$$

where for $2 \leq j \leq n-2$, L_j are the operators from L^2 into $\mathcal{D}_{\hat{R}_{j-1}}$ verifying $L_j\hat{V}_j = \hat{S}L_j$.

It results

$$\begin{aligned} M_{f_n}\hat{h} &= P_{L^2}\hat{V}^n\hat{h} = \hat{R}_1 M_{f_{n-1}}\hat{h} + \cdots + D_{R_1^*}\cdots D_{R_{n-2}^*}L_{n-2}\hat{h} \\ &\quad + D_{\hat{R}_1^*}\cdots D_{R_{n-1}^*}\hat{R}_n D_{\hat{R}_{n-1}}\cdots D_{\hat{R}_1}\hat{h}. \end{aligned}$$

This clearly implies $\hat{R}_n\hat{S}|\mathcal{D}_{\hat{R}_{n-1}} = \hat{S}\hat{R}_n$, and the assertion a) is proved.

From

$$T_n h = P_{H^2}V^n h = P_{H^2}P_{L^2}\hat{V}^n h = P_{H^2}M_{f_n}h = T_{f_n}h$$

it results b).

Since L^2 reduces $\hat{U}$ to $\hat{S}$ it is clear that H^2 is invariant to $\hat{U}$ and $\hat{U}|H^2 = S$. Clearly then $\hat{U}|K = U$. Moreover

$$\begin{aligned}\bigvee_{n\geq 0} \hat{U}^{*n}K &= \bigvee_{n\geq 0} \hat{U}^{*n} \bigvee_{k\geq 0} \hat{V}^k H^2 = \bigvee_{n\geq 0} \hat{V}^k \bigvee_{n\geq 0} \hat{U}^{*n} H^2 \\ &= \bigvee_{k\geq 0} \hat{V}^k \bigvee_{n\geq 0} \hat{S}^{*n} H^2 = \bigvee_{k\geq 0} V^k L^2 = \hat{K}.\end{aligned}$$

Hence $[\hat{K}, \hat{U}]$ is the minimal unitary extension (cf.[10]) of $[K, U]$ and the assertion c) is proved.

Let us prove the assertion d). Clearly the embedding of the space $K_n = K_{n-1} \oplus \mathcal{D}_{R_{n-1}}$ into $\hat{K}_{n-1} \oplus \mathcal{D}_{\hat{R}_{n-1}}$ is given by a triangular matrix as in (3.5).

For $n = 2$ we have

$$Z_2(R_1 h \oplus D_{R_1} h) = Z_2 V h = \hat{V} h = \hat{R}_1 h \oplus D_{\hat{R}_1} h.$$

Hence

$$a_1 D_{R_1} h = 0, b_1 D_{R_1} h = D_{\hat{R}_1} h, h \in H^2.$$

For $n \geq 3$ and $h \in H^2$ we have:

$$\begin{aligned}&Z_n(V_{n-1}V^{n-2}h \oplus D_{R_{n-1}} \cdots D_{R_1} h) = Z_n V^{n-1} h = \hat{V}^{n-1} h \\ &\quad = \hat{V}_{n-1} \hat{V}^{n-2} h \oplus D_{\hat{R}_{n-1}} \cdots D_{\hat{R}_1} h.\end{aligned}$$

Hence

$$Z_n(0 \oplus D_{R_{n-1}} \cdots D_{R_1} h) = \hat{V}_{n-1} \hat{V}^{n-2} h - Z_n V_{n-1} V^{n-2} h \oplus D_{\hat{R}_{n-1}} \cdots D_{\hat{R}_1} h.$$

It results

$$b_{n-1} D_{R_{n-1}} \cdots D_{R_1} h = D_{\hat{R}_{n-1}} \cdots D_{\hat{D}_{\hat{R}_1}} h$$

and

$$\begin{aligned}&a_{n-1} D_{R_{n-1}}, \ldots, D_{R_1} h \\ &\quad = (\hat{V}_{n-1} Z_{n-1} - Z_{n-1} V_{n-1}) V^{n-2} h \\ &\quad = (\hat{V}_{n-1} Z_{n-1} - Z_{n-1} V_{n-1})(P_{K_{n-2}} V^{n-2} D_{R_{n-2}} \cdots D_{R_1} h) \\ &\quad = [\hat{V}_{n-1} Z_{n-1} - Z_{n-1} V_{n-1}] k_{n-1},\end{aligned}$$

with $k_{n-1} = 0 \oplus D_{R_{n-2}} \cdots D_{R_1} h \in K_{n-1}$. We used the fact that on K_{n-2} according to (1.7) we have

$$\begin{aligned}&(\hat{V}_{n-1} Z_{n-1} - Z_{n-1} V_{n-1}) k_{n-2} = \hat{V}_{n-1} Z_{n-2} k_{n-2} - Z_n V_{n-1} k_{n-2} \\ &\quad = \hat{V} Z_{n-2} k_{n-2} - Z_{n-2} V k_{n-2} = 0.\end{aligned}$$

The assertion d) is proved.

As in the proof of c) we can show that $\hat{U}_n$ is the minimal unitary extension of U_n. Clearly $\hat{U}_n\hat{V}_n = \hat{V}_n\hat{U}_n$ and for any $k, k' \in K_n$ we have

$$(\hat{V}_n k, k') = (P_{\hat{K}_n}\hat{V}k, k') = (\hat{V}k, k') = (Vk, k') = (P_{K_n}Vk, k') = (V_n k, k').$$

Hence $V_n = P_{K_n}\hat{V}_n | K_n$ which proves e).

The theorem is completely proved. □

Corollary 2 *For any $n \geq 2$ the following assertions are equivalent*

(i) *U_n is diagonal,*

(ii) *Z_n is diagonal,*

(iii) *b_n is an isometry.*

Proof: U_n is diagonal if and only if $U_{n-1}V_{n-1} = V_{n-1}U_{n-1}$ (see Theorem 1). This happens if and only if the U_{n-1} - Toeplitz operator V_{n-1} is a commutant of U_{n-1} which is equivalent to the fact that the symbol $\hat{V}_{n-1}$ of V_{n-1} extends it. According to the Theorem 2 this happens if and only if $a_{n-1} = 0$ i.e. if and only if b_{n-1} is an isometry. □

Working with $(\hat{R}_n)_{n\geq 1}$ instead of $(R_n)_{n\geq 1}$ in producing positive definite sequences of analytic Toeplitz operators we have the advantage that $\hat{R}_n$ are practically multiplication operators. Using Theorem 2 we can describe, in a similar way as in [8], the obstructions which appear at each step n, in choosing $\hat{R}_n$. But, although some simplifications appear, the difficulties in producing recurrently $(\hat{R}_n)_{n\geq 1}$ remain, essentially, the same as in the general case.

The next result is an illustration of the above remarks.

Theorem 4 *Let $f \in H^\infty$, $\|f\| \leq 1$. The sequence $(T_{f^n})_{n\geq 1} = (T_f^n)_{n\geq 1}$ is the only positive definite sequence $(T_{f_n})_{n\geq 1}$ of analytic Toeplitz operators verifying $f_1 = f$ if and only if f is an extreme point in the unit ball of H^∞, i.e., if and only if*

$$\int_{\mathbf{T}} \log\,(1 - |f|^2)dt = -\infty. \tag{3.10}$$

Proof: Suppose firstly that (3.10) holds. Let $(T_{f_n})_{n\geq 1}$ be a positive definite sequence of analytic Toeplitz operators with $f_1 = f$ and let $[K, V]$ be its isometric dilation. Then for any $n \geq 1$ and any $h \in H^2$ we have

$$\begin{aligned}
1/2\pi \int_{\mathbf{T}} |f_{n+1} - f_n f_1|^2 |h|^2 dt &= \|(f_{n+1} - f_n f_1)h\|^2_{H^2} \\
&= \|(T_{f_{n+1}} - T_{f_n}T_{f_1})h\|^2_{H^2} = \|(P_{H^2}V^{n+1} - P_{H^2}V^n P_{H^2}V)h\|^2_{H^2} \\
&= \|P_{H^2}V^n(I - P_{H^2})Vh\|^2_K \leq \|(I - P_{H^2})Vh\|^2_K \\
&= \|Vh\|^2_K - \|P_{H^2}Vh\|^2_K = \|h\|^2_{H^2} - \|T_{f_1}h\|^2_{H^2} = 1/2\pi \int_{\mathbf{T}} (1 - |f_1|^2)|h|^2 dt.
\end{aligned}$$

It results that

$$\int_{\mathbf{T}} |f_{n+1} - f_n f|^2 \varphi \, dt \leq \int_{\mathbf{T}} (1 - |f_1|^2)\varphi dt,$$

for any $\varphi =$ square of the modulus of any analytic polynomial, firstly, then for any $\varphi =$ square of the modulus of any trigonometric polynomial and finally for any positive continuous function φ on $\mathbf{T}$.

We conclude that

$$|f_{n+1} - f_n f_1|^2 \leq 1 - |f_1|^2$$

a.e. on $\mathbf{T}$ which implies

$$\int_{\mathbf{T}} log|f_{n+1} - f_n f_1|^2 dt \leq \int_{\mathbf{T}} log(1 - |f_1|^2)dt = -\infty.$$

Since $f_{n+1} - f_n f_1 \in H^\infty$ this can happen if and only if $f_{n+1} = f_n f_1, \quad n \geq 1$. It results $f_n = f_1^n = f^n$.

Suppose now that

$$\int_{\mathbf{T}} log\ (1 - |f|^2)dt > -\infty. \tag{3.11}$$

We shall construct a class of choice sequences $(\hat{R}_n)_{n\geq 1}$ of the form $(M_{r_n})_{n\geq 1}, r_n \in L^\infty$, which produce positive definite sequences $(T_{f_n})_{n\geq 1}$ of analytic Toeplitz operators having $r_1 = f_1 = f$.

Take $r_1 = f_1 = f$. From (3.11), using the known theorems of Beurling and Szegö, it results there exists a function $r_2 \in L^\infty, r_2 \neq 0$ such that

$$g = r_2(1 - |f|^2) \in H^\infty.$$

Clearly we can take r_2 of arbitrary small norm. Choose $0 < \varepsilon < 1$ such that

$$\varepsilon < \prod_{j=0}^{\infty}(1 - \varepsilon^{2j}),$$

and take r_2 as above such that

$$0 < |r_2| < \varepsilon^2.$$

Define now recurrently, for any $n \geq 3$

$$r_n = (1 - |r_{n-1}|^2)^{-1} \cdots (1 - |r_2|^2)^{-1} r_{n-1} r_2 \overline{f}. \tag{3.12}$$

An easy induction show that $|r_n| \leq \varepsilon^n$ for any $n \geq 2$. Taking $\hat{R}_n = M_{r_n}$ we obtain a choice sequence $(\hat{R}_n)_{n\geq 1}$ on L^2 such that for any $n \geq 1, \quad \mathcal{D}_{\hat{R}_n} = L^2$.

The corresponding $[\hat{K}, \hat{V}]$ is

$$\hat{K} = L^2 \oplus L^2 \oplus L^2 \oplus \cdots$$

and

$$\hat{V} = \begin{bmatrix} f & (1-|f|^2)^{1/2} r_2 & (1-|f|^2)^{1/2}(1-|r_2|^2))^{1/2} r_3 & \cdots \\ (1-|f|^2)^{1/2} & -\overline{f} r_2 & -\overline{f}(1-|r_2|^2)^{1/2} r_3 & \cdots \\ 0 & (1-|r_2|^2)^{1/2} & -\overline{r}_2 r_3 & \cdots \\ \cdot & \cdot & \cdot & \cdots \\ \cdot & \cdot & \cdot & \cdots \\ \cdot & \cdot & \cdot & \cdots \end{bmatrix}.$$

(3.13)

Clearly $\hat{T}_n = P^{\hat{K}}_{L^2} \hat{V}^n | L^2$ is by the form M_{f_n} with $f_n \in L^\infty$.
We have

$$f_n = \hat{T}_n 1 = P^{\hat{K}}_{L^2} \hat{V}^n \begin{bmatrix} 1 \\ 0 \\ 0 \\ \vdots \end{bmatrix}.$$

Using the form (3.13) of $\hat{V}$ and the recurrent formula (3.12) of r_n we obtain

$$\hat{V} \begin{bmatrix} 1 \\ 0 \\ 0 \\ \vdots \end{bmatrix} = \begin{bmatrix} f \\ (1-|f|^2)^{1/2} \\ 0 \\ \vdots \end{bmatrix} = \begin{bmatrix} f_1 \\ (1-|f|^2)^{1/2} \\ 0 \\ \vdots \end{bmatrix},$$

with $f_1 = f \in H^\infty$;

$$\hat{V}^2 \begin{bmatrix} 1 \\ 0 \\ 0 \\ \vdots \end{bmatrix} = \hat{V} \begin{bmatrix} f_1 \\ (1-|f|^2)^{1/2} \\ 0 \\ \vdots \end{bmatrix} = \begin{bmatrix} f f_1 + (1-|f|^2) r_2 \\ (1-|f|^2)^{1/2} f_1 - (1-|f|^2)^{1/2} r_2 \overline{f} \\ (1-|r_2|^2)^{1/2}(1-|f|^2)^{1/2} \\ 0 \\ \vdots \end{bmatrix}$$

$$= \begin{bmatrix} f_2 \\ (1-|f|^2)^{1/2} f_1 - (1-|f|^2)^{1/2} r_2 \overline{f} \\ (1-|r_2|^2)^{1/2}(1-|f|^2)^{1/2} \\ 0 \\ \vdots \end{bmatrix},$$

with $f_2 = f f_1 + (1-|f|^2) r_2 = f f_1 + g \in H^\infty$;

$$\hat{V}^3 \begin{bmatrix} 1 \\ 0 \\ 0 \\ \vdots \end{bmatrix} = \hat{V} \begin{bmatrix} f_2 \\ (1-|f|^2)^{1/2} f_1 - (1-|f|^2)^{1/2} r_2 \overline{f} \\ (1-|r_2|^2)^{1/2}(1-|f|^2)^{1/2} \\ 0 \\ \vdots \end{bmatrix} = \begin{bmatrix} f_3 \\ k_{31} \\ k_{32} \\ k_{33} \\ 0 \\ \vdots \end{bmatrix},$$

with

$$
\begin{aligned}
f_3 &= ff_2 + (1-|f|^2)r_2 f_1 - (1-|f|^2)r_2^2\overline{f} - (1-|r_2|^2)(1-|f|^2)r_3 \\
&= ff_2 + gf_1 \in H^\infty, \\
k_{31} &= (1-|f|^2)^{1/2} - (1-|f|^2)^{1/2} r_2 \overline{f} f_1 + (1-|f|^2)^{1/2}\overline{f}^2 r_2^2 \\
&\quad -(1-|f|^2)^{1/2}\overline{f} r_3 \\
&= (1-|f|^2)^{1/2} f_2 - (1-|f|^2)^{1/2} r_2 \overline{f} f_1 + (1-|f|^2)^{1/2}\overline{f}^2 r_2^2 \\
&\quad -(1-|f|^2)^{1/2}\overline{f}^2 r_2^2 \\
&= (1-|f|^2)^{1/2} f_2 - (1-|f|^2)^{1/2} r_2 \overline{f} f_1. \\
k_{32} &= (1-|f|^2)^{1/2}(1-|r_2|^2)^{1/2} f_1 - (1-|f|^2)^{1/2}(1-|r_2|^2)^{1/2}\overline{f} r_2 \\
&\quad -(1-|f|^2)^{1/2}(1-|r_2|^2)^{1/2}\overline{r}_2 r_3 \\
&= (1-|f|^2)^{1/2}(1-|r_2|^2)^{1/2} f_1 - (1-|f|^2)^{1/2}(1-|r_2|^2)^{1/2}\overline{f} r_2 \\
&\quad -(1-|f|^2)^{1/2}(1-|r_2|^2)^{1/2}|r_2|^2 r_2 \overline{f} \\
&= (1-|f|^2)^{1/2}(1-|r_2|^2)^{1/2} f_1 \\
&\quad -(1-|f|^2)^{1/2}(1-|r_2|^2)^{-1/2}(1-|r_2|^2+|r_2|^2) r_2 \overline{f} \\
&= (1-|f|^2)^{1/2}(1-|r_2|^2)^{1/2} f_1 \\
&\quad -(1-|f|^2)^{1/2}(1-|r_2|^2)^{1/2} r_2 \overline{f}. \\
k_{33} &= (1-|f|^2)^{1/2}(1-|r_2|^2)^{1/2}(1-|r_3|^2)^{1/2}.
\end{aligned}
$$

Using the same type of calculus we can prove by induction that for any $n \geq 3$ we have

$$
\hat{V}^n \begin{bmatrix} 1 \\ 0 \\ 0 \\ \vdots \end{bmatrix}
$$

$$
= \begin{bmatrix}
f_n \\
(1-|f|^2)^{1/2} f_{n-1} - (1-|f|^2)^{1/2} r_2 \overline{f} f_{n-2} \\
(1-|f|^2)^{1/2}(1-|r_2|^2)^{1/2} f_{n-2} - (1-|f|^2)^{1/2}(1-|r_2|^2)^{-1/2} r_2 \overline{f} f_{n-3} \\
\vdots \\
\prod_{j=2}^{n-2}(1-|f|^2)^{1/2}(1-|r_j|^2)^{1/2} f_1 - \prod_{j=2}^{n-2}(1-|f|^2)^{1/2}(1-|r_j|^2)^{-1/2} r_2 \overline{f} \\
\prod_{j=2}^{n-1}(1-|f|^2)^{1/2}(1-|r_j|^2)^{1/2} \\
0 \\
\vdots
\end{bmatrix}
$$

with

$$f_n = ff_{n-1} + (1 - |f|^2)r_2 f_{n-2} = ff_{n-1} + gf_{n-2} \in H^\infty.$$

Hence the choice sequence $(M_{r_n})_{n\geq 1}$ produces a positive definite sequence of analytic Toeplitz operators $(T_{f_n})_{n\geq 1}$ with $f_1 = f$ and $f_n \neq f^n$ for $n \geq 2$.

The theorem is proved. □

Corollary 3 *For $f \in H^\infty$, $\|f\| \leq 1$, either $(T_f^n)_{n\geq 1}$ is the only positive definite sequence of analytic Toeplitz operators starting with f or there exists a bijection from the unit ball of H^∞ into the set of such sequences. The second case appears when*

$$\int_{\mathbf{T}} log\ (1 - |f|^2)dt > -\infty.$$

Proof: Set

$$M = \{r \in L^2 : r(1 - |f|^2) \in H^2\} = qH^2, q \in L^\infty, \quad |q| = 1, \quad a.e.$$

if it is the case. Then for any $u \in H^\infty$, $\|u\| \leq 1, r(u) = \varepsilon qu$ with ε as in the Theorem, is good for starting the construction of the sequence $(r_n)_{n\geq 2}$ as before. Note also that the sequence $(f_n)_{n\geq 1}$ of the choice u is obtained by the following algorithm :

$$f_1 = f$$

$$f_n = ff_{n-1} + gf_{n-2}, \quad n \geq 2$$

with $f_0 = 1$ and $g = g(u) = (1 - |f|^2)r(u)$.

In the next Section we shall show that the case of maximum entropy truncable positive definite sequence of analytic Toeplitz operators corresponds to the case of U - diagonal Ando dilations considered in [8]. The form of the parameters and the algorithm of producing $(\hat{R}_n)_{n\geq 1}$ for such sequences have a precise function theoretic counterpart. □

4 Maximum Entropy Truncable Sequences

Let $R = (R_n)_{n\geq 1}$ be a choice sequence on H^2 and for any $p \geq 1$, let $R^{p|0} = (R_1, \ldots, R_p, 0, 0 \cdots)$ be its p -truncation. In general, the choice sequence $R^{p|0}$ does not produce positive definite sequence of analytic Toeplitz operators even when R does.

We say that a positive definite sequence (T_{f_n}) of analytic Toeplitz operators is *maximum entropy truncable* if its choice sequence $(R_n)_{n\geq 1}$ has the property that for any $p \geq 1$ the truncated choice sequence $R^{p|0}$ of R produces positive definite sequence of analytic Toeplitz operators.

Recall that for a choice sequence $R = (R_n)_{n\geq 1}$ on H^2 we denoted by $[K, V] = [K(R), V(R)]$ the minimal isometric dilation of the corresponding positive definite sequence $(T_n)_{n\geq 1}$. If $(R_n)_{n\geq 1}$ produces positive definite sequence of analytic Toeplitz operators then we denote by $U = U(R)$ the isometry on $K = K(R)$

attached to R as in the Theorem 1, i.e. the unique isometry U an K such that $UV = VU$ and H^2, as the subspace of K, reduces U to the shift S. We shall also use the notations: $K^{p|0} = K(R^{p|0})$, $V^{p|0} = V(R^{p|0})$, $U^{p|0} = U(R^{p|0})$, etc.

The following geometric characterization of the maximum entropy truncable sequences will play the central roll in the identification of the system of free parameters and in the recursive construction of such sequences.

Theorem 5 *Let $R = (R_n)_{n\geq 1}$ be a choice sequence on H^2, $(T_n)_{n\geq 1}$ be its attached positive definite sequence and $[K, V]$ be the minimal isometric dilation of $(T_n)_{n\geq 1}$. The following assertions are equivalent:*

(i) *$(T_n)_{n\geq 1}$ is a maximum T entropy truncable positive definite sequence of analytic Toeplitz operators.*

(ii) *$(T_n)_{n\geq 1}$ is a positive definite sequence of analytic Toeplitz operators and the isometry $U = U(R)$ on $K = K(R)$ attached to $(T_n)_{n\geq 1}$ as in the Theorem 1 is diagonal.*

Proof: Suppose that (i) holds, i.e. for any $p \geq 1$ the truncated sequence $R^{p|0}$ of R produces positive definite sequence of analytic Toeplitz operators.

Since for any $n \geq 1$ the string $(T_1, \ldots, T_n)$ depends only on $(R_1, \ldots, R_n)$ it results that, for any $n \geq 1$, T_n is analytic Toeplitz say $T_n = T_{f_n}$, $f_n \in H^\infty$.

It results that there exists $U = U(R)$ as in the Theorem 1. Since for any $n \geq 2$, $U_n(R)$ depends only on $(R_1, \ldots, R_{n-1})$ it results that $U_n(R) = U_n(R^{p|0})$ for any n, $2 \leq n \leq p+1$.

So in order to prove (ii) it is sufficient to show that for any $p \geq 1$, $U^{p|0}$ is diagonal.

Recall first that for any $p \geq 1$ the subspace H^2 of $K^{p|0}$ reduces $U^{p|0}$ to the shift S. Hence for $p = 1$ we have

$R^{1|0} = (R_1, 0, 0, \ldots)$

$K^{1|0} = H^2 \oplus \mathcal{D}_{R_1} \oplus \mathcal{D}_{R_1} \oplus \cdots$

and

$$V^{1|0} = \begin{bmatrix} R_1 & 0 & 0 & 0 & \cdots \\ D_{R_1} & 0 & 0 & 0 & \cdots \\ 0 & I & 0 & 0 & \cdots \\ 0 & 0 & I & 0 & \cdots \\ \cdot & \cdot & \cdot & \cdot & \cdots \end{bmatrix} \qquad U^{1|0} = \begin{bmatrix} S & 0 & 0 & 0 & \cdots \\ 0 & Y_2 & U_{23} & U_{24} & \cdots \\ 0 & 0 & Y_3 & U_{34} & \cdots \\ 0 & 0 & 0 & Y_4 & \cdots \\ \cdot & \cdot & \cdot & \cdot & \cdots \end{bmatrix}$$

Equalizing (i, j) entries in the matrix equality $U^{1|0}\ V^{1|0} = V^{1|0}U^{1|0}$ we obtain the following $(i \times j)$ equalities:

$$\begin{array}{ll} (1 \times 1) & SR_1 = R_1 S, \\ (2 \times 2) & Y_1 D_{R_1} = D_{R_1} S, \\ (2 \times j),\ j \geq 2 & U_{2,j} R_1 = 0, \\ (j \times j),\ j \geq 3 & Y_j = Y_{j-1}, \\ (i \times j),\ 3 \leq i < j & U_{i,j} = U_{i-1,\,j-1}. \end{array}$$

It results that $U^{1|0}$ is diagonal.

Clearly $R_1 = T_1 = T_{f_1}$. In the case when

$$\int_{\mathbf{T}} log\,(1 - |f_1|^2)dt = -\infty, \tag{4.1}$$

then, according to Theorem 4, we have $R = R^{1|0},\ V = V^{1|0}, U = U^{1|0}$.

Hence U is diagonal.

Suppose now that

$$\int_{\mathbf{T}} log\,(1 - |f_1|^2)\,dt > -\infty.$$

In this case (cf.[13]) there exists an outer function $g \in H^\infty$ such that

$$1 - |f_1|^2 = |g|^2$$

a.e. on $\mathbf{T}$.

Define the operator X from $\mathcal{D}_{R_1}$ into H^2 by

$$XD_{R_1}h = gh, \quad h \in H^2. \tag{4.2}$$

From

$$\begin{aligned}\|XD_{R_1}h\|^2 &= \|gh\|^2 = 1/2\pi \int_{\mathbf{T}} |g|^2|h|^2 dt = 1/2\pi \int_{\mathbf{T}} (1 - |f_1|^2)|h|^2 dt \\ &= \|h\|^2 - \|T_{f_1}h\|^2 = \|h\|^2 - \|R_1 h\|^2 = \|D_{R_1}h\|^2,\end{aligned}$$

it results that (4.2) defines an isometry from $\mathcal{D}_{R_1}$ into H^2. Since g is outer X is unitary.

From

$$XY_2D_{R_1}h = XD_{R_1}Sh = gSh = Sgh = SXD_{R_1}h$$

it results that Y_2 is a unilateral shift of multiplicity one.

Remark also that ker $D_{R_1} = \{0\}$, hence $\mathcal{D}_{R_1} = \mathbf{H}^2$.

Summing up, at the step 1, we proved :

$1)_1$ the isometry $U^{1|0}$ is diagonal

$2)_1$ either $R = R^{1|0}$ or Y_2 is a unilateral shift of multiplicity one on $\mathcal{D}_{R_1} = H^2$.

We shall prove, by induction, that, for any $p \geq 1$, we have:

$1)_p$ On $K^{p|0} = H^2 \oplus \mathcal{D}_{R_1} \oplus \cdots \oplus \mathcal{D}_{R_p} \oplus \mathcal{D}_{R_p} \oplus \cdots$, the isometry $U^{p|0}$ is diagonal of the form:

$$U^{p|0} = \begin{bmatrix} S & 0 & \cdots & 0 & 0 & \cdots \\ 0 & Y_2 & \cdots & 0 & 0 & \cdots \\ \cdot & \cdot & \cdots & \cdot & \cdots & \\ \cdot & \cdot & \cdots & \cdot & \cdots & \\ 0 & 0 & \cdots & Y_{p+1} & 0 & \cdots \\ 0 & 0 & \cdots & 0 & Y_{p+1} & \cdots \\ \cdot & \cdot & \cdots & \cdot & \cdot & \cdots \end{bmatrix}. \tag{4.3}$$

$2)_p$ either $R = R^{p|0}$ or Y_{p+1} is a unilateral shift of multiplicity one on $\mathcal{D}_{R_p} = H^2$.

This clearly will prove also the implication $(i) \to (ii)$.

Suppose that for some $p \geq 1$, $1)_p$ and $2)_p$ hold.

If $R_{p+1} = 0$ then $K^{p+1|0} = K^{p|0}$, $V^{p+1|0} = V^{p|0}$. It results $U^{p+1|0} = U^{p|0}$ and consequently $1)_{p+1}$ and $2)_{p+1}$ clearly hold.

So we can suppose $R_{p+1} \neq 0$. From $2)_p$ it results that $Y_2, \ldots, Y_{p+1}$ are unilateral shifts of multiplicity one on $\mathcal{D}_{R_1} = \cdots = \mathcal{D}_{R_p} = H^2$.

Moreover on

$$K^{p+1|0} = H^2 \oplus \mathcal{D}_{R_1} \oplus \cdots \oplus \mathcal{D}_{R_p} \oplus \mathcal{D}_{R_{p+1}} \oplus \cdots \oplus \mathcal{D}_{R_{p+1}} \oplus \cdots$$

$U^{p+1|0}$ has the matrix of the form

$$U^{p+1|0} = \begin{bmatrix} S & 0 & \cdots & 0 & \cdots & 0 & 0 & 0 & \cdots & 0 & \cdots \\ 0 & Y_2 & \cdots & 0 & \cdots & 0 & 0 & U_{2,p+2} & \cdots & U_{2,p+j} & \cdots \\ \cdot & \cdot & \cdots & \cdot & \cdots & \cdot & \cdot & \cdot & & \cdot & \cdots \\ \cdot & \cdot & \cdots & \cdot & \cdots & \cdot & \cdot & \cdot & \cdots & \cdot & \cdots \\ \cdot & \cdot & \cdots & \cdot & \cdots & \cdot & \cdot & \cdot & \cdots & \cdot & \cdots \\ 0 & 0 & \ldots & Y_i & \ldots & 0 & 0 & U_{i,p+2} & \cdots & U_{i,p+j} & \cdots \\ \cdot & \cdot & \cdots & \cdot & \cdots & \cdot & \cdot & \cdot & & \cdot & \cdots \\ \cdot & \cdot & \cdots & \cdot & \cdots & \cdot & \cdot & \cdot & & \cdot & \cdots \\ \cdot & \cdot & \cdots & \cdot & \cdots & \cdot & \cdot & \cdot & & \cdot & \cdots \\ 0 & 0 & \cdots & 0 & \cdots & Y_p & 0 & U_{p,p+2} & \cdots & U_{p,p+j} & \cdots \\ 0 & 0 & \cdots & 0 & \cdots & 0 & Y_{p+1} & U_{p+1,p+2} & \cdots & U_{p+1,p+j} & \cdots \\ 0 & 0 & \cdots & 0 & \cdots & 0 & 0 & Y_{p+2} & \cdots & U_{p+2,p+j} & \cdots \\ \cdot & \cdot & \cdots & \cdot & \cdots & \cdot & \cdot & \cdot & & \cdot & \cdots \\ \cdot & \cdot & \cdots & \cdot & \cdots & \cdot & \cdot & \cdot & & \cdot & \cdots \\ \cdot & \cdot & \cdots & \cdot & \cdots & \cdot & \cdot & \cdot & & \cdot & \cdots \\ 0 & 0 & \ldots & 0 & \ldots & 0 & 0 & 0 & \cdots & Y_{p+j} & \cdots \\ \cdot & \cdot & \cdots & \cdot & \cdots & \cdot & \cdot & \cdot & & \cdot & \cdots \end{bmatrix}$$

where, for $2 \leq i \leq p+1$, Y_i is as in (4.3).

Let us write also the matrix of $V^{p+1|0}$ as follows:

$$V^{p+1|0} =$$

$$\begin{bmatrix} R_1 & D_{R_{1}c^*} R_2 & \cdots & D_{R_1^*} \cdots D_{R_{p-1}^*} R_p & D_{R_1^*} \cdots D_{R_p^*} R_{p+1} & 0 & \cdots \\ D_{R_1} & -R_1^* R_2 & \cdots & -R_1^* D_{R_2^*} \cdots D_{R_{p-1}^*} R_p & -R_1^* D_{R_2^*} \cdots D_{R_p^*} R_{p+1} & 0 & \cdots \\ 0 & D_{R_2} & \cdots & -R_2^* D_{R_3^*} \cdots D_{R_{p-1}^*} R_p & -R_2^* D_{R_3^*} \cdots D_{R_p^*} R_{p+1} & 0 & \cdots \\ \cdot & \cdot & \cdots & \cdot & \cdot & \cdot & \cdots \\ 0 & 0 & \cdots & R_{p-1}^* R_p & -R_{p-1}^* D_{R_p^*} R_{p+1} & 0 & \cdots \\ 0 & 0 & \cdots & D_{R_p} & -R_p^* R_{p+1} & 0 & \cdots \\ 0 & 0 & \cdots & 0 & D_{R_{p+1}} & 0 & \cdots \\ 0 & 0 & \cdots & 0 & 0 & I & \cdots \\ \cdot & \cdot & \cdots & \cdot & \cdot & \cdot & \cdots \end{bmatrix}$$

Equalizing (i,j) entries in the matrix equality $V^{p+1|0}\,U^{p+1|0} = U^{p+1|0}V^{p+1|0}$ we obtain the following $(i \times j)$ equalities:

(1×1), $\quad R_1 S = S R_1$

$(1 \times i)$, $\quad 2 \le i \le p+1$, $\quad D_{R_1^*} \cdots D_{R_{i-1}^*} R_i Y_i = S D_{R_1^*} \cdots D_{R_{i-1}^*} R_i$

$(i \times p+j)$, $\quad 2 \le j$, $\quad D_{R_1^*} R_2 U_{2,p+j} + \cdots + D_{R_1^*} \cdots D_{R_p^*} R_{p+1} U_{p+1,p+j} = 0$

$(i \times i-1)$, $\quad 2 \le i$, $\quad D_{R_{i-1}} Y_{i-1} = Y_i D_{R_{i-1}}$

$(i \times i)$, $\quad 2 \le i$, $\quad R_{i-1}^* R_i Y_i = Y_i R_{i-i}^* R_i$

$(p+1 \times p+1)$, $\quad R_p^* R_{p+1} Y_{p+1} = Y_{p+1} R_p^* R_{p+1} - U_{p+1,p+2} D_{R_{p+1}}$

$(p+j, p+j)$, $\quad 2 \le j$, $\quad U_{p+j-1,p+j} = U_{p+j,p+j-1}$

$(i \times j)$, $\quad 2 \le i < j \le p$, $\quad R_{i-1}^* D_{R_i^*} \cdots D_{R_{j-1}^*} R_j Y_j = Y_i R_{i-1}^* D_{R_i^*} \cdots D_{R_{j-1}^*} R_j$

$(i \times p+1), 2 \le i \le p$, $\quad R_{i-1}^* D_{R_i^*} \cdots D_{R_p^*} R_{p+1} Y_{p+1} = Y_i R_{i-1}^* D_{R_i^*} \cdots D_{R_p^*} R_{p+1} - U_{i,p+2} D_{R_{p+1}}$

$(p+1, p+j)$, $\quad 2 \le j$, $\quad D_{R_p} U_{p,p+j} - R_p^* R_{p+1} U_{p+1,p+j} = U_{p+1,p+j+1}$

$(i \times p+j), 2 \le i \le p, 2 \le j$, $\quad D_{R_{i-1}} U_{i-1,p+j} - R_{i-1}^* R_i U_{i,p+j} - \cdots - R_{i-1}^* D_{R_i^*} \ldots D_{R_p^*} R_{p+1} U_{p+1,p+j} = U_{i,p+j+1}$

$(p+i \times p+j)$, $\quad 2 \ge i < j$, $\quad D_{R_{p+i}} U_{p+i-1,p+j} = U_{p+i,p+j+1}$

In case $p = 1$, from (1×2), we obtain

$$D_{R_1^*} R_2 Y_2 = S D_{R_1^*} R_2.$$

Since Y_2 and S are unilateral shifts of multiplicity one we deduce

$$\ker D_{R_1^*} R_2 = \{0\}.$$

From $(1 \times 1 + j)$ it results

$$D_{R_1^*} R_2 U_{2,\,j+1} = 0, \quad j \ge 2,$$

hence

$$U_{2,j+1} = 0, \qquad j \ge 2. \tag{4.4}$$

From $(1+i \times 1+j)$, $2 \geq j$ we have

$$U_{j,1+j} = U_{1+j,\ 1+j+1}$$

which together (4.4) imply $U_{i,j} = 0$ for any $2 \leq i < j$. Hence $U^{2|0}$ is diagonal.

Suppose now that $p \geq 2$. Since Y_{p+1} and S a unilateral shifts of multiplicity one from $(1 \times p+1)$ it results

$$\ker D_{R_1^*} \cdots D_{R_p^*} R_{p+1} = \{0\}. \tag{4.5}$$

From $(1 \times p+j)$, $j \geq 2$, we obtain

$$D_{R_1^*}(R_2 U_{2,p+j} + \cdots + D_{R_2^*} \cdots D_{R_p^*} R_{p+1} U_{p+1,p+j}) = 0.$$

Since the expression under the bracket take values in $D_{R_1^*}$ we have

$$R_2 U_{2,p+j} + \cdots + D_{R_2^*} \cdots D_{R_p^*} R_{p+1} U_{p+1,p+j} = 0.$$

From $(2 \times p+j)$, $j \geq 0$, we obtain

$$U_{2,p+j+1} = -R_1^*(R_2 U_{2,p+j} + \cdots + D_{R_2^*} \cdots D_{R_p^*} R_{p+1} U_{p+1,p+j}) = 0.$$

Hence

$$U_{2,p+j+1} = 0, \qquad j \geq 2. \tag{4.6}$$

From $(1 \times p+j+1)$, $j \geq 2$, using (4.6), we obtain

$$D_{R_1^*} D_{R_2^*}(R_3 U_{3,p+j+1} + \cdots + D_{R_3^*} \cdots D_{R_p^*} R_{p+1} U_{p+1,p+j+1}) = 0$$

which implies

$$R_3 U_{3,p+j+1} + D_{R_3^*} R_4 U_{4,p+j+1} + \ldots + D_{R_3^*} \cdots D_{R_p^*} R_{p+1} U_{p+1,p+j+1} = 0.$$

From $(3 \times p+j+1)$, we obtain

$$U_{3,p+j+2} = -R_2^*(R_3 U_{3,p+j+1} + \cdots + D_{R_3^*} \cdots D_{R_p^*} R_{p+1} U_{p+1,p+j+1}) = 0.$$

We can prove, by the same type of arguments, making induction on i, that we have

$$U_{i,p+j+i-1} = 0 \qquad 2 \leq i \leq p. \tag{4.7}$$

From $(i \times j)$, $p+1 \leq i < j$ we obtain

$$U_{p+1,p+j+1} = D_{R_{p+1}} U_{p,p+j}.$$

Hence

$$U_{p+1,p+j+p} = D_{R_{p+1}} U_{p,p+j+(p-1)} = 0$$

and by induction

(4.8) $$U_{i,p+j+(i-1)} = 0, \qquad i \geq 1,\ j \geq 2.$$

From $(1 \times p + j)$, $j \geq p+1$, since $U_{p,p+j} = 0$, it results

$$D_{R_1^*} \dots D_{R_p^*} R_{p+1} U_{p+1,p+j} = 0, \quad j \geq p+1$$

which, according to (4.5), will implies

$$U_{p+1,p+j} = 0, \qquad j \geq p+1.$$

Hence

$$U_{p+1,2p+1} = 0$$

and again an induction using $(p+i \times p+j)$, will give

$$U_{p+j,2p+j} = 0, \qquad j \geq 1.$$

Now, from $(1 \times p)$ we obtain

$$\begin{aligned} &D_{R_1^*} R_2 U_{2,2p} + \cdots + D_{R_1^*} \cdots D_{R_{p-2}^*} R_{p-1} U_{p-1,2p} \\ &\quad + D_{R_1^*} \cdots D_{R_{p-1}^*} R_p U_{p,2p} + D_{R_1^*} \cdots D_{R_p^*} R_{p+1} U_{p+1,2p} = 0. \end{aligned}$$

But, $U_{j,2p} = 0$ for $2 \leq j < p-1$, hence

$$D_{R_1^*} \cdots D_{R_{p-1}^*} [R_p U_{p,2p} + D_{R_p^*} R_{p+1} U_{p+1,2p}] = 0,$$

which implies

(4.9) $$R_p U_{p,2p} + D_{R_p^*} R_{p+1} U_{p+1,2p} = 0.$$

From $(p+1 \times p+1)$ we obtain

(4.10) $$D_{R_p} U_{p,2p} - R_p^* R_{p+1} U_{p+1,2p} = U_{p+1,2p+1} = 0.$$

The relations (4.9) and (4.10) give

$$\begin{pmatrix} R_p & D_{R_p^*} \\ D_{R_p} & -R_p^* \end{pmatrix} \begin{pmatrix} U_{p,2p} \\ R_{p+1} U_{p+1,2p} \end{pmatrix} = 0$$

hence $U_{p,2p} = 0$, and $R_{p+1} U_{p+1,2p} = 0$. The last equality and (4.5) imply $U_{p+1,2p} = 0$.

Thus we have

(4.11) $$U_{p,2p} = 0, \quad U_{p+1,2p} = 0.$$

From $(p+1 \times p+p-1)$, we obtain

$$D_{R_p}U_{p,p+p-1} - R_p^* R_{p+1}U_{p+1,p+p-1} = U_{p+1,2p} = 0 \tag{4.12}$$

and from $(1 \times p+p-1)$ we obtain

$$D_{R_1^*} \dots D_{R_{p-1}^*}(R_p U_{p,2p-1} + D_{R_p^*} R_{p+1} U_{p+1,2p-1}) = 0$$

which implies

$$R_p U_{p,2p-1} + D_{R_p^*} R_{p+1} U_{p+1,2p-1} = 0. \tag{4.13}$$

From (4.12) and (4.14) we obtain

$$\begin{pmatrix} R_p & D_{R_p^*} \\ D_{R_p} & -R_p^* \end{pmatrix} \begin{pmatrix} U_{p,\,2p-1} \\ R_{p+1}U_{p+1,2p-1} \end{pmatrix} = 0$$

and again

$$U_{p,2p-1} = 0 \qquad U_{p+1,\,2p-1} = 0. \tag{4.14}$$

In the same way we can prove, by induction, that

$$U_{p,p+j} = 0 \qquad U_{p+1,p+j+1} = 0, \qquad j \geq 2.$$

From $(p+1 \times p+1)$ it results

$$R_p^* R_{p+1} Y_{p+1} = Y_{p+1} R_p^* R_{p+1}.$$

Using now $(i \times p+j), 2 \leq i \leq p$ and $(p+1, p+j)$, $2 \leq i < j$, we obtain $U_{i,j} = 0$ for any $i < j$. This means that $U^{p+1|0}$ is diagonal.

We proved the assertion $1)_{p+1}$.

Let us prove the assertion $2)_{p+1}$. Suppose that $R \neq R^{p+1|0}$. This means that there exists $n \geq p+2$ such that $R_n \neq 0$. Take n to be the smallest with this property. Since $R_q = 0$ for $p+2 \leq q < n$, it results, as we remarked before, that the isometry $U^{q|0} = U^{n+1|0}$ is diagonal and Y_{q+1} is a unilateral shift of multiplicity one on $\mathcal{D}_{R_q} = H^2$. So we can suppose $n = p+2$. Since $U_{p+2} = U_{p+2}^{p+2|0}$ is diagonal it results that $U_n V_n = V_n U_n$ for any $n \leq p+2$. Also for $n \leq p+1$ $\mathcal{D}_{R_n} = H^2$, $\mathcal{D}_{\hat{R}_n} = L^2$ and the map from $\mathcal{D}_{R_n} = H^2$ into $\mathcal{D}_{\hat{R}_n} = L^2$ defined by $b_n D_{R_n} h = D_{\hat{R}_n} h$, $h \in H^2$ is an isometry (see Corollary 2). It results that the restriction of $\hat{S}$ to the invariant subspace $\overline{\mathcal{D}_{\hat{R}_{p+1}} H^2}$ is an isometry unitary equivalent with Y_{p+2}. But, by Beurling Theorem on invariant subspace in L^2, the above restriction is either a unilateral shift of multiplicity one or a unitary operator.

We conclude that Y_{p+2} is either unilateral shift of multiplicity one on $D_{R_{p+1}} = H^2$ or a unitary operator.

In case Y_{p+2} is unitary on $\mathcal{D}_{R_{p+1}}$ it results that for any $h \in \mathcal{D}_{R_{p+1}}$ and any $n \geq 1$ there exists $g_n \in \mathcal{D}_{R_{p+1}}$ such that $h = Y_{p+2}^n g_n$. Since

$$D_{R_1^*} \cdots D_{R_{p+1}^*} R_{p+2} Y_{p+2} = S D_{R_1^*} \cdots D_{R_{p+1}^*} R_{p+2},$$

it results

$$\begin{aligned} D_{R_1^*}\cdots D_{R_{p+1}^*}R_{p+2}h &= D_{R_1^*}\cdots D_{R_{p+1}^*}R_{p+2}Y_{p+2}^n g_n \\ &= S^n D_{R_1^*}\cdots D_{R_{p+1}^*}R_{p+2}g_n. \end{aligned}$$

Hence

$$D_{R_1^*}\ldots D_{R_{p+1}^*}R_{p+2}h \in \bigcap_{n\geq 0} S^n H^2 = \{0\}.$$

It results $R_{p+2}h = 0$ for any $h \in \mathcal{D}_{R_{p+1}}$ which is in contradiction with our assumption that $R_{p+2} \neq 0$. This proves $2)_{p+1}$ and our induction is complete.

Hence the implication $(i) \rightarrow (ii)$ is completely proved.

Suppose now that the assumption (ii) holds, i.e. U is diagonal and let

$$U = \begin{bmatrix} S & 0 & 0 & \ldots \\ 0 & Y_2 & 0 & \ldots \\ 0 & 0 & Y_3 & \ldots \\ \ldots & \ldots & \ldots & \ldots \end{bmatrix}$$

be its matrix. Then for any $n \geq 1$, K_n reduces U. This implies $U_n V_n = V_n U_n$ for any $n \geq 1$.

If we take now on

$$K^{p|0} = H^2 \oplus \mathcal{D}_{R_1} \oplus \cdots \oplus \mathcal{D}_{R_p} \oplus \mathcal{D}_{R_p} \oplus \cdots$$

the isometry

$$U^{p|0} = \begin{bmatrix} S & 0 & \cdots & 0 & 0 & \cdots \\ 0 & Y_2 & \cdots & 0 & 0 & \cdots \\ 0 & 0 & \cdots & Y_{p+1} & 0 & \cdots \\ 0 & 0 & \cdots & 0 & Y_{p+1} & \cdots \\ \cdot & \cdot & \cdots & \cdot & \cdot & \cdots \end{bmatrix}$$

it is easy to see that $U^{p|0}V^{p|0} = V^{p|0}U^{p|0}$. From Theorem 1 it results that $R^{p|0}$ produces positive definite sequence of analytic Toeplitz operators and the assertion (i) is proved.

The Theorem is completely proved. □

Corollary 4 *A choice sequence $(\hat{R}_n)_{n\geq 1}$ on L^2 produces a maximum entropy truncable positive definite sequence of analytic Toeplitz operators if and only if:*

1) $\hat{R}_1 = M_{r_1}$ *with* $r_1 \in H^\infty$, $\|r_1\| \leq 1$,

2) *for any* $p \geq 2$, *either* $\hat{R}_n = 0$ *for any* $n \geq p$ *or* $\mathcal{D}_{\hat{R}_{p-1}} = L^2$ *and* $\hat{R}_p = M_{r_p}$ *with* $r_p \in L^\infty$, $\|r_p\| \leq 1$, *verifying*

$$\prod_{j=1}^{p-1}(1 - |r_j|^2)\bar{r}_{k-1}r_p \in H^\infty, \quad 1 \leq k \leq p, \tag{4.15}$$

with $r_0 = 1$.

Proof. Suppose that $(\hat{R}_n)_{n\geq 1}$ produces a maximum entropy truncable positive definite sequence of analytic Toeplitz operators $(T_{f_n})_{n\geq 1}$. Then $\hat{R}_1 = M_{f_1} = M_{r_1}$ with $r_1 = f_1 \in H^\infty$, $\|r_1\| \leq 1$.

According to Theorem 4 and Corollary 2 the embedding of K into $\hat{K}$ has a diagonal matricial form. Hence K is the subspace of $\hat{K}$ given by

$$K = H^2 \oplus \overline{\mathcal{D}_{\hat{R}_1} H^2} \oplus \overline{\mathcal{D}_{R_2}\mathcal{D}_{R_1} H^2} \oplus \cdots \tag{4.16}$$

As in the proof of the Theorem 4, for any $p \geq 1$, we have either $\hat{R}_n = 0$ for any $n > p$ or for any j, $1 \leq j \leq p$, $\mathcal{D}_{\hat{R}_j} = L^2$, $\hat{Y}_{j+1} = \hat{S}$ and $\hat{R}_j\hat{S} = \hat{S}R_j$. So in the last case $\hat{R}_n = M_{r_n}$, $r_n \in L^\infty$, $\|r_n\| \leq 1$. The relations (4.15) come from (4.16) and $VK \subset K$.

Conversely, if $\hat{R}_n = M_{r_n}$, $r_n \in L^\infty$, $\|r_n\| \leq 1$, with $r_1 \in H^\infty$ and, for any p, $p \geq 2$ and any k, $1 \leq k \leq p$, the relations (4.15) hold, then a direct computation shows, by induction, that $(\hat{R}_n)_{n\geq 1}$ produces a positive definite sequence of analytic Toeplitz operators. The relations (4.15) also imply that $\hat{V}_n K_n \subset K_n$ which implies, as we know, that U is diagonal.

Using the Corollary 4 we shall give now an algorithm to produce, by free parameters, the set of all maximum entropy truncable positive definite sequences of analytic Toeplitz operators.

We shall produce recurrently a sequence $(r_n)_{n\geq 1}$ of functions $r_n \in L^\infty$ with the properties as in Corollary 3.

For a function $\varphi \in L^\infty$ we shall denote by H_φ the Hankel operator of symbol φ i.e. the operator from H^2 into $L^2 \ominus H^2$ defined by

$$H_\varphi = P_{L^2\ominus H^2} M_\varphi | H^2.$$

Clearly the subspace $\ker H_\varphi$ of H^2 is a closed subspace invariant to the shift. If φ, ψ are functions in L^∞ clearly the set

$$M = \{r \in L^2 : \psi r \in \ker H_\varphi\}$$

is a closed subspace in L^2, invariant to the bilateral shift $\hat{S}$. According to Beurling Theorem, either M is reducing for $\hat{S}$ or there exists $q \in L^\infty$, $|q| = 1$, a.e. on $\mathbf{T}$, such that $M = qH^2$.

Our algorithm is the following:

Step 1:
Choose a function $u_1 \in H^\infty$, $\|u_1\| \leq 1$. Set $r_1 = u_1$ and

$$M_1 = \{r \in L^\infty : (1 - |r_1|^2) r \in \ker H_{\overline{r}_1}\}.$$

If M_1 is not reducing for $\hat{S}$ set $M_1 = q_1 H^2$

Step: $p + 1$, $p \geq 1$

If M_p is reducing for $\hat{S}$ take $r_n = 0$ for any $n \geq p+1$. End.
If $M_p = q_p H^2$, choose an arbitrary $u_{p+1} \in H^\infty$, $\|u_{p+1}\| \leq 1$, and set $r_{p+1} = q_p u_{p+1}$, and

$$M_{p+1} = \left\{ r \in L^\infty : \prod_{j=1}^{p+1} (1 - |r_j|^2) r \in \bigcap_{j=1}^{p+1} \ker H_{\overline{r_j}} \right\}$$

If M_{p+1} is not reducing for $\hat{S}$ set $M_{p+1} = q_{p+1} H^2$.
Continue.

It is clear that in this way we obtain all the sequences $(r_n)_{n\geq 1}$ of functions in L^∞ verifying the conditions in the Corollary 4. The free parameters are the sequences $(u_n)_{n\geq 1}$ of functions $u_n \in H^\infty$, $\|u_n\| \leq 1$, under the only restriction that $u_n = 0$ for $n \geq p+1$ in case M_p as above is reducing for $\hat{S}$.

An example of a sequence $(u_n)_{n\geq 1}$ which produces, by the above algorithm, a long sequence $(r_n)_{n\geq 1}$ as in the Corollary 3 is the following: take $u_n = \lambda_n \theta_1 \cdot \theta_2 \cdots \theta_n$, where for any $n \geq 1$, θ_n is an inner function and $\lambda_n \in \mathbf{C}$, $|\lambda_n| < 1$. Here $q_n = 1$ hence $r_n = u_n$.

Let us remark that, in the simplest case when $u_n(z) = \lambda z^n$, $|\lambda| < 1$, although the matricial form of $[\hat{K}, \hat{V}]$ in terms of the choice sequence $(\hat{R}_n)_{n\geq 1}$ initiated on L^2 is simple, the corresponding choice sequence $(R_n)_{n\geq 1}$ initiated on H^2 and the corresponding matricial form of $[K, V]$ look sufficiently complicate. This shows that the L^2-extension presented in Section 3, simplifies the recursive constructions in our interpolation problem.

5 Application to a Moment Problem on $\mathbf{Z}^2$

Consider the bitorus $\mathbf{T}^2 = \mathbf{T}\times\mathbf{T}$ as the dual group of the discrete group $\mathbf{Z}^2 = \mathbf{Z}\times\mathbf{Z}$. A regular Borel measure μ on $\mathbf{T}^2$ is well defined by the matrix $[c_{n,m}(\mu)]_{\substack{-\infty<n<\infty \\ -\infty<m<\infty}}$ of its Fourier coefficients:

$$c_{n,m}(\mu) = \int e^{-int} e^{-ims} d\mu(t,s), \quad (n,m) \in \mathbf{Z}^2. \tag{5.1}$$

A *representing measure* on the bitorus is a probability (positive of total mass 1) measure μ on $\mathbf{T}^2$ verifying

$$c_{n,m}(\mu) = 0, \; (n,m) \in Z^2, \; n,m \geq 0, \; n+m > 0. \tag{5.2}$$

Let us denote by $\mathcal{M}_0$ the set of all such measures. The elements of $\mathcal{M}_0$ represent the origin as a point in the maximal ideal space of the bidisc algebra.

The moment problem considered here is the following:

Let $[a_{n,m}]_{\substack{1\leq n<p\leq\infty \\ 1\leq m<q\leq\infty}}$ *be a* $p-1 \times q-1$ *matrix of complex numbers. Give necessary and sufficient conditions in order there exists a representing measure* μ

on the bitorus such that

$$a_{n,m} = c_{-n,m}(\mu), \quad 1 \le n < p, \quad 1 \le m < q. \tag{5.3}$$

In the case the set of all solutions is non void, find a system of free parameters which label it and give algorithms to produce recurrently, by successive choice of a free parameter, a desired solution.

The matrix of the Fourier coefficients of a solution μ looks like

$$\begin{array}{cccc|c|ccc} & \vdots & \vdots & \vdots & \vdots & \vdots & \\ \cdots & \bar{a}_{22} & \bar{a}_{21} & 0 & 0 & 0 & \cdots \\ \cdots & \bar{a}_{12} & \bar{a}_{11} & 0 & 0 & 0 & \cdots \\ \hline \cdots & 0 & 0 & 1 & 0 & 0 & \cdots \\ \hline \cdots & 0 & 0 & 0 & a_{11} & a_{12} & \cdots \\ \cdots & 0 & 0 & 0 & a_{21} & a_{22} & \cdots \\ & \vdots & \vdots & \vdots & \vdots & \vdots & \end{array} \tag{5.4}$$

The problem, in this general formulations, seems to be very difficult, as any extension problem concerning positive definiteness on the group $\mathbf{Z}^2$ is.

We shall treat here the case when either p or q is ∞.

The next result will show that, in this case, the above moment problem can be reduced to a problem of extension to a positive definite sequence of analytic Toeplitz operators considered in the preceding sections.

For a matrix $[a_{n,m}]_{\substack{1\le n<\infty \\ 1\le m<\infty}}$ let us denote formally

$$f_n(z) = \sum_{m=1}^{\infty} a_{n,m} z^m, \quad z \in \mathbf{D}, \quad n \ge 1. \tag{5.5}$$

Theorem 6 *The matrix* $[a_{n,m}]_{\substack{1\le n<\infty \\ 1\le m<\infty}}$ *generates by* (5.3) *a representing measure* μ *on the bitorus if and only if for any* $n \ge 1$, (5.5) *defines a function* f_n *in* H^∞, $\|f_n\|_\infty \le 1$ *and the sequence* $(T_{f_n})_{n\ge 1}$, *is a positive definite sequence of analytic Toeplitz operators.*

Proof: Suppose the matrix $[a_{n,m}]_{\substack{1\le n<\infty \\ 1\le m<\infty}}$ generates by (5.3) a representing measure μ on the bitorus. Let U, V be the isometries defined on the Hilbert space $K = H^2(d\mu)$ - the closure in $L^2(d\mu)$ of the set of all analytic polynomials, by multiplication by the coordinate functions e^{it}, e^{is} respectively.

Let

$$K_0 = \bigvee_{m \geq 0} U^m \mathbf{C},\ U_0 = U|K_0,\ T_n = P^K_{K_0} V^n | K_0.$$

Using the form (5.4) of μ we can identify K_0 with H^2 such that U_0 becomes the shift S. Clearly H^2 reduces U to S and consequently $T_n = T_{f_n}$ with f_n verifying (5.5).

Conversely, if (5.5) defines, for any $n \geq 1$, a function f_n in H^∞ such that the sequence $(T_{f_n})_{n \geq 1}$ of analytic Toeplitz operators is positive definite, then, constructing K, V, U as in Theorem 1, we can extend the commuting pair $[U, V]$ of the isometries on K to its minimal unitary extension -a unitary representation of the group $\mathbf{Z}^2$ on a supra space $\tilde{K}$ of K generated by a pair $[\tilde{U}, \tilde{V}]$ of commuting unitary operators on $\tilde{K}$, verifying $\tilde{U}|K = U$, $\tilde{V}|K = V$ (cf.[]). Let E be the spectral measure, supported on $\mathbf{T}^2$, attached to this representation. Setting for any Borel subset σ of $\mathbf{T}^2$, $\mu(\sigma) = (E(\sigma)1, 1)_{\tilde{K}}$, 1 being here the constant function 1 in H^2, as a vector in $\tilde{K}$, we obtain a positive measure μ on $\mathbf{T}^2$. For any $n \geq 0$ and $m \in \mathbf{Z}$ we have

$$\begin{aligned} c_{n,m}(\mu) &= \int e^{-int} e^{-ims} d\mu(t,s) = \int e^{-int} e^{-ims} d(E(t,s)1, 1)_{\tilde{K}} \\ &= (\tilde{U}^m V^n 1, 1)_{\tilde{K}} = (V_n e^{imt}, 1)_K = (T_{f_n} e^{imt}, 1)_{H^2} \\ &= 1/2\pi \int e^{imt} f_n(e^{it}) dt, \end{aligned}$$

with $f_0 = 1$. This clearly implies $\mu \in \mathcal{M}_0$ and verifies (5.3).

The Theorem is completely proved. □

The algorithms following Corollary 3 and Corollary 4 can be applied to produce representing measures on the bitorus.

References

[1] V.M. Adamian, D.Z. Arov and M.G. Krein, Infinite Hankel block matrices and related extension problems *Izv. Acad. Nauk Armjan SSR, Matematika* **6** (1971), pp. 87–112.

[2] T. Ando, On a pair of commutative contractions, *Acta Sci. Math.* **24** (1963), pp. 88–90.

[3] Gr. Arsene, Z. Ceauşescu and C. Foiaş, On intertwining dilation VIII, *J. Operator Theory* **4** (1980), pp. 55–91.

[4] I.A. Ball and I. Gohberg, A commutant lifting theorem for triangular matrices with diverse applications, *Integral Eq. and Operator Theory* **8** (1985), pp. 205–267.

[5] J.P. Burg, *Maximum Entropy Spectral Analysis*, Ph. D. thesis, Department of Geophysics, Stanford University, (1975).

[6] Z. Ceauşescu and C. Foiaş, On intertwining dilations V, *Acta Sci. Math.* **40** (1978), pp. 9–32.

[7] Z. Ceauşescu and I. Suciu, Isometric dilations of commuting contractions I, *J. Operator Theory* **12** (1984), pp. 65–88.

[8] Z. Ceauşescu and I. Suciu, Isometric dilations of commuting contractions IV, *Advances in invariant subspaces and other results of operator theory, Operator Theory; Advances and Applications* **17** (1986) Birkhäuser Verlag, Basel, pp. 55–77.

[9] T. Constantinescu, A maximum entropy principle for contractive intertwining dilations, *Operators in indefinite metric spaces, scattering theory and other topics, Operator Theory; Advances and Applications* **24** (1987) Birkhäuser Verlag Basel, pp. 69–85.

[10] R.G. Douglas, On the operator equation $S^*XT = T$ and related topics, *Acta Sci. Math.* **30** (1969), pp. 19–32.

[11] C. Foiaş and A.E. Frazho, *The Commutant Lifting Approach to Interpolation Problems*, Birkhäuser Verlag, Basel-Boston-Berlin, 1990.

[12] H. Helson, *Lectures on Invariant Subspaces,* New York: Academic Press, 1964.

[13] K. Hoffman, *Banach spaces of analytic functions*, Englewood cliffs, Prentince Hall, New Jersey, 1962.

[14] I. Suciu, Operatorial extrapolations and prediction, *Special classes of linear operators and other topics; Operator Theory: Advances and Applications* **28** (1988) Birkhäuser Verlag, Basel-Boston-Berlin, pp. 291–301.

[15] I. Suciu, On the matrix of the Fourier coefficients of a representing measures on the bitorus, *Preprint IMAR* No. **22** 1994.

[16] B. Sz.-Nagy, C. Foiaş, *Harmonic Analysis of Operators on Hilbert Space*, North-Holland Publishing Co., Amsterdam-Budapest, 1970.

Ion Suciu
Institute of Mathematics
Romanian Academy
P.O. Box 1–764, RO–70700
Bucharest
Romania

Operator Theory:
Advances and Applications, Vol. 115

Subdecomposable Operators and Rationally Invariant Subspaces

Lawrence R. Williams

We develop a 2x2 matrix model for a decomposable operator T relative to a certain open covers $\{U_1, U_2\}$ of the spectrum of T. This model is used to construct new and different proofs of the facts such as decomposable and subdecomposable operators have the single-valued extension property and satisfy Dunford's Condition C. We discuss rationally invariant subspaces, and given an operator T and an invariant subspace $\mathcal{M}$, we categorize those complex numbers λ for which $\mathcal{M}$ is invariant for $(T-\lambda)^{-1}$. Finally, we give an example of a pure cyclic hyponormal T operator and a nonzero vector w for which the local spectrum of w with respect to T is not equal to the spectrum of T. This answers negatively the question of whether the local spectrum of each vector with respect to T is equal to the spectrum of T for a pure cyclic hyponormal operator T.

1 Introduction

Let $\mathcal{H}$ be a Hilbert space, and $\mathcal{L}(\mathcal{H})$ denote the algebra of all bounded linear operators on $\mathcal{H}$. For T $\in \mathcal{L}(\mathcal{H})$, let σ(T) denote the spectrum of T and $\rho(T)$ denote the resolvent set of T. Suppose that T $\in \mathcal{L}(\mathcal{H})$ and x $\in \mathcal{H}$. Let f(λ) = (T-λ)$^{-1}$x for each λ in $\rho(T)$. The function f is analytic on ρ(T), and f may have an analytic extension $\hat{f}$ to an open subset U of the complex plane $\mathcal{C}$ containing $\rho(T)$ such that (T-λ) $\hat{f}(\lambda)$ = x for all λ in U. If each two such extensions of f agree on their common domains, then T is said to have the *single-valued extension property*. In this case f must have a maximum analytic extension which we denote by $\hat{x}\,(\cdot)$. Let ρ_T(x) be the domain of $\hat{x}$ and σ_T(x) = $\mathcal{C}$ - ρ_T(x). The sets ρ_T(x) and σ_T(x) are called the *local resolvent* and the *local spectrum* of x with respect to T, respectively. Observe that σ_T(x) is a subset of σ(T) that is compact and that σ_T(x) = ∅ if and only x = 0. (More information about the local spectra of operators can be found in [D].)

Suppose that T $\in \mathcal{L}(\mathcal{H})$ and T has the single-valued extension property. The operator T is said to satisfy Dunford's Condition C (DCC) if the linear submanifold $X_T(F) \doteq \{x \in \mathcal{H} : \sigma_T(x) \subseteq F\}$ is closed for each closed subset F of $\mathcal{C}$. It is known that hyponormal operators have the single-valued extension property, and Stampfli [S] and Radjabalipour [R2] have shown that hyponormal operators satisfy DCC. (Recall that an operator T is said to be hyponormal if $T^*T - TT^* \geq 0$.)

An operator T in $\mathcal{L}(\mathcal{H})$ is said to be *decomposable* if for each pair $\{U_1, U_2\}$ of open subsets of $\mathcal{C}$ satisfying $\sigma(T) \subseteq U_1 \cup U_2$, there exist invariant subspaces $\mathcal{K}_1$ and $\mathcal{K}_2$ for T such that $\mathcal{K}_1 + \mathcal{K}_2 = \mathcal{H}$, $\sigma(T \mid \mathcal{K}_1) \subseteq U_1$, and $\sigma(T \mid \mathcal{K}_2) \subseteq U_2$. An operator S is said to be *subdecomposable* if S has a decomposable extension. It is known that each hyponormal operator is similar to a subdecomposable operator. (See [P]. The result can also be found in [EP] and in [MP].) This similarity has been exploited to prove some results about hyponormal operators. For example, Liming Yang [Y] was able to prove that quasisimilar hyponormal operators have equal essential spectra by showing that a similar theorem is true for subdecomposable operators.

The class of subdecomposable operators contains a variety of operators. Normal operators are evidently decomposable. Hence the class of subdecomposable operators contains the class of subnormal operators. Also the Riesz functional calculus shows that operators with countable spectra are decomposable. It follows that the classes of decomposable and subdecomposable operators contain the algebraic operators, the nilpotent and quasinilpotent operators, and the compact operators. (Recall that an operator is subnormal if it is the restriction of a normal operator to an invariant subspace and an operator is algebraic if it satisfies a polynomial equation.)

Let F be a compact subset of $\mathcal{C}$. For a bounded function $f : F \to \mathcal{C}$, define $\|f\|_F = sup\{|f(z)| : z \in F\}$. The *polynomial convex hull* of F is the set $\hat{F} = \{z \in \mathcal{C} : |p(z)| \leq \|p\|_F$ for each polynomial $p\}$. We say that F is *polynomially convex* if $F = \hat{F}$. We can show that $\mathcal{C}$ -$\hat{F}$ is the unbounded component of $\mathcal{C} - \mathcal{F}$. It follows that F is polynomially convex if and only if $\mathcal{C} - \mathcal{F}$ is connected. (See [C] for a discussion of polynomially convex sets.) Recall that a *hole* in F is a bounded component of $\mathcal{C}$ - F. Hence $\hat{F}$ is equal to F union the collection of all the holes in F.

Suppose that T belongs to $\mathcal{L}(\mathcal{H})$, $\mathcal{M}$ is an invariant subspace for T, and $T_1 = T \mid \mathcal{M}$. Since $\partial\sigma(T_1) \subseteq \sigma(T)$, then $\sigma(T_1)$ is a subset of $\sigma(T)$ union a subcollection of the holes in $\sigma(T)$. That is, $\sigma(T_1) \subseteq \widehat{\sigma(T)}$. This fact shall be used in the sequel.

2 A Matrix Model for Decomposable Operators

In this section, we first present a 2x2 matrix model for a decomposable operator T relative to certain open covers $\{U_1, U_2\}$ of $\sigma(T)$. We then show that this model can be used to obtain results such as decomposable and subdecomposable operators satisfy Dunford's Condition C. The following theorem gives a 2x2 matrix model for decomposable operators. Its proof is based on a proof given in [Y]. Corollary 2.5 will show that the invariant subspace $\mathcal{M}$ for T in Theorem 2.1 can be taken to be a rationally invariant subspace. (See Section 3 for the definition of a rationally invariant subspace for an operator.)

Theorem 2.1 *Suppose that T is a decomposable operator in $\mathcal{L}(\mathcal{H})$, $\{U_1, U_2\}$ is a pair of open subsets of $\mathcal{C}$, where U_2 is convex (or just simply connected), and $\sigma(T) \subseteq U_1 \cup U_2$. Then there exists an invariant subspace $\mathcal{M}$ for T such that*

$$T = \begin{bmatrix} A & B \\ 0 & C \end{bmatrix}$$

with respect to the decomposition $\mathcal{H} = \mathcal{M} \oplus \mathcal{M}^{\perp}$, *where* $\sigma(A) \subseteq U_1$ *and* $\sigma(C) \subseteq U_2$.

Proof: Suppose that T is a decomposable operator, and $\{U_1, U_2\}$ is a pair of open subsets of C, where U_2 is convex (or simply connected), and $\sigma(T) \subseteq U_1 \cup U_2$. There exist invariant subspaces $\mathcal{K}_1$ and $\mathcal{K}_2$ for T such that $\mathcal{K}_1 + \mathcal{K}_2 = \mathcal{H}$, $\sigma(T \mid \mathcal{K}_1) \subseteq U_1$, and $\sigma(T \mid \mathcal{K}_2) \subseteq U_2$. Let $\mathcal{M} = \mathcal{K}_1$. Then

$$T = \begin{bmatrix} A & B \\ 0 & C \end{bmatrix}$$

with respect to the decomposition $\mathcal{H} = \mathcal{M} \oplus \mathcal{M}^{\perp}$. Clearly, $\sigma(A) \subseteq U_1$. Let's show that $\sigma(C) \subseteq U_2$.

Define a linear transformation W: $\mathcal{M}^{\perp} \to \mathcal{H}/\mathcal{M}$ by Wz $=$ z $+ \mathcal{M}$ for each z in $\mathcal{M}^{\perp}$. For each z in $\mathcal{M}^{\perp}$, $\| Wz \| = \| z + \mathcal{M} \| = inf\{\| z + t \| : t \in \mathcal{M}\} = inf\{\sqrt{\| z \|^2 + \| t \|^2} : t \in \mathcal{M}\} = \| z \|$. Suppose that z $+\mathcal{M} \in \mathcal{H}/\mathcal{M}$ for some z in $\mathcal{H}$. Then z $= z_1 + z_2$ for $z_1 \in \mathcal{M}$ and $z_2 \in \mathcal{M}^{\perp}$. Hence W$z_2 = z_2 + \mathcal{M} = z + \mathcal{M}$, since $z - z_2 = z_1$ belongs to $\mathcal{M}$. Hence W is an isometric isomorphism of $\mathcal{M}^{\perp}$ onto $\mathcal{H}/\mathcal{M}$.

Define $T_0 : \mathcal{H}/\mathcal{M} \to \mathcal{H}/\mathcal{M}$ by $T_0(z + \mathcal{M}) =$ (Tz $+\mathcal{M}$) for each z in $\mathcal{H}$. Since $\mathcal{M}$ is an invariant subspace for T, T_0 is a well-defined linear operator on $\mathcal{H}/\mathcal{M}$. Let P : $\mathcal{H} \to \mathcal{H}$ be the orthogonal projection onto $\mathcal{M}^{\perp}$. Then Cz $=$ PTz for each z in $\mathcal{M}^{\perp}$. Let z $\in \mathcal{M}^{\perp}$. Then WCz $=$ W(PTz) $=$ (PTz $+ \mathcal{M}$) $=$ (Tz $+ \mathcal{M}$) $=$ T_0(z $+ \mathcal{M}$) $=$ T_0Wz . It follows that WC $=$ T_0W, and thus C and T_0 are similar.

Define V : $\mathcal{H}/\mathcal{M} \to \mathcal{K}_2/(\mathcal{M} \cap \mathcal{K}_2)$ by V(z $+\mathcal{M}$) $= z_2 + (\mathcal{M} \cap \mathcal{K}_2)$, where z $= z_1 + z_2$ for z in $\mathcal{H}$, z_1 in $\mathcal{M}$, and z_2 in $\mathcal{K}_2$. Suppose that z $= z_1 + z_2 = w_1 + w_2$, where z_1 and w_1 belong to $\mathcal{M}$ and z_2 and w_2 belong to $\mathcal{K}_2$. Then $z_1 - w_1 = w_2 - z_2$ belongs to $\mathcal{M} \cap \mathcal{K}_2$. Hence $z_2 + (\mathcal{M} \cap \mathcal{K}_2) = w_2 + (\mathcal{M} \cap \mathcal{K}_2)$. It follows that V is well-defined. Clearly, V is linear and ker(V) $= 0$. Suppose that $z + (\mathcal{M} \cap \mathcal{K}_2) \in \mathcal{K}_2/(\mathcal{M} \cap \mathcal{K}_2)$. Then $z \in \mathcal{K}_2$, and V(z $+ \mathcal{M}$) $=$ z $+ (\mathcal{M} \cap \mathcal{K}_2)$. Therefore, V is also surjective.

Now suppose that $z \in \mathcal{H}$. Write z $= z_1 + z_2$ for some z_1 in $\mathcal{M}$ and z_2 in $\mathcal{K}_2$. Observe that

$$\| z + \mathcal{M} \| = inf\{\| z + t \| : t \in \mathcal{M}\} = inf\{\| z - z_1 + t \| : t \in \mathcal{M}\}$$

$$= inf\{\| z_2 + t \| : t \in \mathcal{M}\} \leq inf\{\| z_2 + t \| : t \in \mathcal{M} \cap \mathcal{K}_2\} = \| z_2 + \mathcal{M} \cap \mathcal{K}_2 \| .$$

It follows that V^{-1} is a bounded linear transformation. By the Open Mapping Theorem, V is also a bounded linear transformation.

Define $T_1 : \mathcal{K}_2/(\mathcal{M} \cap \mathcal{K}_2) \to \mathcal{K}_2/(\mathcal{M} \cap \mathcal{K}_2)$ by $T_1(z + (\mathcal{M} \cap \mathcal{K}_2)) = Tz + (\mathcal{M} \cap \mathcal{K}_2)$, for each z in $\mathcal{K}_2$. Clearly, T_1 is a bounded linear operator, and $(VT_0)(z + \mathcal{M}) =$ V(Tz $+ \mathcal{M}$) $= Tz_2 + (\mathcal{M} \cap \mathcal{K}_2) = T_1(z_2 + (\mathcal{M} \cap \mathcal{K}_2)) = (T_1 V)(z + \mathcal{M})$ for z $= z_1 + z_2$, where $z \in \mathcal{H}$, $z_1 \in \mathcal{M}$, and $z_2 \in \mathcal{K}_2$. It follows that T_0, and therefore C, are similar to T_1.

We now show that $\sigma(C) = \sigma(T_1)$ is a subset of U_2. Define $T_2 : \mathcal{K}_2 \to \mathcal{K}_2$ by $T_2 z = \mathrm{Tz}$ for each z in $\mathcal{K}_2$, that is $T_2 = T \mid \mathcal{K}_2$. Define $T_3 : \mathcal{M} \cap \mathcal{K}_2 \to \mathcal{M} \cap \mathcal{K}_2$ by $T_3 z = \mathrm{Tz}$ for each z in $\mathcal{M} \cap \mathcal{K}_2$, that is $T_3 = T_2 \mid \mathcal{M} \cap \mathcal{K}_2$. We know that $\sigma(T_2) \subseteq U_2$. Also, $\sigma(T_3) \subseteq \widehat{\sigma(T_2)}$, and $\widehat{\sigma(T_2)} \subseteq U_2$. The latter inclusion holds because $\sigma(T_2) \subseteq U_2$ and U_2 is simply connected.

To finish the proof, let's show that $\sigma(T_1) \subseteq \sigma(T_2) \cup \sigma(T_3)$. Suppose that λ belongs to $\mathcal{C}$ and both $(T_2 - \lambda)$ and $(T_3 - \lambda)$ are invertible. Suppose that $w \in \mathcal{K}_2$ and $(T_1 - \lambda)(w + (\mathcal{M} \cap \mathcal{K}_2) = 0$. Then $(T - \lambda)w \in \mathcal{M} \cap \mathcal{K}_2$. There exists z in $\mathcal{M} \cap \mathcal{K}_2$ such that $(T_3 - \lambda)z = (T - \lambda)w$. It follows that $(T_2 - \lambda)z = (T_2 - \lambda)w$. Since $(T_2 - \lambda)$ is one-to-one, we have w = z. Hence $w \in \mathcal{M} \cap \mathcal{K}_2$, and so $w + \mathcal{M} \cap \mathcal{K}_2 = 0$. Hence $(T_1 - \lambda)$ is one-to-one.

Now suppose that $w \in \mathcal{K}_2$. Then there exists z in $\mathcal{K}_2$ such that $(T_2 - \lambda)z = \mathrm{w}$. So $z + (\mathcal{M} \cap \mathcal{K}_2) \in \mathcal{K}_2/(\mathcal{M} \cap \mathcal{K}_2)$, and $(T_1 - \lambda)(z + (\mathcal{M} \cap \mathcal{K}_2) = (T - \lambda)z + \mathcal{M} \cap \mathcal{K}_2 = w + \mathcal{M} \cap \mathcal{K}_2$. Hence, $(T_1 - \lambda)$ is surjective also. It follows by the Open Mapping Theorem that $(T_1 - \lambda)$ is invertible. □

I. Colojoară and C. Foias [CF] showed that decomposable operators have the single-valued extension property and satisfy Dunford's Condition C. In Proposition 2.2 and Theorem 2.4, we use the matrix model of Theorem 2.1 to give different proofs of the results of Colojoară and Foias.

Proposition 2.2 *Suppose that T is a decomposable operator in $\mathcal{L}(\mathcal{H})$. Then T satisfies the single-valued extension property.*

Proof: Suppose that $\mathcal{O}$ is an open subset of $\mathcal{C}$ and f: $\mathcal{O} \to \mathcal{H}$ is analytic function satisfying $(T - \lambda)f(\lambda) = 0$ for each λ in $\mathcal{O}$. We will show that $\mathrm{f} \equiv 0$ on $\mathcal{O}$. Suppose that $\lambda_0 \in \mathcal{O}$. There exist $\delta > 0$ such that $B(\lambda_0, \delta) \doteq \{\lambda \in \mathcal{C} : \mid \lambda - \lambda_0 \mid < \delta\} \subseteq \mathcal{O}$. Let $U_2 = B(\lambda_0, \delta/2)$ and $U_1 = \mathcal{C} - \overline{B(\lambda_0, \delta/4)}$. Hence, $\sigma(T) \subseteq U_1 \cup U_2$. (Observe that U_2 is convex, and thus simply connected.) By Theorem 2.1, there exists an invariant subspace $\mathcal{M}$ for T such that

$$T = \begin{bmatrix} A & B \\ 0 & C \end{bmatrix}$$

with respect to the decomposition $\mathcal{H} = \mathcal{M} \oplus \mathcal{M}^{\perp}$, where $\sigma(A) \subseteq U_1$ and $\sigma(C) \subseteq U_2$. For each $\lambda \in \mathcal{O}$, we have $\mathrm{f}(\lambda) = f_1(\lambda) \oplus f_2(\lambda)$ with respect to the decomposition $\mathcal{H} = \mathcal{M} \oplus \mathcal{M}^{\perp}$. Observe that f_1 and f_2 are also analytic on $\mathcal{O}$, and that

$$(T - \lambda)f(\lambda) = \begin{bmatrix} (A - \lambda)f_1(\lambda) + Bf_2(\lambda) \\ (C - \lambda)f_2(\lambda) \end{bmatrix} = \begin{bmatrix} 0 \\ 0 \end{bmatrix}$$

for each λ in $\mathcal{O}$. We have $(C - \lambda)f_2(\lambda) = 0$ for each λ in $\mathcal{O}$. Since $\sigma(C) \subseteq U_2$, it follows that $f_2(\lambda) = (C - \lambda)^{-1}0 = 0$ for each λ in $B(\lambda_0, \delta) - B(\lambda_0, \delta/2)$. Since f_2 is analytic on $B(\lambda_0, \delta)$, it follows that $f_2 \equiv 0$ on $B(\lambda_0, \delta)$. Hence $(A - \lambda)f_1(\lambda) = 0$ for each $\lambda \in B(\lambda_0, \delta)$. Since $B(\lambda_0, \delta/4) \subseteq \rho(A)$, we have

$f_1(\lambda) = (A-\lambda)^{-1}0 = 0$ for each λ in $B(\lambda_0, \delta/4)$. In particular, $f_1(\lambda_0) = 0$. Therefore, $f(\lambda_0) = 0$. Since λ_0 is an arbitrary point in $\mathcal{O}$, we have $f \equiv 0$ on $\mathcal{O}$. It follows that T has the single-valued extension property. □

Corollary 2.3 *Suppose that S is a subdecomposable operator. Then S has the single-valued extension property.*

Theorem 2.4 *Suppose that T is a decomposable operator in $\mathcal{L}(\mathcal{H})$. Then T satisfies Dunford's Condition C.*

Proof: Suppose that F is a closed subset of $\mathcal{C}$, and let $X_T(F) = \{w \in \mathcal{H} : \sigma_T(w) \subseteq F\}$. We wish to show that $X_T(F)$ is closed in $\mathcal{H}$. Suppose that $\{w_n\}_{n=1}^{\infty}$ is a sequence in $X_T(F)$, $w \in \mathcal{H}$, and $w_n \to w$. Let's show that $w \in X_T(F)$.

Suppose that $\lambda_0 \in \mathcal{C}$ - F. Then there exists $\delta > 0$ such that $B(\lambda_0, \delta) \subseteq \mathcal{C}$ - F. Let $U_2 = B(\lambda_0, \delta/2)$ and $U_1 = \mathcal{C} - \overline{B(\lambda_0, \delta/4)}$. Then $\sigma(T) \subseteq U_1 \cup U_2$. (Observe that U_2 is simply connected.) By Theorem 2.1, there exists an invariant subspace $\mathcal{M}$ for T such that

$$T = \begin{bmatrix} A & B \\ 0 & C \end{bmatrix}$$

with respect to the decomposition $\mathcal{H} = \mathcal{M} \oplus \mathcal{M}^{\perp}$, where $\sigma(A) \subseteq U_1$ and $\sigma(C) \subseteq U_2$.

For each n in N, there exists a function $\hat{w}_n$ that is analytic on $\mathcal{C}$ - F such that $(T-\lambda)\hat{w}_n(\lambda) = w_n$, for each $\lambda \in \mathcal{C}$ - F. Now $w_n = u_n \oplus v_n$ and $\hat{w}_n(\lambda) = f_n(\lambda) \oplus g_n(\lambda)$ with respect to the decomposition $\mathcal{H} = \mathcal{M} \oplus \mathcal{M}^{\perp}$, for each n in N and each λ in $\mathcal{C}$ - F.

Fix n in N. Observe that f_n and g_n are analytic on $\mathcal{C}$ - F. We have

$$(T-\lambda)\hat{w}_n(\lambda) = \begin{bmatrix} (A-\lambda)f_n(\lambda) + Bg_n(\lambda) \\ (C-\lambda)g_n(\lambda) \end{bmatrix} = \begin{bmatrix} u_n \\ v_n \end{bmatrix}$$

for each λ in $\mathcal{C}$ - F. In particular, $(C-\lambda)g_n(\lambda) = v_n$ for each λ in $\mathcal{C}$ - F. Let $h_n(\lambda) = (C-\lambda)^{-1}v_n$ for each λ in $\rho(C)$. Note that h_n is analytic on $\rho(C)$. Since $\rho(C) \cap (\mathcal{C} - F) \neq \emptyset$, and $g_n(\lambda) = (C-\lambda)^{-1}v_n = h_n(\lambda)$ for each λ in $\rho(C) \cap (\mathcal{C} - F)$, it follows that we can extend g_n to an analytic function on $\rho(C) \cup (\mathcal{C} - F)$, which we continue to call g_n. But $\rho(C) \cup (\mathcal{C} - F) = \mathcal{C}$ because $F \subseteq \rho(C)$. Thus g_n is analytic on $\mathcal{C}$. For $|\lambda| > \| C \|$, we have

$$\| (C-\lambda)^{-1} \| = \left\| \sum_{n=0}^{\infty} C^n/\lambda^{n+1} \right\| \leq \sum_{n=0}^{\infty} \| C \|^n / |\lambda|^{n+1} = \frac{1}{|\lambda| - \| C \|}.$$

Hence

$$\| g_n(\lambda) \| \leq \frac{1}{|\lambda| - \| C \|} \| v_n \|$$

for $|\lambda| > \| C \|$. It follows that g_n is entire, bounded, and $lim_{\lambda \to \infty} \| g_n(\lambda) \| = 0$. Hence, by Liouville's Theorem, $g_n \equiv 0$ on $\mathcal{C}$, and thus $v_n = 0$, for each n in N.

Since $w_n \to$ w and $w_n = u_n \oplus 0$, we have w $=$ u $\oplus$ 0 with respect to the decomposition $\mathcal{H} = \mathcal{M} \oplus \mathcal{M}^{\perp}$. An easy argument shows that $\sigma_T(w) \subseteq \sigma_A(u) \subseteq \sigma(A) \subseteq U_1$. Since $\lambda_0 \notin U_1$, then $\lambda_0 \notin \sigma_T(w)$, that is, $\lambda_0 \in \rho_T(w)$. We have shown that $\mathcal{C}$ - F $\subseteq \rho_T(w)$. Hence, $\sigma_T(w) \subseteq$ F. Therefore, $w \in X_T(F)$, and $X_T(F)$ is closed. □

Corollary 2.5 *In the matrix model for decomposable operators in Theorem 2.1, the subspace $\mathcal{M}$ can be taken to be a rationally invariant subspace.*

Proof: Let $\mathcal{K}_1$ be the Hilbert space in the proof of Theorem 2.1, and let $\mathcal{K}_1^+ = X_T(\sigma(T \mid \mathcal{K}_1))$. Since T satisfies Dunford's Condition C, $\mathcal{K}_1^+$ is a closed subspace of $\mathcal{H}$. It follows that $\mathcal{K}_1^+$ is a hyperinvariant subspace for T and $\mathcal{K}_1 \subseteq \mathcal{K}_1^+$. Hence we can replace $\mathcal{K}_1$ by $\mathcal{K}_1^+$ in the proof of Theorem 2.1. It follows that $\mathcal{M}$ is a hyperinvariant, and thus a rationally invariant subspace for T. (See Section 3 for the definition of a rationally invariant subspace for T.) □

We use our matrix model to prove the following proposition that can be used to show that subdecomposable operators satisfy Dunford's Condition C. In fact, the following proposition is a special case of a result of E. Albrecht and J. Eschmeier [AE] that decomposable operators have Bishop's property (β). Indeed, Albrecht and Eschmeier show that Bishop's property (β) characterizes the class of subdecomposable operators.

Proposition 2.6 *Suppose that T is a decomposable operator in $\mathcal{L}(\mathcal{H})$, F is a closed subset of $\mathcal{C}$, $\{w_n\}$ is a sequence of vectors in $X_T(F)$, $w \in X_T(F)$, and $w_n \to w$. Then $\hat{w}_n(\lambda) \to \hat{w}(\lambda)$ for each λ in $\mathcal{C}$ - F.*

Proof: Suppose that $\lambda_0 \in \mathcal{C}$ - F. Let δ, U_1, and U_2 be as in the proof of Theorem 2.4. By Theorem 2.1, there exists an invariant subspace $\mathcal{M}$ for T such that

$$T = \begin{bmatrix} A & B \\ 0 & C \end{bmatrix}$$

with respect to the decomposition $\mathcal{H} = \mathcal{M} \oplus \mathcal{M}^{\perp}$, where $\sigma(A) \subseteq U_1$ and $\sigma(C) \subseteq U_2$. For each n in N, there exists a function $\hat{w}_n$ that is analytic on $\mathcal{C}$ - F such that $(T - \lambda)\hat{w}_n(\lambda) = w_n$, for each $\lambda \in \mathcal{C}$ - F. Also, there exists a function $\hat{w}$ that is analytic on $\mathcal{C}$ - F such that $(T - \lambda)\hat{w}(\lambda) =$ w for each λ in $\mathcal{C}$ - F. Now $w_n = u_n \oplus v_n$ and $\hat{w}_n(\lambda) = f_n(\lambda) \oplus g_n(\lambda)$ with respect to the decomposition $\mathcal{H} = \mathcal{M} \oplus \mathcal{M}^{\perp}$, for each n in N and each λ in $\mathcal{C}$ - F. Likewise, w $=$ u $\oplus$ v and $\hat{w}(\lambda) = f(\lambda) \oplus g(\lambda)$ for each λ in $\mathcal{C}$ - F. As in the proof of Theorem 2.4, $g_n \equiv 0$, and thus $v_n = 0$, for each n in N. Also, $g \equiv 0$, and v $=$ 0. It follows that for each n in N and λ in $\mathcal{C} - F$, $(A - \lambda) f_n(\lambda) = u_n$. Also (A- $\lambda) f(\lambda) = u$ for each λ in $\mathcal{C}$ - F. Since $B(\lambda_0, \delta/4) \subseteq \rho(A)$, we have

$$f_n(\lambda) = (A - \lambda)^{-1} u_n \longrightarrow (A - \lambda)^{-1} u = f(\lambda)$$

for each λ in $B(\lambda_0, \delta/4)$. Therefore,

$$\hat{w}_n(\lambda) = f_n(\lambda) \oplus 0 \longrightarrow f(\lambda) \oplus 0 = \hat{w}(\lambda)$$

for each λ in $B(\lambda_0, \delta/4)$. In particular, $\hat{w}_n(\lambda_0) \to \hat{w}(\lambda_0)$. The result holds since λ_0 is an arbitrary point in $\mathcal{C}$ - F. □

The following theorem follows from Proposition 2.6 and the fact that decomposable operators satisfy Dunford's Condition C. We note that this result appears in [R1] with a slightly different proof.

Theorem 2.7 *Subdecomposable operators satisfy Dunford's Condition C.*

Proof: Suppose that S is a subdecomposable operator in $\mathcal{L}(\mathcal{H})$. There exists a Hilbert space $\mathcal{H}_0$ and linear transformations A and B such that

$$T = \begin{bmatrix} S & A \\ 0 & B \end{bmatrix},$$

with respect to the direct sum $\mathcal{H} \oplus \mathcal{H}_0$, is a decomposable operator. Suppose that F is a closed subset of $\mathcal{C}$, $\{u_n\}_{n=1}^{\infty}$ is a sequence of vectors in $X_S(F)$, $u \in \mathcal{H}$, and $u_n \to u$. For each n in N, there exists a function $\hat{u}_n$ that is analytic on $\mathcal{C}$ - F such that $(S - \lambda)\hat{u}_n(\lambda) = u_n$, for each λ in $\mathcal{C}$ - F. In $\mathcal{H} \oplus \mathcal{H}_0$, let $w_n = u_n \oplus 0$ for each n in N, and let $w = u \oplus 0$. Also, let $h_n(\lambda) = \hat{u}_n(\lambda) \oplus 0$ for each n in N and λ in $\mathcal{C}$ - F. Observe that for each n in N,

$$(T - \lambda)h_n(\lambda) = (S - \lambda)\hat{u}_n(\lambda) \oplus 0 = u_n \oplus 0 = w_n$$

for each λ in $\mathcal{C}$ - F. It follows that $\sigma_T(w_n) \subseteq F$ for each n in N, that is, $w_n \in X_T(F)$ for each n in N. Since $w_n \to w$ and T satisfies Dunford's Condition C, then $w \in X_T(F)$. Hence, there is a function $\hat{w}$ that is analytic on $\mathcal{C}$ - F such that $(T - \lambda)\hat{w}(\lambda) = w$ for each λ in $\mathcal{C}$ - F. We may write $\hat{w}(\lambda) = f(\lambda) \oplus g(\lambda)$ for each λ in $\mathcal{C}$ - F. By Proposition 2.6, $\hat{w}_n(\lambda) \to \hat{w}(\lambda)$ for each λ in $\mathcal{C}$ - F. Since $\hat{u}_n(\lambda) \oplus 0 \to f(\lambda) \oplus g(\lambda)$ for each λ in $\mathcal{C}$ - F, we have $g \equiv 0$ on $\mathcal{C}$ - F. Hence $(S - \lambda)f(\lambda) = u$ for each λ in $\mathcal{C}$ - F. Hence, $\sigma_S(u) \subseteq F$, and so $u \in X_S(F)$. It follows that $X_S(F)$ is closed. Therefore, S satisfies Dunford's Condition C. □

The next corollary follows because each hyponormal operator is similar to a subdecomposable operator. This result was first obtained by Stampfli [S] and Radjabalipour [R2].

Corollary 2.8 *Hyponormal operators satisfy Dunford's Condition C.*

3 Rationally Invariant Subspaces

Suppose that $T \in \mathcal{L}(\mathcal{H})$ and $\mathcal{M}$ is an invariant subspace for T. In some cases $\sigma(T) \subseteq \sigma(T \mid \mathcal{M})$. (For example, if T is the bilateral shift and $T| \mathcal{M}$ is the unilateral shift, then $\sigma(T)$ is the unit circle and $\sigma(T \mid \mathcal{M})$ is the closed unit disk.) In other, cases we have $\sigma(T \mid \mathcal{M}) \subseteq \sigma(T)$. (For example, this is the case when $\mathcal{M}$ is a reducing subspace for T.) However, as observed previously, we have in every case that $\sigma(T \mid \mathcal{M}) \subseteq \widehat{\sigma(T)}$.

The following proposition characterizes those invariant subspaces $\mathcal{M}$ for T for which $\sigma(T \mid \mathcal{M}) \subseteq \sigma(T)$. We first state a definition.

Suppose that T $\in \mathcal{L}(\mathcal{H})$ and $\mathcal{M}$ is a subspace of $\mathcal{H}$. Then $\mathcal{M}$ is said to be a *rationally invariant subspace* for T if $\mathcal{M}$ is invariant for r(T) for each rational function r with poles off $\sigma(T)$.

Proposition 3.1 *Suppose that $T \in \mathcal{L}(\mathcal{H})$ and $\mathcal{M}$ is an invariant subspace for T. Then $\sigma(T \mid \mathcal{M}) \subseteq \sigma(T)$ if and only if $\mathcal{M}$ is a rationally invariant subspace for T.*

Proof: Suppose that $\mathcal{M}$ is an invariant subspace for T. Observe that $\mathcal{M}$ is rationally invariant for T if and only if $\mathcal{M}$ is invariant for (T - $\lambda)^{-1}$ for each λ in $\rho(T)$. Let us write

$$T = \begin{bmatrix} A & B \\ 0 & C \end{bmatrix}$$

with respect to $\mathcal{H} = \mathcal{M} \oplus \mathcal{M}^{\perp}$. Suppose that λ belongs to $\rho(T)$. Write

$$(T - \lambda)^{-1} = \begin{bmatrix} X & Y \\ Z & W \end{bmatrix}$$

with respect to $\mathcal{H} = \mathcal{M} \oplus \mathcal{M}^{\perp}$. Then we have

$$(T - \lambda)^{-1}(T - \lambda) = \begin{bmatrix} X(A - \lambda) & XB + Y(C - \lambda) \\ Z(A - \lambda) & ZB + W(C - \lambda) \end{bmatrix} = \begin{bmatrix} 1 & 0 \\ 0 & 1 \end{bmatrix}$$

and

$$(T - \lambda)(T - \lambda)^{-1} = \begin{bmatrix} (A - \lambda)X + BZ & (A - \lambda)Y + BW \\ (C - \lambda)Z & (C - \lambda)W \end{bmatrix} = \begin{bmatrix} 1 & 0 \\ 0 & 1 \end{bmatrix}.$$

It follows that $\mathcal{M}$ is invariant for $(T - \lambda)^{-1}$ if and only if Z $= 0$ if and only $(A - \lambda)$ is invertible if and only if $\lambda \in \rho(A)$. Hence $\mathcal{M}$ is rationally invariant for T if and only if $\sigma(A) \subseteq \sigma(T)$. □

Corollary 3.2 *Suppose that T belongs to $\mathcal{L}(\mathcal{H})$ and $\mathcal{M}$ is a rationally invariant subspace for T. Write*

$$T = \begin{bmatrix} A & B \\ 0 & C \end{bmatrix}$$

with respect to $\mathcal{H} = \mathcal{M} \oplus \mathcal{M}^{\perp}$. Then $\sigma(T) = \sigma(A) \cup \sigma(C)$.

Proof: Use the proof of Proposition 3.1. We see that if $(T - \lambda)$ is invertible, then $Z = 0$, and thus $(A - \lambda)$ and $(C - \lambda)$ are invertible. So $\sigma(A) \cup \sigma(C) \subseteq \sigma(T)$. On the other hand, suppose that $(A - \lambda)$ and $(C - \lambda)$ are invertible. Let $Z = 0$, $X = (A - \lambda)^{-1}$, $W = (C - \lambda)^{-1}$, and $Y = -(A - \lambda)^{-1} B (C - \lambda)^{-1}$. A simple matrix calculation shows that $(T - \lambda)$ is invertible. So $\sigma(T) \subseteq \sigma(A) \cup \sigma(C)$. Therefore, $\sigma(T) = \sigma(A) \cup \sigma(C)$. □

Observe that Corollary 3.2 does not hold if we assume that $\mathcal{M}$ is only an invariant subspace for T. (Take T to be the bilateral shift and A the unilateral shift.) We remark also that in light of Corollary 2.5 and Corollary 3.2, we can assume that $\mathcal{M}$ is a rationally invariant subspace for T and $\sigma(T) = \sigma(A) \cup \sigma(C)$ in the matrix model of decomposable operators given in Theorem 2.1.

Suppose that T belongs to $\mathcal{L}(\mathcal{H})$ and $\mathcal{M}$ is an invariant subspace for T that is not necessarily rationally invariant for T. In the following, we investigate the question of which complex numbers λ in $\rho(T)$ is it true that $\mathcal{M}$ is invariant for $(T - \lambda)^{-1}$? To facilitate the discussion, let's make the following definitions:

$$\gamma(T, \mathcal{M}) \doteq \{\lambda \in \rho(T) : (T - \lambda)^{-1}\mathcal{M} \subseteq \mathcal{M}\}$$

and

$$\delta(T, \mathcal{M}) \doteq \mathcal{C} - \gamma(T, \mathcal{M}).$$

The set $\gamma(T, \mathcal{M})$ is called the *rationally invariant resolvent of T with respect to* $\mathcal{M}$ and $\delta(T, \mathcal{M})$ is called the *rationally invariant spectrum of T with respect to* $\mathcal{M}$.

Proposition 3.3 *Suppose that T belongs to $\mathcal{L}(\mathcal{H})$ and $\mathcal{M}$ is an invariant subspace for T. Then $\gamma(T, \mathcal{M}) = \rho(T \mid \mathcal{M}) \cap \rho(T)$ and $\delta(T, \mathcal{M}) = \sigma(T \mid \mathcal{M}) \cup \sigma(T)$.*

Proof: Let $\lambda \in \rho(T)$. By the proof of Proposition 3.1, $\mathcal{M}$ is invariant for $(T - \lambda)^{-1}$ if and only if $\lambda \in \rho(T \mid \mathcal{M})$. Hence, the proposition follows. □

Corollary 3.4 *Suppose that T belongs to $\mathcal{L}(\mathcal{H})$ and $\mathcal{M}$ is an invariant subspace for T. Then $\gamma(T, \mathcal{M})$ is open and $\delta(T, \mathcal{M})$ is compact.*

The following proposition and corollary show interesting relations between the connected components of the resolvent of an operator and its rationally invariant resolvents (and rationally invariant spectra).

Proposition 3.5 *Suppose that T belongs to $\mathcal{L}(\mathcal{H})$, $\mathcal{M}$ is an invariant subspace for T, H is a connected component of $\rho(T)$, and $H \cap \gamma(T, \mathcal{M}) \neq \emptyset$. Then $H \subseteq \gamma(T, \mathcal{M})$.*

Proof: Let $\lambda_0 \in H \cap \gamma(T, \mathcal{M})$. Since $H \cap \gamma(T, \mathcal{M})$ is open, there exists $\epsilon > 0$ such that $B(\lambda_0, \epsilon) \subseteq H \cap \gamma(T, \mathcal{M})$. Suppose w belongs to $\mathcal{M}$. Let

$h(\lambda) = (T - \lambda)^{-1}w$ for each λ in $\rho(T)$. Write $h(\lambda) = f(\lambda) \oplus g(\lambda)$ for each λ in $\rho(T)$, and write

$$T = \begin{bmatrix} A & B \\ 0 & C \end{bmatrix}$$

with respect to the decomposition $\mathcal{H} = \mathcal{M} \oplus \mathcal{M}^{\perp}$. Observe that the functions h, f, and g are analytic on $\rho(T)$. We have

$$\begin{bmatrix} f(\lambda) \\ g(\lambda) \end{bmatrix} = \begin{bmatrix} (A - \lambda)^{-1} & Y_\lambda \\ 0 & (C - \lambda)^{-1} \end{bmatrix} \begin{bmatrix} w \\ 0 \end{bmatrix} = \begin{bmatrix} (A - \lambda)^{-1}w \\ 0 \end{bmatrix}$$

for each λ in $B(\lambda_0, \epsilon)$. It follows that $g \equiv 0$ on $B(\lambda_0, \epsilon)$. Since H is connected, we have $g \equiv 0$ on H. Hence $h(\lambda) = (T - \lambda)^{-1}w = f(\lambda) \oplus 0$ belongs to $\mathcal{M}$ for each λ in H and for each w in $\mathcal{M}$. Therefore, $H \subseteq \gamma(T, \mathcal{M})$. □

Corollary 3.6 *Suppose that T belongs to $\mathcal{L}(\mathcal{H})$ and $\mathcal{M}$ is an invariant subspace for T. If H is the unbounded component of $\rho(T)$, then $H \subseteq \gamma(T, \mathcal{M})$.*

Proof: Choose λ_0 in $\mathcal{C}$ such that $|\lambda_0| > \|T\|$. Suppose that w belongs to $\mathcal{M}$. Then λ_0 belongs to H and

$$(T - \lambda_0)^{-1}w = -\sum_{n=0}^{\infty} \frac{T^n}{\lambda_0^{n+1}} w \in \mathcal{M}.$$

It follows that $\mathcal{M}$ is invariant for $(T - \lambda_0)^{-1}$. Hence $H \cap \gamma(T, \mathcal{M}) \neq \emptyset$. By Proposition 3.5, $H \subseteq \gamma(T, \mathcal{M})$. □

Note that Corollary 3.6 also follows from Proposition 3.5 and the fact that $\delta(T, \mathcal{M})$ is compact.

Suppose that S is a subdecomposable operator in $\mathcal{L}(\mathcal{H})$. There exists a Hilbert space $\mathcal{H}_0$ and linear transformations A and B such that

$$T = \begin{bmatrix} S & A \\ 0 & B \end{bmatrix},$$

with respect to the direct sum $\mathcal{H} \oplus \mathcal{H}_0$, is decomposable. Suppose that F is a closed subset of $\mathcal{C}$. Example 3.7 shows that it is not always the case that $X_S(F) = X_T(F) \cap \mathcal{H}$.

Example 3.7 Let S be the unilateral shift and T the bilateral shift. Since T is normal, S is subdecomposable. Theorem 2.5 of [W1] shows that $\sigma_S(x)$ is equal to the closed unit disk for each nonzero x in $\mathcal{H}$. Let F be the unit circle. Then $X_S(F) = \{0\}$. Since $\sigma(T) = F$, we have $X_T(F) \cap \mathcal{H} = \mathcal{H}$. Hence $X_S(F) \neq X_T(F) \cap \mathcal{H}$. Because $\sigma(S) \not\subseteq \sigma(T)$, then note that $\mathcal{H}$ is not a rationally invariant subspace for T.

Question 3.8 *What are necessary and sufficient conditions on F in order that $X_S(F) = X_T(F) \cap \mathcal{H}$?*

The following theorem gives a sufficient condition on F so that $X_S(F) = X_T(F) \cap \mathcal{H}$ in the case that $\mathcal{H}$ is a rationally invariant subspace for T. Example 3.7 shows that Theorem 3.9 is not true if $\mathcal{H}$ is assumed to be only an invariant subspace for T.

Theorem 3.9 *Suppose that S is a subdecomposable operator in $\mathcal{L}(\mathcal{H})$. There exists a Hilbert space $\mathcal{H}_0$ and linear transformations A and B such that*

$$T = \begin{bmatrix} S & A \\ 0 & B \end{bmatrix},$$

with respect to the direct sum $\mathcal{H} \oplus \mathcal{H}_0$, is decomposable. Suppose that $\mathcal{H}$ is a rationally invariant subspace for T and F is a closed subset of $\mathcal{C}$. If $\mathcal{C}$ - F has no connected component H such that $H \subseteq \sigma(T)$, then $X_S(F) = X_T(F) \cap \mathcal{H}$.

Proof: Since clearly $\sigma_T(w) \subseteq \sigma_S(w)$ for each w in $\mathcal{H}$, we have $X_S(F) \subseteq X_T(F) \cap \mathcal{H}$. So suppose that $w \in X_T(F) \cap \mathcal{H}$. Then there is a function $\hat{w}$ that is analytic on $\mathcal{C}$ - F such that $(T - \lambda)\hat{w}(\lambda) =$ w for each λ in $\mathcal{C}$ - F. Write $\hat{w}(\lambda) = f(\lambda) \oplus g(\lambda)$ for each λ in $\mathcal{C}$ - F. Observe that the functions f and g are analytic on $\mathcal{C}$ - F. Suppose that $\lambda_0 \in \mathcal{C}$ - F. Then λ_0 belongs to a connected component H of $\mathcal{C}$ - F. Observe that $H \cap \rho(T)$ is a nonempty open subset of $\mathcal{C}$ and that $\mathcal{H}$ is invariant for $(T - \lambda)^{-1}$ for each λ in $H \cap \rho(T)$. Hence

$$\begin{bmatrix} f(\lambda) \\ g(\lambda) \end{bmatrix} = \hat{w}(\lambda) = (T - \lambda)^{-1} w = \begin{bmatrix} (S - \lambda)^{-1} & Y_\lambda \\ 0 & (B - \lambda)^{-1} \end{bmatrix} \begin{bmatrix} w \\ 0 \end{bmatrix}$$
$$= \begin{bmatrix} (S - \lambda)^{-1} w \\ 0 \end{bmatrix}.$$

for each λ in $H \cap \rho(T)$. It follows that $g(\lambda) = 0$ for each λ in $H \cap \rho(T)$. Since g is analytic on H and H is connected and open, it follows that $g \equiv 0$ on H. In particular, $g(\lambda_0) = 0$. Hence $g \equiv 0$ on $\mathcal{C}$ - F. Therefore, $(T - \lambda)\hat{w}(\lambda) = (S - \lambda) f(\lambda) \oplus 0 = w \oplus 0$ for each λ in $\mathcal{C}$ - F. It follows that $w \in X_S(F)$. Hence, $X_S(F) = X_T(F) \cap \mathcal{H}$. □

4 Pure Cyclic Hyponormal Operators

Suppose that T is a pure cylic hyponormal operator on $\mathcal{H}$ with cyclic vector x. (Recall that T is said to be pure if T has no nonzero reducing subspace on which T restricted to that subspace is normal.) Then $\mathcal{H} = \vee\{x, Tx, T^2x, T^3x, ...\} = \overline{\{p(T)x : p \text{ is a complex polynomial}\}}$. The following theorem appears in [W2]

Theorem A *If S belongs to $\mathcal{L}(\mathcal{H})$, S commutes with T, and kernel(S^*) is finite dimensional, then $\sigma_T(Sx) = \sigma(T)$.*

By taking S = p(T) for a polynomial p, we can see that $\sigma_T(w) = \sigma(T)$ for each w in a dense subset of $\mathcal{H}$. The following question also appears in [W2].

Question A *Does $\sigma_T(w) = \sigma(T)$ for each nonzero w in $\mathcal{H}$?*

The following example shows that the condition that kernel(S^*) is finite dimensional is necessary in Theorem A and that the answer to Question A is negative.

Example 4.1 Let $G = G_1 \cup G_2$, where $G_1 = \{\lambda \in \mathcal{C} :| \lambda + 2 |< 1\}$ and $G_2 = \{\lambda \in \mathcal{C} :| \lambda - 2 |< 1\}$. Observe that $G_1 \cap G_2 = \emptyset$. Consider $L^2(G, \mathcal{M}, \mu)$, where $\mathcal{M}$ is the σ-algebra of Lebesque measurable subsets of G and μ is area measure restricted to G. Let $L^2_a(G)$ be the subspace of $L^2(G, \mathcal{M}, \mu)$ consisting of those functions that are analytic on G. The Bergman operator T for G is the operator multiplication by z on $L^2_a(G)$. (See [C] for a discussion of Bergman operators.) Note that T is a pure subnormal operator, and therefore a pure hyponormal operator. Since $\mathcal{C}$ - G is connected, Runge's Theorem implies that T is cyclic with cyclic vector e where e(z) =1 for each z in G. Note also that $\sigma(T) = \bar{G} = \bar{G_1} \cup \bar{G_2}$.

Let $f(z) = 1$ for each z in a neighborhood of G_1 and $f(z) = 0$ for each z in a neighborhood of G_2. Then f belongs to $L^2_a(G)$ and f is not the zero vector. Let $U = \mathcal{C}$ - $\overline{G_1}$. Define a function $\hat{f} : U \to L^2_a(G)$ for λ in U by $(\hat{f}(\lambda)(z) = \frac{1}{z-\lambda}$ for z in G_1 and $\hat{f}(\lambda)(z) = 0$ for z in G_2. So $\hat{f}(\lambda) \in L^2_a(G)$ and $(T - \lambda)$ $\hat{f}(\lambda) =$ f for each λ in U. Observe that $\hat{f}$ is analytic on U. (In fact, for each λ in U $\hat{f}'(\lambda)(z) = \frac{-1}{(z-\lambda)^2}$ for z in G_1 and $\hat{f}'(\lambda)(z) = 0$ for each z in G_2.) It follows that $\sigma_T(f) \subseteq \overline{G_1}$. In particular, $f \neq 0$ and $\sigma_T(f) \neq \sigma(T)$. This shows that the answer to Question A is negative.

Note that f(T) is defined by the Riesz functional calculus. Let $S = f(T)$. Then S commutes with T and f = Se, where e is the cyclic vector for T defined above and f is the function in $L^2_a(G)$ defined above. Hence $\sigma_T(Se) \neq \sigma(T)$. Theorem A implies that kernel(S^*) is infinite dimensional. Hence the condition that kernel(S^*) is finite dimensional is necessary in Theorem A.

References

[CF] I. Colojoară and C. Foias, *The Theory of Generalized Spectral Operators*, Gordan and Breach, New York, 1968.

[C] J. Conway, *Subnormal Operators*, Research Notes in Mathematics Series, vol. 51, Pitman, Boston, 1981.

[D] N. Dunford, *Spectral II. Resolutions of the identity*, Pacific J. Math. **2** (1952), 559–614.

[EP] J. Eschmeir and M. Putinar, *Bishop's condition (β) and rich extension of linear operators*, Indiana U. Math. J. **3** (1988), 325–347.

[AE] E. Albrecht and J. Eschmeir, *Analytic functional models and local spectral theory*, Proc. London Math. Soc. (3) **75** (1997), 323–348.

[MP] M. Martin and M. Putinar, *"Lectures in Hyponormal Operators,"* Operator Theory Advances and Applications, vol. 39, Birkhäuser, Basel.

[P] M. Putiner, *Hyponormal operators are subscalar*, J. Operator Theory **12** (1984), 385–395.

[R1] M. Radjabailpour, *On subnormal operators*, Trans. Amer. Math. Soc. **211** (1975), 377–389.

[R2] M. Radjabalipour, *Ranges of hyponormal operators*, Illinois J. Math. **21** (1977), 70–75.

[S] J.T. Stampfli, *A local spectra theory for operators, V. Spectral subspaces for hyponormal operators*, Trans. Amer. Math. Soc. **217** (1976), 285–296.

[W1] L.R. Williams, *The local spectra of pure quasinormal operators*, J. Math. Anal. and Appls. **187** (1994), 842–850.

[W2] L.R. Williams, *Bounded point evaluations and the local spectra of cyclic hyponormal operators*, Dynamic Systems and Appl. **3** (1994), 103–112.

[Y] Liming Yang, *Quasisimilarity of hyponormal and subdecomposable operators*, J. Functional Anal. **112** (1993), 204–217.

Division of Mathematics and Statistics
The University of Texas at San Antonio
San Antonio, Texas 78249
williams@math.utsa.edu

GPSR Compliance

The European Union's (EU) General Product Safety Regulation (GPSR) is a set of rules that requires consumer products to be safe and our obligations to ensure this.

If you have any concerns about our products, you can contact us on ProductSafety@springernature.com

In case Publisher is established outside the EU, the EU authorized representative is:

Springer Nature Customer Service Center GmbH
Europaplatz 3
69115 Heidelberg, Germany

Batch number: 10732312

Printed by Printforce, the Netherlands